Farzin Asadi and Kei Eguchi
Power Electronics Circuit Analysis with PSIM®

Also of Interest

Farzin Asadi and Kei Eguchi

Power Electronics Circuit Analysis with PSIM®

—

DE GRUYTER

Authors
Dr. Farzin Asadi
Department of Electrical and Electronics Engineering
Maltepe university
34857 Istanbul
Turkey
farzinasadi@maltepe.edu.tr

Dr. Kei Eguchi
Department of Information Electronics
Fukuoka Institute of Technology
Higashi-Ku 811-0295
Japan
eguti@fit.ac.jp

ISBN 978-3-11-074063-9
e-ISBN (PDF) 978-3-11-074065-3
e-ISBN (EPUB) 978-3-11-074091-2

Library of Congress Control Number: 2021937997

Bibliographic information published by the Deutsche Nationalbibliothek
The Deutsche Nationalbibliothek lists this publication in the Deutsche Nationalbibliografie;
detailed bibliographic data are available on the Internet at http://dnb.dnb.de.

© 2021 Walter de Gruyter GmbH, Berlin/Boston
Cover image: Ladislav Kubeš / iStock / Getty Images Plus
Typesetting: Integra Software Services Pvt. Ltd.
Printing and binding: CPI books GmbH, Leck

www.degruyter.com

In loving memory of
my father Moloud Asadi and
my mother Khorshid Tahmasebi,
always on my mind, forever in my heart.

Farzin Asadi

Dedicated to my lovely family.

Kei Eguchi

Preface

Power electronic systems are widely used today to provide power processing for applications ranging from computing and communications to medical electronics, appliance control, transportation, and high-power transmission. The associated power levels range from milliwatts to megawatts.

Modeling and simulation are essential ingredients of the analysis and design process in power electronics. They help a design engineer gain an increased understanding of circuit operation. With this knowledge the designer can, for a given set of specifications, choose a topology, select appropriate circuit component types and values, estimate circuit performance, and complete the design by ensuring that the circuit performance will meet specifications even with the anticipated variations in operating conditions and circuit component values. Power electronic systems are nonlinear variable structure systems. They involve passive components such as resistors, capacitors, and inductors, semiconductor switches such as thyristors and MOSFETs, and circuits for control. The analysis and design of such systems present significant challenges. Fortunately, increased availability of powerful computer and simulation programs makes the analysis/design process much easier.

PSIM® is an electronic circuit simulation software package, designed specifically for use in power electronics and motor drive simulations but can be used to simulate any electronic circuit. With fast simulation speed and user-friendly interface, PSIM provides a powerful simulation environment to meet the user simulation and development needs.

This book shows how to simulate the power electronic converter circuits in PSIM environment. The prerequisite for this book is a first course on power electronics. This book is composed of eight chapters.

Chapter 1 is an introduction to PSIM. This chapter introduces the different parts of PSIM environment.

Chapter 2 shows the fundamentals of circuit simulation with PSIM. You will learn how to draw the circuit schematic, connect the components together, run the simulation, measure what you want, and many other important tasks in this chapter.

Chapter 3 introduces Simview™. Simview is PSIM's waveform display and postprocessing program. This chapter studies the most important measurements that can be done with Simview.

Chapter 4 introduces the most commonly used components of PSIM.

https://doi.org/10.1515/9783110740653-202

Chapter 5 shows how PSIM can be used for analysis of power electronic circuits. Many examples are studied in this chapter. The studied examples are selected among the most fundamental circuits of power electronics. All the details are shown, so you can follow the examples easily.

Chapter 6 shows how you can simulate motors and mechanical loads in PSIM.

Chapter 7 introduces SimCoupler™. SimCoupler fuses PSIM with Simulink® by providing an interface for co-simulation. With the aid of SimCoupler, you have access to all of the Simulink blocks in your simulation.

Chapter 8 introduces the SmartCtrl®. SmartCtrl is a controller design software specifically geared toward power electronic applications. It features a user-friendly interface, a simple workflow, and a comprehensible display of control loop stability and performance. SmartCtrl facilitates the simple and fast designing of controllers for various power converters. In this chapter, the controller design process is shown for a buck converter.

We want to acknowledge the Powersim Inc. for providing access to PSIM during the writing of this book. We hope that this book will be useful to readers, and we welcome comments on the book. Enjoy the world of PSIM!

Farzin Asadi
Kei Eguchi

Contents

Preface —— VII

Chapter 1
An overview of PSIM® —— 1
1.1   Introduction —— 1
1.2   PSIM software —— 1
1.3   Licensing —— 3
1.4   PSIM environment —— 3
1.5   Version of PSIM and installed modules —— 10
1.6   PSIM's help menu —— 11

Chapter 2
Basics of PSIM —— 16
2.1   Introduction —— 16
2.2   Analysis of a simple resistive circuit —— 16
2.3   Simulating the circuit —— 35
2.4   Measurement of circuit current —— 39
2.5   Calculation of power —— 43
2.6   Voltage and current sensors —— 46
2.7   Measurement with current flag and voltage flag —— 51
2.8   Calculation of power using voltage/current flag —— 54
2.9   Color of connections —— 57
2.10   Exporting the drawn schematic —— 62
2.11   Addition of text to the schematic —— 63
2.12   Addition of project information to the schematic —— 65
2.13   Saving the schematic —— 67
2.14   Protection of schematic files with password —— 70
2.15   Compatible files for older versions of PSIM —— 73
2.16   Enabling and disabling a component —— 74
2.17   Shortcut keys —— 78
2.18   List of used elements —— 80
2.19   Removing the grids —— 81
2.20   Backup files —— 84
2.21   Automatic code generation blocks —— 85
2.22   Utilities menu —— 88
2.23   Simulation control block —— 100
2.24   Free-run mode —— 109
2.25   Change of elements' values during the simulation —— 116
2.26   AC and DC voltmeters and ammeters —— 119

Chapter 3
Simview™ —— **128**
3.1   Introduction —— **128**
3.2   A sample simulation —— **128**
3.3   Addition of label to the waveforms —— **134**
3.4   Changing the color of waveforms and addition of new axis to the
       graph —— **136**
3.5   Splitting the window —— **140**
3.6   Zoom in, zoom out, and move —— **143**
3.7   Customizing the X- and Y-axes —— **144**
3.8   Cursors —— **146**
3.9   Calculation of average and root mean square —— **151**
3.10   Calculation of power factor, active power,
       and apparent power —— **153**
3.11   Calculation of total harmonic distortion —— **155**
3.12   Fast Fourier transform analysis —— **159**
3.13   Exporting the waveforms to Excel® and MATLAB® —— **162**
3.14   Reading the exported file in MATLAB environment —— **164**
3.15   Exporting the graphic file of waveforms —— **166**

Chapter 4
PSIM's elements —— **167**
4.1   Introduction —— **167**
4.2   Resistor, inductor, and capacitor —— **167**
4.3   Saturable inductor block —— **178**
4.4   Coupled inductor block —— **179**
4.5   DC Load block —— **188**
4.6   Nonlinear elements —— **196**
4.7   Filters —— **202**
4.8   Computational blocks —— **207**
4.9   Transfer function block —— **220**
4.10   Logic elements —— **224**
4.11   Math function —— **226**
4.12   Switches —— **234**
4.13   Transformers —— **243**
4.14   Op amp —— **247**
4.15   IC models —— **252**

Chapter 5
Simulation of power electronic converters —— 254
5.1   Introduction —— **254**
5.2   Example 1: simulation of a simple RC circuit —— **254**
5.3   Example 2: effect of capacitor value on output waveform —— **258**
5.4   Example 3: nonlinear elements —— **262**
5.5   Example 4: switching the MOSFET —— **266**
5.6   Example 5: R_switch_on and R_switch_off —— **272**
5.7   Example 6: MOSFET conduction loss —— **286**
5.8   Example 7: calculation of total harmonic distortion —— **289**
5.9   Example 8: THD block —— **293**
5.10   Example 9: Math Function voltage source block —— **298**
5.11   Example 10: Wattmeter block —— **300**
5.12   Example 11: average power measurement in presence of
         harmonics —— **307**
5.13   Example 12: VAR meter block —— **310**
5.14   Example 13: reactive power calculation in presence of harmonics —— **318**
5.15   Example 14: VA/power factor meter block —— **323**
5.16   Example 15: half-wave rectifier —— **333**
5.17   Example 16: measurement of source power and diode losses for
         half-wave rectifier circuit —— **338**
5.18   Example 17: single-phase full-wave uncontrolled rectifier —— **342**
5.19   Example 18: diode bridge block —— **348**
5.20   Example 19: three-phase uncontrolled rectifier —— **351**
5.21   Example 20: single-phase half-wave controlled rectifier —— **358**
5.22   Example 21: single-phase full-wave controlled rectifier —— **363**
5.23   Example 22: measurement of power factor for different triggering
         angles —— **374**
5.24   Example 23: three-phase controlled rectifier —— **380**
5.25   Example 24: buck converter —— **388**
5.26   Example 25: effect of time step on edges of signals —— **396**
5.27   Example 26: making subcircuit —— **398**
5.28   Example 27: PWM generation with carrier PWM controller block —— **407**
5.29   Example 28: efficiency of SEPIC converter —— **409**
5.30   Example 29: limiter blocks —— **419**
5.31   Example 30: frequency response of control to output voltage for buck
         converter —— **424**
5.32   Example 31: frequency response of control to inductor current for buck
         converter —— **432**
5.33   Example 32: frequency response of electric circuit —— **433**

5.34   Example 33: frequency response of filters —— **438**
5.35   Example 34: input impedance of buck converter —— **442**
5.36   Example 35: output impedance of buck converter —— **450**
5.37   Example 36: closed-loop control of buck converter —— **459**
5.38   Example 37: closed-loop control of buck converter with simplified C code block —— **468**
5.39   Example 38: multiplexer block —— **472**
5.40   Example 39: simulation of flyback and push–pull converter —— **474**
5.41   Example 40: single-phase inverter —— **476**
5.42   Example 41: three-phase inverter —— **484**
5.43   Example 42: cascaded inverters —— **488**
5.44   Example 43: multilevel inverters and SV PWM block —— **496**
5.45   Example 44: magnetic components —— **499**
5.46   Example 45: code generation and thermal modules —— **511**

Chapter 6
Electrical machines —— 514
6.1   Introduction —— **514**
6.2   Simulation of a loaded DC motor —— **514**
6.3   Reference direction of mechanical systems —— **519**
6.4   Measurement of motor efficiency —— **521**
6.5   Generator simulation —— **522**
6.6   Efficiency of generator —— **528**

Chapter 7
SimCoupler™ —— 532
7.1   Introduction —— **532**
7.2   Setting up the SimCoupler —— **532**
7.3   Preparing the PSIM model —— **535**
7.4   Preparing the Simulink model —— **539**

Chapter 8
SmartCtrl —— 546
8.1   Introduction —— **546**
8.2   Power stage of the converter —— **546**
8.3   Obtaining the frequency response of the power stage —— **550**
8.4   Designing the controller —— **552**
8.5   Parameters of designed controller —— **560**
8.6   Exporting the parameters of designed controller —— **561**
8.7   Addition of the controller to the power stage —— **563**

8.8   Testing the controller —— 570
8.9   Designing with another type of controller —— 573
8.10   Automatic design of power stage and controller —— 579
8.11   Controller design for common type of DC–DC converters —— 582
8.12   Sample designs for SmartCtrl —— 589

References for further study —— 591

Index —— 593

Chapter 1
An overview of PSIM®

1.1 Introduction

This chapter introduces PSIM, different parts of PSIM environment, and its help system.

1.2 PSIM software

PSIM, which is developed by Powersim Inc. (https://powersimtech.com/), is an electronic circuit simulation software package, designed specifically for use in power electronics and motor drive simulations but can be used to simulate any electronic circuit. With fast simulation speed and user-friendly interface, PSIM provides a powerful simulation environment to meet the user simulation and development needs. PSIM is used by industries for research and product development and it is used by educational institutions for research and teaching.

PSIM includes the basic package as well as the following add-on modules (modules, extended PSIM functionality into specific areas of circuit simulation, and design):

Motor drive: Built-in electric machine models and mechanical load models for motor drive system studies.

Digital control: Discrete library elements such as zero-order hold, z-domain transfer function blocks, quantization blocks, and digital filters for digital control system analysis.

Thermal: Library elements and functions calculate semiconductor device losses and inductor losses.

Renewable energy: Library elements, such as solar module, wind turbine, battery, and ultracapacitor models for renewable energy applications.

Power Supply Design Suite: Pre-built design templates for resonant LLC converters.

EMI Design Suite: Design templates for EMI analysis and EMI filter design.

Motor control design suite: Pre-built templates for induction motor and linear/nonlinear PMSM drives.

Hybrid electric vehicle design suite: Pre-built templates for hybrid electric vehicle powertrain system design.

SimCoupler: Interface between PSIM and MATLAB/Simulink for co-simulation.

SPICE: Functions to link to LTspice.

MagCoupler: Interface between PSIM and the electromagnetic field analysis software JMAG for co-simulation.

MagCoupler-RT: Link between PSIM and JMAG-RT data files.

https://doi.org/10.1515/9783110740653-001

ModCoupler: Interface between PSIM and ModelSim for co-simulation. There are two versions of the interface: ModCoupler-VHDL that supports VHDL code, and ModCoupler-Verilog that supports Verilog code.

SimCoder: Function for automatic code generation.

F2833x Target: Library elements for automatic code generation for TI F2833x series DSP.

F2803x Target: Library elements for automatic code generation for TI F2803x series DSP.

F2802x Target: Library elements for automatic code generation for TI F2802x series DSP.

F2806x Target: Library elements for automatic code generation for TI F2806x series DSP.

F2837x Target: Library elements for automatic code generation for TI F2837x series DSP.

F28004x Target: Library elements for automatic code generation for TI F28004x series DSP.

PE-Expert4 Target: Library elements for automatic code generation for Myway PE-Expert4 hardware platform.

Processor in the loop: Interface between PSIM and TI DSP hardware boards for processor-in-the-loop simulation. It also includes the function block to support TI's InstaSPIN® motor control algorithm.

In addition, PSIM integrates the DSIM® engine into its simulation environment. DSIM is known for its incredible speed and accuracy in solving very large power converter systems, and for solving detailed switching transients. DSIM shares the same schematic capture and waveform processing environment as PSIM.

Also, PSIM links with the software SmartCtrl® for control loop design. SmartCtrl is designed specifically for power converter applications.

PSIM also provides the function to export the power stage to Typhoon® HIL's real-time simulator for hardware-in-the-loop (HIL) simulation.

With these product lineups, Powersim provides a complete platform from design to simulation, to hardware implementation. The overall environment is shown in Fig. 1.1.

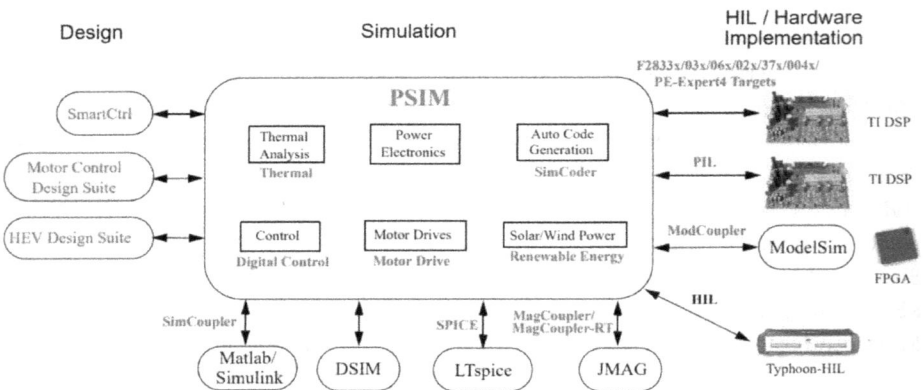

Fig. 1.1.

1.3 Licensing

Powersim provides four types of license for users: FREE DEMO, 30-DAY TRIAL, PSIM
STANDARD, and PSIM PROFESSIONAL. Students and educational institutes can get
a discount.

The free demo version of PSIM can be downloaded form Products section of
https://powersimtech.com/ website (Fig. 1.2).

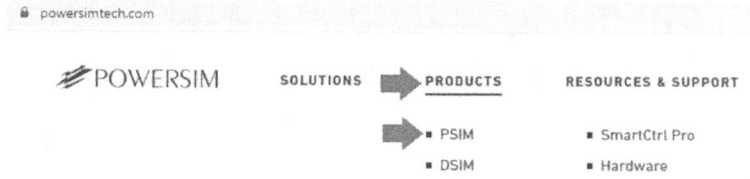

Fig. 1.2.

The demo version does not expire but is limited in component count and allowed
circuit complexity. When you use the demo version, you cannot have a waveform with
more than 6,000 points in the Simview environment (Fig. 1.3). Simview is a tool for
analyzing the obtained waveform.

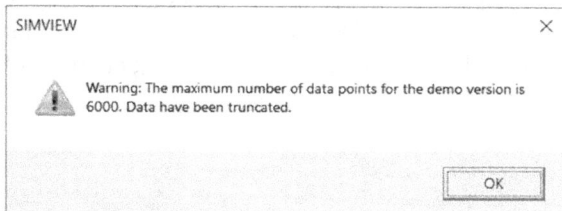

Fig. 1.3.

1.4 PSIM environment

In this section, we will take a look at different parts of PSIM. PSIM environment
(Fig. 1.4) is composed of four parts: Menu bar, Library Browser window, commonly
used components, and schematic capture.

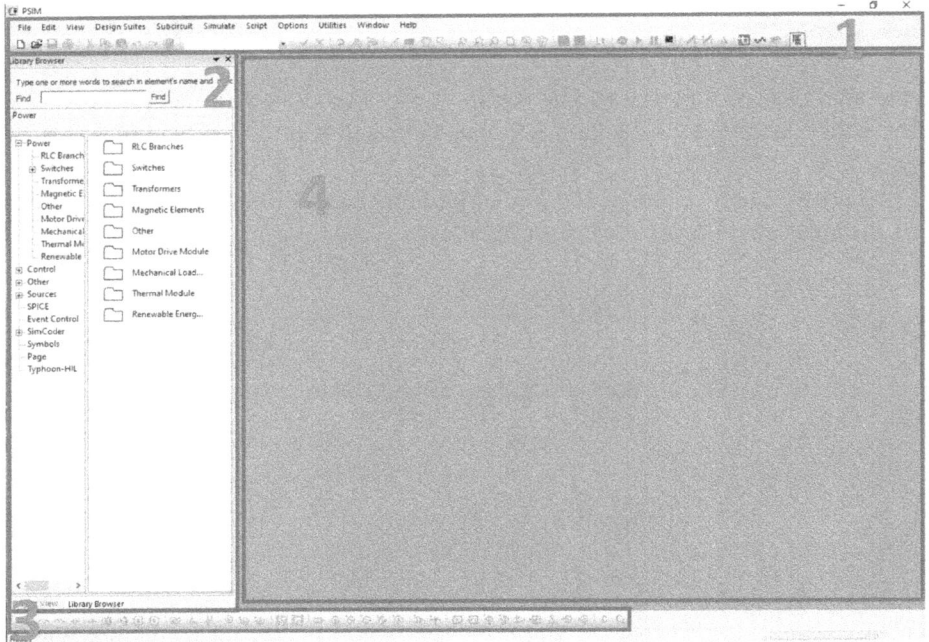

Fig. 1.4.

Like any other Windows® program, PSIM's menu bar (Fig. 1.5) is composed of some menus and some shortcut icons that act as an interface between the user and PSIM. These menus and their shortcut icons will be studied in the forthcoming chapters of this book.

Fig. 1.5.

PSIM has many components. The components are categorized into different groups under the Elements menu (Fig. 1.6).

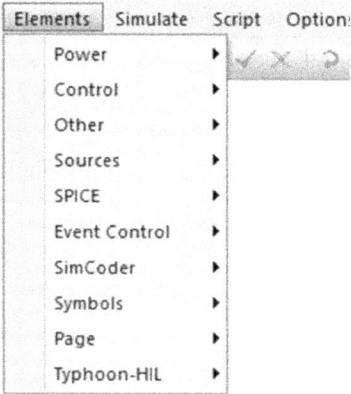

Elements	Simulate	Script	Option:
Power	▶		
Control	▶		
Other	▶		
Sources	▶		
SPICE	▶		
Event Control	▶		
SimCoder	▶		
Symbols	▶		
Page	▶		
Typhoon-HIL	▶		

Fig. 1.6.

For instance, assume that you want to add a MOSFET to your schematic using the Elements menu. In order to do this, you need to open the Elements menu, then open the Power, then open the Switches and finally click the MOSFET (Fig. 1.7).

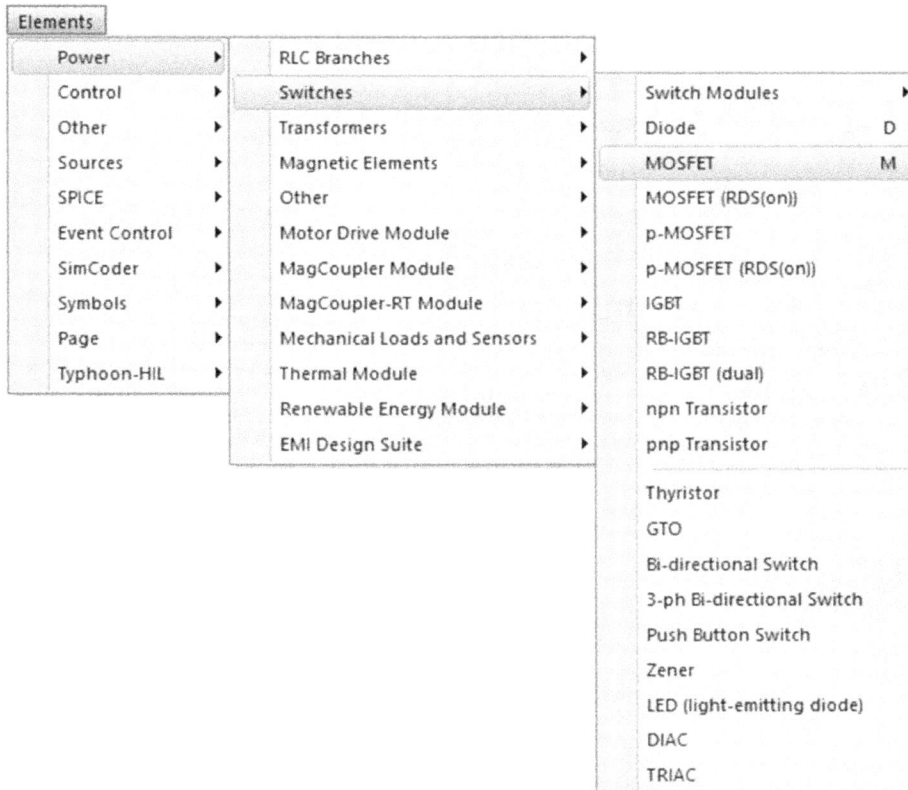

Elements

Power	▶	RLC Branches	▶	
Control	▶	Switches	▶	Switch Modules ▶
Other	▶	Transformers	▶	Diode D
Sources	▶	Magnetic Elements	▶	MOSFET M
SPICE	▶	Other	▶	MOSFET (RDS(on))
Event Control	▶	Motor Drive Module	▶	p-MOSFET
SimCoder	▶	MagCoupler Module	▶	p-MOSFET (RDS(on))
Symbols	▶	MagCoupler-RT Module	▶	IGBT
Page	▶	Mechanical Loads and Sensors	▶	RB-IGBT
Typhoon-HIL	▶	Thermal Module	▶	RB-IGBT (dual)
		Renewable Energy Module	▶	npn Transistor
		EMI Design Suite	▶	pnp Transistor

Thyristor

GTO

Bi-directional Switch

3-ph Bi-directional Switch

Push Button Switch

Zener

LED (light-emitting diode)

DIAC

TRIAC

Fig. 1.7.

Library browser (Fig. 1.8), is a search tool for components. It is the easiest way to add a component to your schematic. For instance, if you type MOSFET in the Find box and press the Enter key, then all the related components will appear in the list (Fig. 1.9).

Now click on the component that you want and the component will be selected. After selecting the component, click on the schematic and the selected component will be added to it.

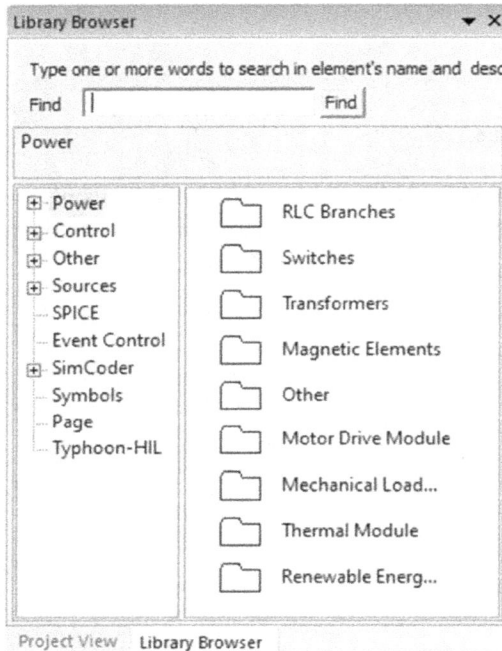

Library Browser ▼ ✕

Type one or more words to search in element's name and desc

Find | Find

Power

- ⊞ Power
- ⊞ Control
- ⊞ Other
- ⊞ Sources
- ⋯ SPICE
- ⋯ Event Control
- ⊞ SimCoder
- ⋯ Symbols
- ⋯ Page
- ⋯ Typhoon-HIL

📁 RLC Branches

📁 Switches

📁 Transformers

📁 Magnetic Elements

📁 Other

📁 Motor Drive Module

📁 Mechanical Load...

📁 Thermal Module

📁 Renewable Energ...

Project View   Library Browser

Fig. 1.8.

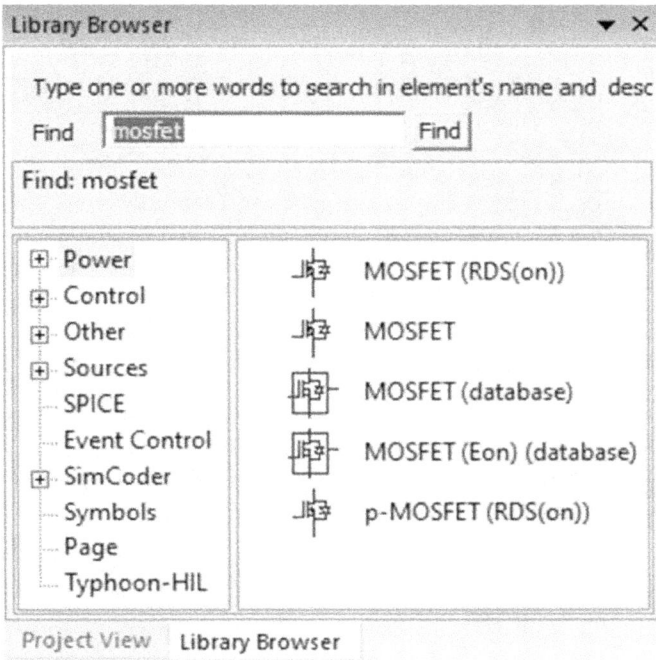

Fig. 1.9.

If you cannot see the Library Browser window, then click on the Library Browser icon (Fig. 1.10) in order to activate the Library Browser window. You can use the View> Library Browser as another way to activate the Library Browser window (Fig. 1.11).

Fig. 1.10.

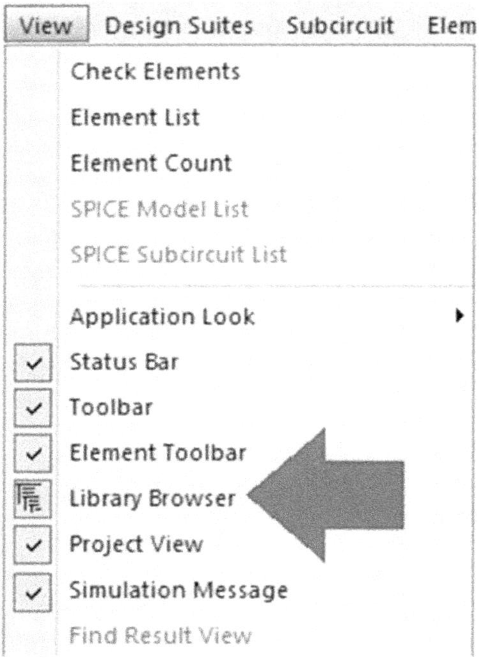

Fig. 1.11.

Some of the most commonly used components are added to the bottom of the screen (Fig. 1.12). You can click on the desired component and add it to schematic with one click.

Fig. 1.12.

When the mouse pointer is put on a component, its name is shown in the left bottom side (Fig. 1.13).

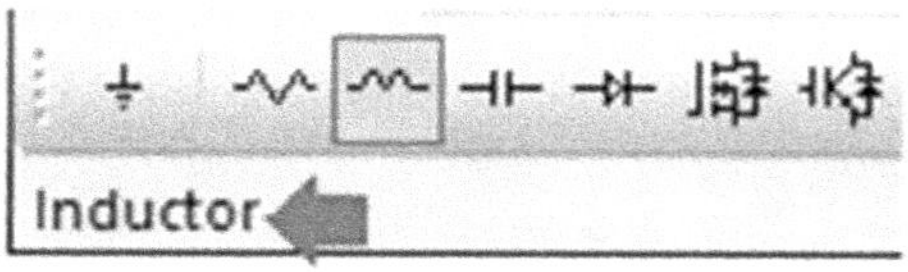

Fig. 1.13.

PSIM shows the simulation progress with a progress bar (Fig. 1.14).

Fig. 1.14.

If your simulation has an error, then PSIM informs you using the Simulation Message window (Fig. 1.15).

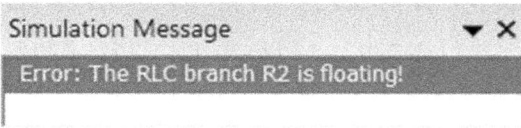

Fig. 1.15.

You can see the use the View menu to open/close a window (Fig. 1.16).

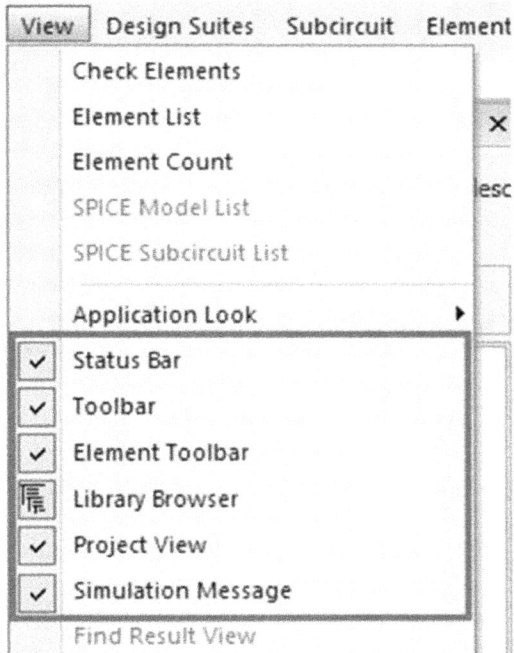

Fig. 1.16.

1.5 Version of PSIM and installed modules

You can see the version of PSIM and installed modules with the aid of Help> About (Figs. 1.17 and 1.18).

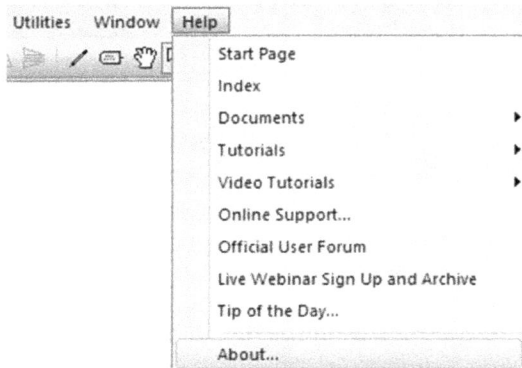

Fig. 1.17.

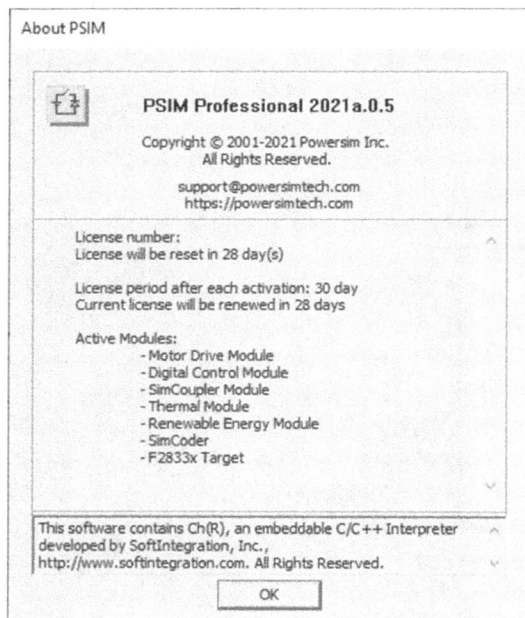

Fig. 1.18.

1.6 PSIM's help menu

PSIM has a powerful help system. You can see the PSIM User Manual in the Help>
Documents section (Fig. 1.19).

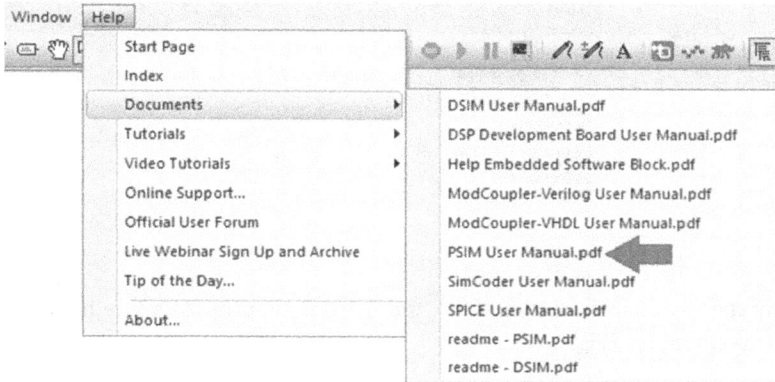

Fig. 1.19.

The Tutorial (Fig. 1.20) and Video Tutorial (Fig. 1.21) sections contain many useful
documents and videos.

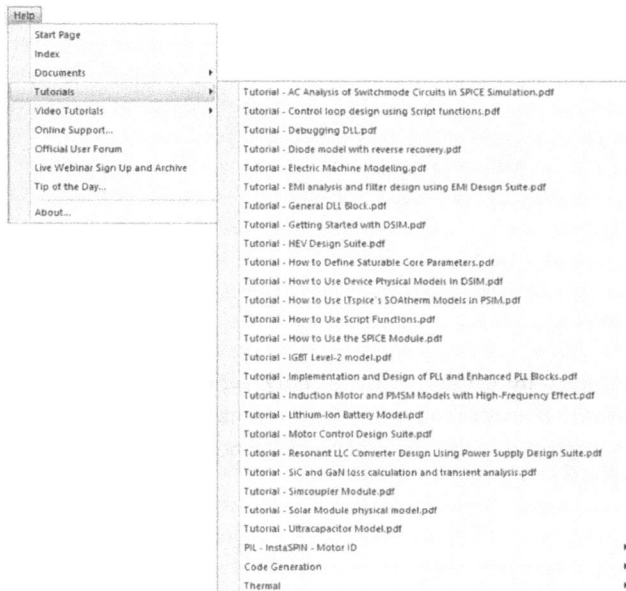

Fig. 1.20.

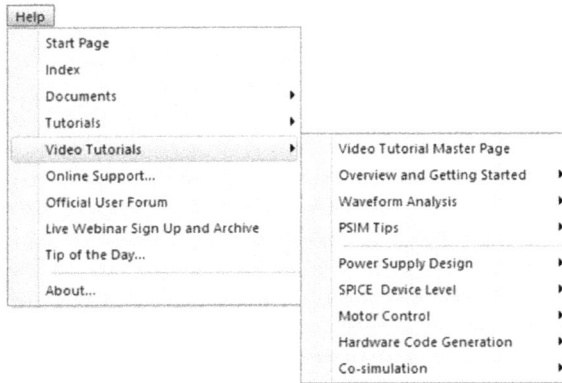

Fig. 1.21.

PSIM has many ready-to-use examples. Click the Open Examples from File menu in order to see these examples (Fig. 1.22).

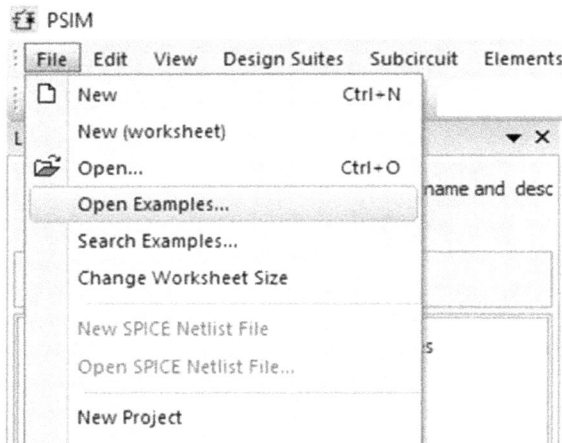

Fig. 1.22.

After clicking the Open Examples, the window shown in Fig. 1.23 appears. Now you can open the folders and there are plenty of inspiring simulations there. For instance, if you want to see simulation related to rectifiers, you need to open the ac–dc folder.

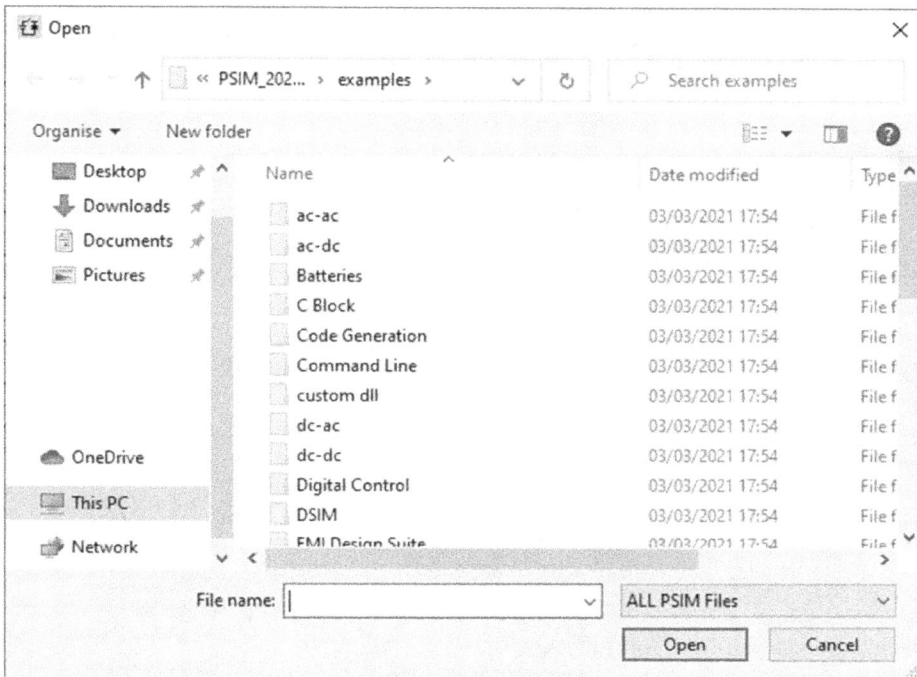

Fig. 1.23.

You can use the File> Search Examples (Fig. 1.24) in order to search in the examples.

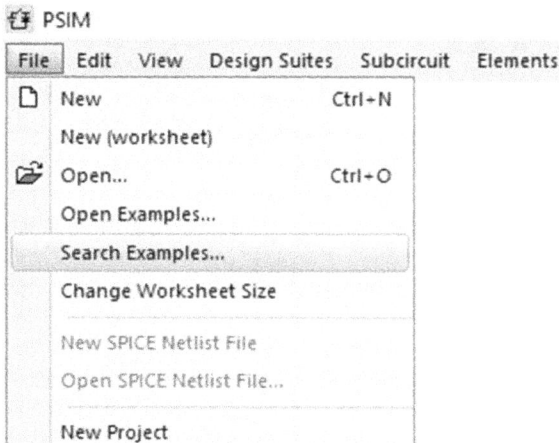

Fig. 1.24.

After clicking the Search Examples, enter the desired search term in the Find box (Fig. 1.25).

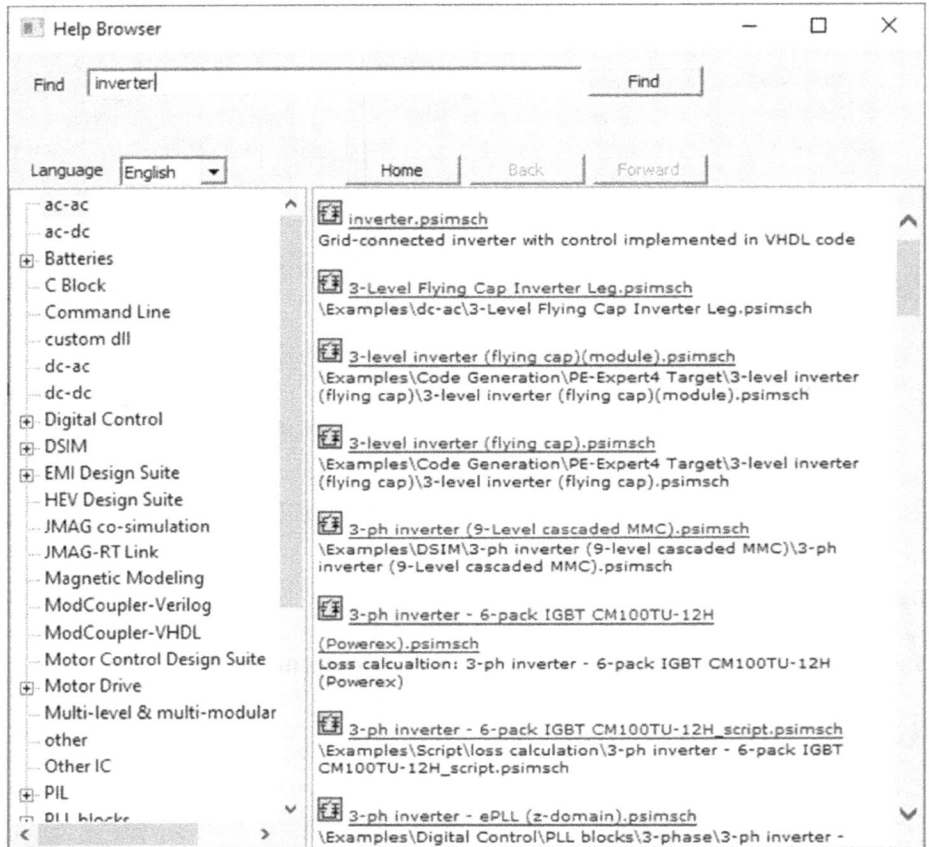

Fig. 1.25.

When you double click on any component placed on schematic, you can see the Help button for that component (Fig. 1.26). If you click the Help button, the help file for that component is shown.

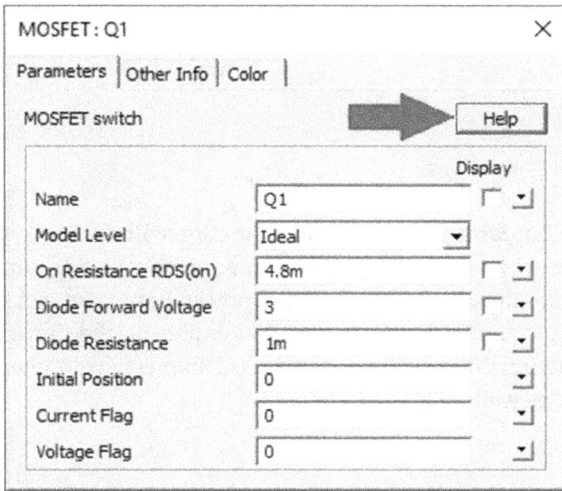

MOSFET : Q1   ✕

Parameters | Other Info | Color |

MOSFET switch   ➡ | Help |

		Display
Name	Q1	☐ ▾
Model Level	Ideal ▾	
On Resistance RDS(on)	4.8m	☐ ▾
Diode Forward Voltage	3	☐ ▾
Diode Resistance	1m	☐ ▾
Initial Position	0	▾
Current Flag	0	▾
Voltage Flag	0	▾

Fig. 1.26.

Chapter 2
Basics of PSIM

2.1 Introduction

In this chapter, you will learn the fundamentals of circuit simulation with PSIM. You will learn how to draw the circuit schematic, connect the components together, run the simulation, and measure the circuit voltages, current, and power. Important tasks such as enable/disable a block, addition of text to schematic, exporting the drawn schematic, Free Run Mode, Simulation Control block, and AC/DC ammeter/voltmeter blocks are studied in this chapter, as well.

2.2 Analysis of a simple resistive circuit

Consider a simple resistive circuit like Fig. 2.1. The voltage across resistor R2 must be $\frac{10}{90+10} \times 100 = 10 \ V$. We want to analyze this simple circuit with the aid of PSIM.

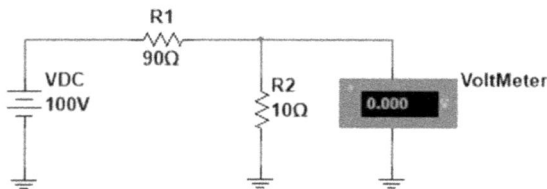

Fig. 2.1.

First of all, we need to make a new file. In order to do this, click the new file icon (Fig. 2.2).

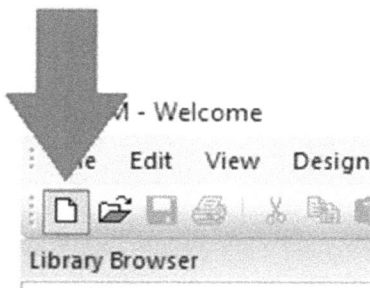

Fig. 2.2.

https://doi.org/10.1515/9783110740653-002

After clicking the new file icon, the PSIM opens a new file for you (Fig. 2.3). You can draw the schematic in this file.

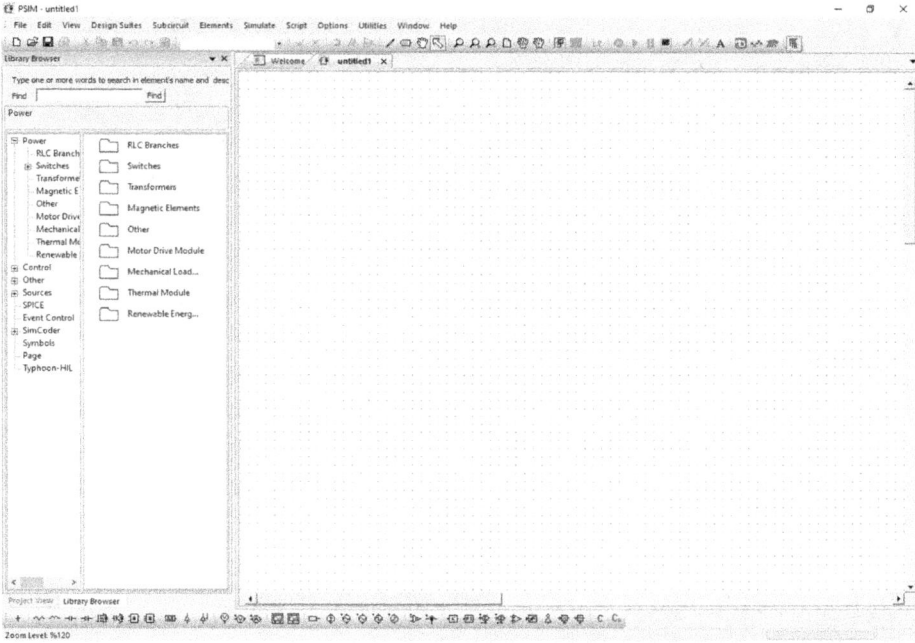

Fig. 2.3.

Select a DC source from Elements>Sources>Voltage>DC (Fig. 2.4).

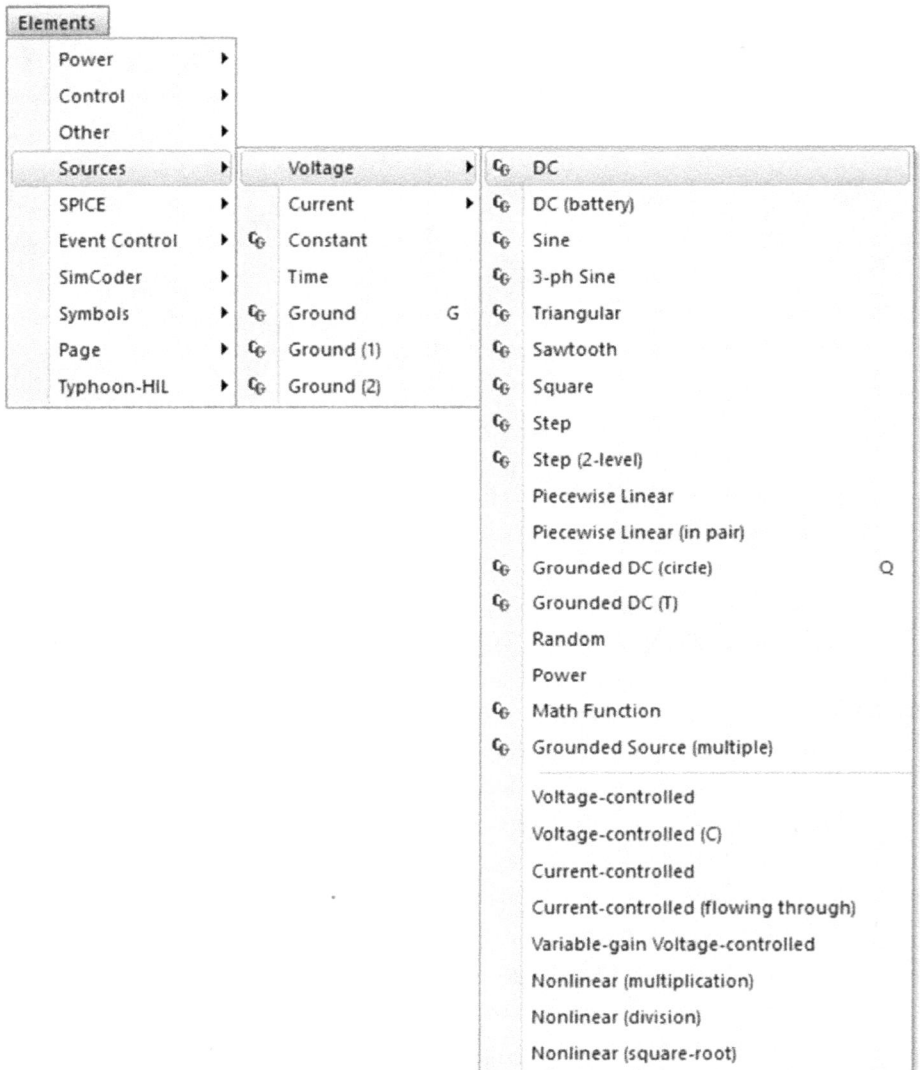

Elements		
Power ▶		
Control ▶		
Other ▶		
Sources ▶	Voltage ▶	DC
SPICE ▶	Current ▶	DC (battery)
Event Control ▶	Constant	Sine
SimCoder ▶	Time	3-ph Sine
Symbols ▶	Ground G	Triangular
Page ▶	Ground (1)	Sawtooth
Typhoon-HIL ▶	Ground (2)	Square
		Step
		Step (2-level)
		Piecewise Linear
		Piecewise Linear (in pair)
		Grounded DC (circle) Q
		Grounded DC (T)
		Random
		Power
		Math Function
		Grounded Source (multiple)
		Voltage-controlled
		Voltage-controlled (C)
		Current-controlled
		Current-controlled (flowing through)
		Variable-gain Voltage-controlled
		Nonlinear (multiplication)
		Nonlinear (division)
		Nonlinear (square-root)

Fig. 2.4.

You can use the Library Browser in order to select the DC source. To do this, enter the "dc source" in the Library Browser text box and press the Enter key. As shown in Fig. 2.5, a list of related blocks will be shown. Now you can select the desired one from the shown list.

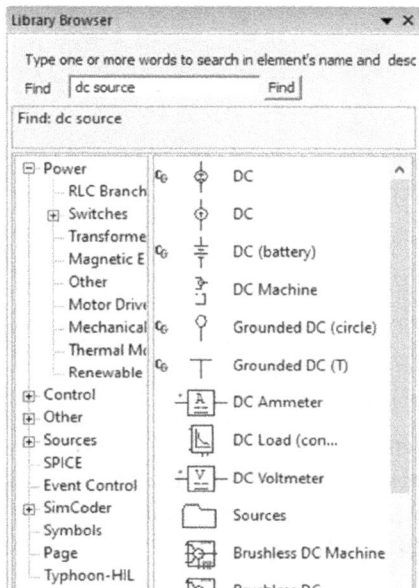

Fig. 2.5.

When you select the DC voltage source block, your mouse pointer will be changed into a DC voltage source in order to show that the DC voltage source block is selected (Fig. 2.6).

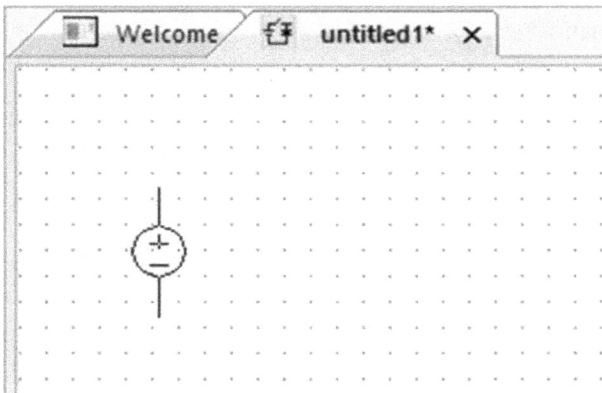

Fig. 2.6.

If you press the mouse right button, then the block will be rotated in clockwise direction. If you press the mouse left button, then the block will be added to the schematic (Fig. 2.7).

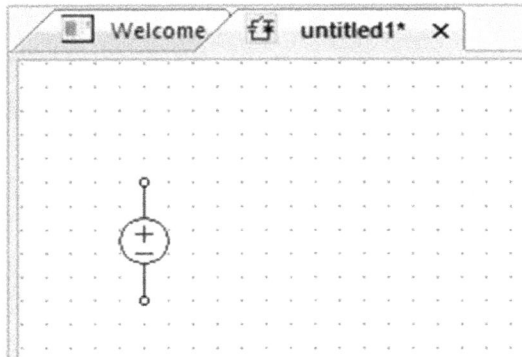

Fig. 2.7.

After placing the DC voltage source to the schematic, press the Esc key on the keyboard or click the Select icon (Fig. 2.8). Otherwise, in each click on the schematic, a DC voltage source will be placed there. If you want to remove a block from the schematic, left click on it and press the Delete key on the keyboard.

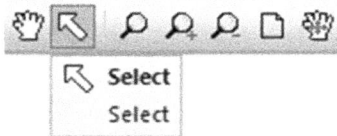

Fig. 2.8.

Add two resistors to the schematic (Fig. 2.9).

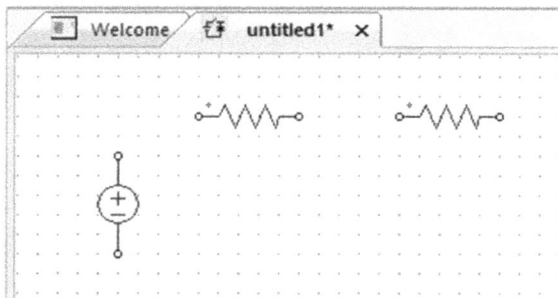

Fig. 2.9.

The resistor block can be found in the Elements>Power>RLC Branches>Resistor (Fig. 2.10). You can search for "Resistor" in the Library Browser search box as well (Fig. 2.11). The third method for selecting the resistor block is to press the R key of your keyboard.

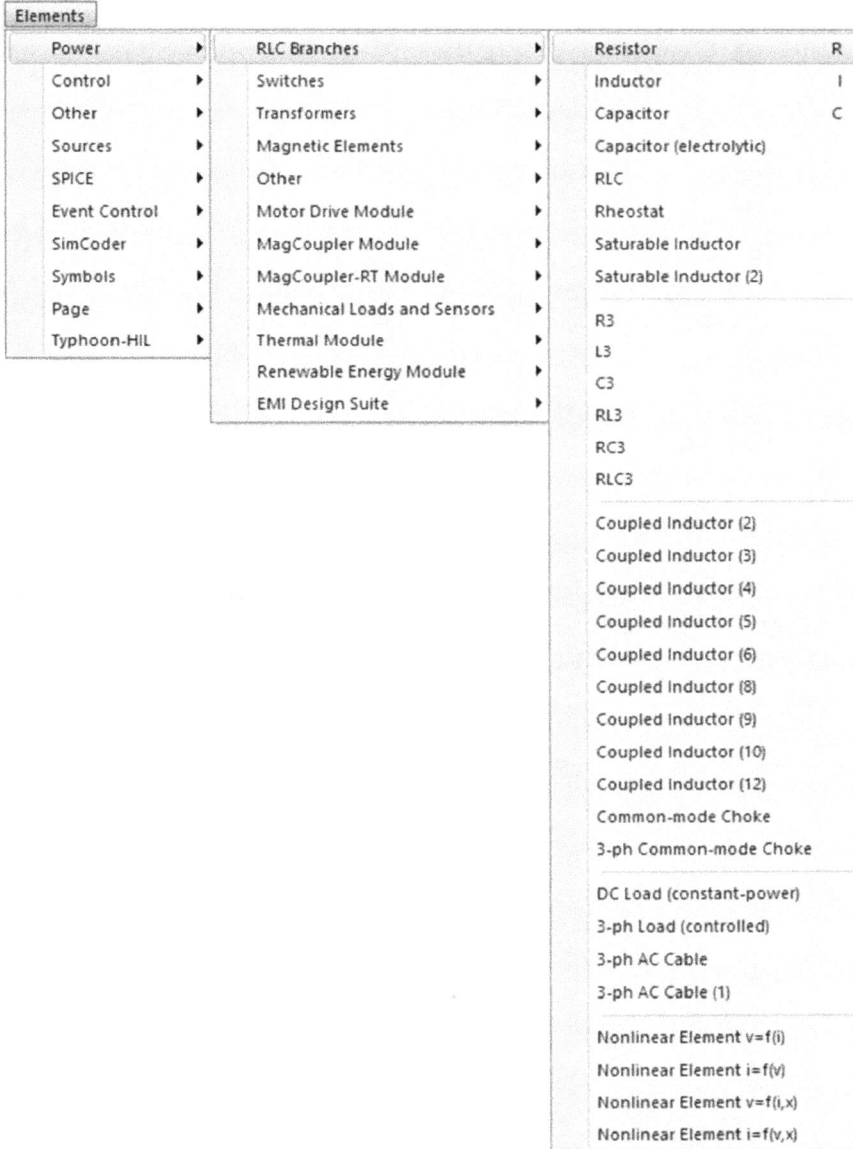

Elements							
Power	▶	RLC Branches	▶	Resistor		R	
Control	▶	Switches	▶	Inductor		I	
Other	▶	Transformers	▶	Capacitor		C	
Sources	▶	Magnetic Elements	▶	Capacitor (electrolytic)			
SPICE	▶	Other	▶	RLC			
Event Control	▶	Motor Drive Module	▶	Rheostat			
SimCoder	▶	MagCoupler Module	▶	Saturable Inductor			
Symbols	▶	MagCoupler-RT Module	▶	Saturable Inductor (2)			
Page	▶	Mechanical Loads and Sensors	▶				
Typhoon-HIL	▶	Thermal Module	▶	R3			
		Renewable Energy Module	▶	L3			
		EMI Design Suite	▶	C3			
				RL3			
				RC3			
				RLC3			
				Coupled Inductor (2)			
				Coupled Inductor (3)			
				Coupled Inductor (4)			
				Coupled Inductor (5)			
				Coupled Inductor (6)			
				Coupled Inductor (8)			
				Coupled Inductor (9)			
				Coupled Inductor (10)			
				Coupled Inductor (12)			
				Common-mode Choke			
				3-ph Common-mode Choke			
				DC Load (constant-power)			
				3-ph Load (controlled)			
				3-ph AC Cable			
				3-ph AC Cable (1)			
				Nonlinear Element v=f(i)			
				Nonlinear Element i=f(v)			
				Nonlinear Element v=f(i,x)			
				Nonlinear Element i=f(v,x)			

Fig. 2.10.

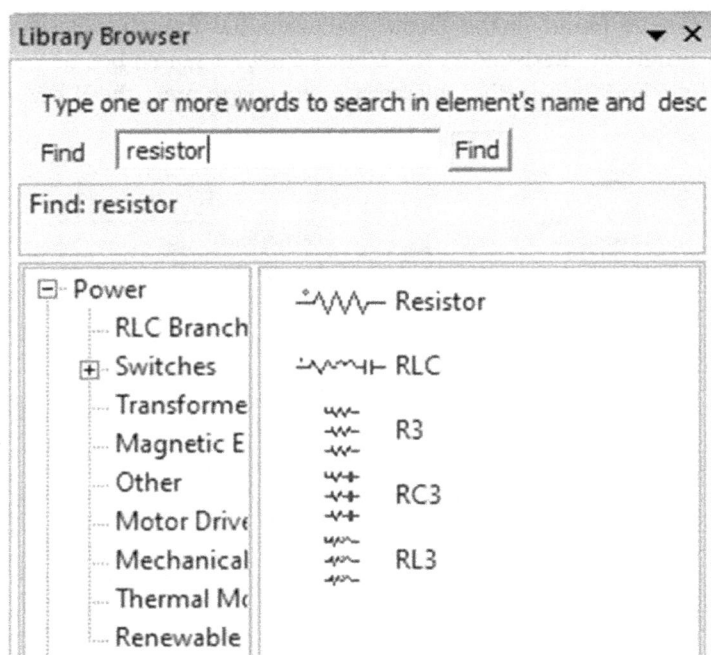

Fig. 2.11.

Click one of the resistors in order to select it (Fig. 2.12). The PSIM draws a rectangle around the resistor in order to show that it is selected. Click the Rotation icon (Fig. 2.13) in order to rotate it (Fig. 2.14).

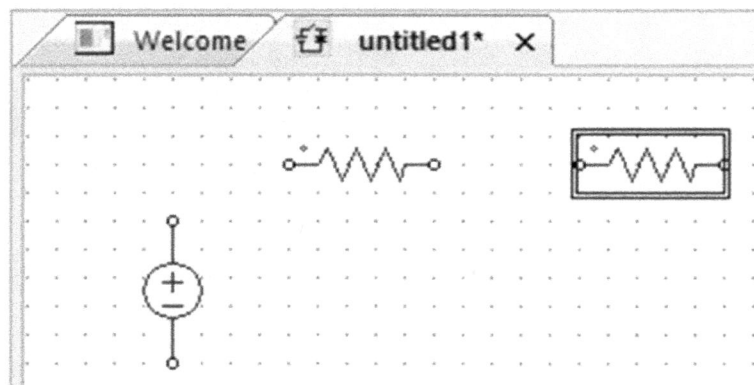

Fig. 2.12.

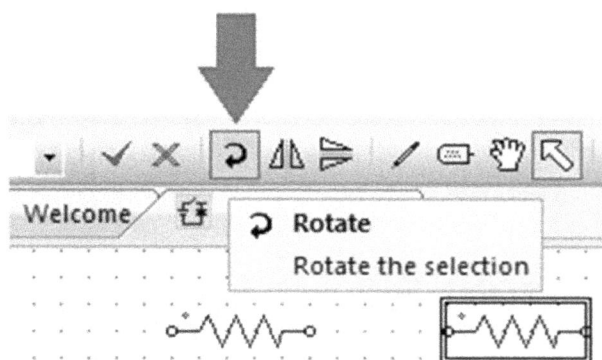

Fig. 2.13.

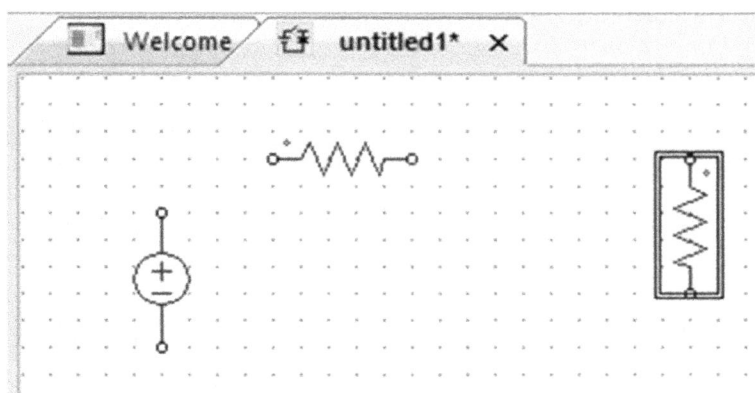

Fig. 2.14.

If you want to flip a block, you need to use the Flip vertical and Flip horizontal icons.

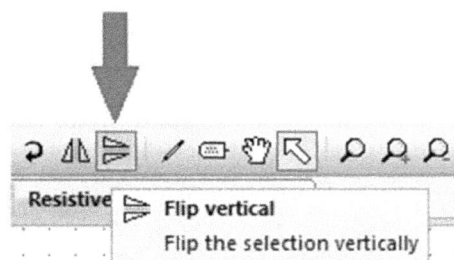

Fig. 2.15.

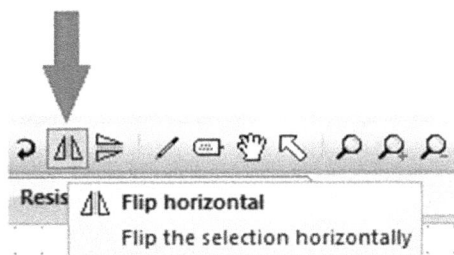

Fig. 2.16.

Use the wire tool (Fig. 2.17) in order to draw the circuit wires. You can click on its icon or you can simply press the W key on your keyboard. After wire tool is activated, the mouse cursor will be changed into a pencil. Use this pencil in order to connect the terminals of blocks together (Fig. 2.18).

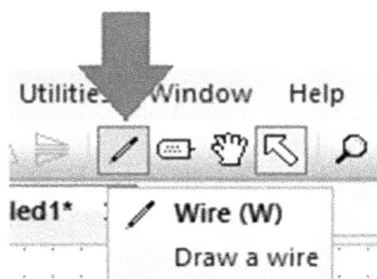

Fig. 2.17.

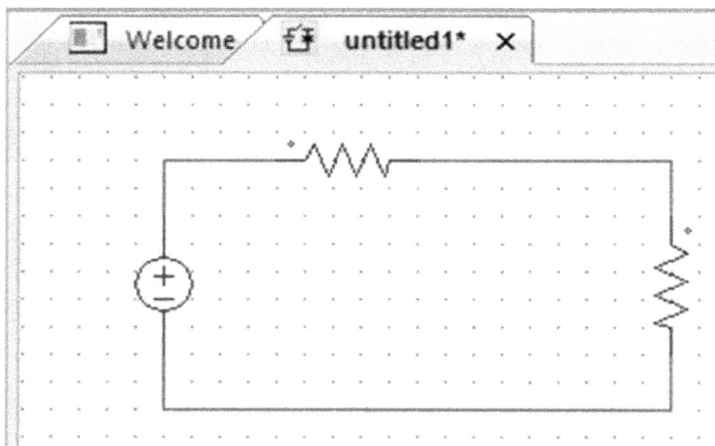

Fig. 2.18.

Your circuits must have a ground. Click the ground icon (or press the G key on your keyboard) to activate the ground block (Fig. 2.19). Place the ground block in the schematic (Fig. 2.20) and use the wire tool in order to connect it to the negative terminal of the DC voltage source (Fig. 2.21).

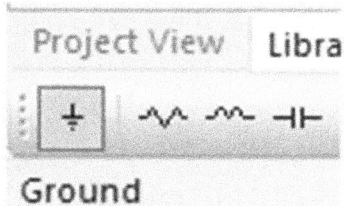

Ground

Fig. 2.19.

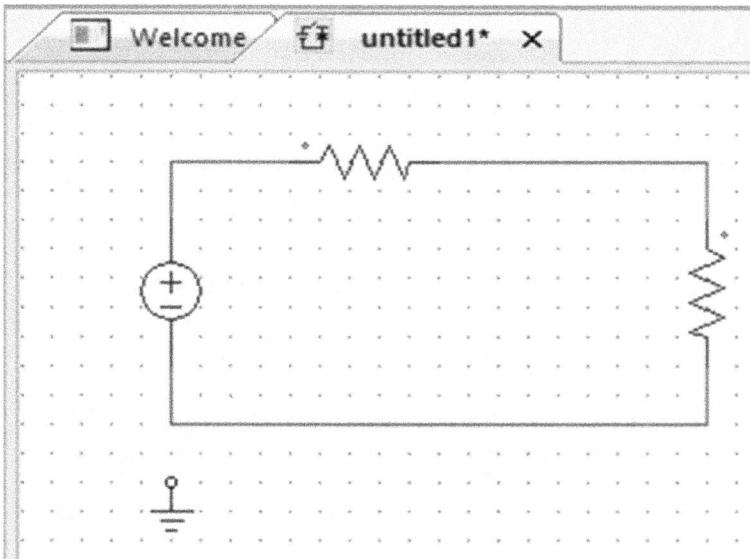

Fig. 2.20.

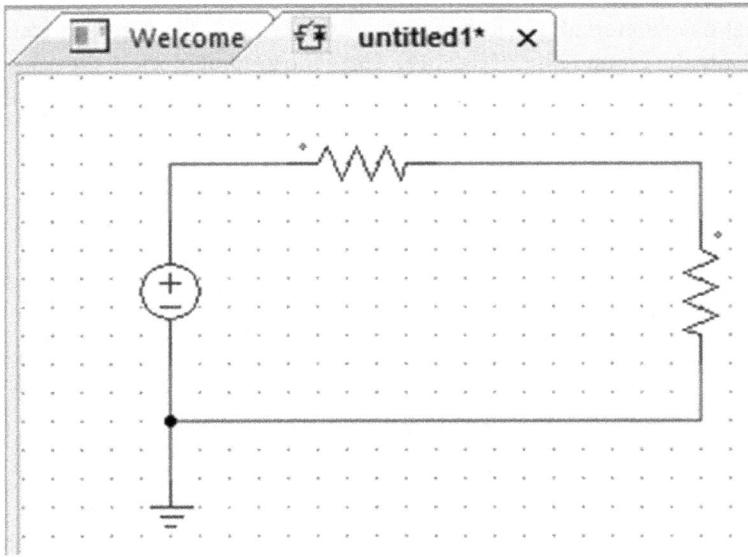

Fig. 2.21.

If you open the Elements>Sources, you can see that PSIM has three ground blocks: Ground, Ground (1), and Ground (2) (Fig. 2.22).

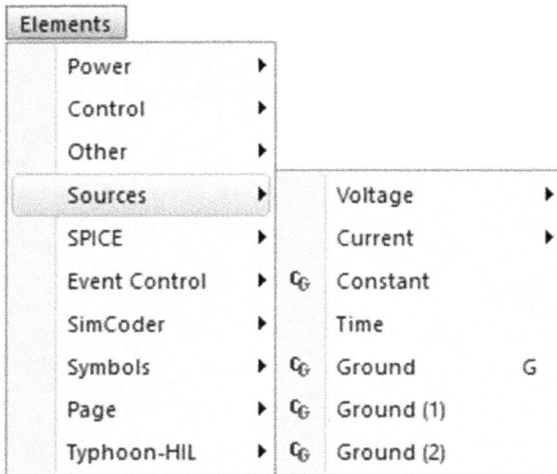

Fig. 2.22.

The symbols for these blocks are shown in Fig. 2.23.

Ground   Ground (1)   Ground (2)

Fig. 2.23.

Although they have different appearance, all of them are the same thing from PSIM point of view (they are electrically connected together). In order to understand this matter, consider the circuit shown in Fig. 2.24. This circuit is equivalent to the one shown in Fig. 2.25 and the current drawn from the source is 5 A. The different shapes of these grounds provide the convenience for user to separate the grounding in different functional sections of the circuits.

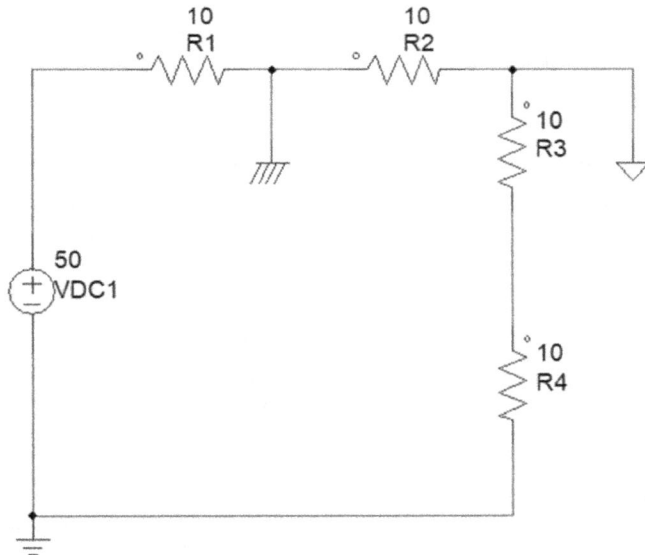

Fig. 2.24.

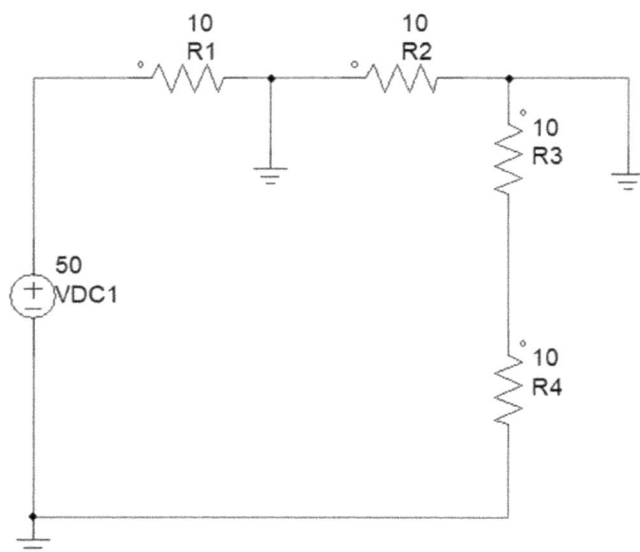

Fig. 2.25.

We need a voltage probe in order to measure the voltage across the resistor. PSIM has two voltage probes. One of them has two terminals (Fig. 2.26) and the other one has only one terminal (Fig. 2.27).

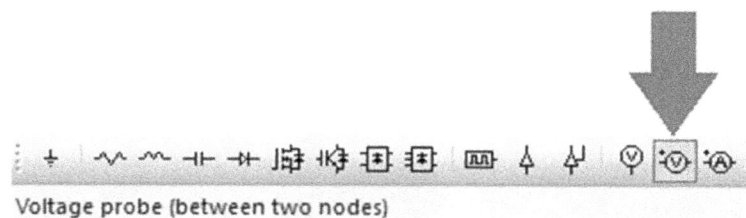

Voltage probe (between two nodes)

Fig. 2.26.

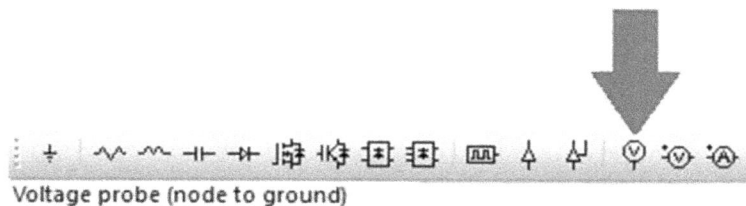

Voltage probe (node to ground)

Fig. 2.27.

You can connect each terminal of two-terminal voltage probe (Fig. 2.26) to the desired node of the circuit. With the aid of this block, you can measure the voltage difference between any two nodes of your circuit. Note that one of the terminals has a small dot behind it (Fig. 2.28). The voltage probe measures the $V_{terminal\ with\ dot} - V_{terminal\ without\ dot}$.

The voltage probe with only one terminal (Fig. 2.27) measures the voltage with respect to the ground.

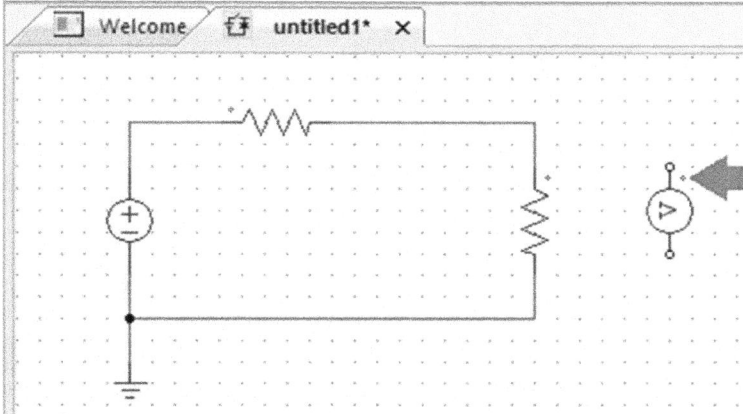

Fig. 2.28.

Add a two-terminal voltage probe to the schematic (Fig. 2.28) and connect it to the resistor (Fig. 2.29).

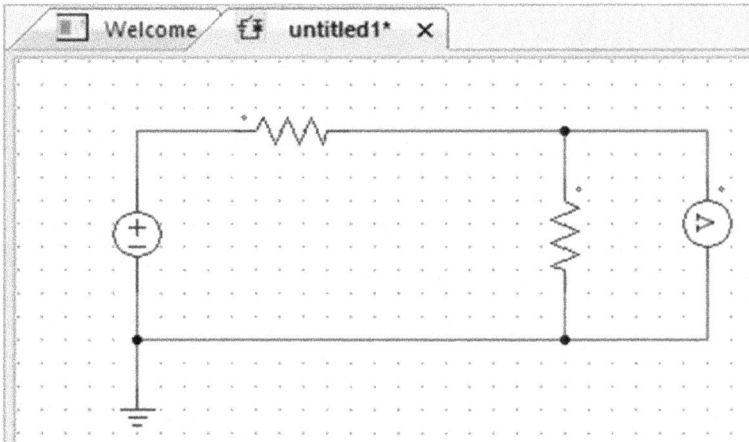

Fig. 2.29.

You can use a single-terminal voltage probe as well (Fig. 2.29), because one of the resistor terminals is grounded. Reading of both probes in Figs. 2.29 and 2.30 are the same.

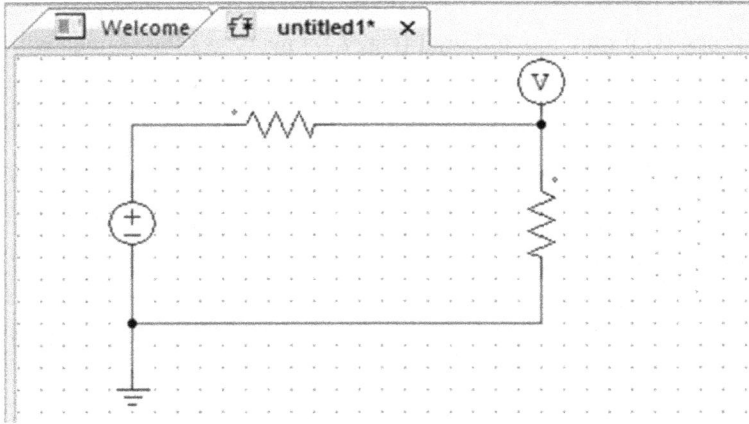

Fig. 2.30.

Now the schematic is ready. It is time to set the components values. Double click the DC voltage source. The window shown in Fig. 2.31 appears. Enter 100 in the Amplitude box. The default value for this box is 100 so there is no need to enter it. The Amplitude box determines the voltage of the source. If the internal impedance of your source is RL type (Fig. 2.32), you can enter the values for series resistor and inductor in the "Series Resistance" and "Series Inductance" boxes, respectively.

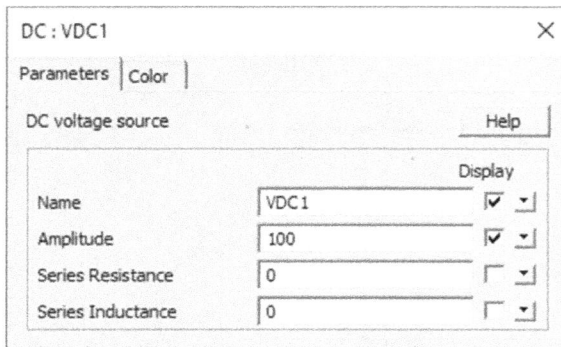

Fig. 2.31.

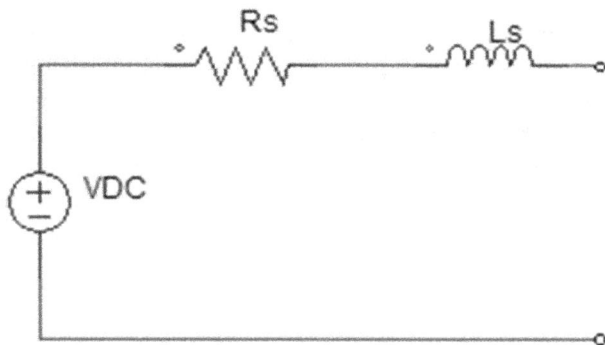

Fig. 2.32.

If you click the Help button in Fig. 2.31, the window shown in Fig. 2.33 will appear. It is highly recommended to read the help page of the blocks. They contain many valuable and important information about the blocks.

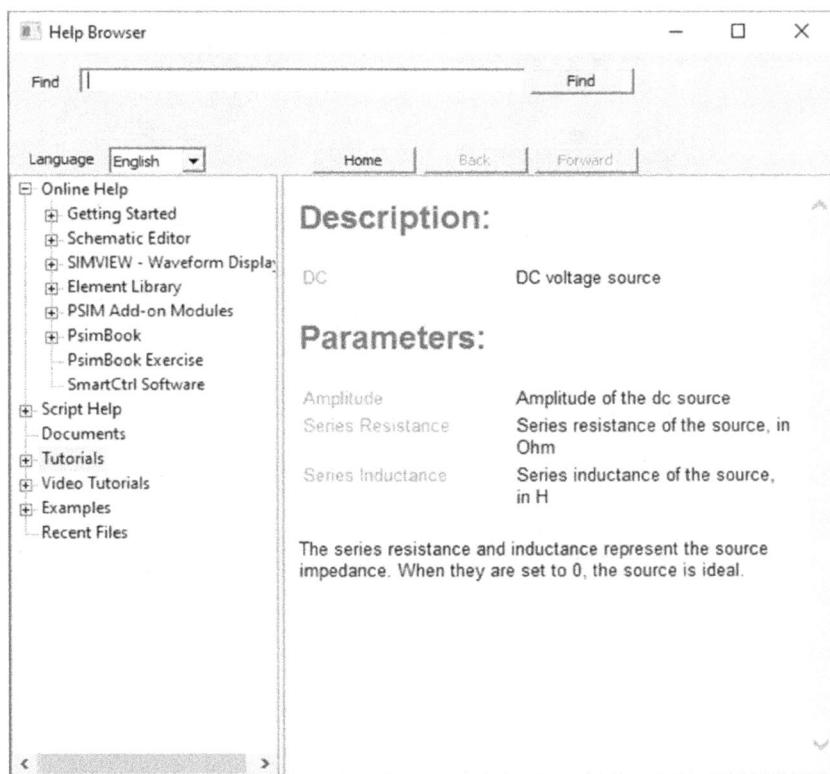

Fig. 2.33.

Double click the resistor blocks and enter their resistances. The default value for Model Level is "Level 1." Do not change that property. When you check the display box (see Fig. 2.34), PSIM show the corresponding property in the schematic environment. For instance, in Fig. 2.35, the display box behind the Name and Resistance boxes are checked. So, PSIM shows the name of component (R2) and its value (10) on the schematic (Fig. 2.36).

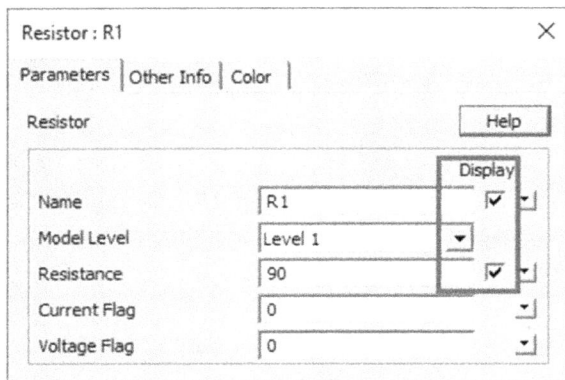

Fig. 2.34.

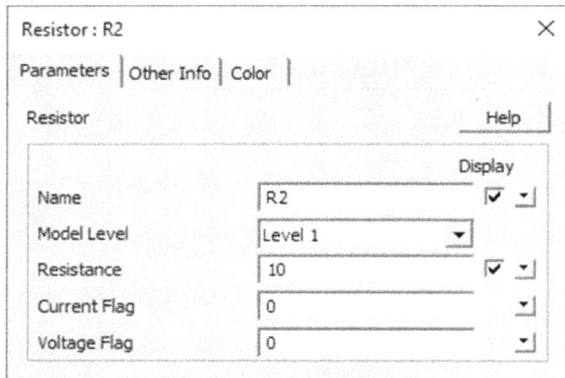

Fig. 2.35.

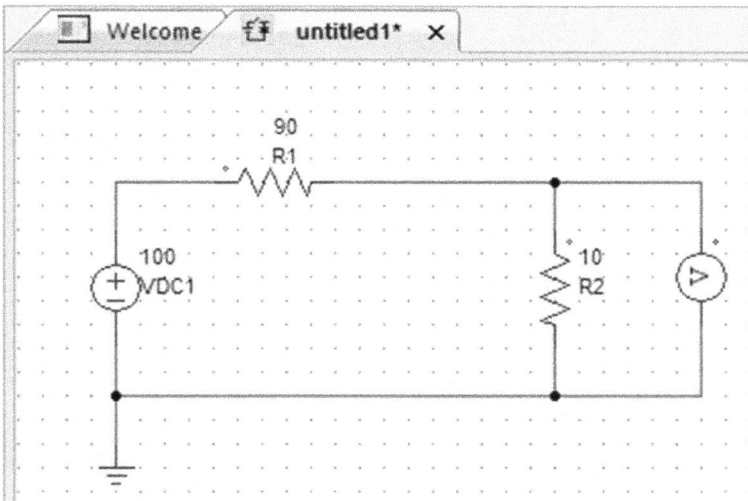

Fig. 2.36.

Double click the voltage probe block and enter the VR2 in the Name box (Fig. 2.37).

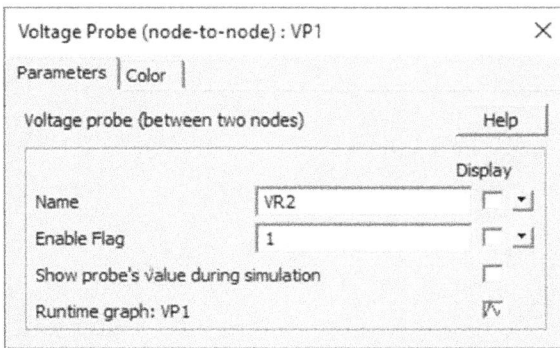

Fig. 2.37.

You can use the View>Fit to Page (Fig. 2.38) in order to automatically fit the schematic into the page (Fig. 2.39). You can zoom in/out with the combination of Ctrl key on your keyboard and middle mouse button. Another way is using the + and – keys on the keyboard. The zoom level of your schematic is shown in the bottom left corner of the screen (Fig. 2.40).

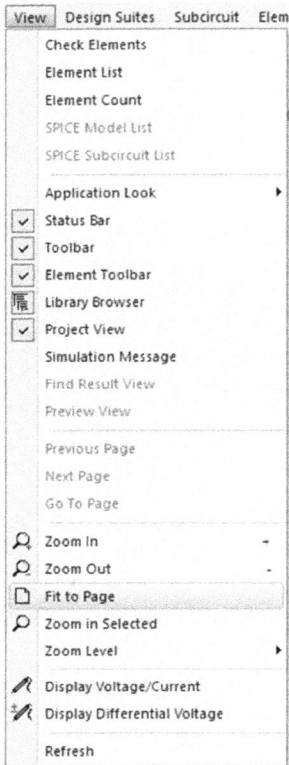

Fig. 2.38.

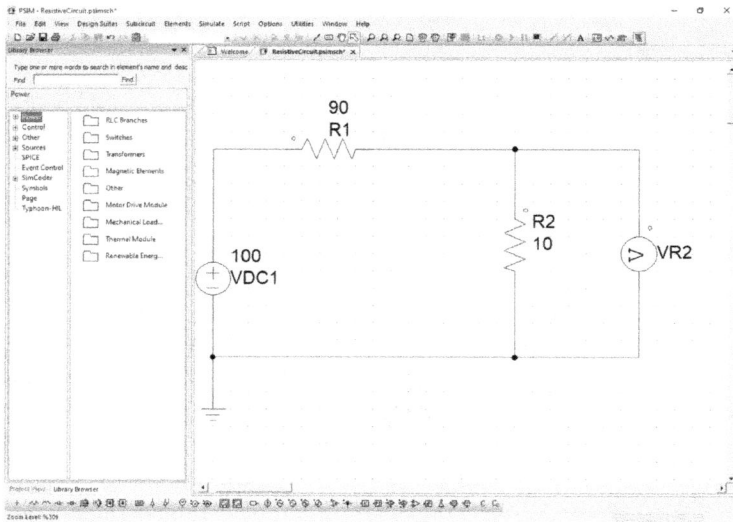

Fig. 2.39.

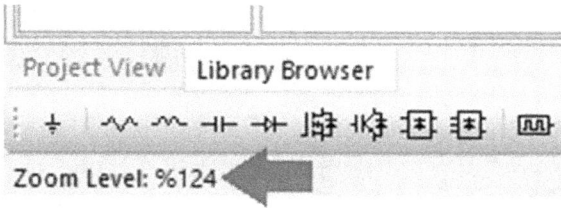

Fig. 2.40.

2.3 Simulating the circuit

You need to add a Simulation Control block to the schematic in order to simulate the circuit. The Simulation Control block can be found in the Simulate menu (Fig. 2.41). Add a Simulation Control block to the schematic (Fig. 2.42)..

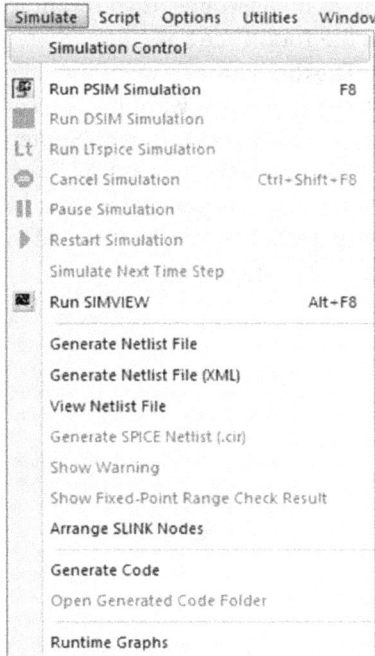

Fig. 2.41.

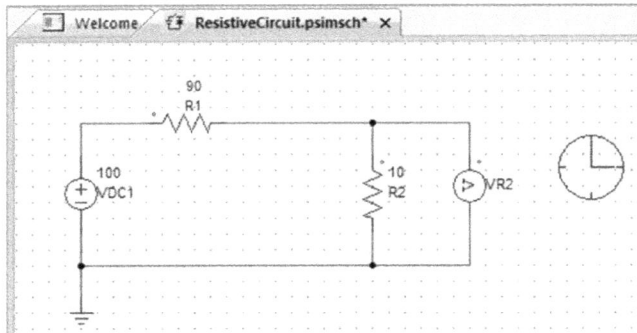

Fig. 2.42.

Double click the Simulation Control block. Enter 0.02 in the Total Time box. The Total Time box determines the duration of simulation. So, 0.02 means that simulate the circuit behavior for [0 s,0.02 s] interval. Instead of 0.02 you can write 20 m. The letter m in PSIM means milli, that is, 10^{-3}, so 20 m means 20 milliseconds. If you want to simulate the circuit for 100 μs, simply enter 100 u in the Total Time box.

Computers do the calculations in discrete time. The difference between the time points can be constant or variable. For instance, a fixed step solver with time step of 10 μs do the analysis at 10 μs, 20 μs, 30 μs, 40 μs, 50 μs, ... However, a variable step solver uses different steps for generating the output. The variable step solvers have two steps: a small step and a big step. The small step is used for transitions, that is, change in the status of switches, or narrow pulses. Figure 2.43 shows the Simulation Control block. The Solver Type is set to fixed step with 10 μs steps.

Fig. 2.43.

Click the Run button (or press the F8 on your keyboard) (Fig. 2.44).

Fig. 2.44.

The PSIM shows the message shown in Fig. 2.45. PSIM decreases the time step automatically in order to increase accuracy. Click OK to continue.

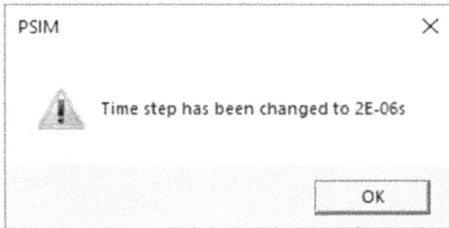

Fig. 2.45.

After clicking the OK, the Simview window shown in Fig. 2.46 will appear. This is waveform measured by VR2 voltage probe. Double click on the VR2. The Properties window shown in Fig. 2.47 will appear.

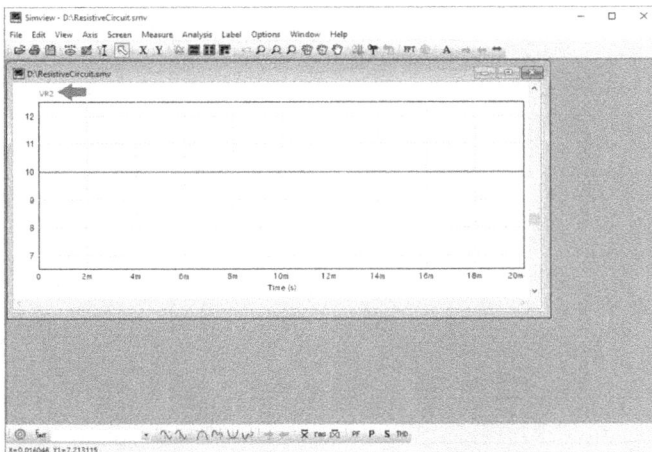

Fig. 2.46.

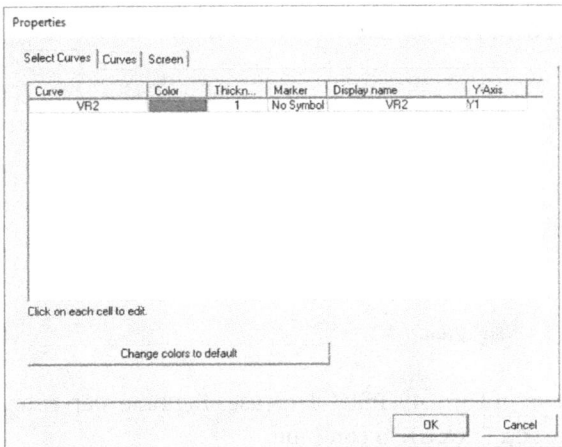

Fig. 2.47.

Click on the Color box in order to select the desired color for the waveform shown in Fig. 2.46. Click the Display name box (Fig. 2.48) in order to change the title of waveform (Fig. 2.49).

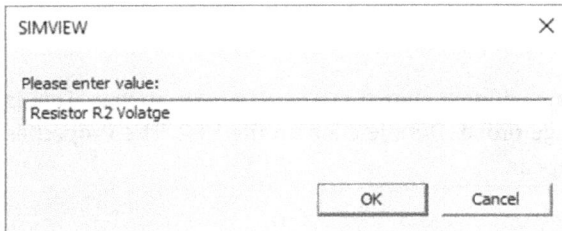

Fig. 2.48.

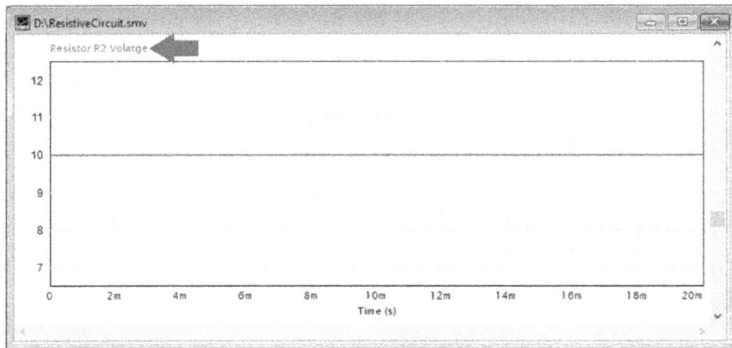

Fig. 2.49.

2.4 Measurement of circuit current

Click on the piece of wire that connects the resistor R1 to the DC source (Fig. 2.50). Press the Delete key on your keyboard in order to remove it (Fig. 2.51).

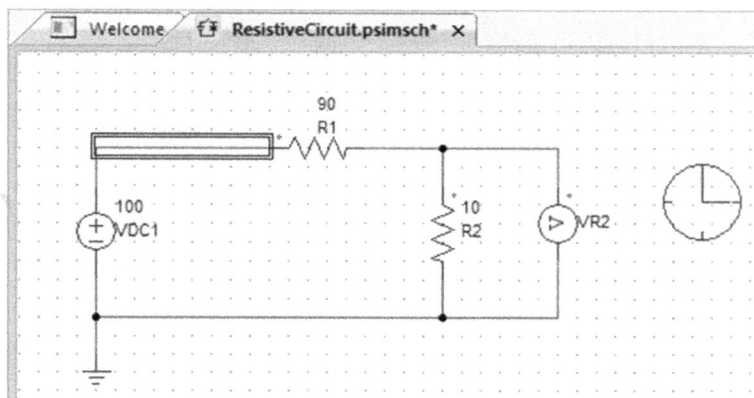

Fig. 2.50.

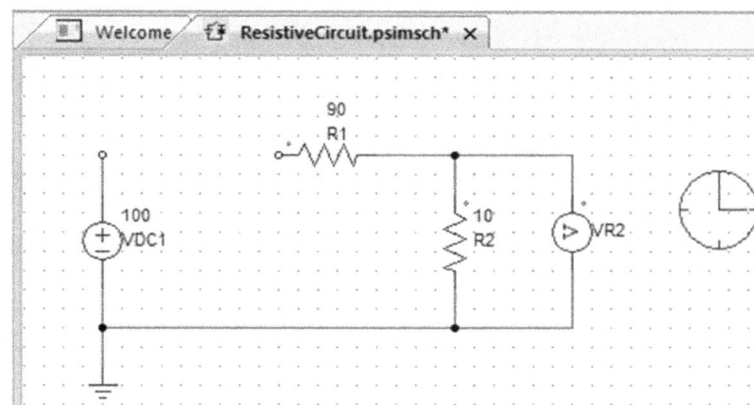

Fig. 2.51.

Select the current probe (Fig. 2.52), add it to the schematic, and connect it to the resistor R1 and DC voltage source (Fig. 2.53).

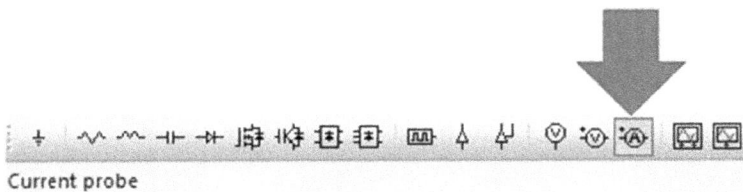

Current probe

Fig. 2.52.

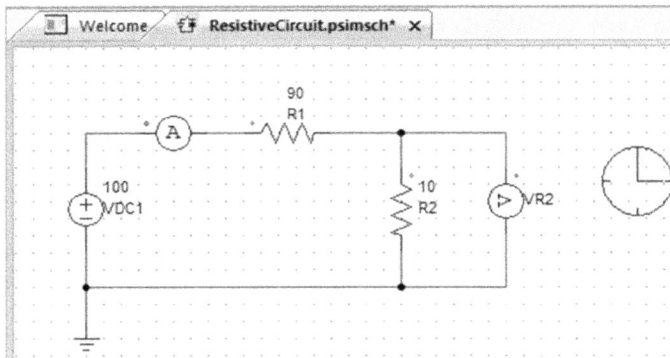

Fig. 2.53.

There is another way to add a current probe to the desired location without removing the wire. If you select and drop a current probe onto a wire, it will be connected automatically to the wire.

The current probe has a terminal with a small dot behind it. Direction of the current which enters the dot is assumed to be positive.

After connecting the current probe to the circuit, double click on it, and rename it to I1. Click the help button. The window shown in Fig. 2.54 will appear.

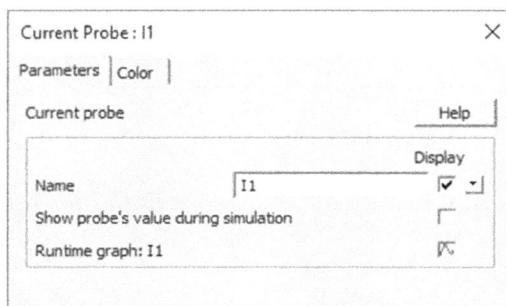

Fig. 2.54.

According to the Help file (Fig. 2.55), the current probe block uses a $1\mu\Omega$ sense resistor in order to produce a very small voltage drop and measures the current using the Ohm's law, that is, $I = \dfrac{V}{R}$.

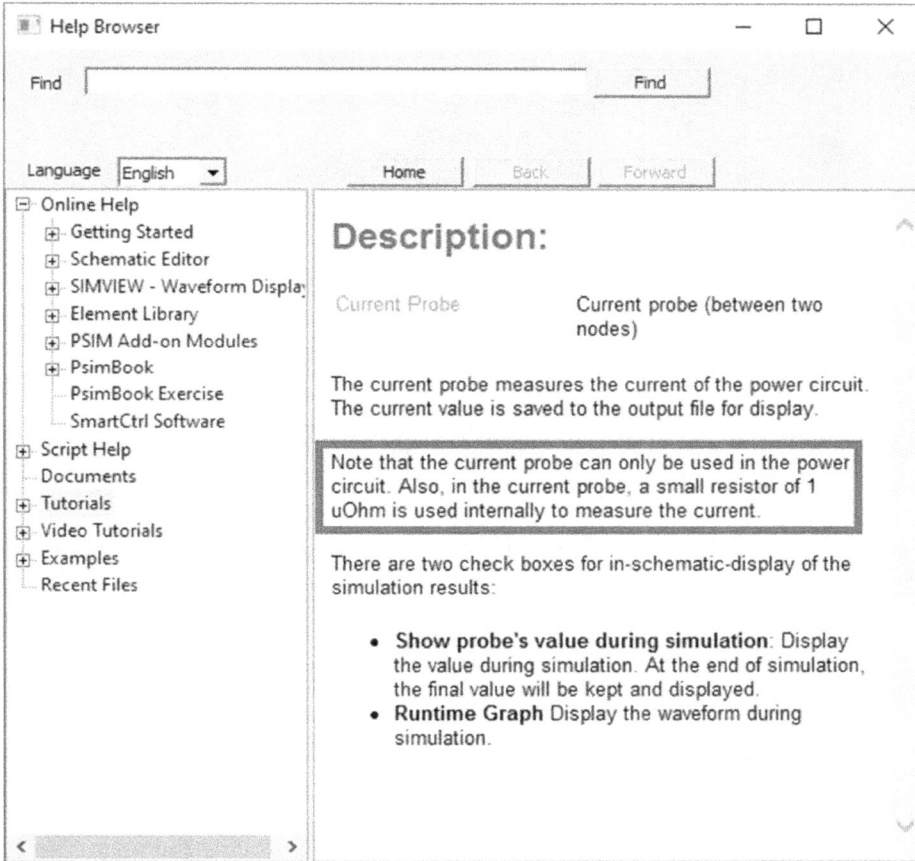

Fig. 2.55.

Now run the simulation. This time the window shown in Fig. 2.56 will appear and asks you to determine the output or outputs that you want to see.

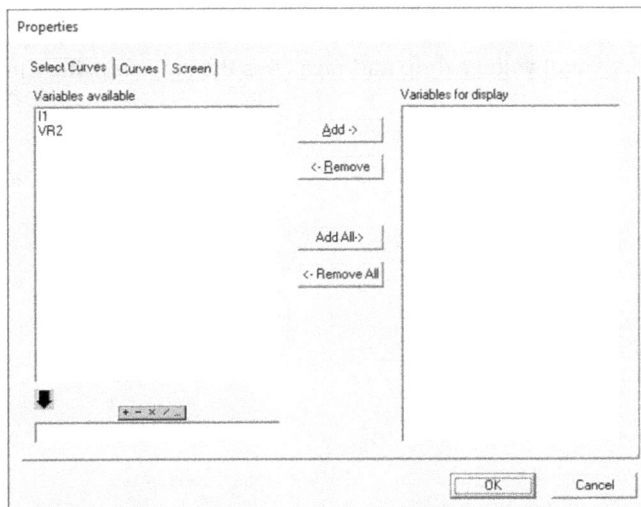

Fig. 2.56.

Select I1 from the left side list by clicking on it and press the Add button to add it to the right list (Fig. 2.57). The right list is the variables that will be displayed after clicking the OK button.

Click the OK button after adding the I1 to the right list. The waveform shown in Fig. 2.58 appears. As expected the current is 1 A.

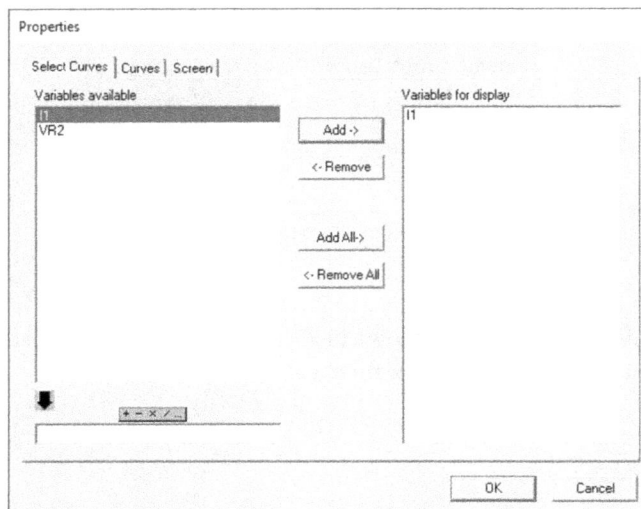

Fig. 2.57.

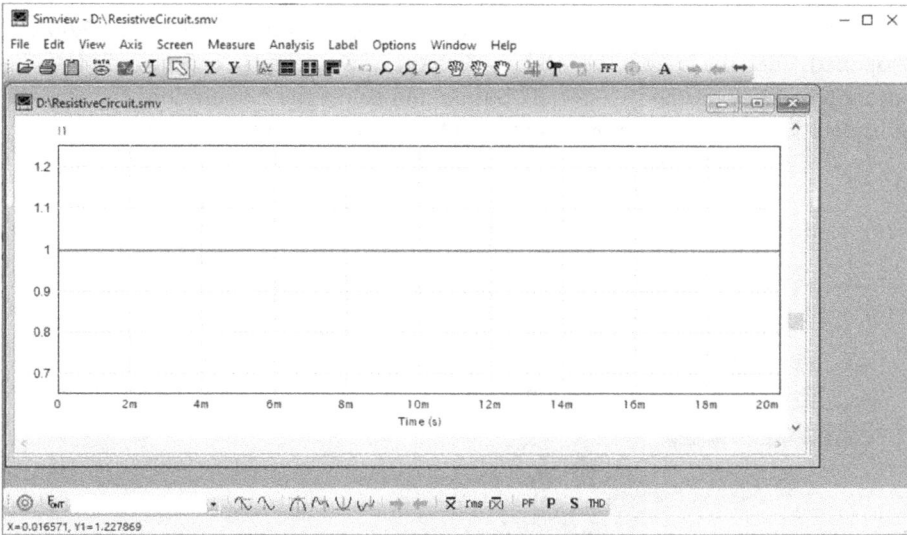

Fig. 2.58.

2.5 Calculation of power

In this section, we want to see the waveform of power dissipated in resistor R2. The power is nothing more than the product of current and voltage. Click the Run SIMVIEW button (Fig. 2.59) in order to ask the PSIM to draw the product of current and voltage.

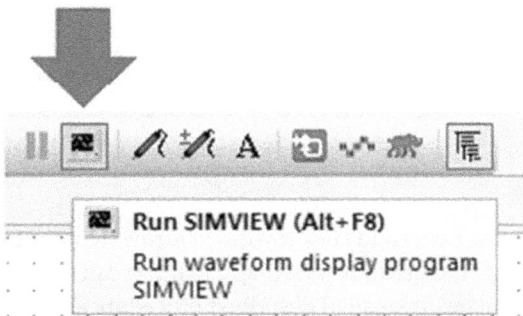

Fig. 2.59.

After clicking the Run SIMVIEW button, the properties window shown in Fig. 2.60 will be opened. Click the I1 from the left list. Click the down arrow in order to add it to the bottom box. Click on the "+ − × / …" button and select the * operation. * shows the multiplication operation. After clicking the *, the bottom box changes to I1*.

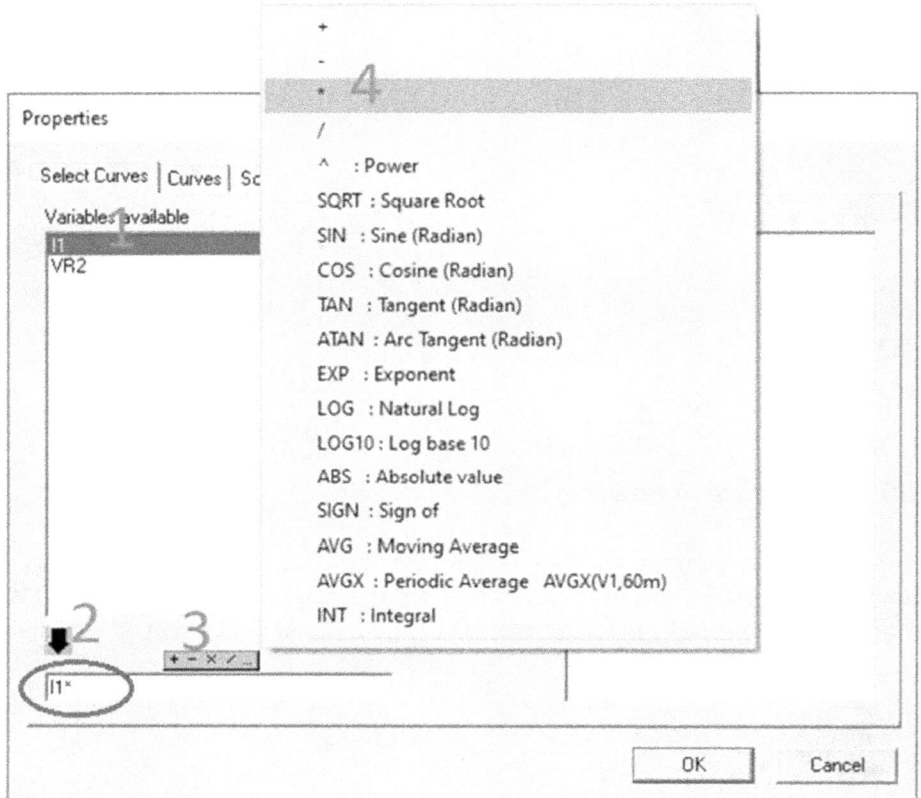

Fig. 2.60.

Now select the VR2 from the left list (click on it) and click the down arrow in order to add it to the bottom box. The bottom box changes to I1*VR2 (Fig. 2.61). You can simply type the I1*VR2 expression in the bottom box instead of described procedure as well.

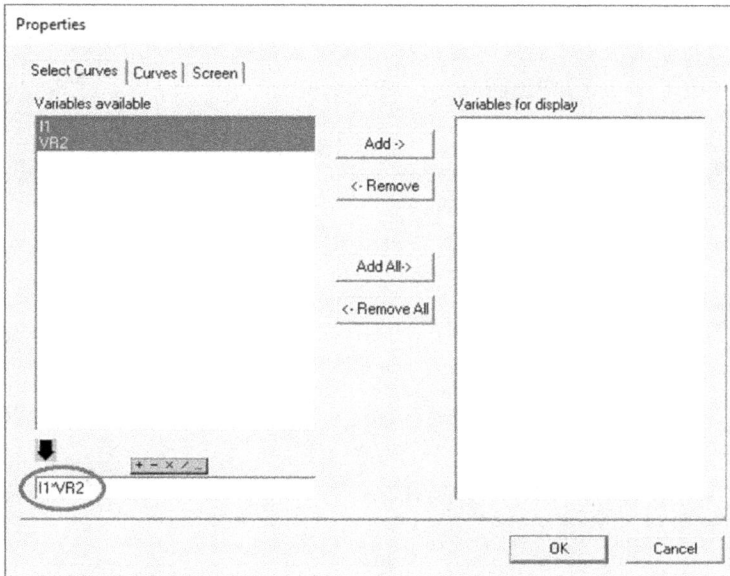

Fig. 2.61.

Click the Add button in order to send the I1*VR2 expression to the right list (Fig. 2.62).

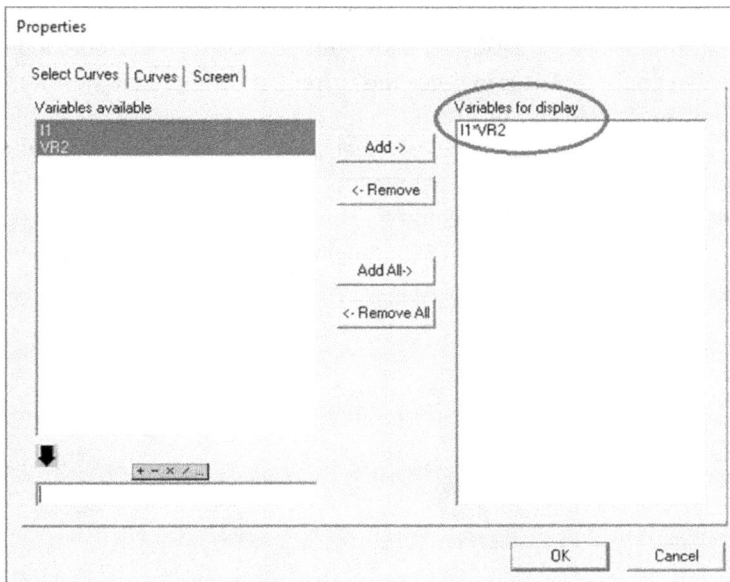

Fig. 2.62.

Click the OK button. The waveform shown in Fig. 2.63 appears. As expected, the power dissipated in resistor R2 is 10 W.

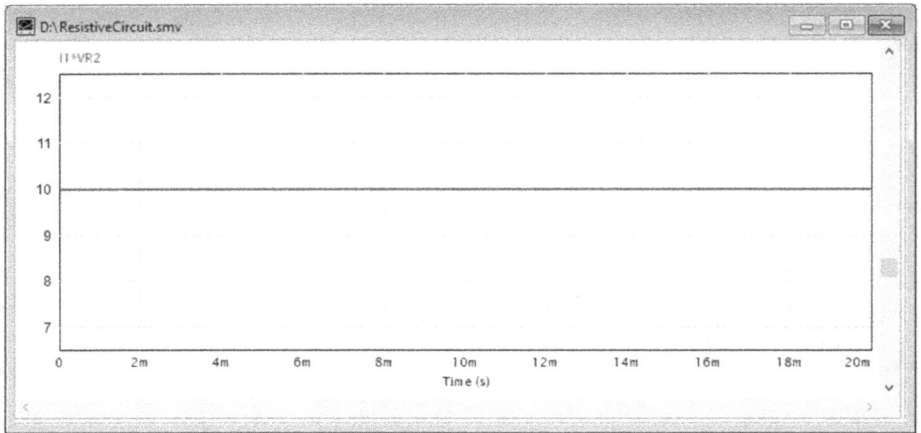

Fig. 2.63.

2.6 Voltage and current sensors

Assume that you want to simulate a DC–DC converter with average current-mode control. In this case, you need a sensor to sense the current of the inductor. Result of measurement will be used in the control circuit.

The voltage/current sensor block in PSIM (Fig. 2.64) measures the voltage/current and permits you to use that measurement where needed. One of the voltage sensor terminals has a small dot behind it. The voltage sensor measures the $V_{terminal\ with\ dot} - V_{terminal\ without\ dot}$. Similar to the voltage sensor, one of the current sensor terminals has a dot. The current that enters this terminal is assumed to be positive.

Fig. 2.64.

According to the Help file (Fig. 2.65), the current sensor block uses a $1\mu\Omega$ sense resistor in order to produce a very small voltage drop and measures the current using the Ohm's law, that is, $I = \dfrac{V}{R}$.

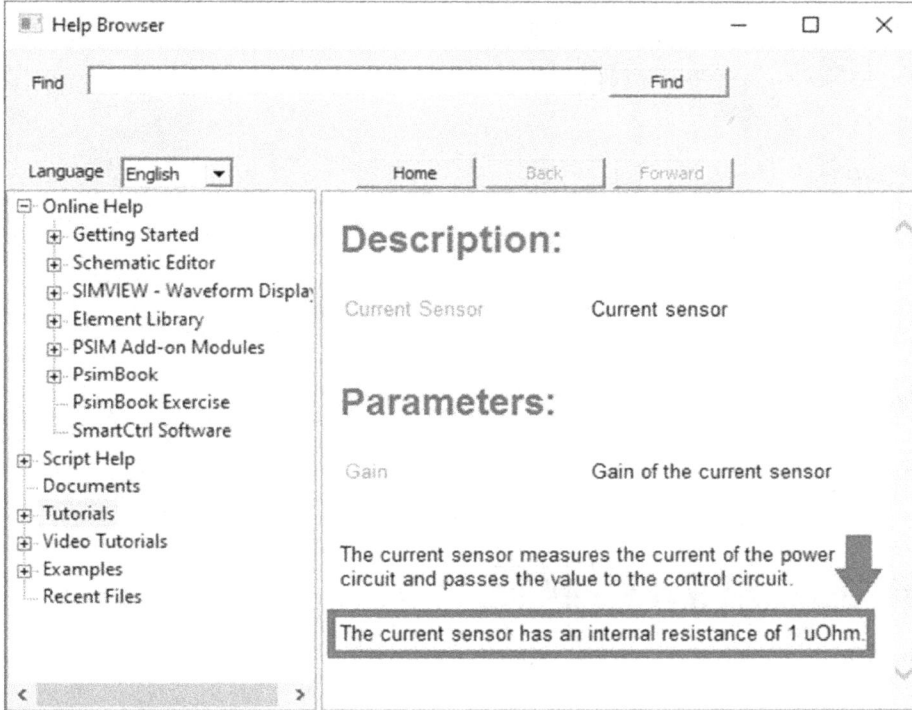

Fig. 2.65.

Assume that we want to measure the resistor R2 power. We can use a voltage sensor to measure the voltage of resistor R2, and a current sensor to measure its current. Then we can use a multiplier block (Fig. 2.66) in order to multiply these two waveforms.

Fig. 2.66.

The schematic shown in Fig. 2.67 calculates the power waveform using the voltage/current sensor blocks. A voltage probe block (Fig. 2.68) is used to save the calculated power waveform in this figure.

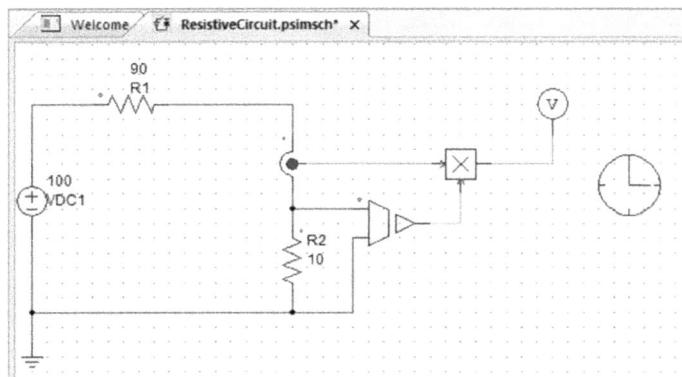

Fig. 2.67.

Fig. 2.68.

Settings of components used in Fig. 2.67 are shown in Figs. 2.69–2.73. The Gain property of the voltage sensor and current sensor is equal to 1. So, the measurement result will not be amplified/attenuated.

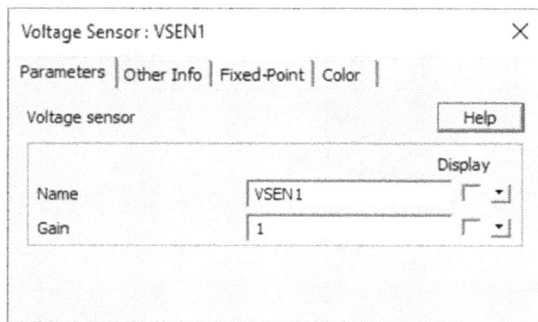

Fig. 2.69.

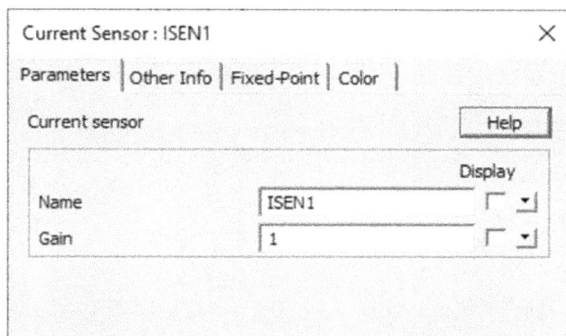

Fig. 2.70.

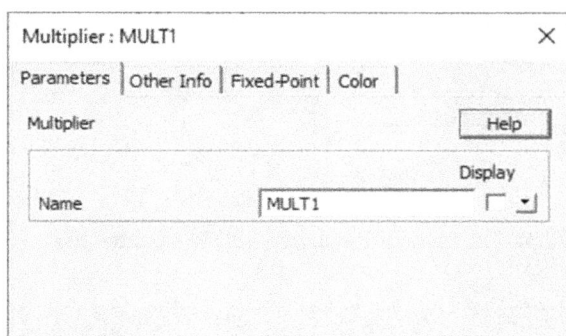

Fig. 2.71.

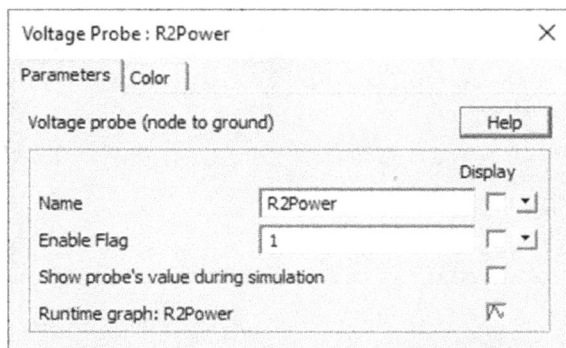

Fig. 2.72.

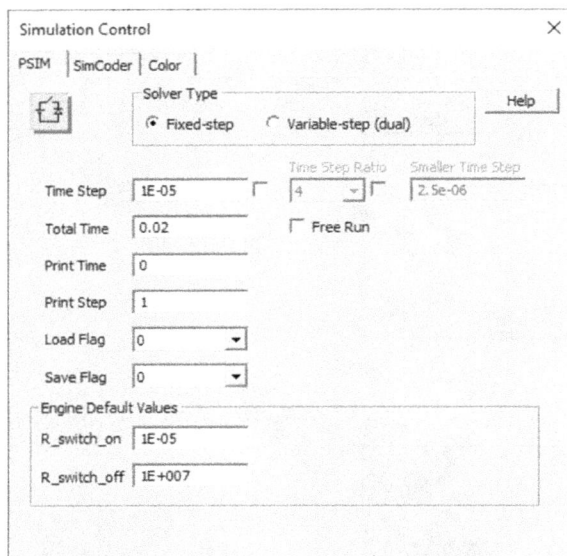

Fig. 2.73.

Simulation result is shown in Fig. 2.74. The resistor R2 power is 10 W as expected.

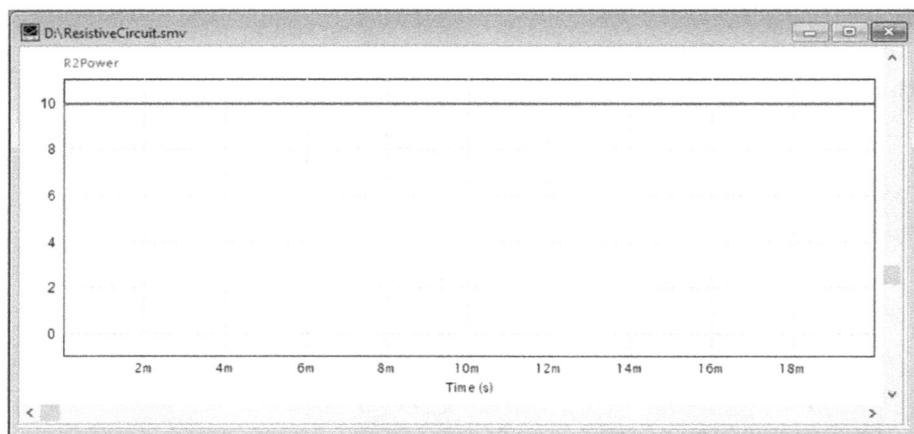

Fig. 2.74.

2.7 Measurement with current flag and voltage flag

Using the voltage/current probe is not the only way to measure the voltage/current of an element. You can measure the voltage/current of an element using the voltage/current flag as well.

For instance, assume that we want to measure the voltage and current of resistor R2 in the schematic shown in Fig. 2.75.

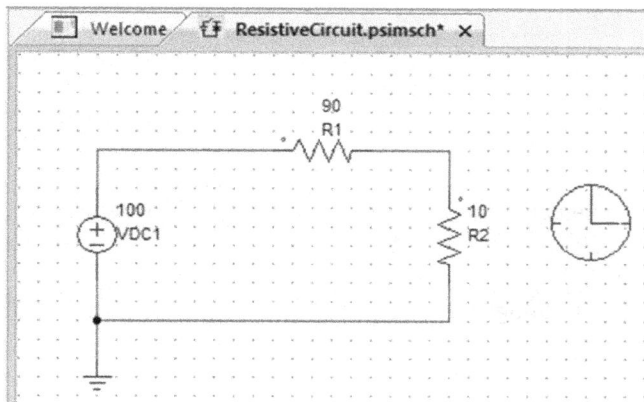

Fig. 2.75.

Double click on the resistor R2. The Current Flag and Voltage Flag boxes default values are zero (Fig. 2.76). Change them to 1 (Fig. 2.77). Note that the resistor has a small dot behind to one of the terminals. The voltage flag measures the $V_{terminal\ with\ dot} - V_{terminal\ without\ dot}$. The current probe measures the current through the component and assumes the current that enters the dot to be positive.

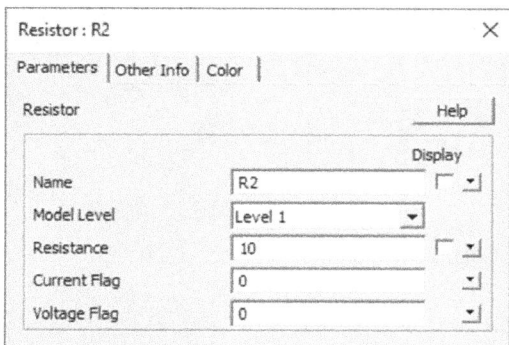

Fig. 2.76.

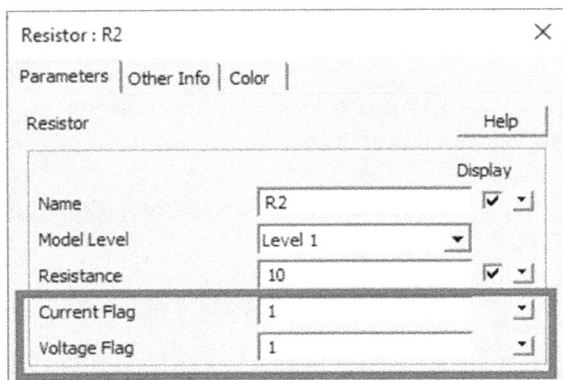

Fig. 2.77.

As shown in Fig. 2.75, the schematic does not contain any voltage/current probe, however if we set the voltage/current flags to 1, then we are able to measure the voltage/current of components.

Run the simulation. After running the simulation, the properties window will appear (Fig. 2.78).

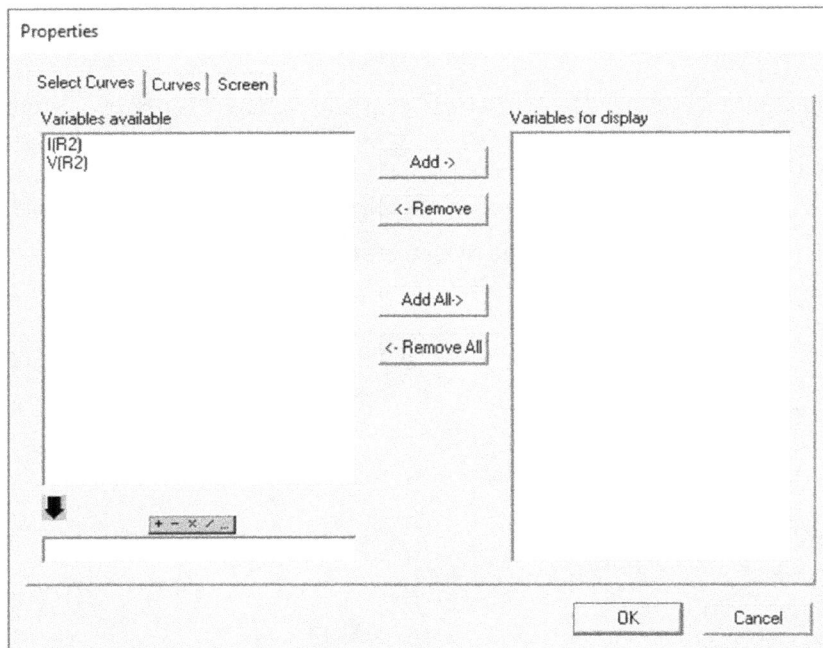

Fig. 2.78.

Click on I(R2) and V(R2) (Fig. 2.79). Then click the Add and OK buttons, respectively (Fig. 2.80).

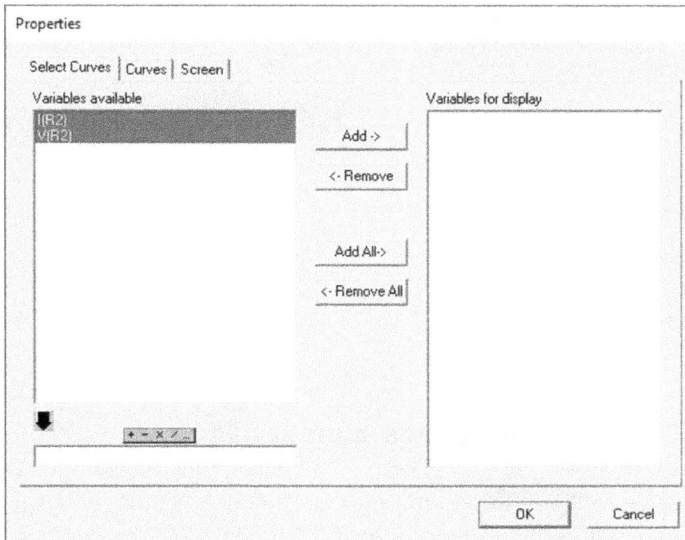

Fig. 2.79.

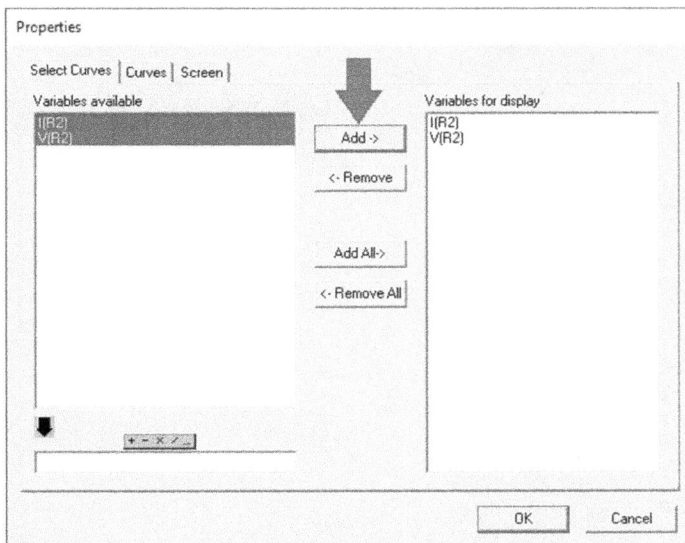

Fig. 2.80.

After clicking on the OK button, the voltage/current waveforms of resistor R2 appears (Fig. 2.81).

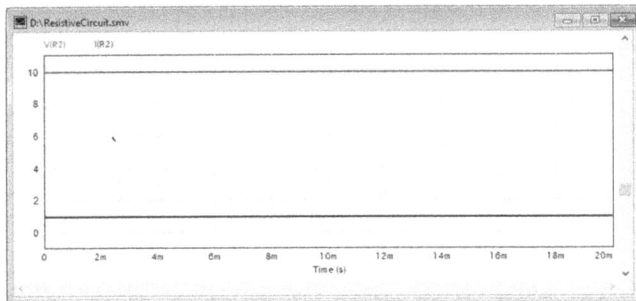

Fig. 2.81.

2.8 Calculation of power using voltage/current flag

You can use the voltage/current flag in order to measure the power of an element. Assume that we want to measure the power of resistor R2 in Fig. 2.75. In order to do this, run the simulation. After seeing the Properties window, select the I(R2) from the left list. Click the down arrow in order to add the I(R2) to the bottom box. After that click the "+ − × / ..." and select the *.

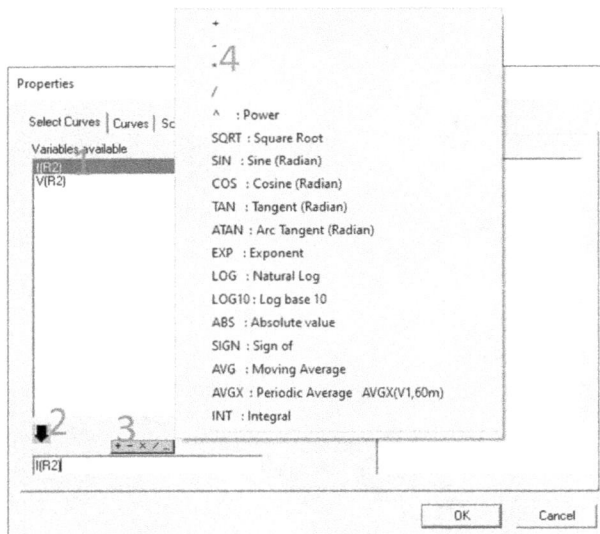

Fig. 2.82.

Now select the V(R2) from the left list. Click the down arrow in order to add the V(R2) to the bottom box. After clicking the arrow, the bottom box will change to "I(R2)*V(R2)." You can simply type "I(R2)*V(R2)" in the bottom box instead of doing the aforementioned procedure.

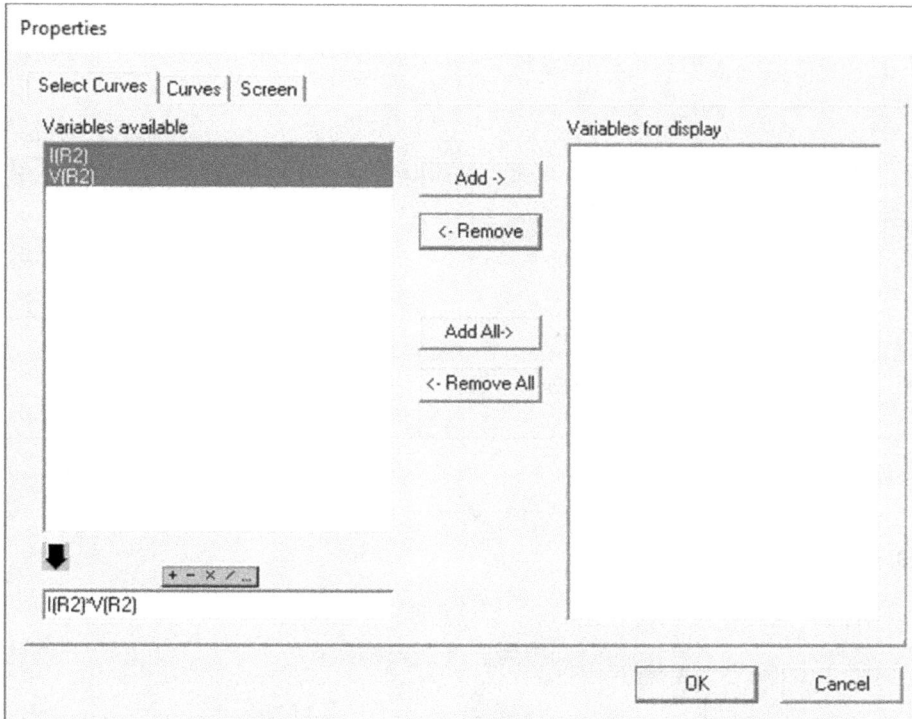

Fig. 2.83.

Click the Add button to add the expression "I(R2)*V(R2)" to the right list (Fig. 2.84). Click the OK button to see the "I(R2)*V(R2)" waveform. As shown in Fig. 2.85, the resistor R2 power is 10 W.

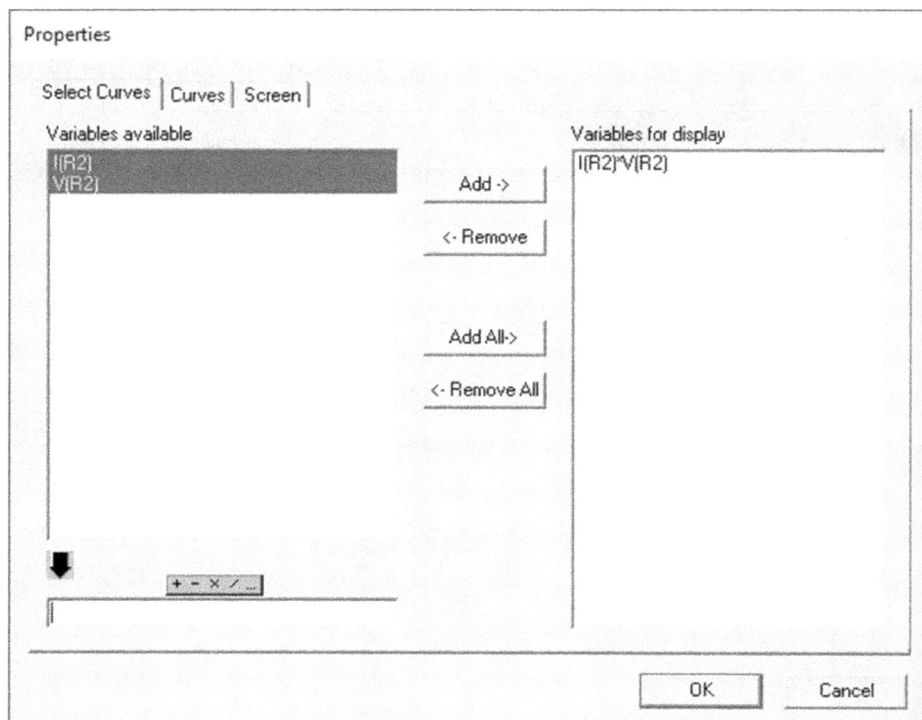

Fig. 2.84.

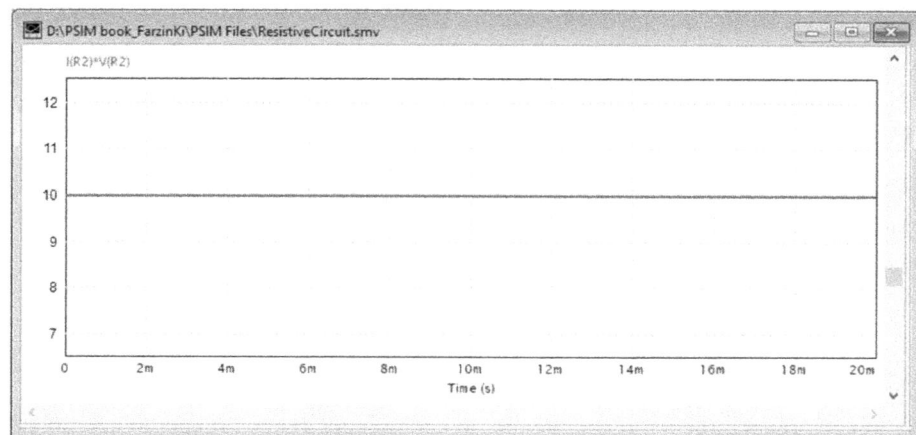

Fig. 2.85.

2.9 Color of connections

In PSIM, different types of connections are shown with different colors. As shown in Fig. 2.86, the wires that connect the electrical components together are red and the color of wires that connect the voltage/current sensor to the multiplier is green. By default, the red color is used for power circuit connections, green color is used for control signals, and maroon is used for mechanical connections (for instance, a load connected to the shaft of a motor).

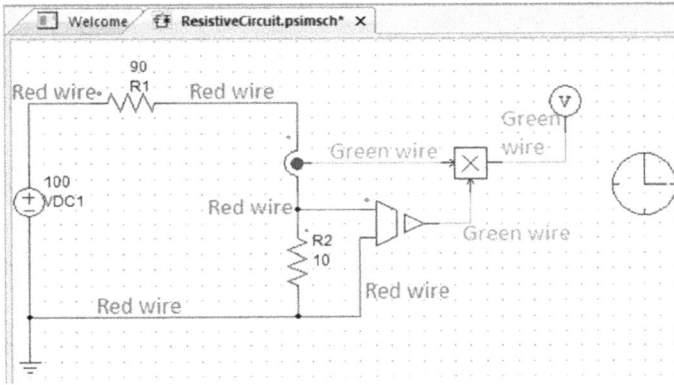

Fig. 2.86.

The power circuit is composed of components which are found in the "RLC Branches," "Switches," "Transformers," and "Magnetic Elements" sections of Power menu (Fig. 2.87). The connection between components of the aforementioned families is shown in red color (Fig. 2.86).

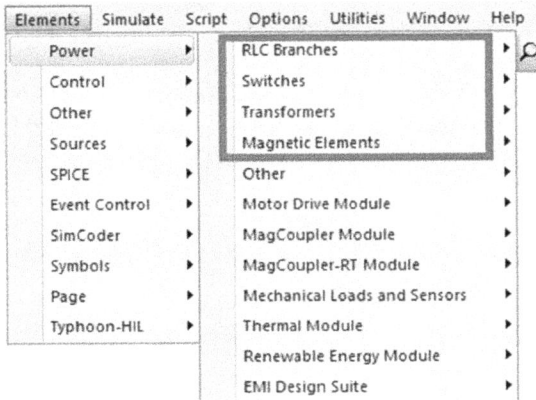

Fig. 2.87.

The control blocks can be found in the "Control" section of Elements menu (Fig. 2.88). By default, color of connections between the blocks of this family is green.

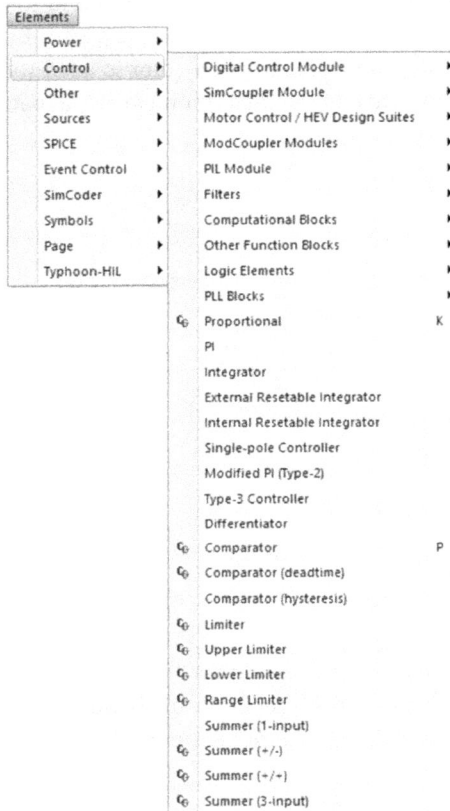

Fig. 2.88.

Mechanical components can be found in the "Mechanical Loads and Sensors" section of Power menu (Fig. 2.89). By default, color of connections between the blocks of this family is maroon.

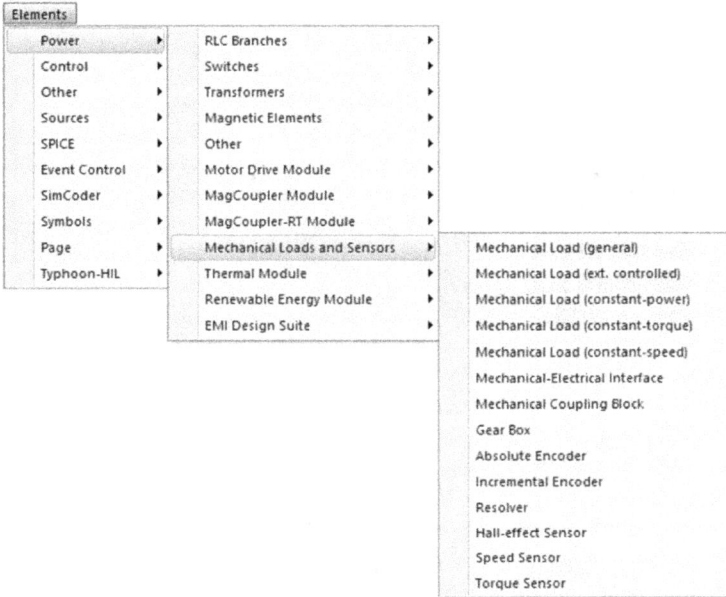

Fig. 2.89.

In order to see the default colors for different types of connections, click the Options>-Settings (Fig. 2.90). The Option window appears. Click the Colors tab (Fig. 2.91). You can change the colors if you like by clicking the "..." button.

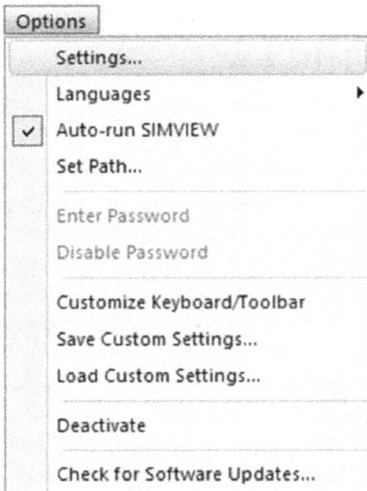

Fig. 2.90.

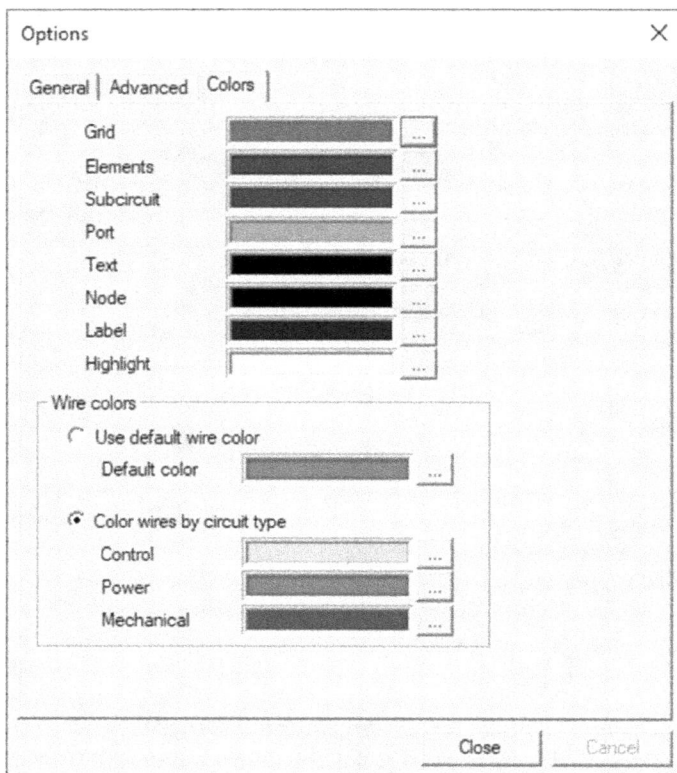

Fig. 2.91.

Note that you cannot connect different types of signals together. In order to understand this matter, consider the simple circuit shown in Fig. 2.92.

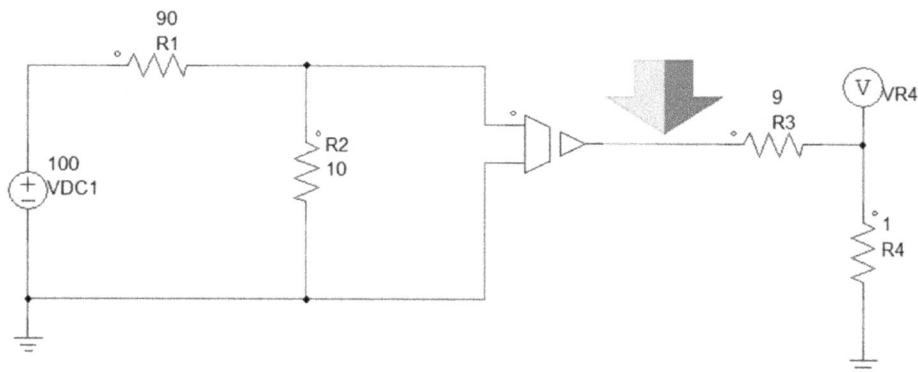

Fig. 2.92.

The output of a voltage sensor (control signal) is connected to a resistor R3 (power circuit element). As shown in Fig. 2.92, a section of the wire that connects the voltage sensor to the resistor R3 is green and another section of that wire is green.

PSIM uses such a color scheme in order to show that a connection between two different families of elements has been established. Sometimes, you can run the simulation despite this visual warning.

You can use the "Control-to-power Interface" block (Fig. 2.93) in order to solve the problem of circuit shown in Fig. 2.92,

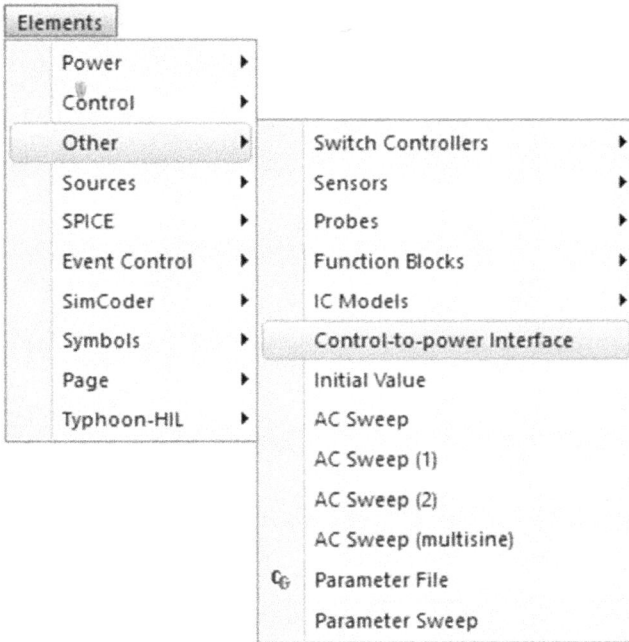

Fig. 2.93.

Add the "Control-to-power Interface" block as shown in Fig. 2.94. Now the piece of wire that connects the voltage sensor to "Control-to-power Interface" block is completely green and the piece of wire that connects the "Control-to-power Interface" to the resistor R3 is completely red. The simulation result for circuit shown in Fig. 2.94 is shown in Fig. 2.95. As expected, the voltage of resistor R4 is 1 V.

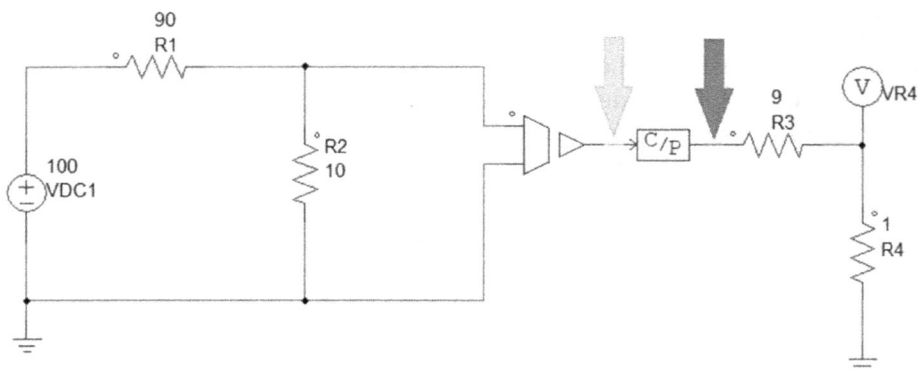

Fig. 2.94.

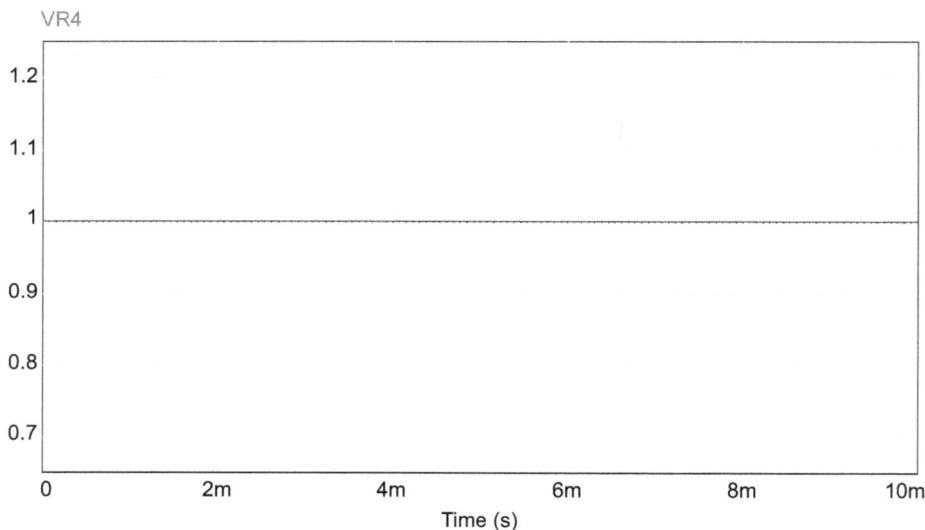

Fig. 2.95.

2.10 Exporting the drawn schematic

You can export the schematic of circuits which you draw in PSIM. In order to do this, click the Color Bitmap (Fig. 2.96).

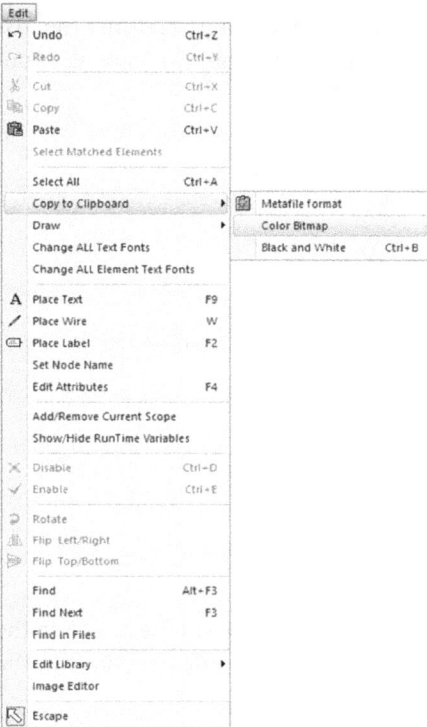

Fig. 2.96.

After clicking the Color Bitmap, the schematic drawn in PSIM will be copied into the clipboard and you can paste it in the desired software like Microsoft Word®, Paint®, and so on.

2.11 Addition of text to the schematic

You can use F9 key or the icon shown in Fig. 2.97 in order to add text to the schematic.

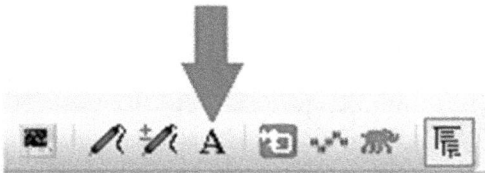

Fig. 2.97.

After pressing the F9 key, the window shown in Fig. 2.98 will appear. Write the text that you want in this box. You can select the desired font for the text using the Font button. You can change the color of text using the "..." button.

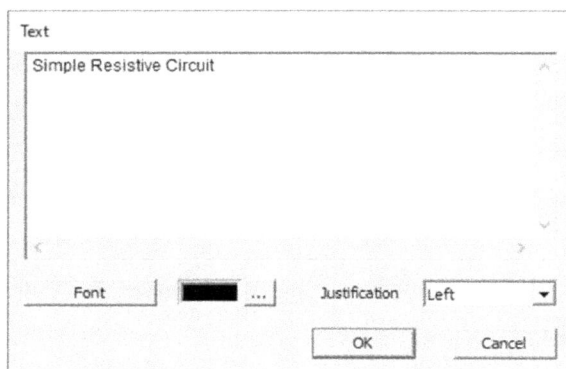

Fig. 2.98.

After clicking the OK button in Fig. 2.98, the text inside the box will be added to the mouse pointer. Click on desired location of the schematic in order to add it to the schematic.

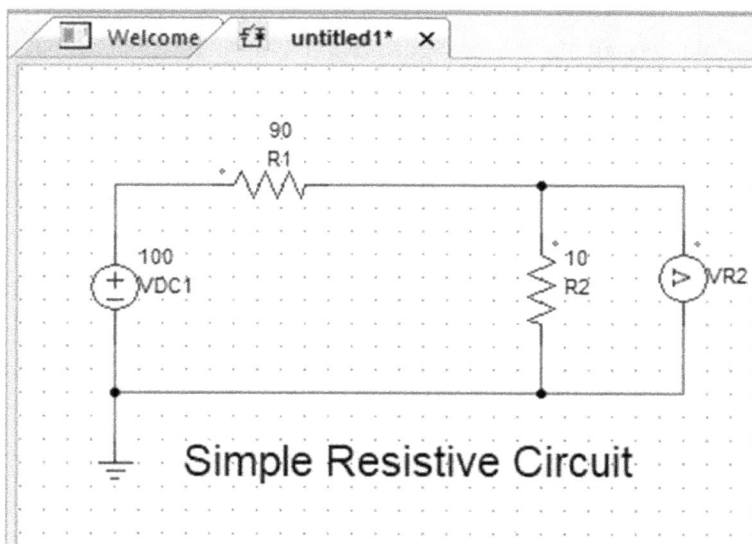

Fig. 2.99.

2.12 Addition of project information to the schematic

You can use the PAGE block (Fig. 2.100) in order to add some useful information like title of circuit, designer, company name, and so on to the schematic. Such a block has no effect on the simulation of circuit. It is a comment for someone who will study the schematic.

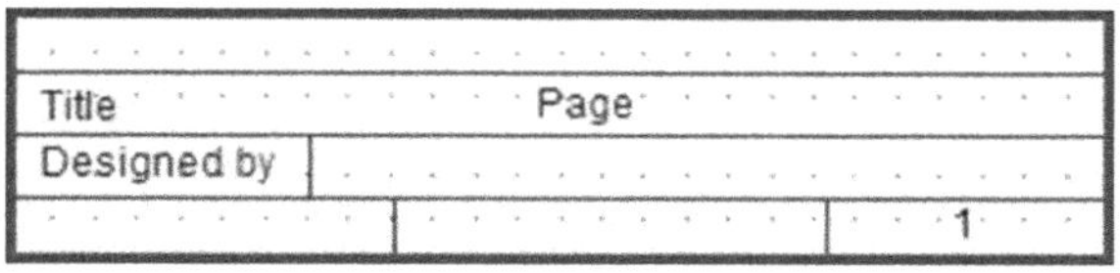

Fig. 2.100.

Search for "page" in the Library Browser window in order to add a PAGE block to your schematic. Click on the PAGE block (shown with arrow in Fig. 2.101) in order to select it. Then click on the desired location of schematic to add the block to the schematic.

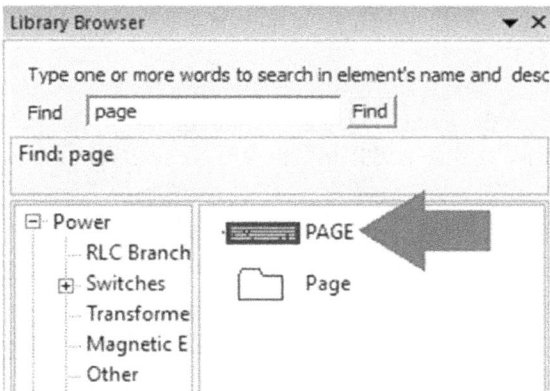

Fig. 2.101.

Double click on the block in order to enter the desired information. For instance, if you fill the boxes as shown in Fig. 2.102, the block will be changed to what shown in Fig. 2.103.

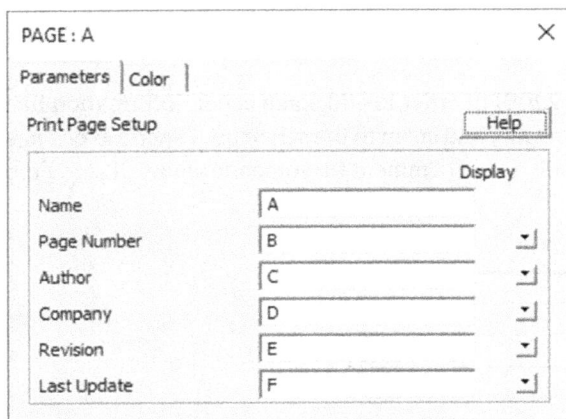

PAGE : A   ✕

Parameters | Color |

Print Page Setup   Help

                                                    Display

Name	A	
Page Number	B	▾
Author	C	▾
Company	D	▾
Revision	E	▾
Last Update	F	▾

Fig. 2.102.

	D	
Title	A	
Designed by	C	
F	E	B

Fig. 2.103.

If you select the New (worksheet) from file menu (Fig. 2.104), then the PAGE block will be added to the schematic automatically (Fig. 2.105) and there is no need for manual addition.

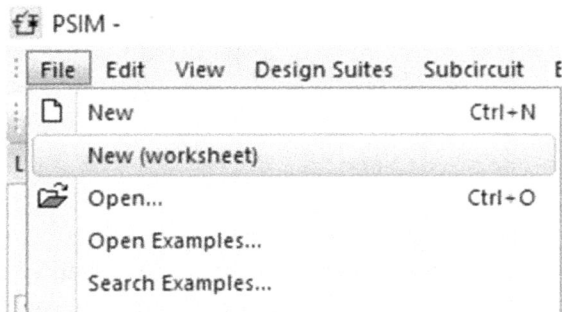

⚡ PSIM -

| File | Edit | View | Design Suites | Subcircuit | E |

☐ New   Ctrl+N

New (worksheet)

📂 Open...   Ctrl+O

Open Examples...

Search Examples...

Fig. 2.104.

Fig. 2.105.

2.13 Saving the schematic

Saving a PSIM schematic is similar to any other Windows® program. Simply click the floppy disk icon and give the desired path.

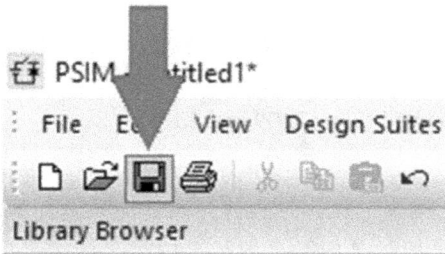

Fig. 2.106.

Consider the schematic shown in Fig. 2.107. As shown in Figs. 2.107 and 2.108, the name of file has a star behind it. This star shows that this file has changed and the changes are not saved yet.

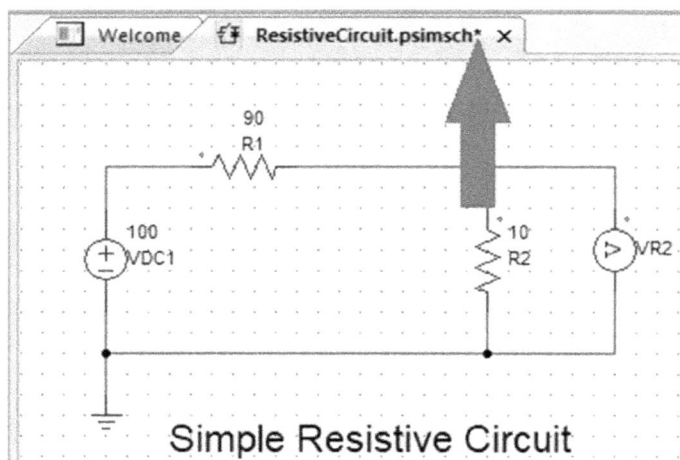

Fig. 2.107.

Fig. 2.108.

As soon as you save the file, the * will be removed (Figs. 2.109 and 2.110).

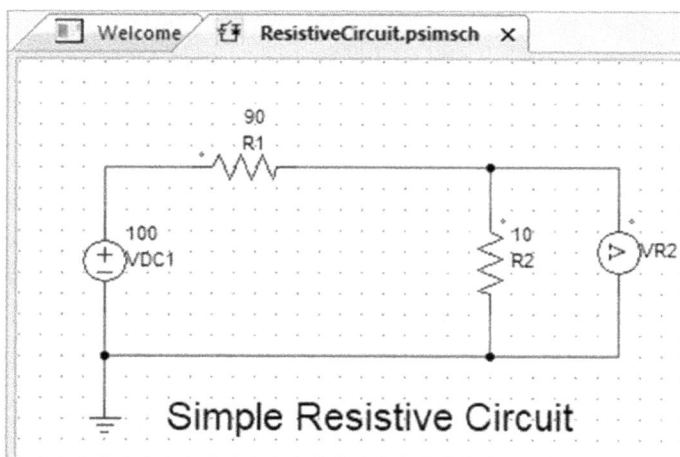

Fig. 2.109.

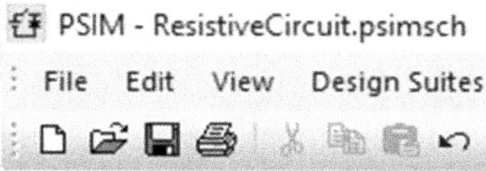

Fig. 2.110.

Another way to save a file is to use the File menu (Fig. 2.111).

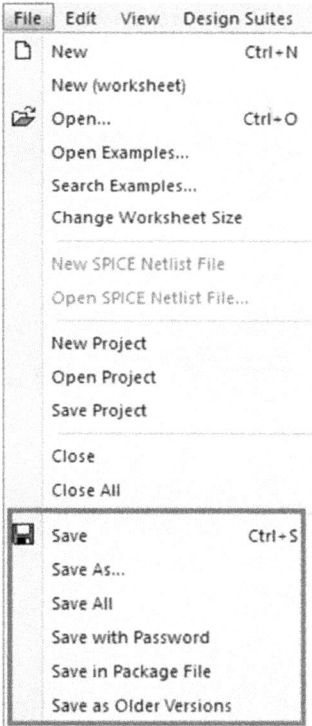

Fig. 2.111.

Use the "Save As …" when you want to make a copy of the current file with a different name.

Use the "Save All" when you want to save all the open files. Sometime you have more than one schematic in PSIM environment. If you click the Save All, as the name suggests, all the open files will be saved.

2.14 Protection of schematic files with password

You can protect your files with password if you prefer. In order to do this, click the "Save with Password" from File menu. The Set Password window will appear (Fig. 2.112). Enter the desired password here and click the Set Password button.

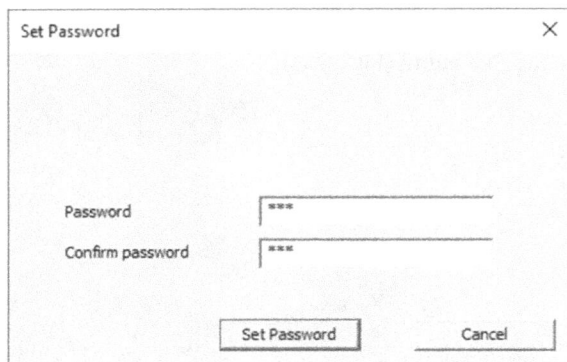

Fig. 2.112.

After clicking the warning message shown in Fig. 2.113 appears. Click the OK button to continue.

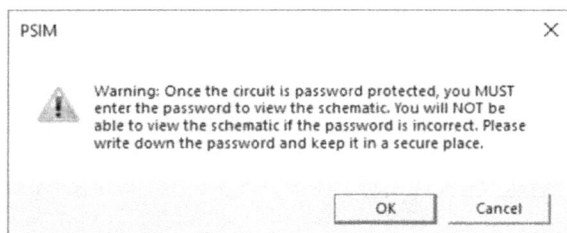

Fig. 2.113.

Now close the file and open it again using File>Open. As shown in Fig. 2.114, the schematic of circuit will not be shown automatically.

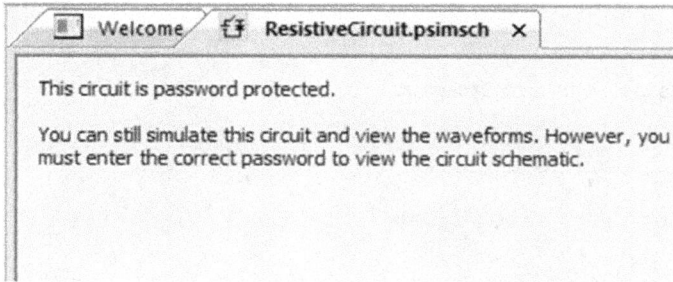

```
▣  Welcome    ↺  ResistiveCircuit.psimsch  ✕

This circuit is password protected.

You can still simulate this circuit and view the waveforms. However, you
must enter the correct password to view the circuit schematic.
```

Fig. 2.114.

In order to see the schematic file, open the Options menu and click the Enter Password button (Fig. 2.115).

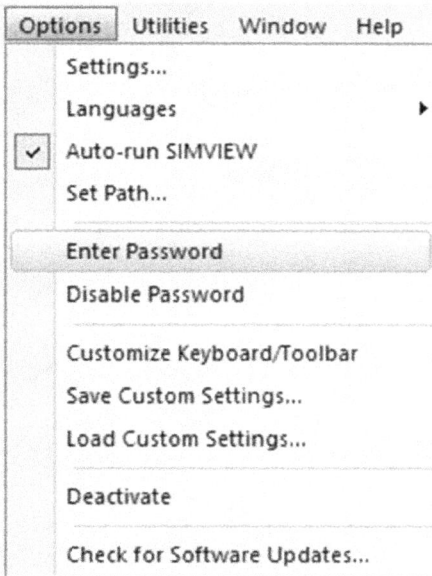

```
Options │ Utilities   Window   Help

      Settings...

      Languages                    ▶

 ✓    Auto-run SIMVIEW

      Set Path...

      Enter Password

      Disable Password

      Customize Keyboard/Toolbar

      Save Custom Settings...

      Load Custom Settings...

      Deactivate

      Check for Software Updates...
```

Fig. 2.115.

After clicking the Enter Password, the Enter Password window will appear. Enter the password here.

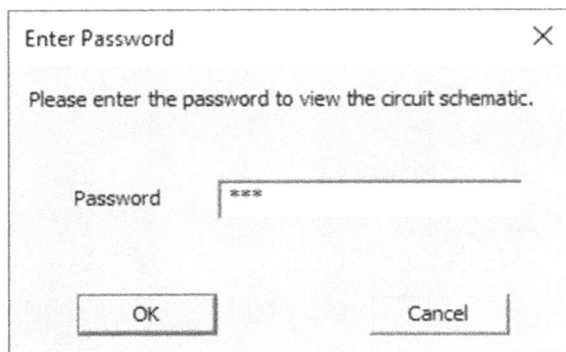

Enter Password   ✕

Please enter the password to view the circuit schematic.

Password   [***]

OK   Cancel

Fig. 2.116.

As soon as you enter the correct password, the schematic will be shown (Fig. 2.117).

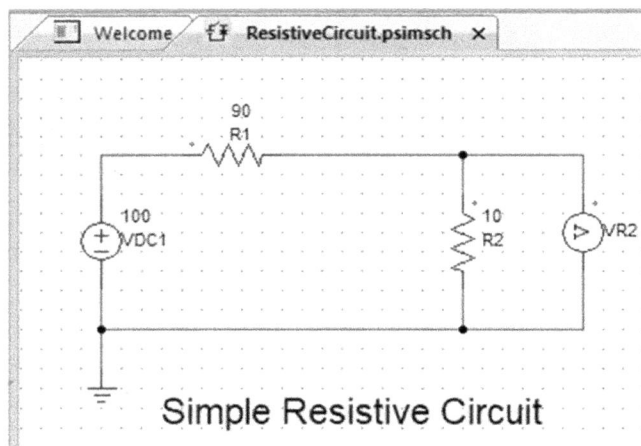

Fig. 2.117.

Click the "Disable Password" (Fig. 2.118) to remove the password from a password protected file. After clicking the "Disable Password" the window shown in Fig. 2.119 will appear and informs you that the file is no longer password protected.

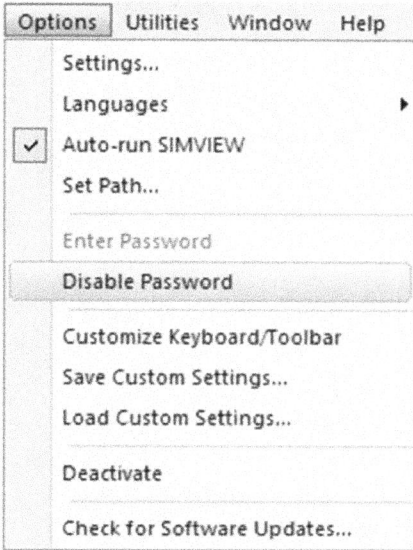

Fig. 2.118.

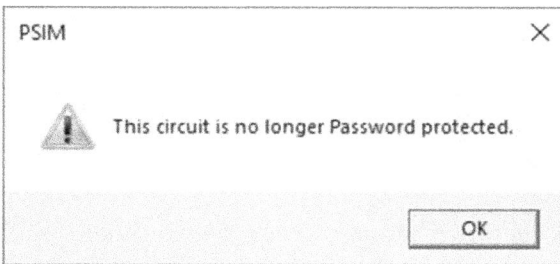

Fig. 2.119.

2.15 Compatible files for older versions of PSIM

You can make compatible file for older version of PSIM. For instance, assume you do some simulation in PSIM 12 and you want to share your work with someone who uses PSIM 10. In this case, click the File>Save as Older Versions and select the PSIM 10.0 Format from save as Type drop down list (Fig. 2.120).

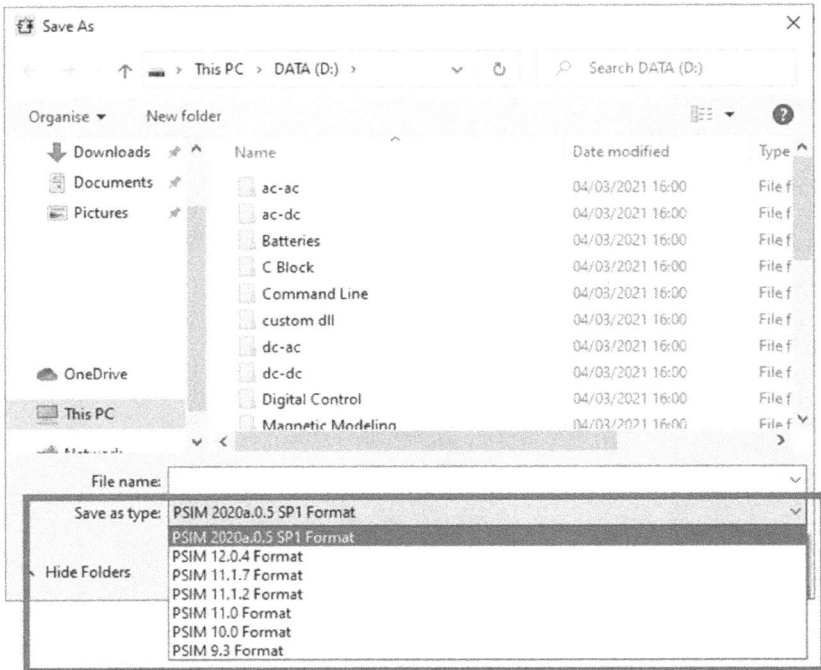

Fig. 2.120.

2.16 Enabling and disabling a component

In PSIM, you can disable a component (so it has no effect on the simulation result) without removing it from the schematic. Assume a schematic like the one shown in Fig. 2.121. This circuit is composed of three resistors. As shown in Fig. 2.122, the voltage across the resistors R2 and R3 is about 5.25 V.

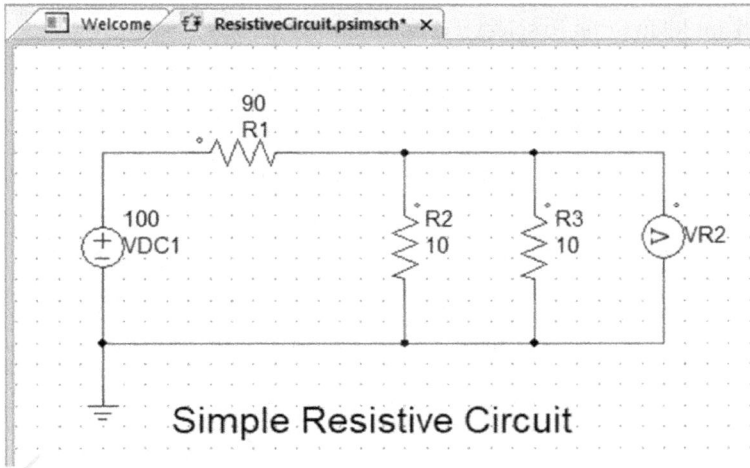

Fig. 2.121.

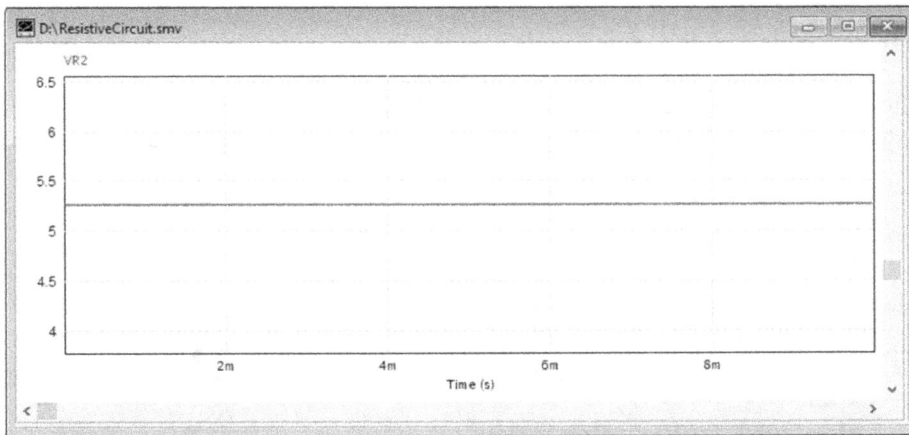

Fig. 2.122.

Now (single) click on R3 in order to select it (Fig. 2.123).

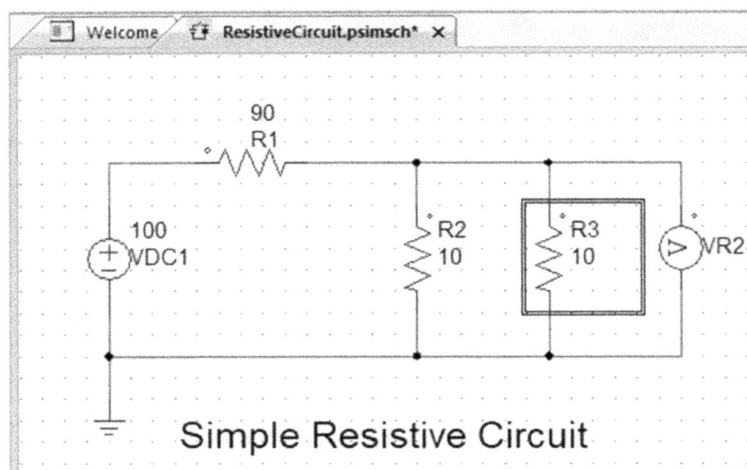

Fig. 2.123.

Click the Disable icon (or press Ctrl+D) (Fig. 2.124).

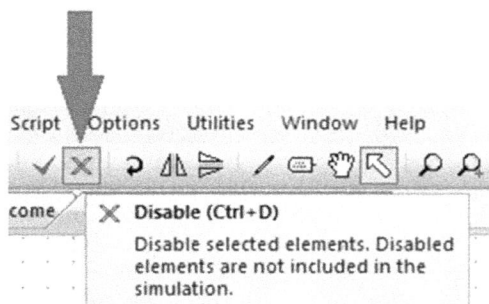

Fig. 2.124.

As shown in Fig. 2.125, the resistor R3 become disabled and has no effect on the simulation. If you run the simulation, the result shown in Fig. 2.126 will appear.

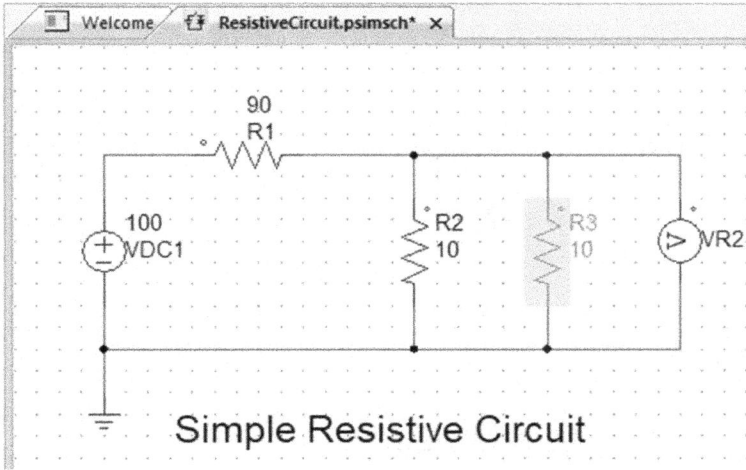

Fig. 2.125.

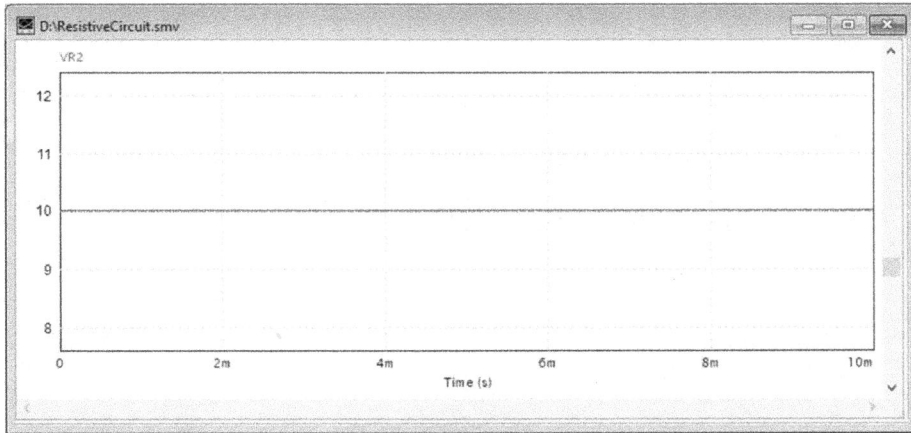

Fig. 2.126.

In order to enable the resistor R3, (single) click on it (Fig. 2.127). Then click the Enable icon (or press Ctrl+E) (Fig. 2.128).

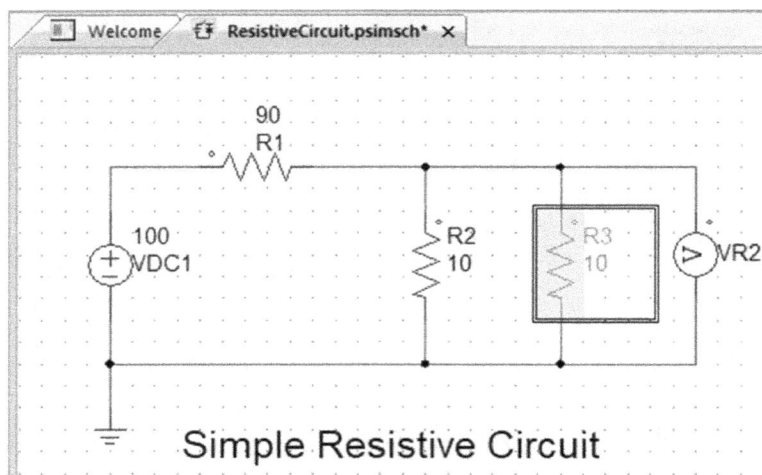

Fig. 2.127.

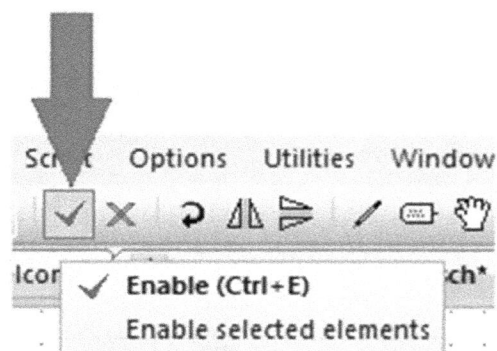

Fig. 2.128.

2.17 Shortcut keys

There is no doubt that shortcut keys make the user work easy. In PSIM, the shortcut keys are accessible using the Options menu (Fig. 2.129).

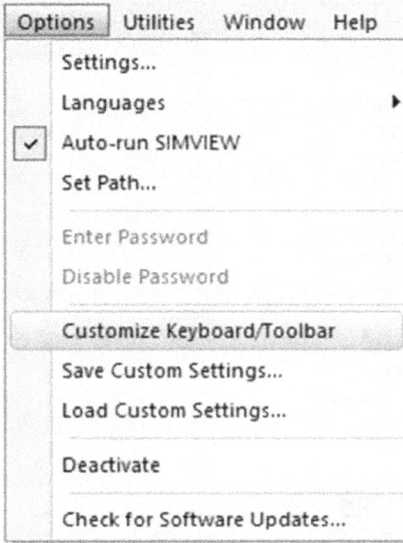

Fig. 2.129.

After clicking the Customize Keyboard/Toolbar, the window shown in Fig. 2.130 will appear. You can define your own shortcuts here. The list of default shortcuts is shown in Fig. 2.131. For instance, according to Fig. 2.131, the shortcut for capacitor is C. This means if you press the C key on your keyboard, then PSIM activates the capacitor for you and you can add a capacitor to your schematic quickly.

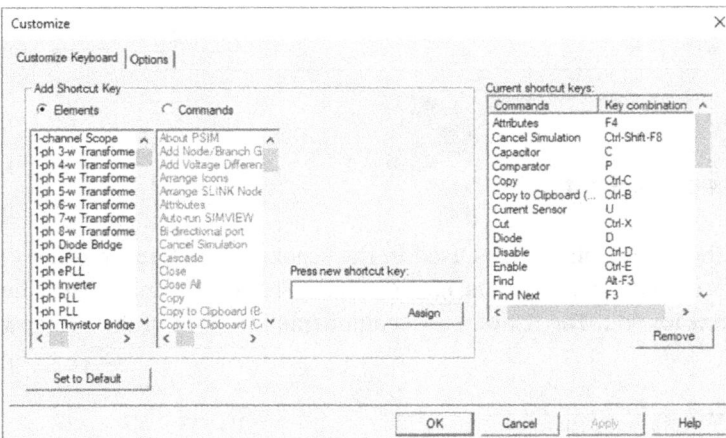

Fig. 2.130.

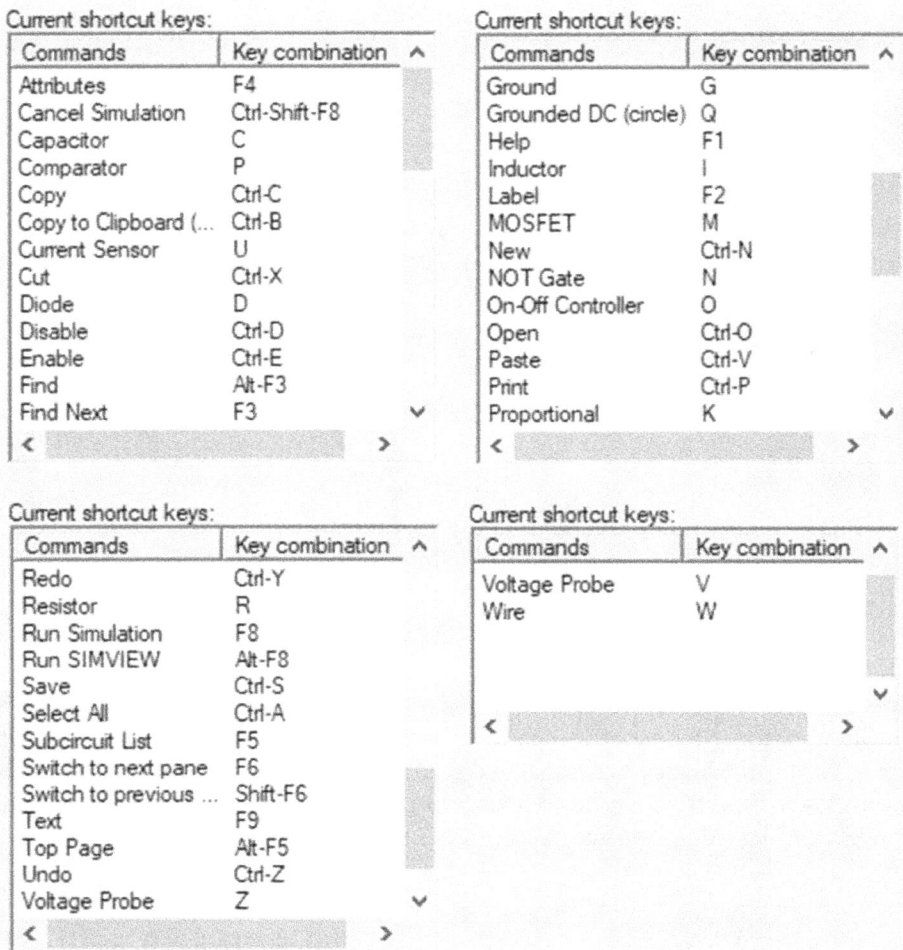

Current shortcut keys:

Commands	Key combination
Attributes	F4
Cancel Simulation	Ctrl-Shift-F8
Capacitor	C
Comparator	P
Copy	Ctrl-C
Copy to Clipboard (...	Ctrl-B
Current Sensor	U
Cut	Ctrl-X
Diode	D
Disable	Ctrl-D
Enable	Ctrl-E
Find	Alt-F3
Find Next	F3

Current shortcut keys:

Commands	Key combination
Ground	G
Grounded DC (circle)	Q
Help	F1
Inductor	I
Label	F2
MOSFET	M
New	Ctrl-N
NOT Gate	N
On-Off Controller	O
Open	Ctrl-O
Paste	Ctrl-V
Print	Ctrl-P
Proportional	K

Current shortcut keys:

Commands	Key combination
Redo	Ctrl-Y
Resistor	R
Run Simulation	F8
Run SIMVIEW	Alt-F8
Save	Ctrl-S
Select All	Ctrl-A
Subcircuit List	F5
Switch to next pane	F6
Switch to previous ...	Shift-F6
Text	F9
Top Page	Alt-F5
Undo	Ctrl-Z
Voltage Probe	Z

Current shortcut keys:

Commands	Key combination
Voltage Probe	V
Wire	W

Fig. 2.131.

2.18 List of used elements

PSIM can extract the list of components used in the schematic for you. In order to extract the list of components, click the Element List (Fig. 2.132). This opens the Element List window for you. The list of used components is shown in this window (Fig. 2.133).

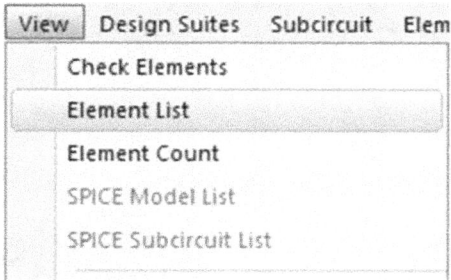

Fig. 2.132.

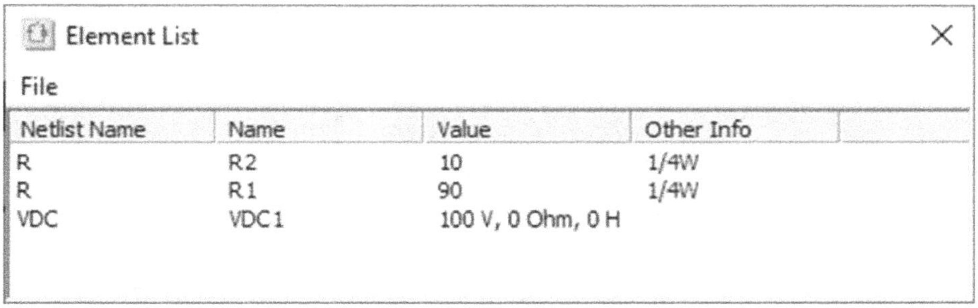

Fig. 2.133.

Use the File menu to export the components list (Fig. 2.134).

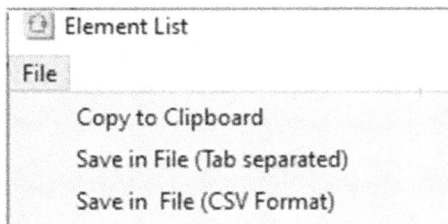

Fig. 2.134.

2.19 Removing the grids

By default, the schematic has some grid points on it (Fig. 2.135). You can remove the grid points from the schematic if you like. In order to do that, click the Settings.

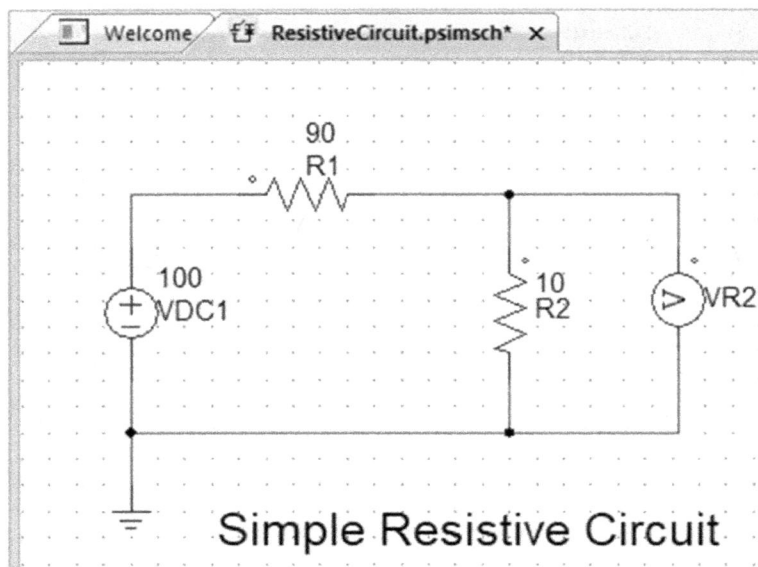

Fig. 2.135.

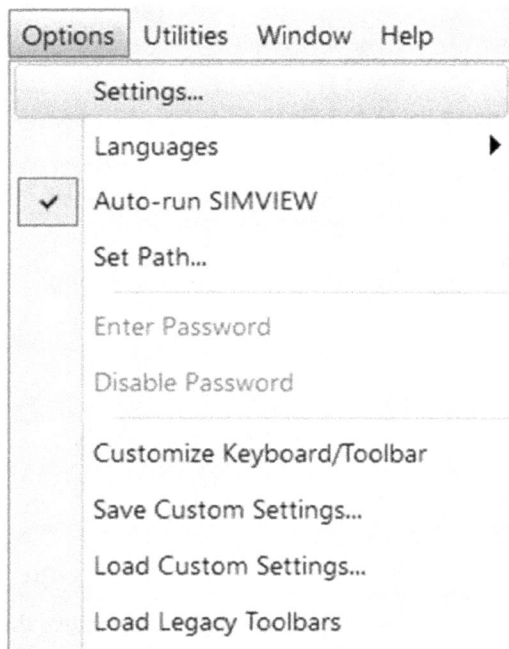

Fig. 2.136.

In the appeared window, uncheck the Display grid box (Fig. 2.137). Now the grid points will be removed from the schematic (Fig. 2.138).

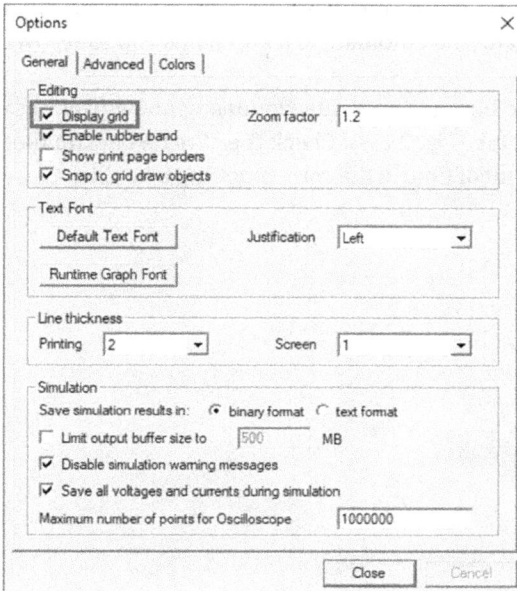

Fig. 2.137.

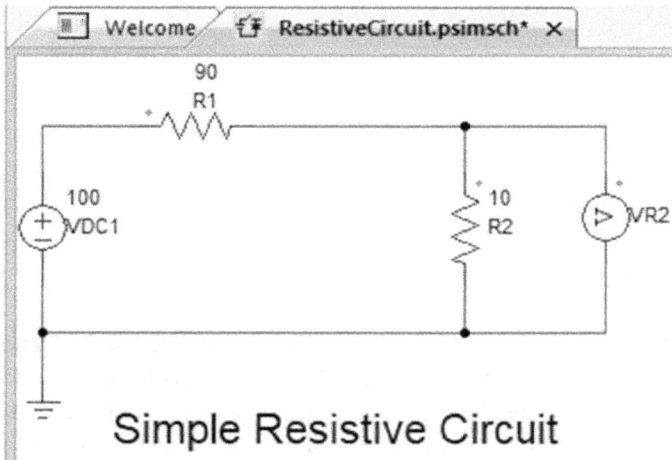

Fig. 2.138.

2.20 Backup files

PSIM can back up your files automatically after a certain amount of time is passed. Such an automatic backup is very useful when you are working on a schematic and suddenly the electricity is disconnected. The automatic backup helps you to recover your work.

In order to activate the automatic backup, open the Options window (Options> Settings …). Then go to the Advanced tab (Fig. 2.139). Check the "Create backup files every" box and enter the desired amount of time in the corresponding box. Click Close button in order to apply the changes.

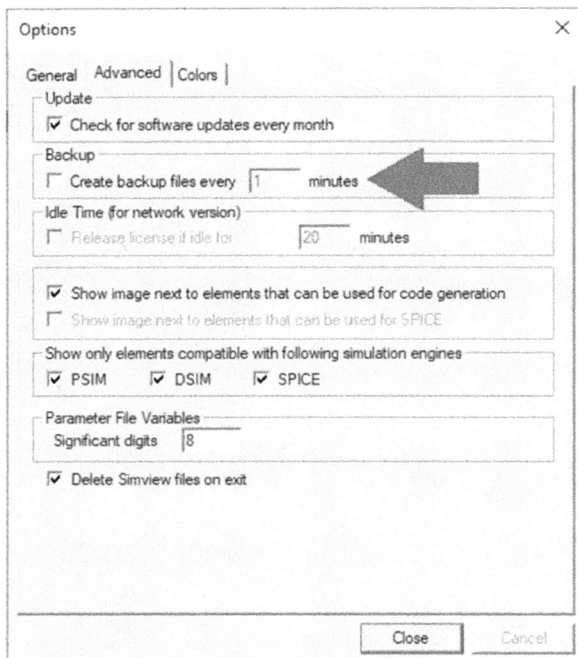

Fig. 2.139.

After activating the automatic backup, PSIM makes another file with the same name as original file name and backup postfix (Fig. 2.140). PSIM uses this backup file to save changes after the amount of time determined by you is passed. The PSIM does not save changes to the original file, that is, the file without the backup postfix. So, the original file is saved when the user clicks the save icon or press the Ctrl+S.

Name	Date modified	Type	Size
ResistiveCircuit (backup)1	04/03/2021 17:55	PSIM Document	16 KB
ResistiveCircuit	04/03/2021 14:45	PSIM Document	16 KB

Fig. 2.140.

2.21 Automatic code generation blocks

Control is an important part of power electronics. Without control, power electronics converters cannot provide the desired voltage/current for the load. PSIM can translate the control circuit of the schematic into the required code for some of the microcontrollers (Fig. 2.141). However, all the components cannot be translated into code. Only components with the CG label (see Fig. 2.142) are allowed. CG stands for code generation.

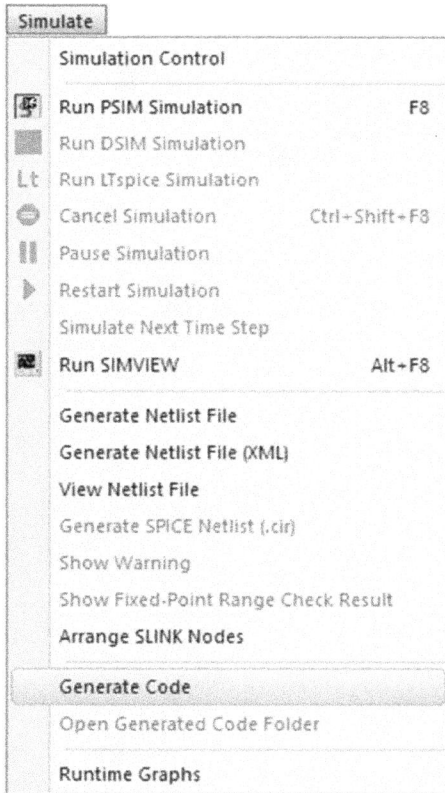

Simulate

Simulation Control

Run PSIM Simulation   F8

Run DSIM Simulation

Lt Run LTspice Simulation

Cancel Simulation   Ctrl+Shift+F8

Pause Simulation

Restart Simulation

Simulate Next Time Step

Run SIMVIEW   Alt+F8

Generate Netlist File

Generate Netlist File (XML)

View Netlist File

Generate SPICE Netlist (.cir)

Show Warning

Show Fixed-Point Range Check Result

Arrange SLINK Nodes

Generate Code

Open Generated Code Folder

Runtime Graphs

Fig. 2.141.

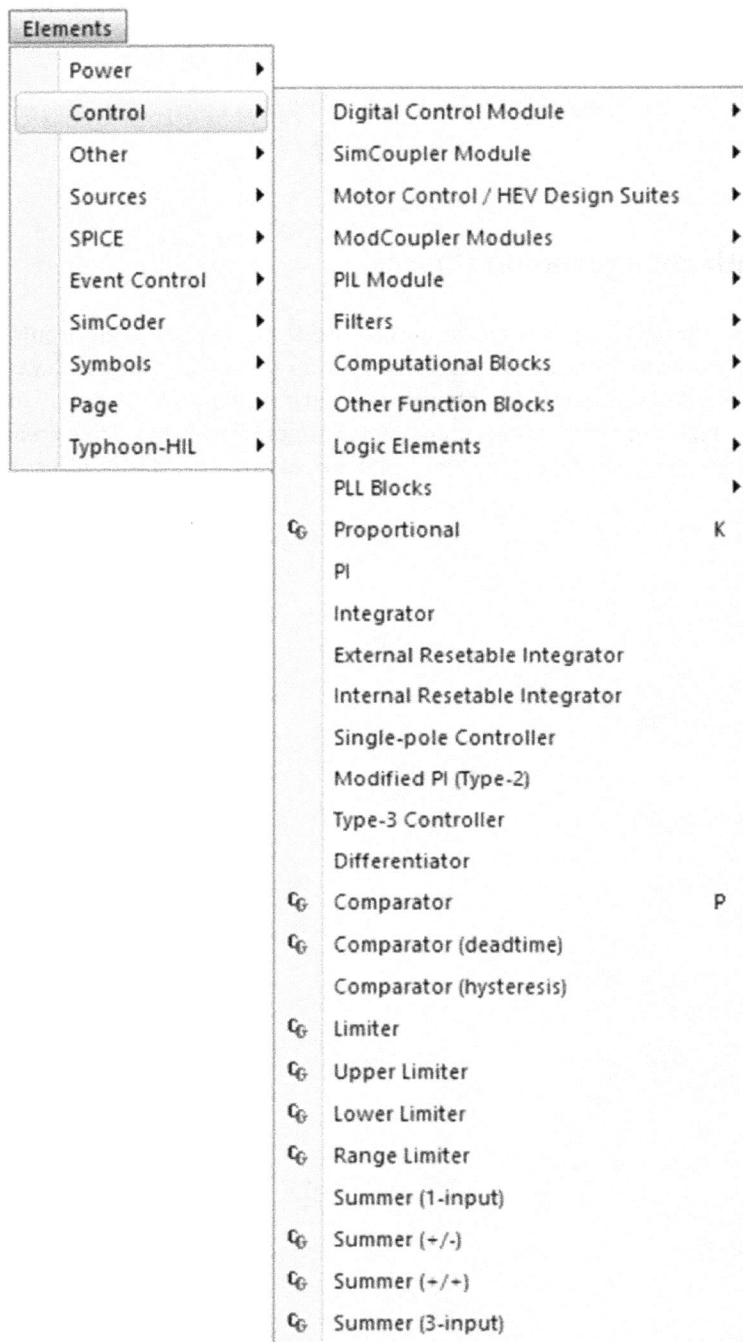

Elements		
Power	▶	
Control	▶	
Other	▶	
Sources	▶	
SPICE	▶	
Event Control	▶	
SimCoder	▶	
Symbols	▶	
Page	▶	
Typhoon-HIL	▶	

Digital Control Module	▶	
SimCoupler Module	▶	
Motor Control / HEV Design Suites	▶	
ModCoupler Modules	▶	
PIL Module	▶	
Filters	▶	
Computational Blocks	▶	
Other Function Blocks	▶	
Logic Elements	▶	
PLL Blocks	▶	
₲ Proportional	K	
PI		
Integrator		
External Resetable Integrator		
Internal Resetable Integrator		
Single-pole Controller		
Modified PI (Type-2)		
Type-3 Controller		
Differentiator		
₲ Comparator	P	
₲ Comparator (deadtime)		
Comparator (hysteresis)		
₲ Limiter		
₲ Upper Limiter		
₲ Lower Limiter		
₲ Range Limiter		
Summer (1-input)		
₲ Summer (+/-)		
₲ Summer (+/+)		
₲ Summer (3-input)		

Fig. 2.142.

Open the Options window (Option>Settings …) in order to ensure that the PSIM adds the CG label to the blocks allowed for code generation. Check the "Show image next to elements that can be used for code generation" box (Fig. 2.143).

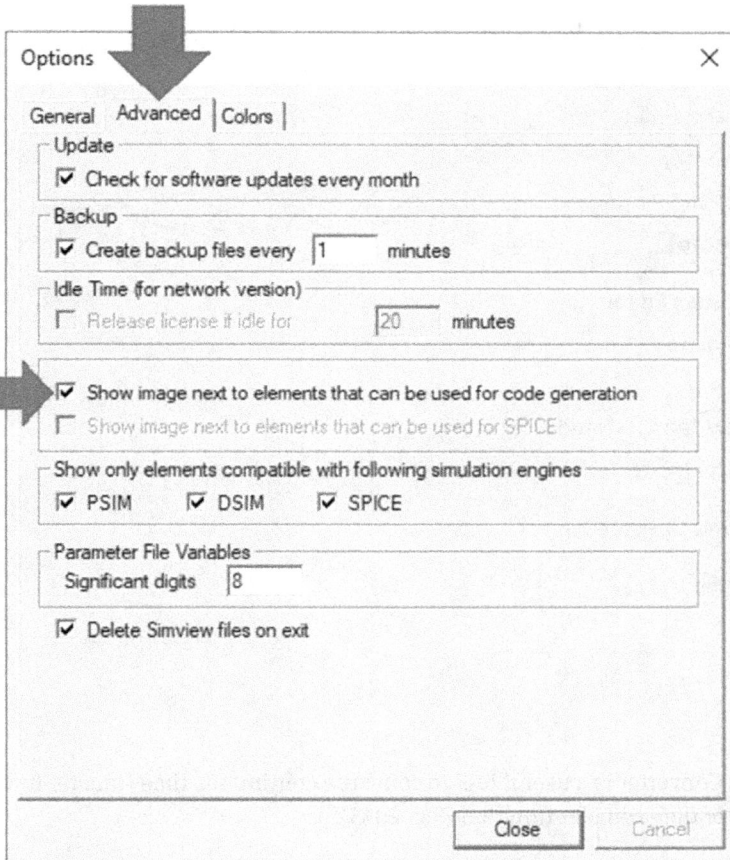

Fig. 2.143.

2.22 Utilities menu

The Utilities menu (Fig. 2.144) contains some very useful tools.

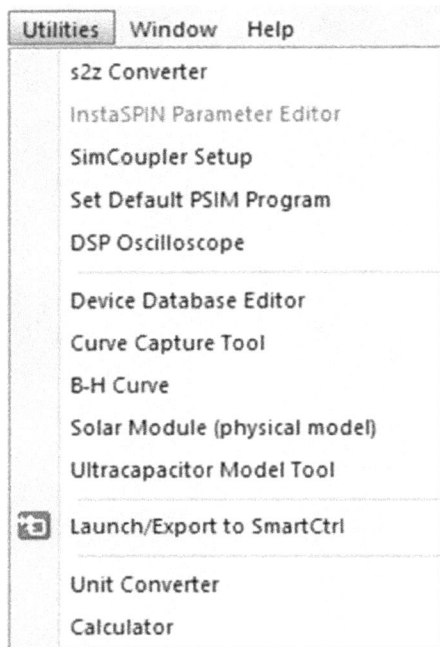

Fig. 2.144.

The Utilities>s2z Converter is a useful tool to convert a continuous time transfer function into a discrete time transfer function (Fig. 2.145).

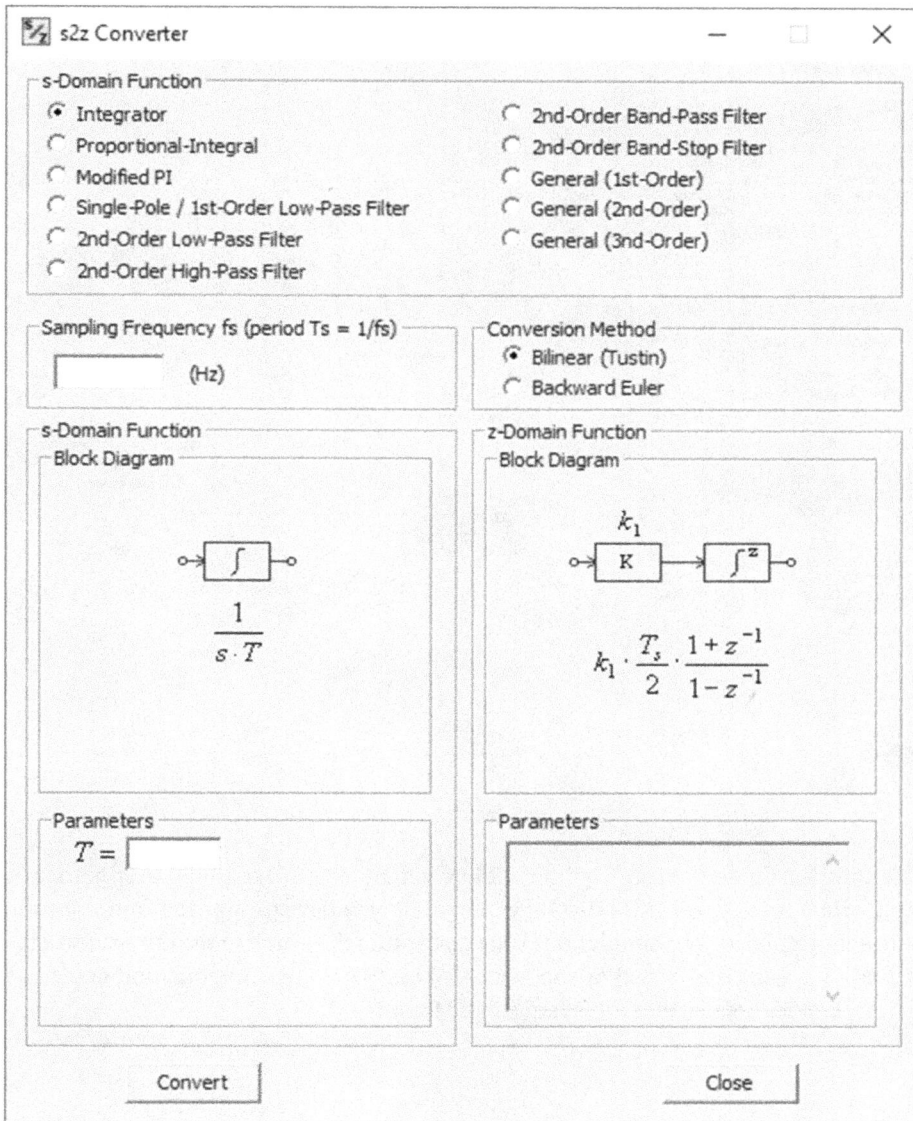

Fig. 2.145.

PSIM's thermal module lets you quickly calculate the switching and conduction losses of switching devices and the core and winding losses of inductors. The Utilities>Device Database Editor (Fig. 2.146) is used for thermal analysis. It provides a convenient way to add new devices and manage existing devices. These devices can then be used in the PSIM schematic and their power losses calculated in the simulation. More details about this tool can be found in [13].

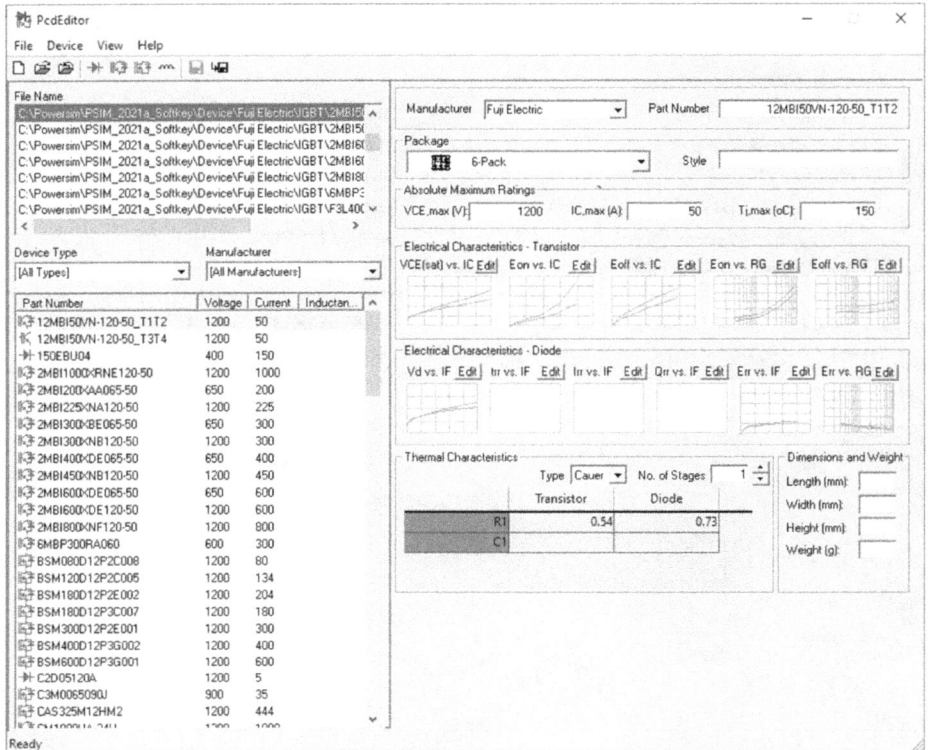

Fig. 2.146.

The Utilities>Curve Capture Tool (Fig. 2.147) generates a set of data point pairs in a *.txt format for a curve. This is useful when you have a device datasheet and you want to use its graphs in your simulation. You can capture the graph in the datasheet using the print screen (PrtScr key) or snapshot tool of PDF Reader program and enter it to this tool. More details about this tool can be found in [14].

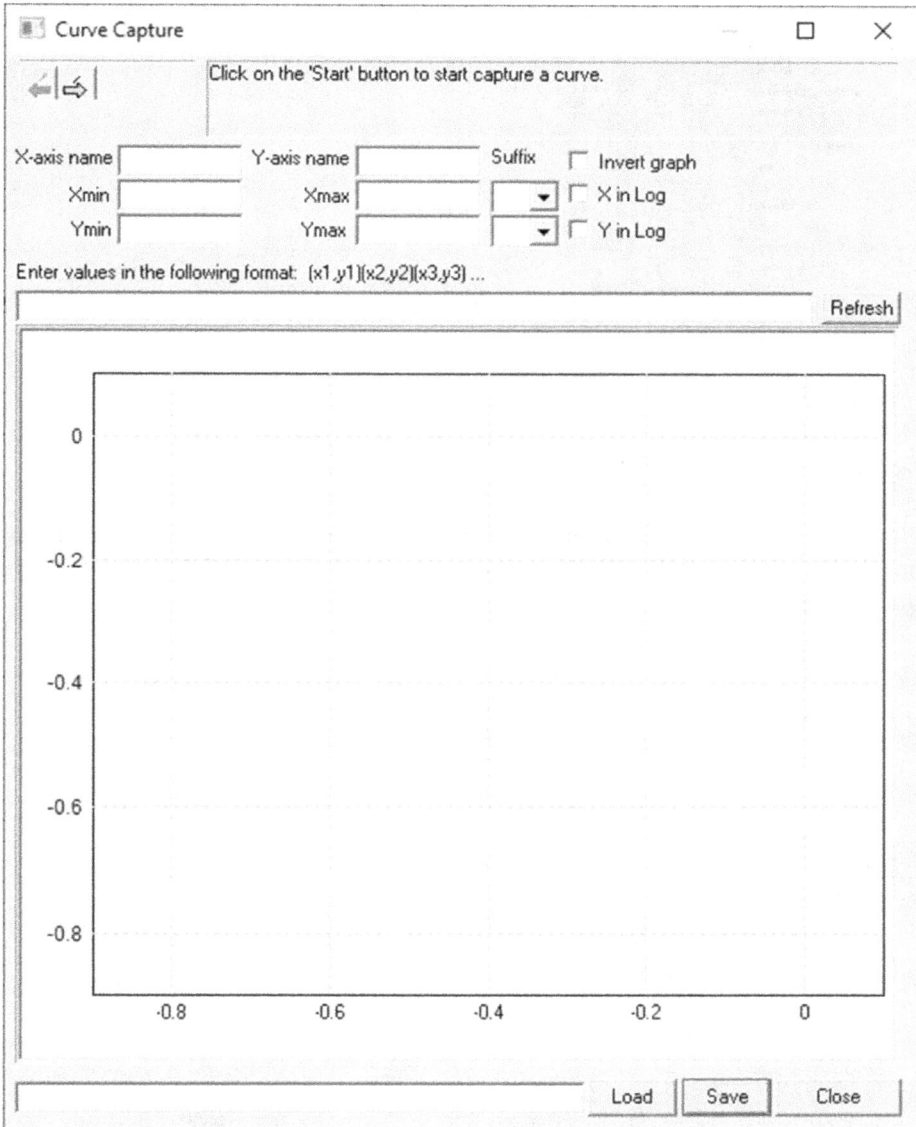

Fig. 2.147.

The tool under the Utilities>B-H Curve provides a convenient way to define the parameters of a saturable core element. The dialog of this tool is shown in Fig. 2.148.

When all the required parameters are entered and the Calculate button is clicked, PSIM will calculate the B-H curve and plot the curve on the right hand side diagram. One can then compare this curve with the curve from the manufacturer's datasheet and readjust the parameters accordingly. More details about this tool can be found in [15].

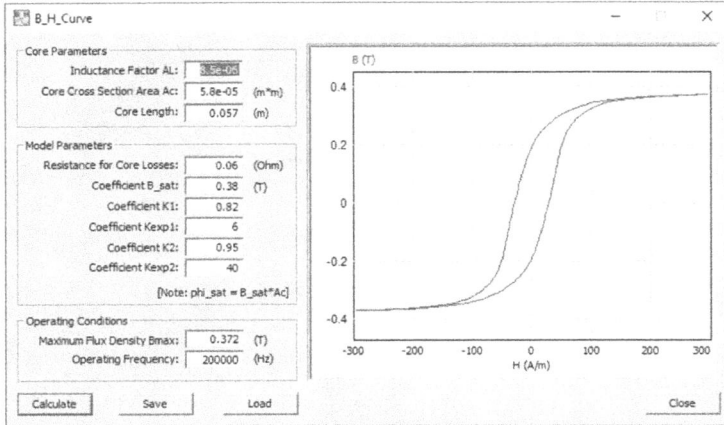

Fig. 2.148.

Photovoltaics (PV) is the conversion of light into electricity using semiconducting materials that exhibit the PV effect. PSIM provides two types of PV models: Functional model and Physical model (Fig. 2.149).

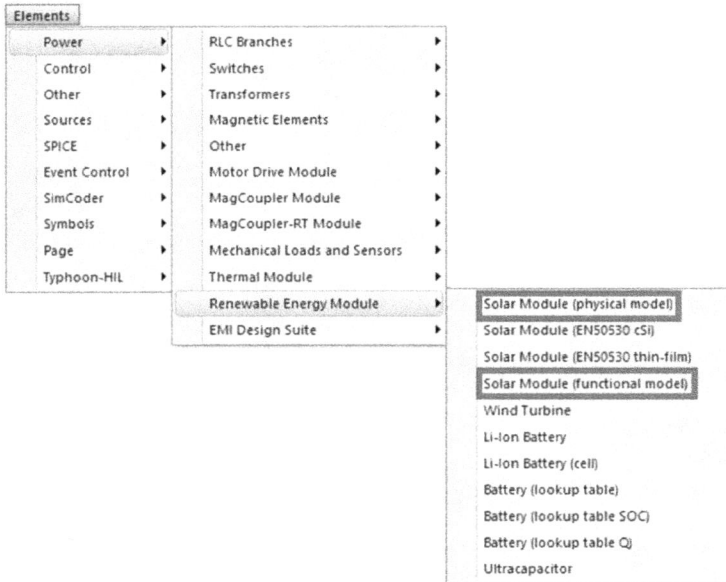

Fig. 2.149.

Functional model is simplified and easy to use. Physical model block takes into account the effects of light intensity and temperature. Physical model block and its setting are shown in Figs. 2.150 and 2.151, respectively.

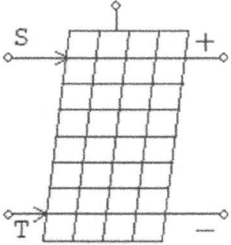

Fig. 2.150.

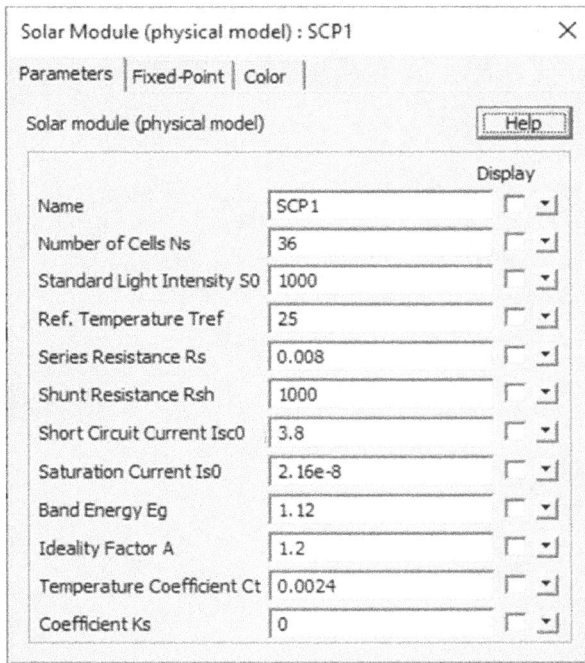

Fig. 2.151.

The physical model allows users to enter detailed parameters from the solar cell's datasheet. In order to make it easier for users to define parameters for a particular solar module, a utility tool called Solar Module (physical model) is provided in the PSIM's Utility menu (Fig. 2.152). This simplifies the process of modeling and analyzing a real-world PV power system.

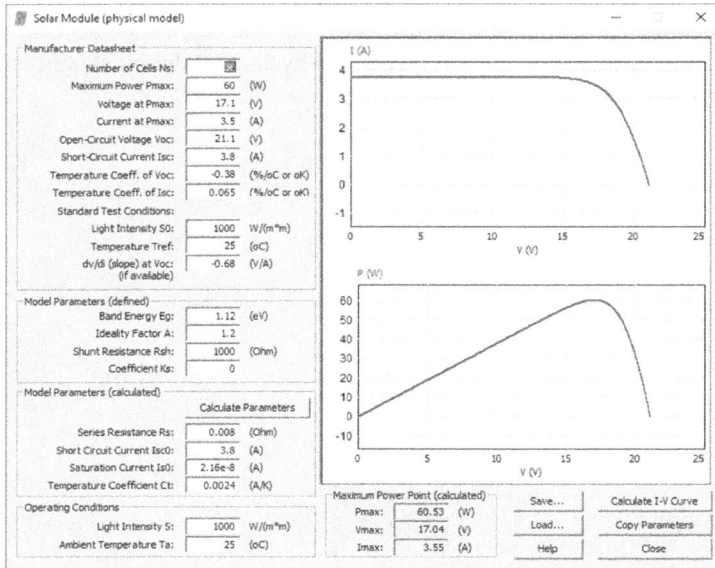

Fig. 2.152.

Enter the parameters of the solar module (from datasheet) into the Solar Module (physical model) tool (Fig. 2.153).

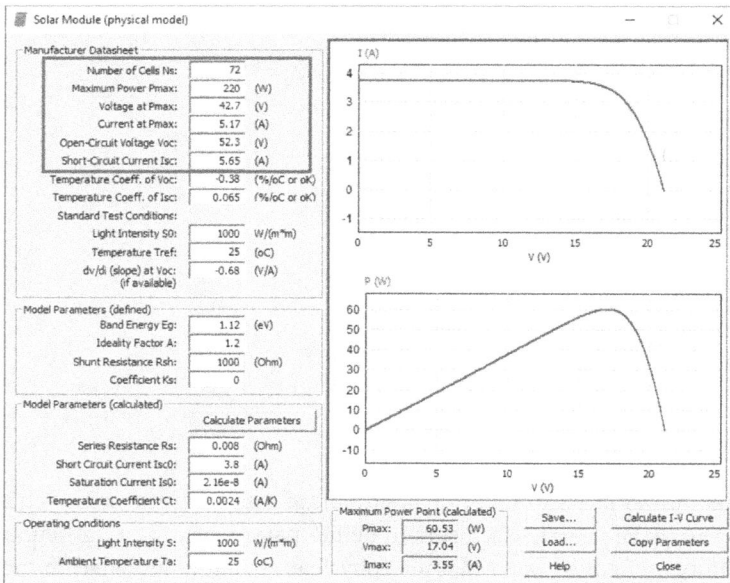

Fig. 2.153.

Click the Calculate Parameters (Fig. 2.154).

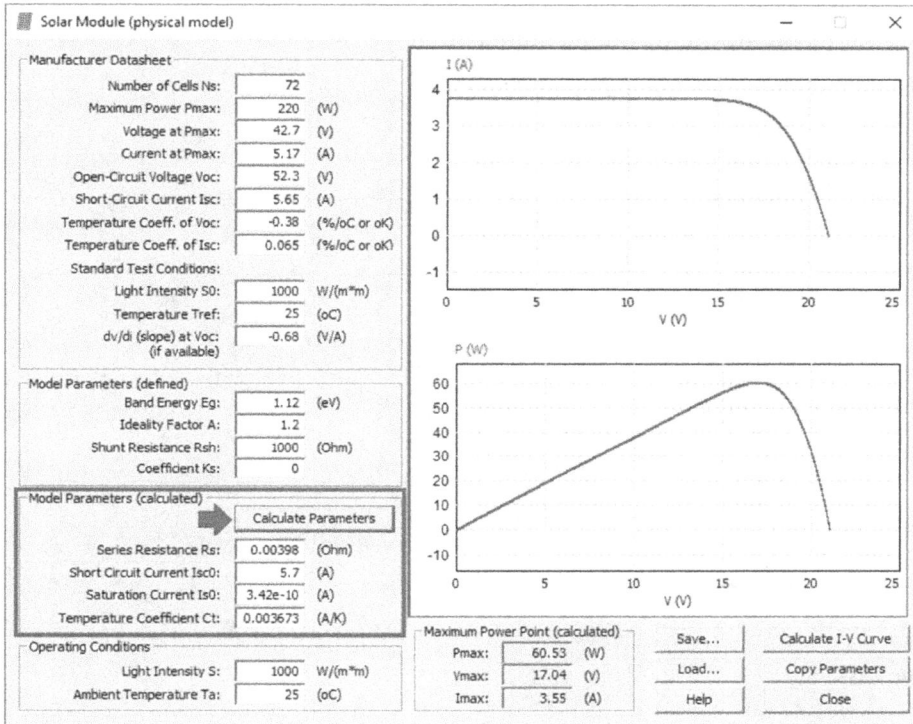

Fig. 2.154.

Click the Calculate I–V Curve (Fig. 2.155) in order to update the I–V curve based on the entered data.

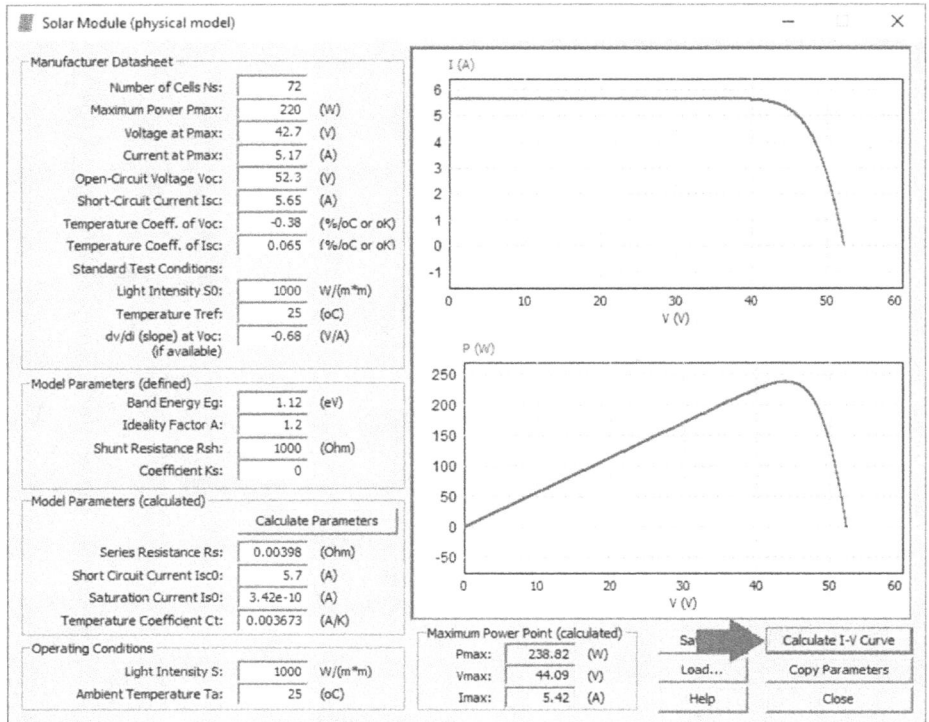

Fig. 2.155.

Click the Copy Parameters button (Fig. 2.156) in order to update the parameters of Solar Module (physical model) block you put on the schematic. With the aid of Copy Parameters button, you transfer the calculated parameters from the tool to the block on the schematic automatically. More details about this tool can be found in [16].

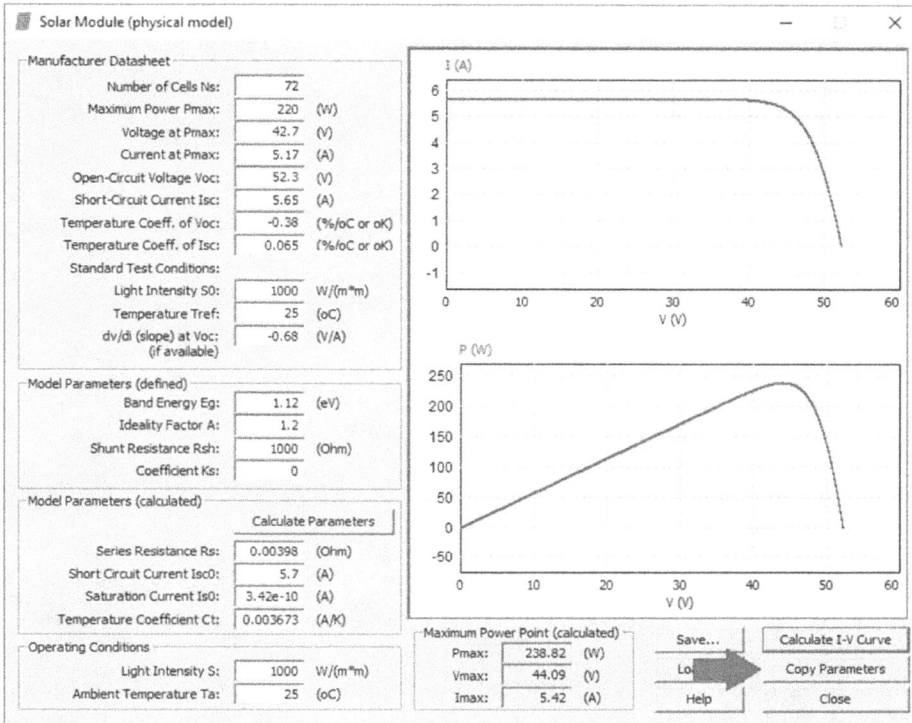

Fig. 2.156.

An ultracapacitor is a high-capacity capacitor with a capacitance value much higher than other capacitors, but with lower voltage limits, that bridges the gap between electrolytic capacitors and rechargeable batteries. It typically stores 10 to 100 times more energy per unit volume or mass than electrolytic capacitors, can accept and deliver charge much faster than batteries, and tolerates many more charge and discharge cycles than rechargeable batteries.

Ultracapacitors are used in applications requiring many rapid charge/discharge cycles, rather than long-term compact energy storage. For instance, in automobiles, buses, trains, cranes and elevators, where they are used for regenerative braking, short-term energy storage, or burst-mode power delivery. Smaller units are used as power backup for static random access memory (SRAM).

To facilitate the use of the ultracapacitor model, a parameter extraction tool (Fig. 2.157) called Ultracapacitor Model Tool is provided. It will adjust the model parameters and do curve fitting so that the simulated results match closely with the experimental data. More details about this tool can be found in [17].

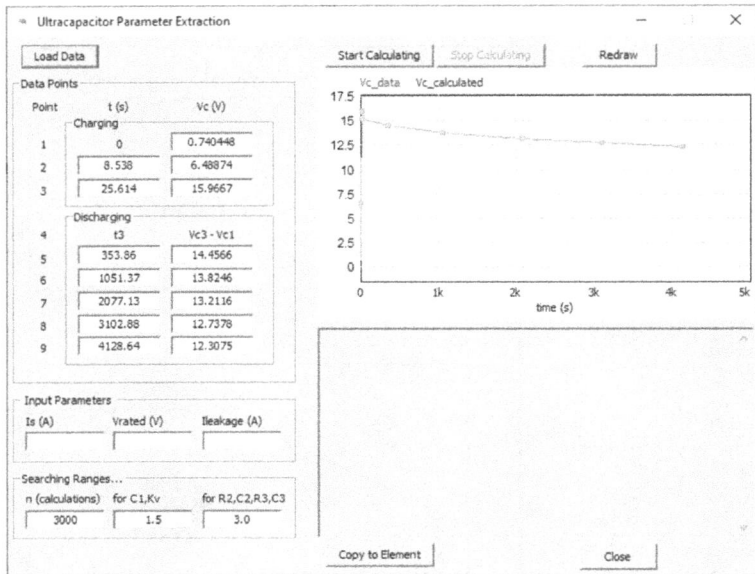

Fig. 2.157.

The Utilities>Launch/Export to SmartCtrl runs the smart control (Fig. 2.158) program. Smart control is an important tool that permits you to design the control loop of power electronics converters. Smart control is studied in Chapter 8. Another way to run smart control is to click the smart control icon (Fig. 2.159).

Fig. 2.158.

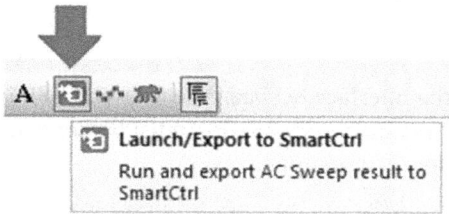

Fig. 2.159.

The Utilities>Unit Converter runs the Unit Converter program (Fig. 2.160). You can convert different units with the aid of this tool.

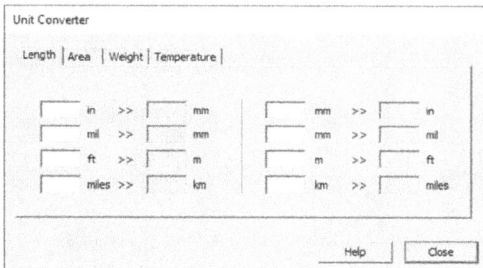

Fig. 2.160.

The Utilities>Calculator runs the PSIM Calculator. You can do the required calculations with the aid of this calculator.

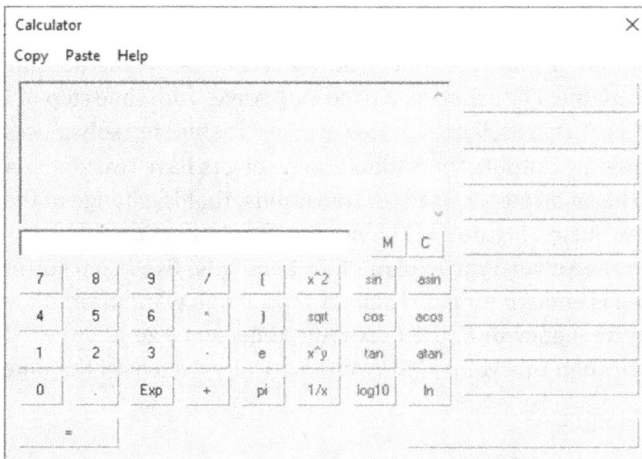

Fig. 2.161.

2.23 Simulation control block

The simulation control block (Fig. 2.162) is the interface between the user and PSIM. With the aid of this block, the user determines how the simulation must be done. In this section, we take a look to some of this block's properties.

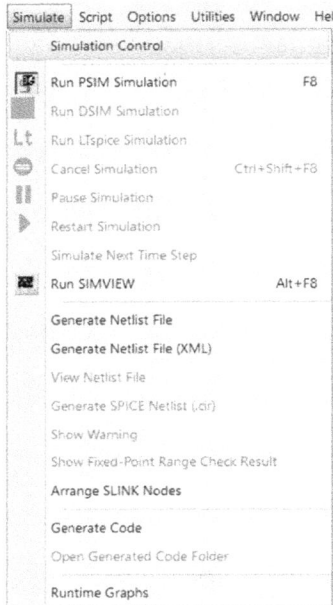

Fig. 2.162.

Computers do the calculations in discrete time. The difference between the time points can be constant or variable. For instance, a fixed step solver with time step of 1 μs do the analysis at $1\,\mu s$, $2\,\mu s$, $3\,\mu s$, $4\,\mu s$, $5\,\mu s$, ... However, a variable step solver uses different steps for generating the output. The variable step solvers have two steps: A small step and a big step. The small step is used for transitions, that is, change in the status of switches, or narrow pulses. Figure 2.163 shows the Simulation Control blocks help page. Pay attention to the Solver Type explanation. Generally, fixed step solver with small enough time step is enough for many simulations. If you want to simulate a converter with switching frequency of F, the maximum time step size is $\frac{1}{10 \times F}$. If you select a time step bigger than this value, PSIM automatically decreases the time step to $\frac{1}{10 \times F}$.

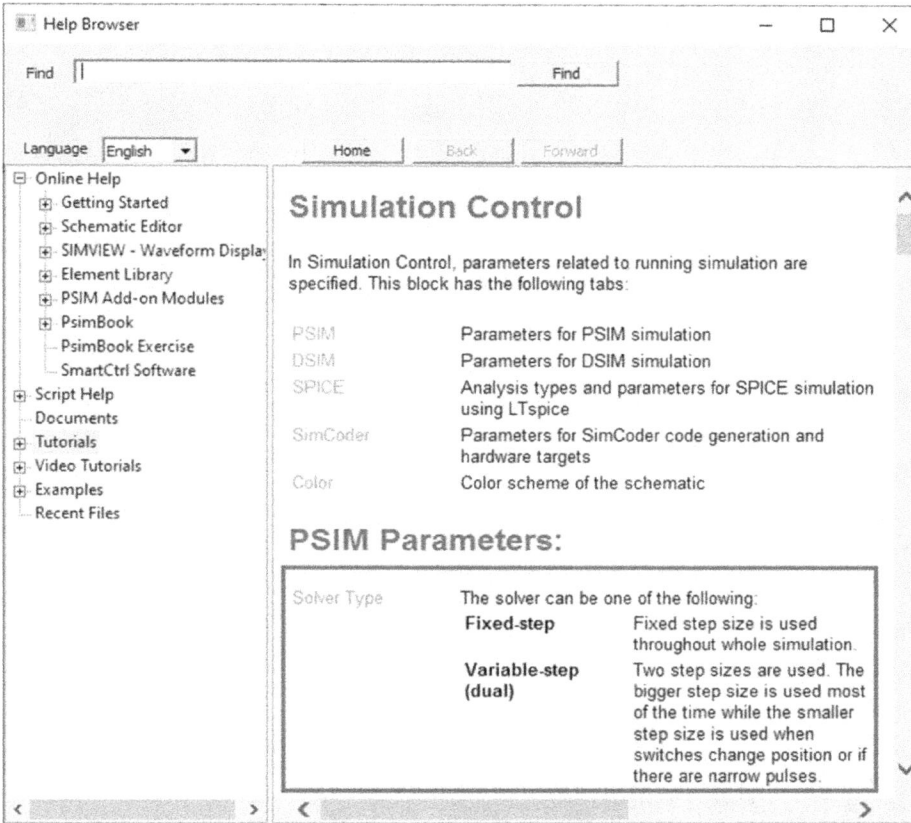

Fig. 2.163.

Consider a simple circuit like the one shown in Fig. 2.164. Location of the elements of this circuit is shown in Fig. 2.165. The settings of elements are shown in Figs. 2.166–2.169.

This circuit is a PWM modulator. The Carrier is a high frequency saw tooth signal like the one shown in Fig. 2.170. The Reference is the modulating signal. In this circuit, it is a DC signal with amplitude of 0.25 V. According to the carrier signal shown in Fig. 2.170, the output must be something like the waveform shown in Fig. 2.171.

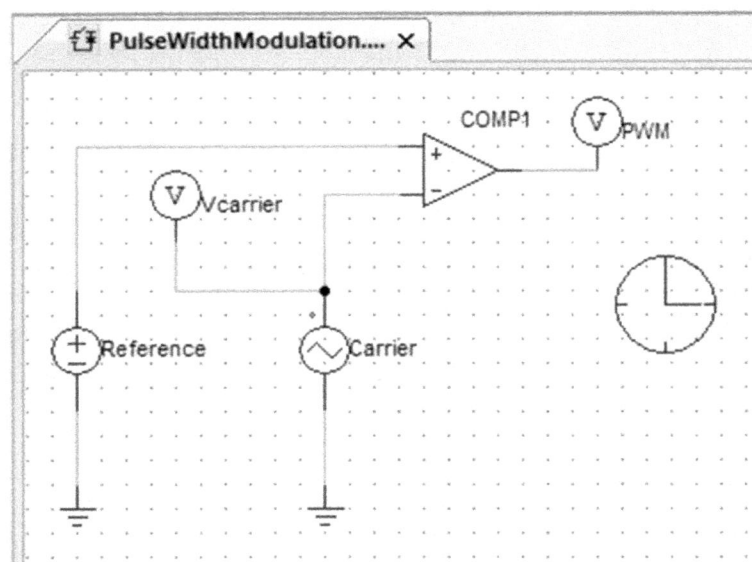

Fig. 2.164.

Fig. 2.165.

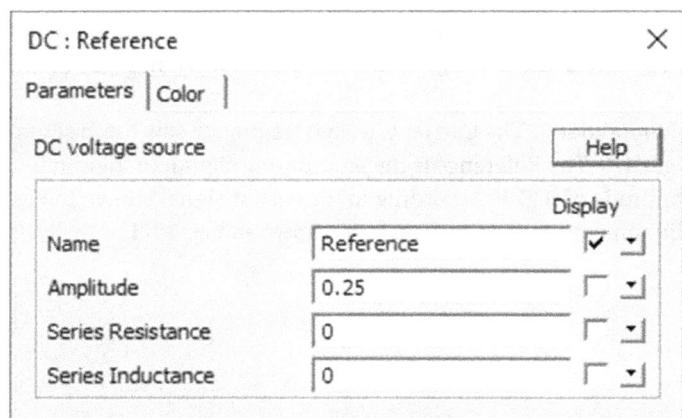

Fig. 2.166.

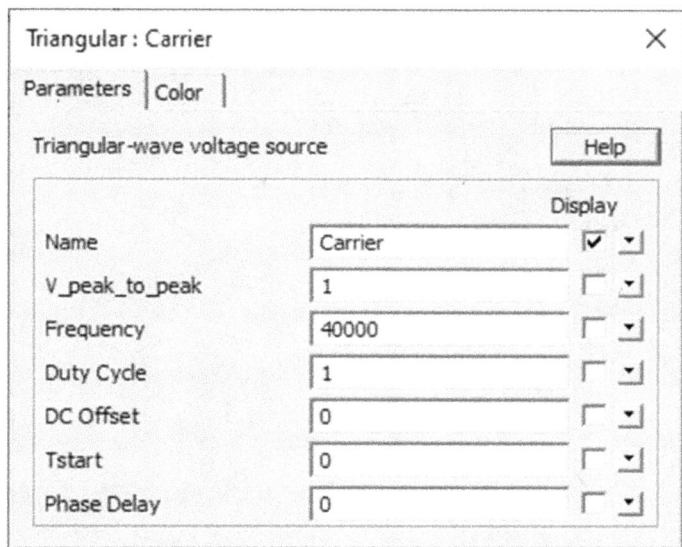

Fig. 2.167.

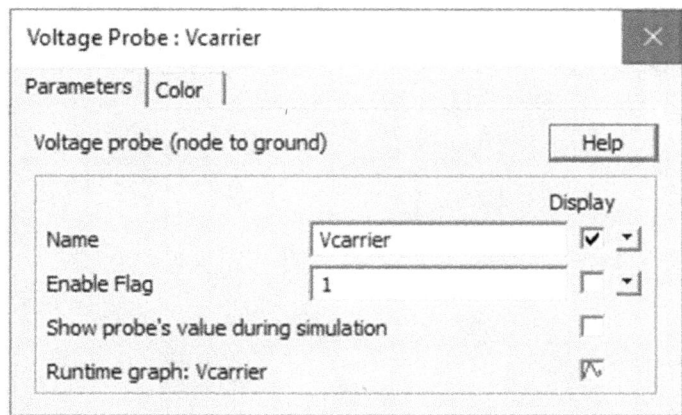

Fig. 2.168.

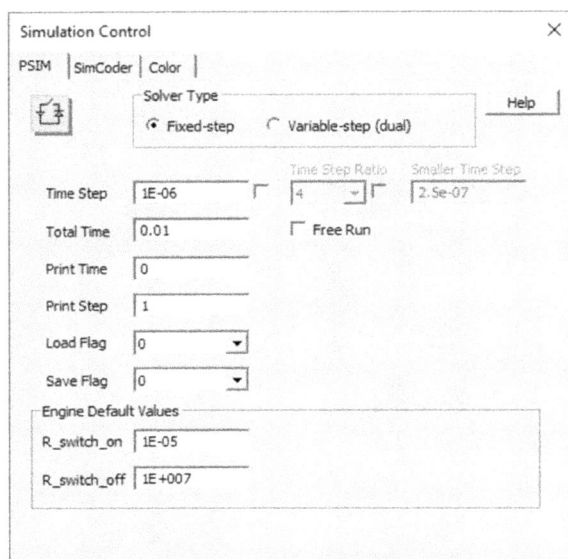

Fig. 2.169.

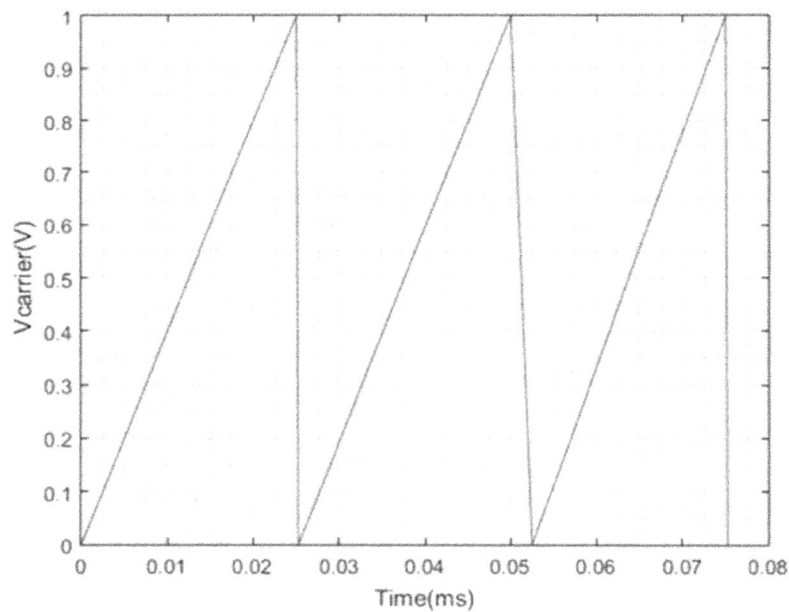

Fig. 2.170.

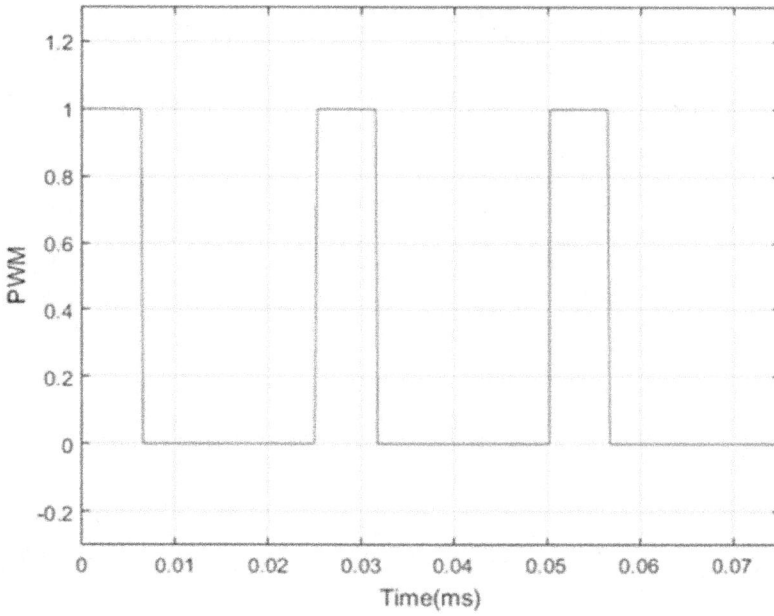

Fig. 2.171.

After running the simulation, the result shown in Figs. 2.172 and 2.173 is obtained. The transitions are not very sharp.

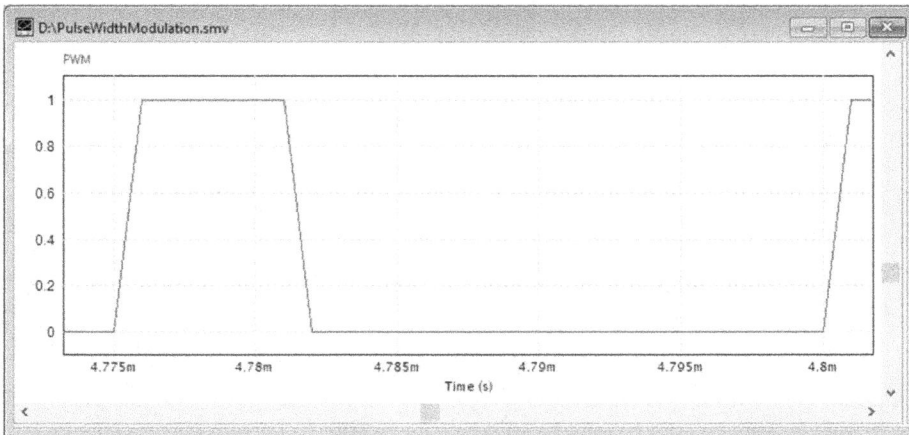

Fig. 2.172.

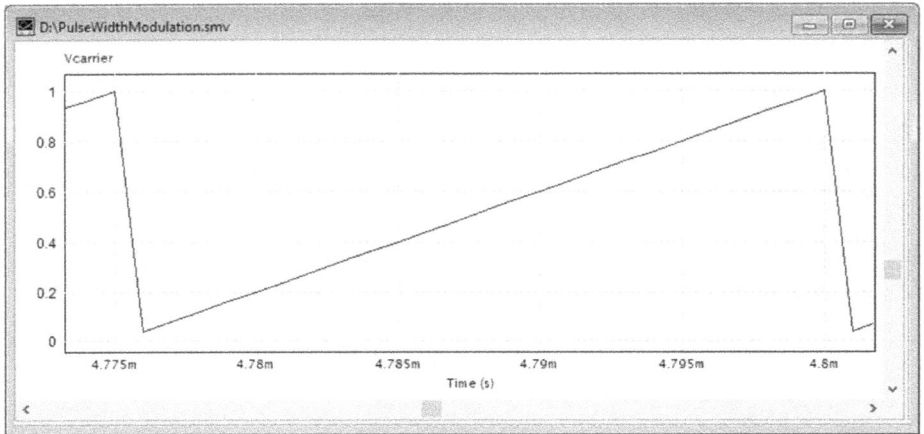

Fig. 2.173.

Change the time step to 1E-07 = 10^{-7} (Fig. 2.174) and run the simulation. The simulation results are shown in Fig. 2.175.

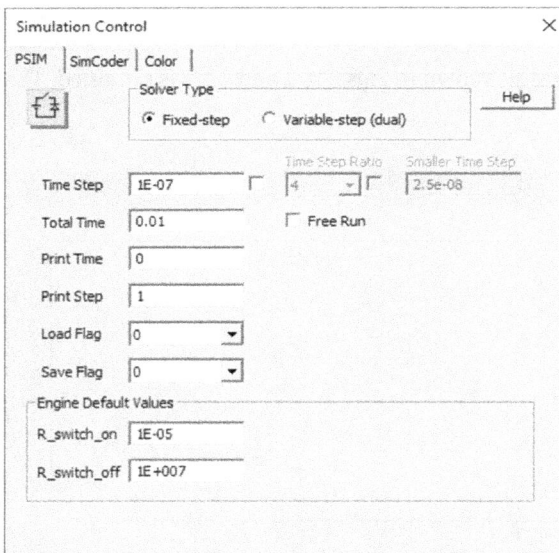

Fig. 2.174.

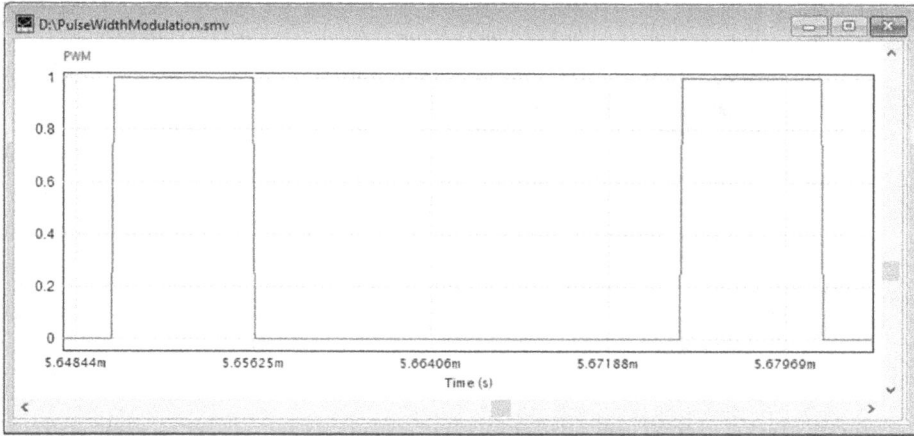

Fig. 2.175.

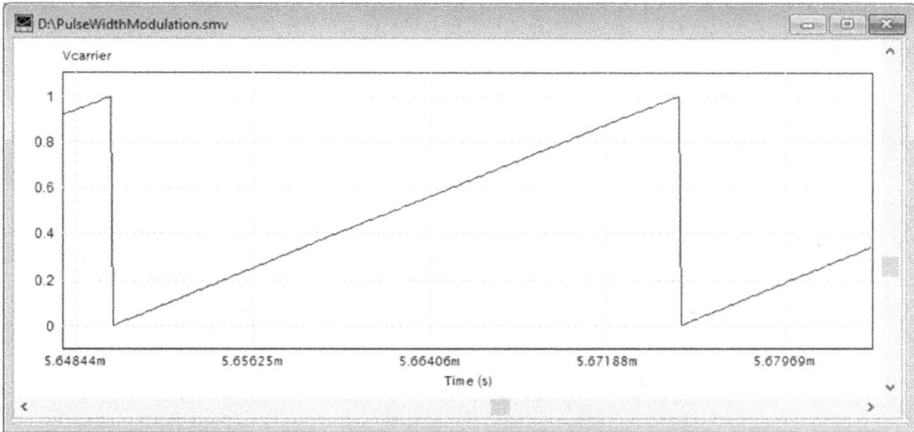

Fig. 2.176.

The Print Time determine the time from which simulation results are saved to the output file (default value is 0). No output is saved before this time. For instance, change the Print Time of previous simulation to 0.002 (Fig. 2.177) and rerun the simulation. The simulation result is shown in Figs. 2.178 and 2.179. As shown, the waveform starts from 0.002 s.

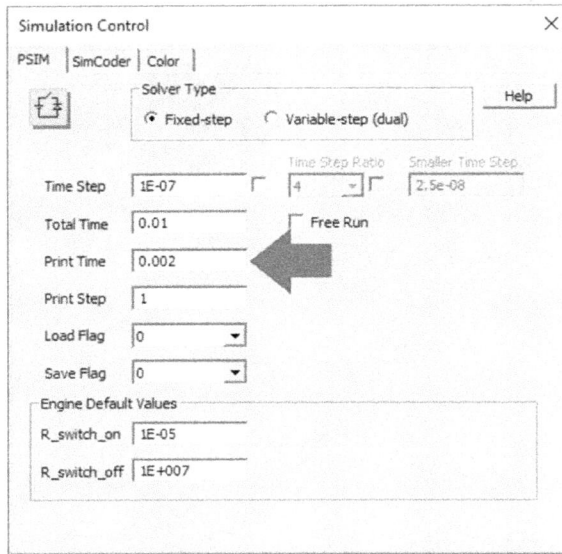

Fig. 2.177.

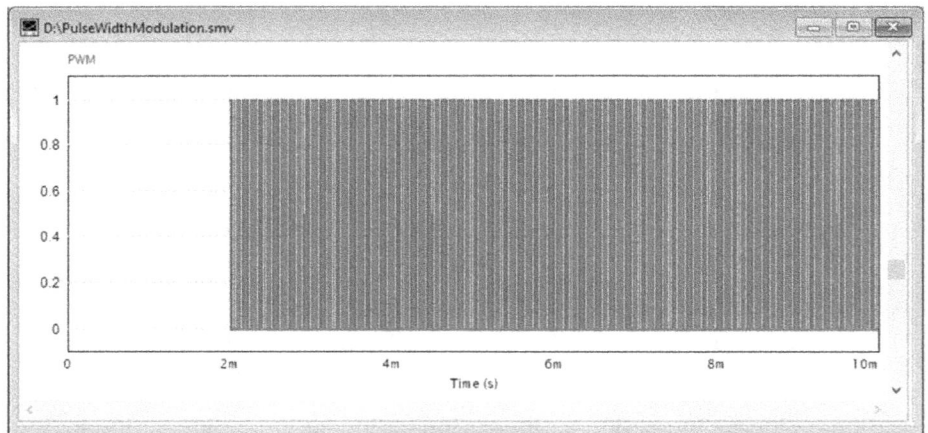

Fig. 2.178.

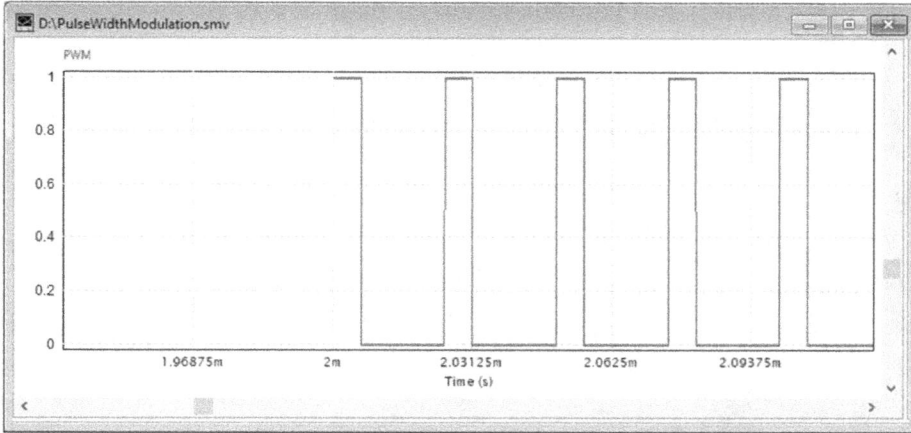

Fig. 2.179.

2.24 Free-run mode

Free-run mode allows a user to run a PSIM simulation and vary components of the simulation much the same as they would on the test bench. Consider a simple rectifier circuit like the one shown in Fig. 2.180. The location of oscilloscope blocks is shown in Figs. 2.181 and 2.182. You can use the library browser (Fig. 2.183) to add oscilloscope block to your schematics as well.

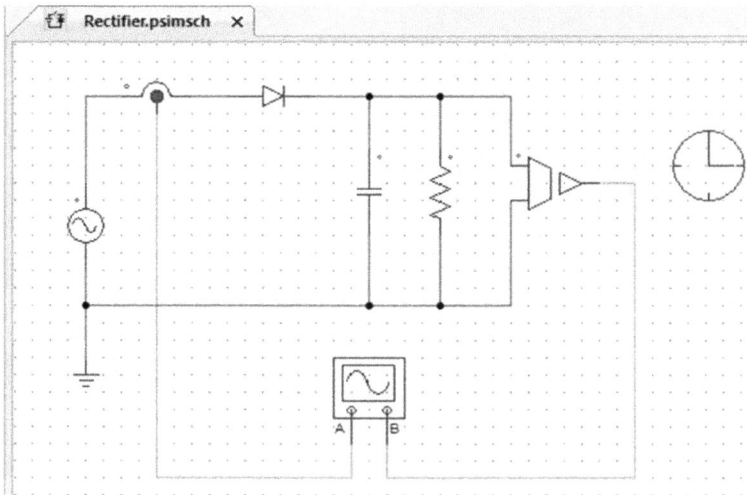

Fig. 2.180.

Fig. 2.181.

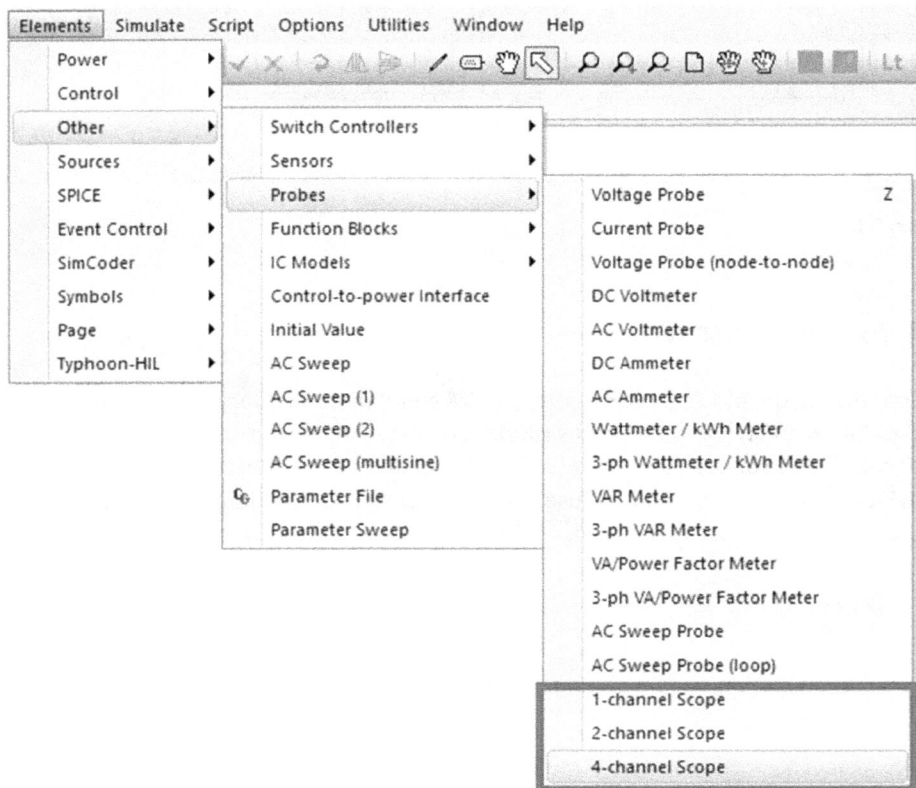

Fig. 2.182.

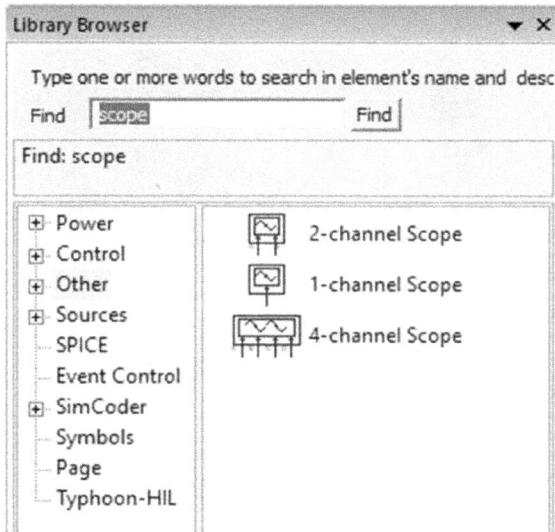

Fig. 2.183.

Settings of the elements used in Fig. 2.180 are shown in Figs. 2.184–2.190.

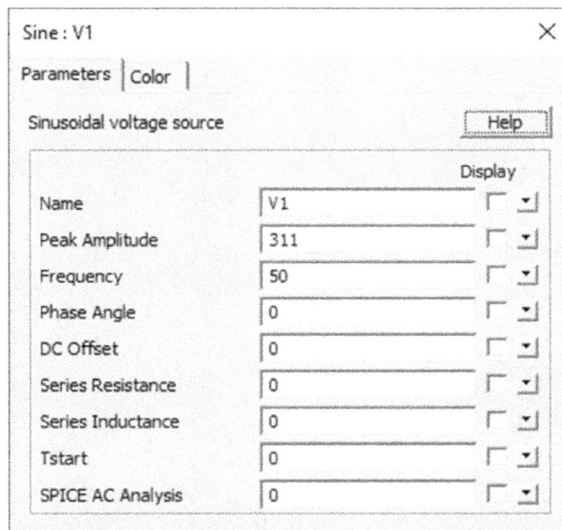

Fig. 2.184.

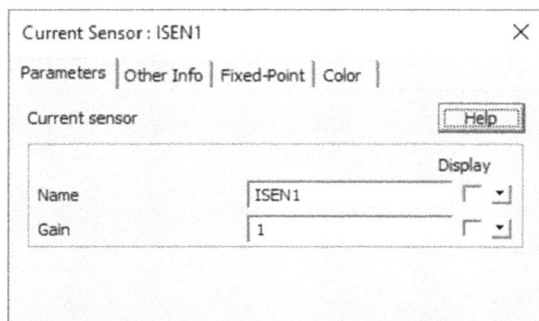

Fig. 2.185.

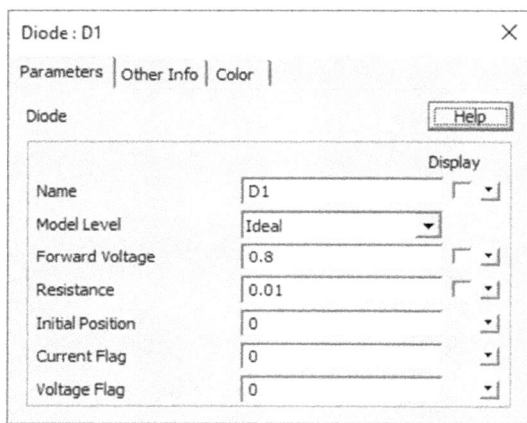

Fig. 2.186.

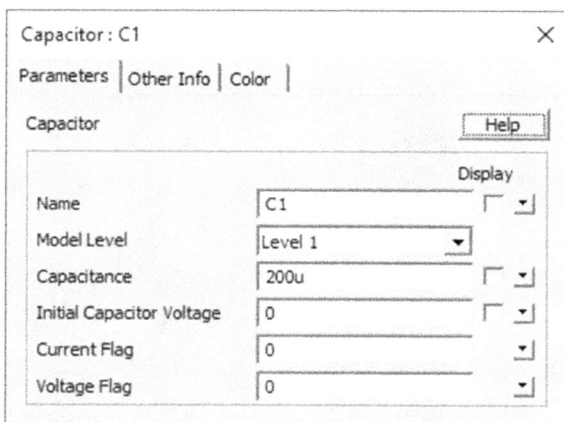

Fig. 2.187.

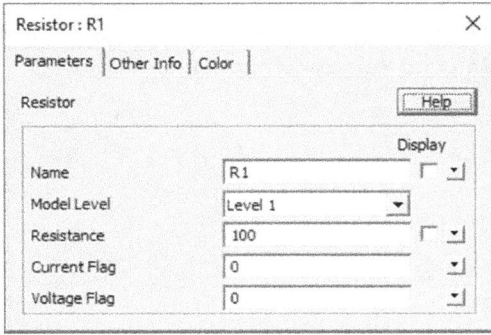

Fig. 2.188.

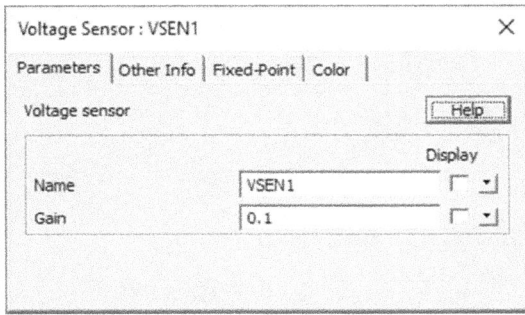

Fig. 2.189.

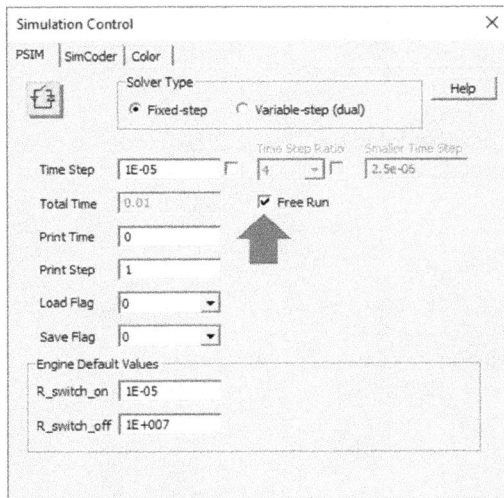

Fig. 2.190.

Run the simulation. The simulation starts and the oscilloscope screen shows the waveforms (Fig. 2.191). If the waveform walks on the screen, click the ON button.

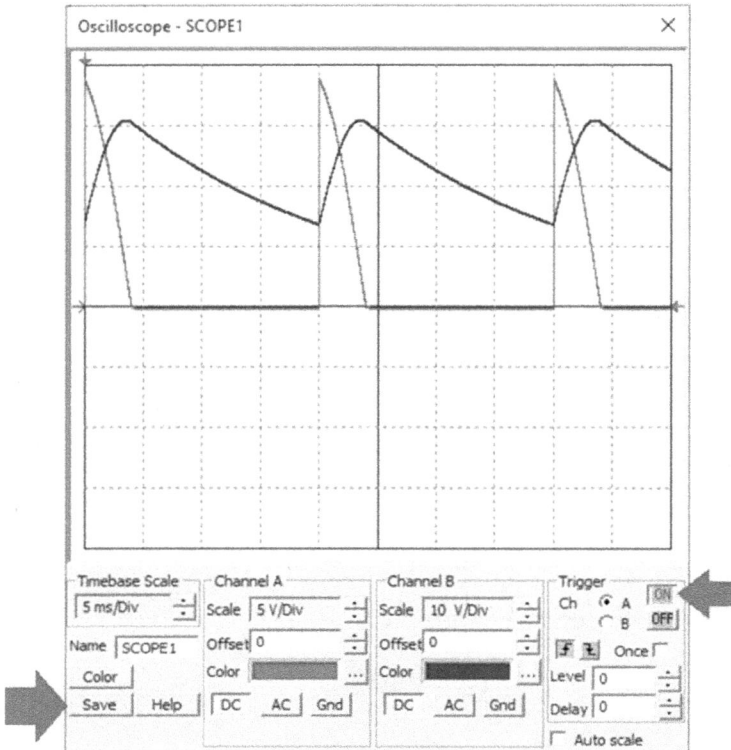

Fig. 2.191.

After stabilizing the waveform, you can use the delay box (Fig. 2.192) in order to move the stopped waveform to right/left.

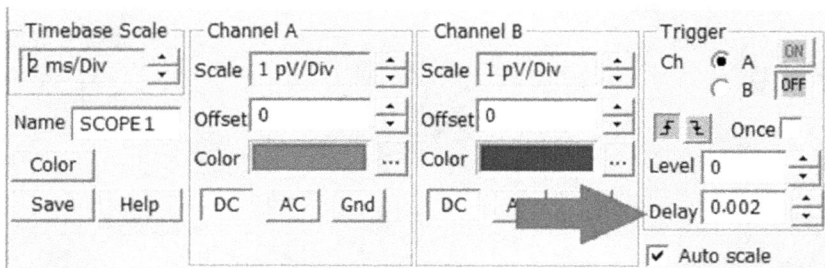

Fig. 2.192.

Click the Save button (Fig. 2.191) in order to save the waveform that you are seeing on the oscilloscope. After saving the file, PSIM automatically opens Simview (Fig. 2.193) and you can the select the waveforms to be shown in the Simview environment (Fig. 2.194).

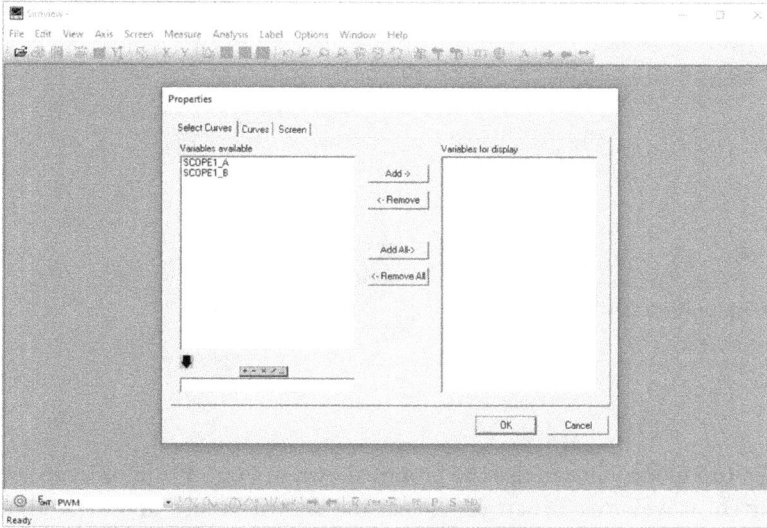

Fig. 2.193.

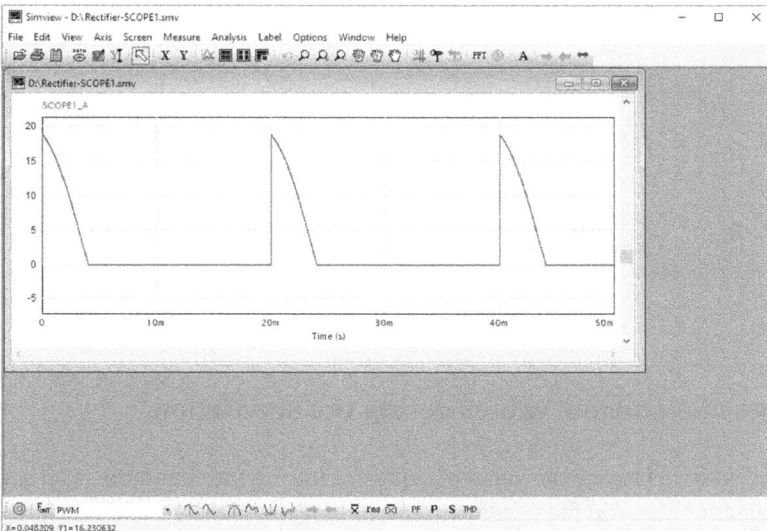

Fig. 2.194.

In order to pause and stop the simulation, use the icons shown in Figs. 2.195 and 2.196. The oscilloscope waveforms will be stopped once the pause button is clicked.

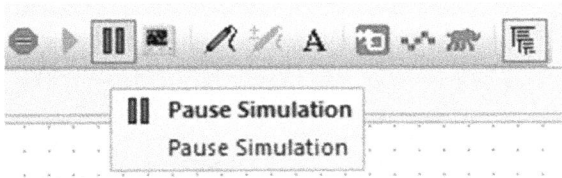

Fig. 2.195.

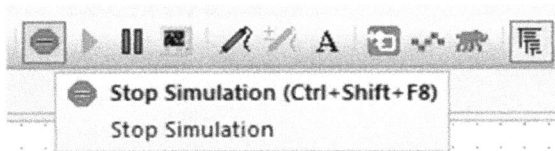

Fig. 2.196.

Note that in free-run mode you can use the probes shown in Fig. 2.197 in order to see the voltage/current waveforms. When you select the shown probes and click on a node, you can see the voltage of that node. When you click on an element, then you can see the current of that element.

Fig. 2.197.

2.25 Change of elements' values during the simulation

Assume that we want to change the value of capacitor during the simulation and see the effect of capacitor value on the circuit current. In order to do this, double click the capacitor and check the Display box in front of the Capacitance (Fig. 2.198). This will add the capacitor value to the schematic as shown in Fig. 2.199.

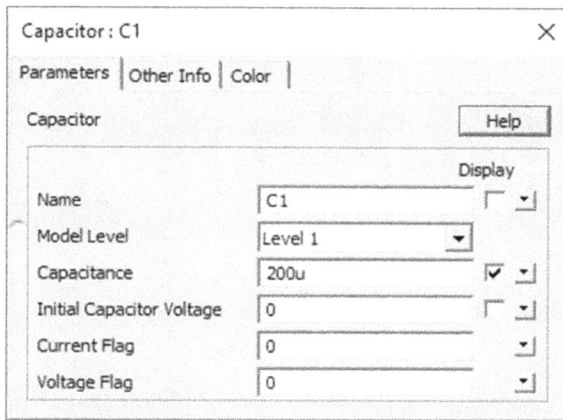

Fig. 2.198.

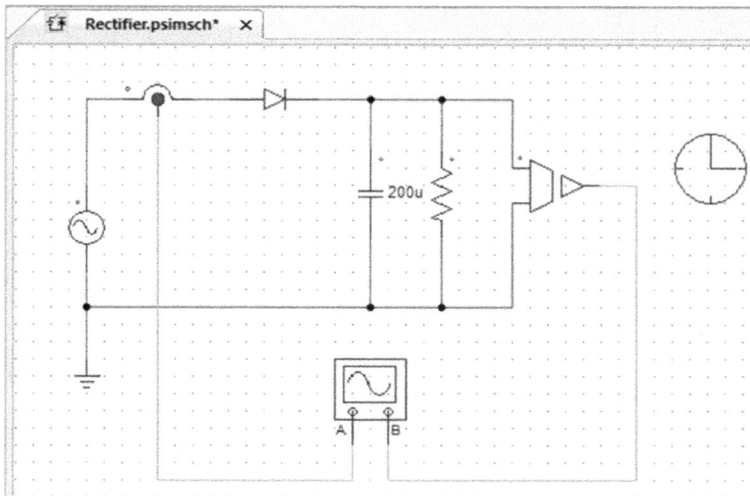

Fig. 2.199.

Run the simulation. Double click the capacitor value (200 u) which is added to the schematic. The Run time variable window will appear (Fig. 2.200). Use the up/down arrows in order to increase/decrease the capacitor value. The waveform shown in Fig. 2.201 is for $C = 400\ \mu F$ capacitor. According to Fig. 2.201, as the capacitor value increases, peak of the current increases as well. For 400 μF capacitor, peak of the current is about 30 A.

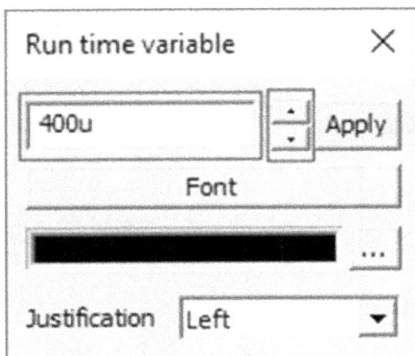

Fig. 2.200.

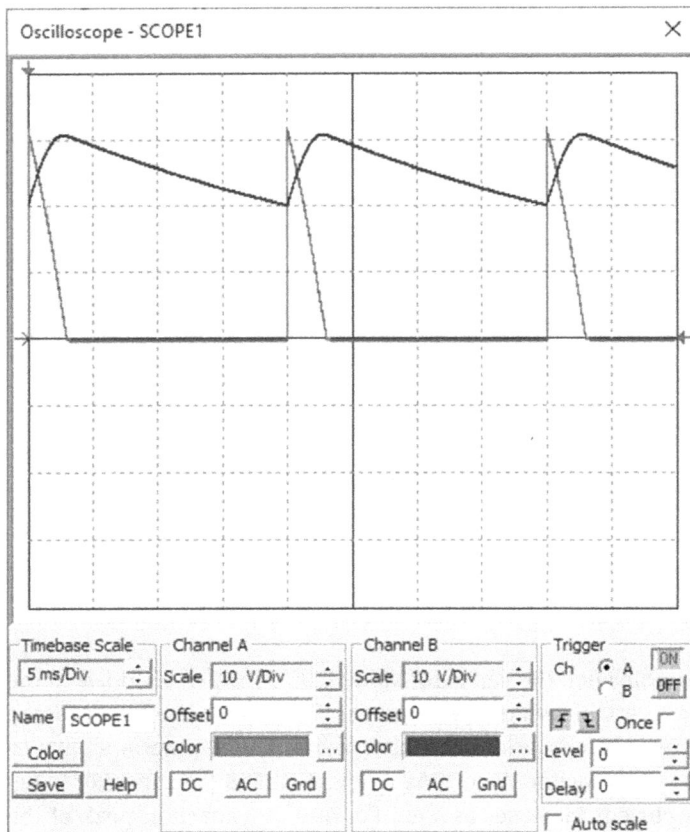

Fig. 2.201.

2.26 AC and DC voltmeters and ammeters

PSIM has four measurement blocks to measure AC/DC currents/voltages. These blocks are shown in Fig. 2.202.

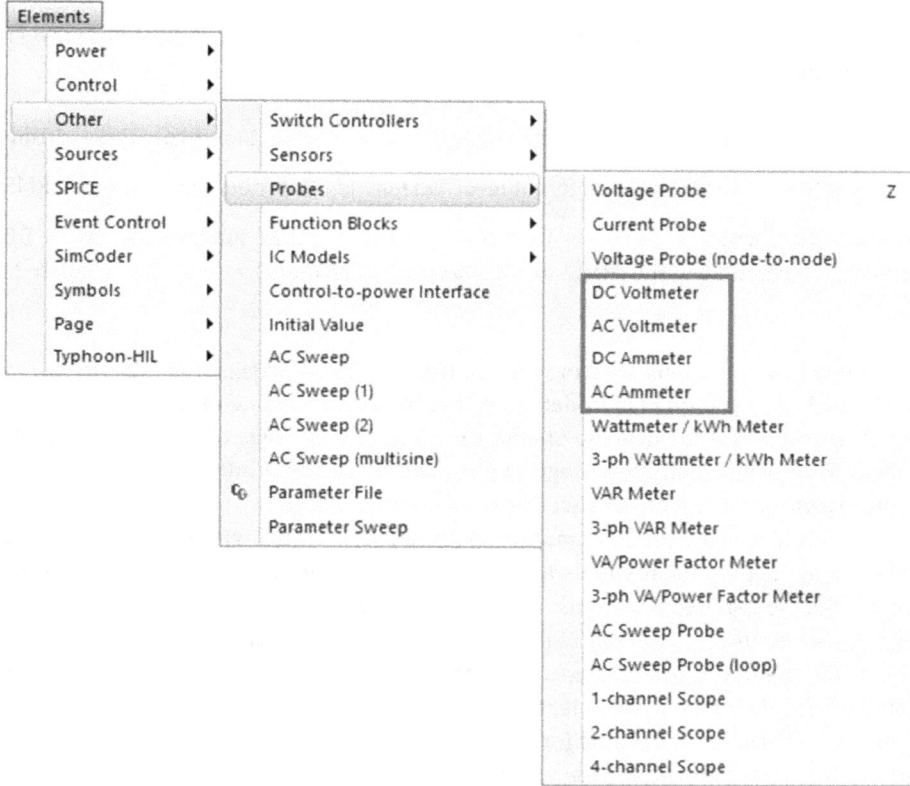

Fig. 2.202.

A DC ammeter measures the average value of a current signal and a DC volt meter measures the average value of a voltage signal. Remember that average value of a periodic signal $f(t + T) = f(t)$ is calculated using the $\frac{1}{T}\int_{t_0}^{t_0 + T} f(\tau)d\tau$.

An AC ammeter measures the Root Mean Square (RMS) value of a current signal and an AC volt meter measures the RMS value of a voltage signal. The RMS value of a periodic signal $f(t + T) = f(t)$ is calculated using the $\sqrt{\frac{1}{T}\int_{t_0}^{t_0 + T} f(\tau)^2 d\tau}$. The AC ammeter/voltmeter block of PSIM ignores the average DC component of signal $f(t)$, that is, removes the DC term and calculates the average for remaining signal.

In order to understand the aforementioned blocks better, consider the circuit shown in Fig. 2.203. The blocks settings are shown in Figs. 2.204–2.213. According to the entered values, the resistor voltage is $V = 200 + 300\sin(\omega_0 t) + 200\sin(2\omega_0 t) + 100\sin(3\omega_0 t)$ where $\omega_0 = 2 \times \pi \times 50 = 314 \frac{\text{Rad}}{\text{s}}$. The average voltage is 200 V and the mathematical RMS is equal to $\sqrt{200^2 + \frac{1}{2}(300^2 + 200^2 + 100^2)} = 479.6 \text{ V}$. However, PSIM ignores the 200 V DC term so it calculates the RMS of $300\sin(\omega_0 t) + 200\sin(2\omega_0 t) + 100\sin(3\omega_0 t)$ which is equal to $\sqrt{\frac{1}{2}(300^2 + 200^2 + 100^2)} = 264.6 \text{ V}$.

According to the Ohm's law, the current is $I = V / R = 20 + 30\sin(\omega_0 t) + 20\sin(2\omega_0 t) + 10\sin(3\omega_0 t)$ The average value of current is 20 A and mathematical RMS is equal to $\sqrt{20^2 + \frac{1}{2}(30^2 + 20^2 + 10^2)} = 47.96 \text{ A}$. However, PSIM ignores the 20 A DC term so it calculates the RMS of $30\sin(\omega_0 t) + 20\sin(2\omega_0 t) + 10\sin(3\omega_0 t)$ which is $\sqrt{\frac{1}{2}(30^2 + 20^2 + 10^2)} = 26.46 \text{ A}$.

When you take a look at an expression like $V = 200 + 300\sin(\omega_0 t) + 200\sin(2\omega_0 t) + 100\sin(3\omega_0 t)$, you quickly understand that the average value of this expression is equal to 200. Since the average operator is linear, the average of sinusoidal signals is zero. However, a computer program does not has such an intelligence. They need some algorithm in order to extract the average value of signal.

The average value (or DC component) is the part of signal with frequency of zero. In mathematical terms, it equals to the term (constant term) of Fourier series of the signal. So, we can use a low-pass filter (LPF) in order to extract the average of the signal. As the cut-off frequency of the LPF decreases, the accuracy of result increases. This is the method that PSIM uses to measure the average value of voltage/currents using the DC Voltameter/Ammeter blocks. When you double click the DC Voltameter/Ammeter blocks, the PSIM asks for the cut-off frequency of the filter. A smaller value for this box increases the accuracy; however, the simulation must be run for a longer time in order to reach the steady-state final value. When you use a DC/AC voltmeter/ammeter, only the steady-state values are important. The transient section of the waveform is not important.

When you double click on AC voltmeter/ammeter block, you need to determine two frequencies: operating frequency and cut-off frequency. Operating frequency determines the value of period, that is, $f(t + T) = f(t)$. Remember that the period is the reciprocal of the frequency. The cut-off frequency determines the cut-off frequency of high-pass filter which is used to remove the DC component of the signal. A smaller value for this box increases the accuracy; however, the simulation must be run for a longer time in order to reach the steady-state final value.

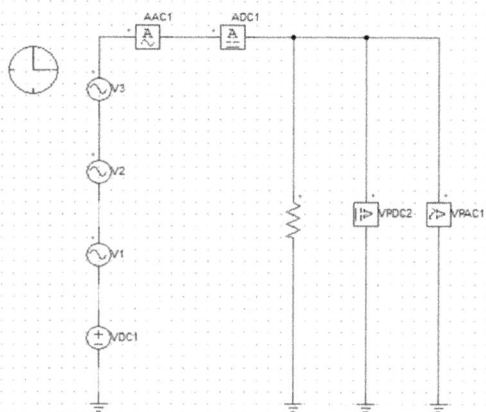

Fig. 2.203.

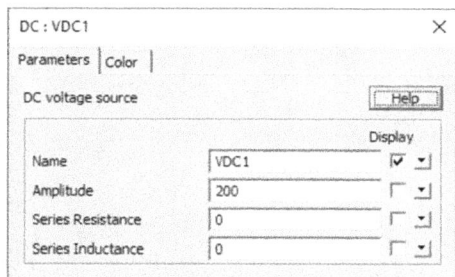

Fig. 2.204.

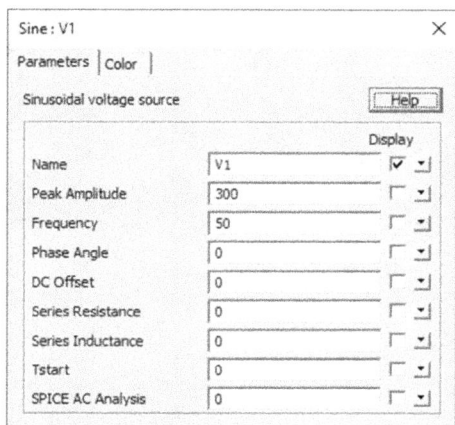

Fig. 2.205.

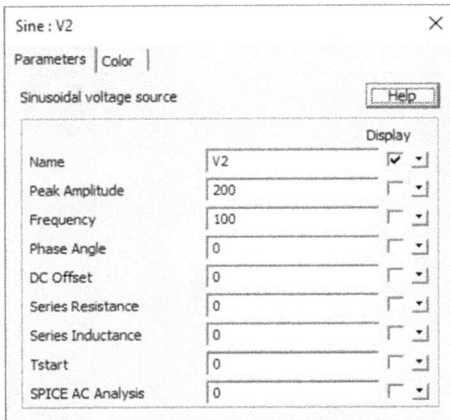

Fig. 2.206.

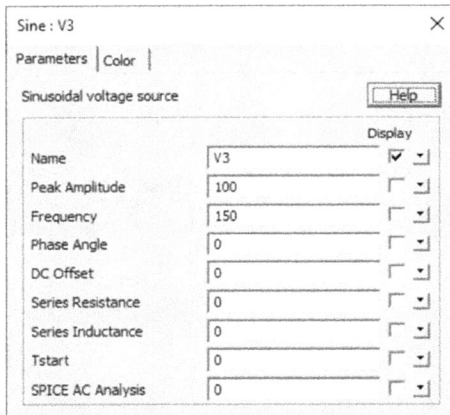

Fig. 2.207.

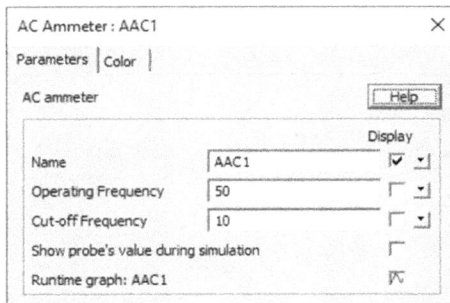

Fig. 2.208.

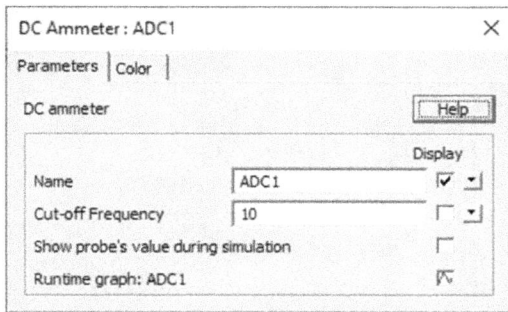

Fig. 2.209.

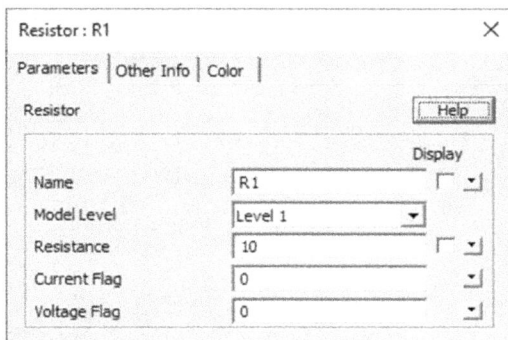

Fig. 2.210.

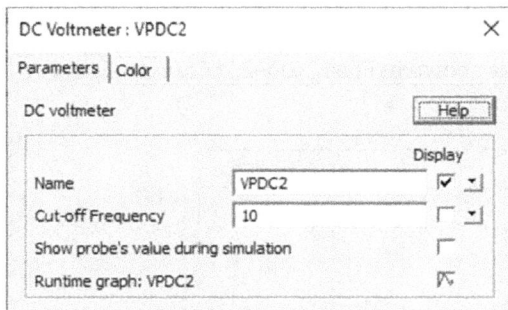

Fig. 2.211.

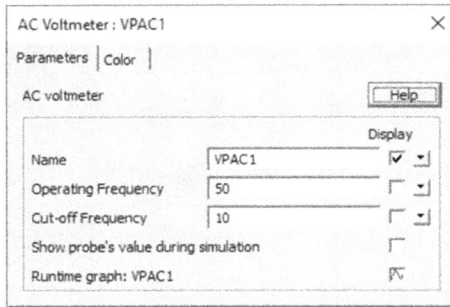

Fig. 2.212.

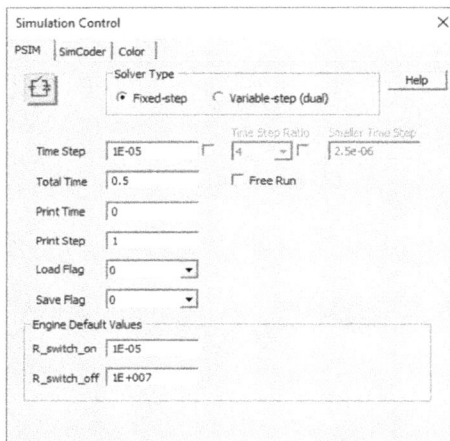

Fig. 2.213.

After running the simulation, the results shown in Figs. 2.214–2.217 are obtained.

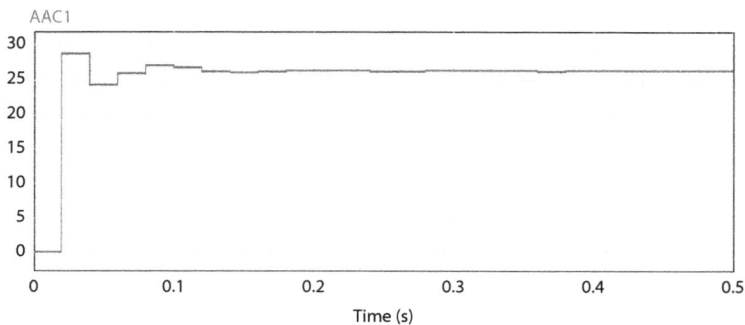

Fig. 2.214.

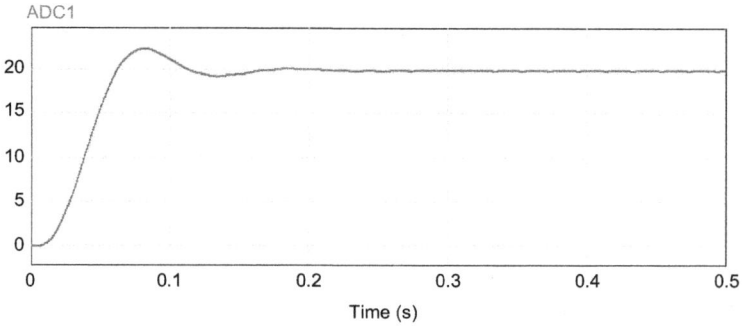

Fig. 2.215.

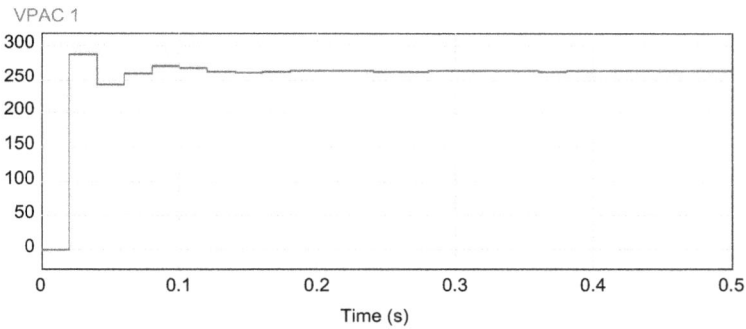

Fig. 2.216.

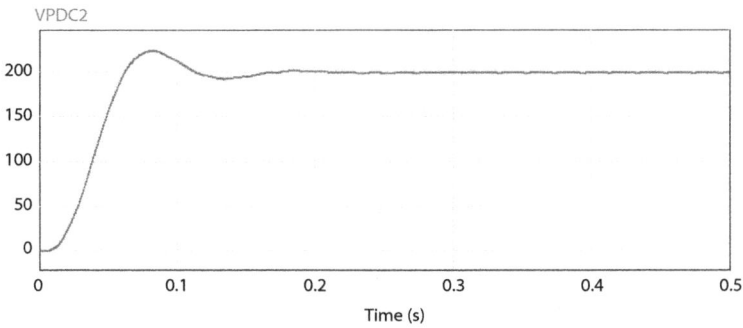

Fig. 2.217.

Use the zoom tool (Fig. 2.218) in order to see steady-state values of measurements. The steady-state values of measurements are shown in Figs. 2.219–2.222. The obtained values are very close to the hand analysis results.

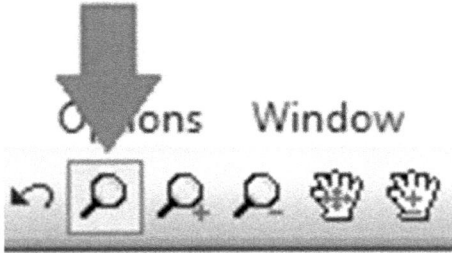

Fig. 2.218.

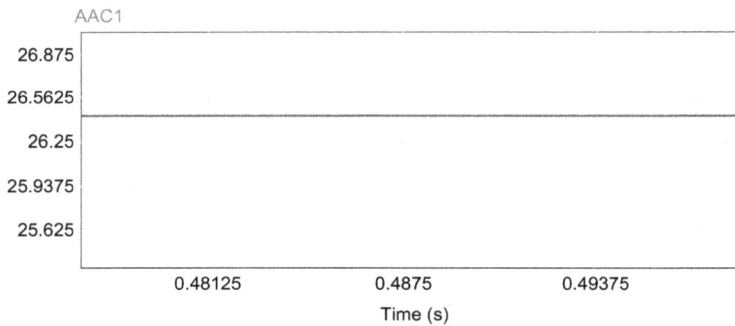

Fig. 2.219.

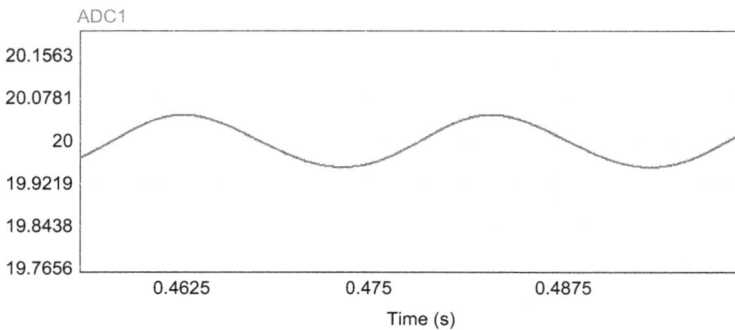

Fig. 2.220.

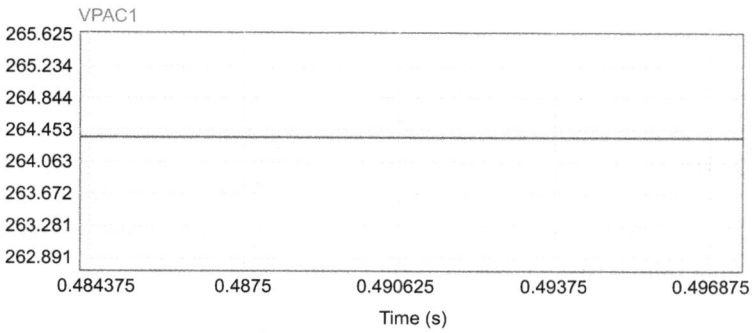

Fig. 2.221.

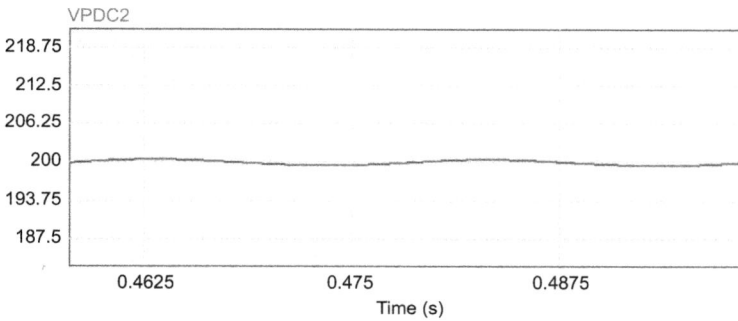

Fig. 2.222.

Chapter 3
Simview™

3.1 Introduction

Simview is PSIM's waveform display and post-processing program. This chapter studies the most important measurements that can be done with Simview.

3.2 A sample simulation

Assume a simple circuit that is shown in Fig. 3.1.

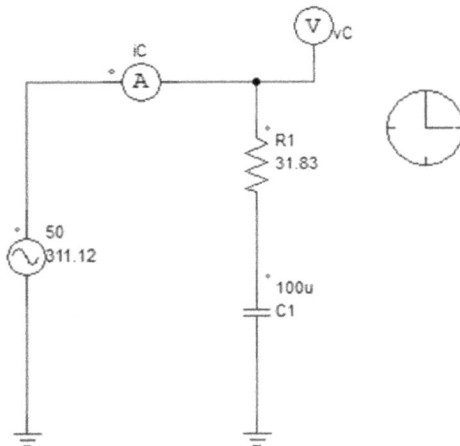

Fig. 3.1.

Settings of the used blocks are shown in Figs. 3.2–3.7.

https://doi.org/10.1515/9783110740653-003

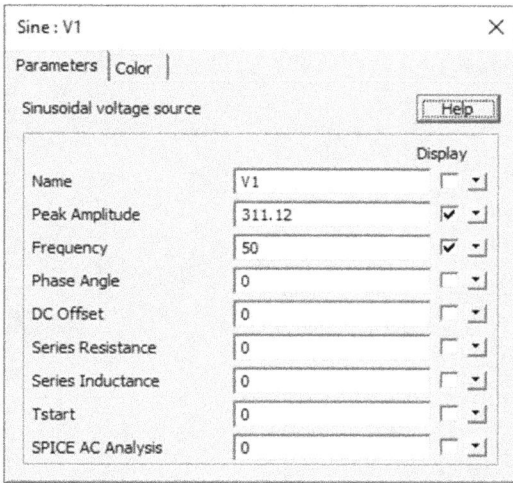

Fig. 3.2.

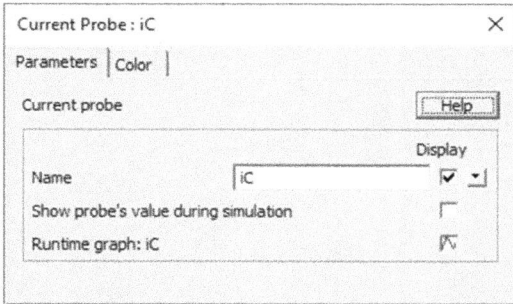

Fig. 3.3.

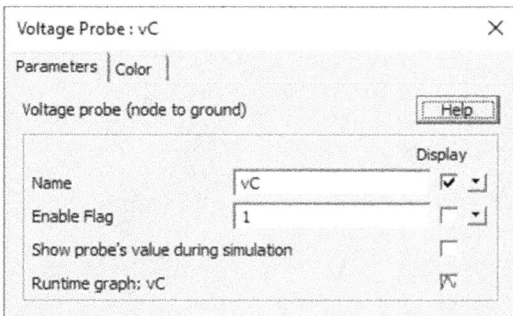

Fig. 3.4.

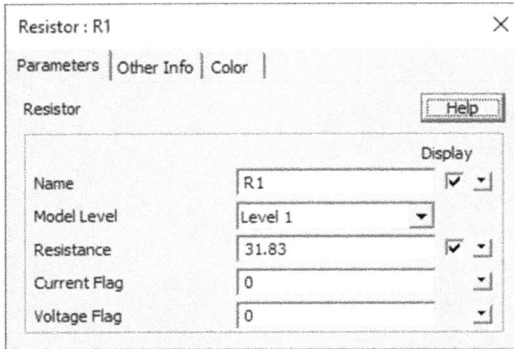

Fig. 3.5.

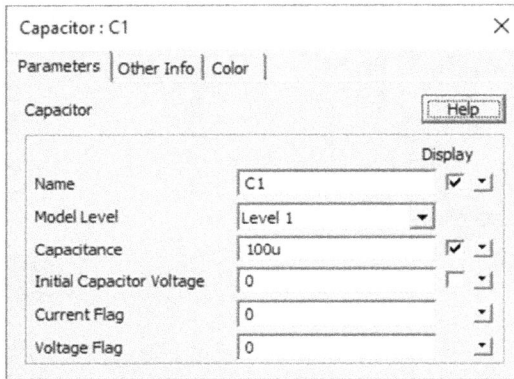

Fig. 3.6.

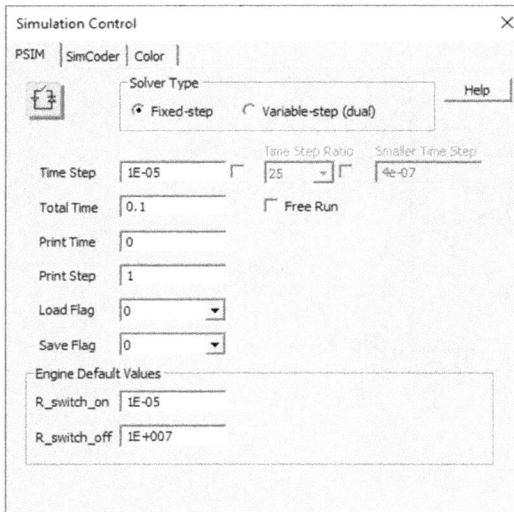

Fig. 3.7.

Run the simulation by clicking on the Run PSIM Simulation icon (Fig. 3.8).

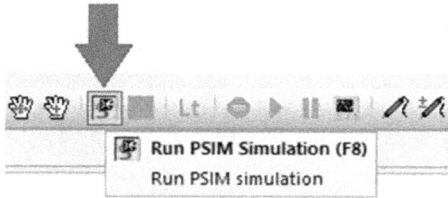

Fig. 3.8.

After simulation is completed, the Properties window appears (Fig. 3.9). The variables that are available in the right-hand side list (Variables for display) will be shown by Simview.

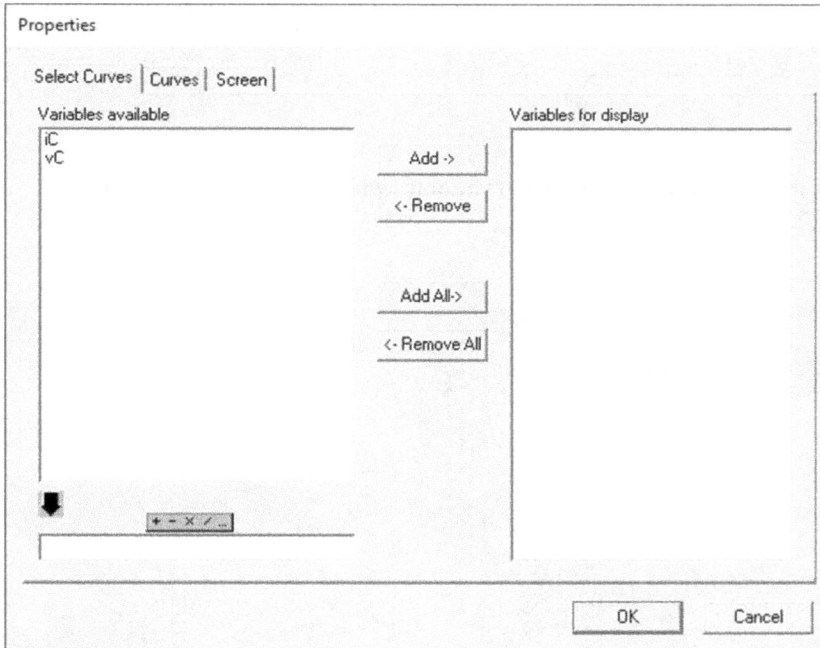

Fig. 3.9.

Assume that we want to see both of iC and vC. In order to do this, click on iC and vC (Fig. 3.10).

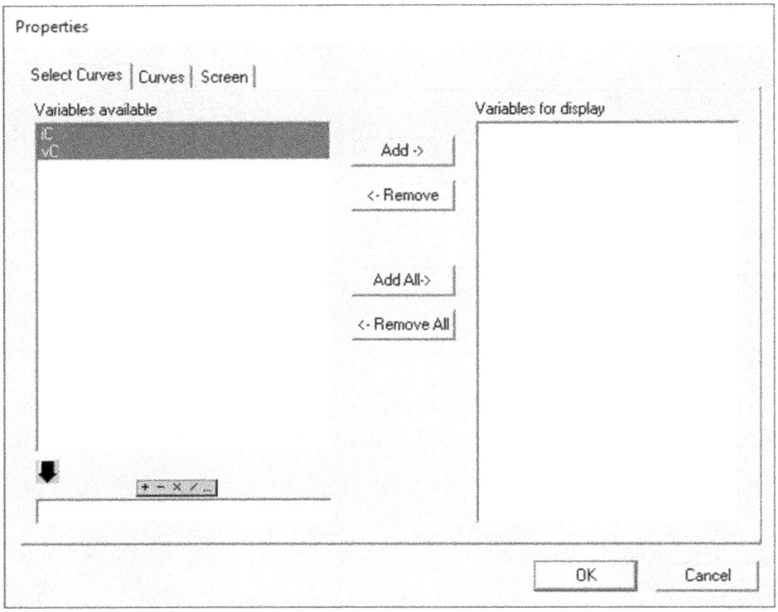

Fig. 3.10.

Then click the add button. The selected variables will be added to the right list.

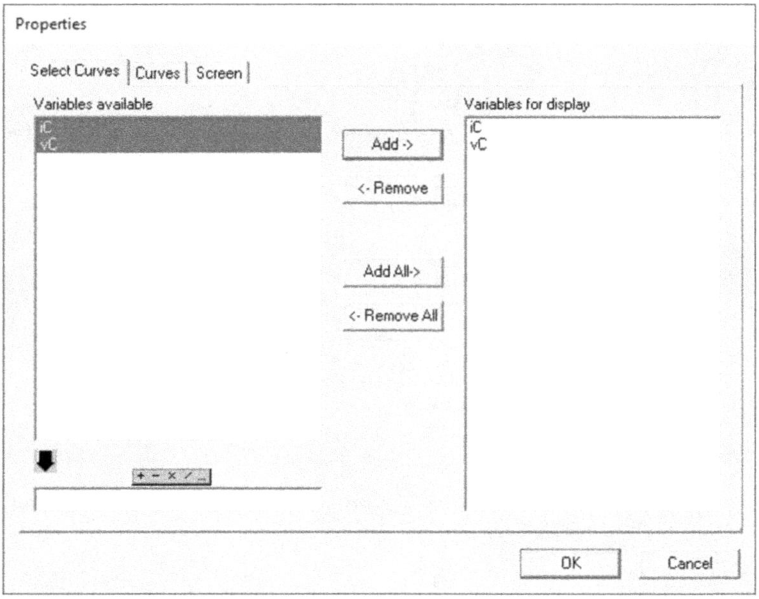

Fig. 3.11.

Click the OK button shown in Fig. 3.11. After clicking the OK button, the waveforms will be shown on the screen (Fig. 3.12).

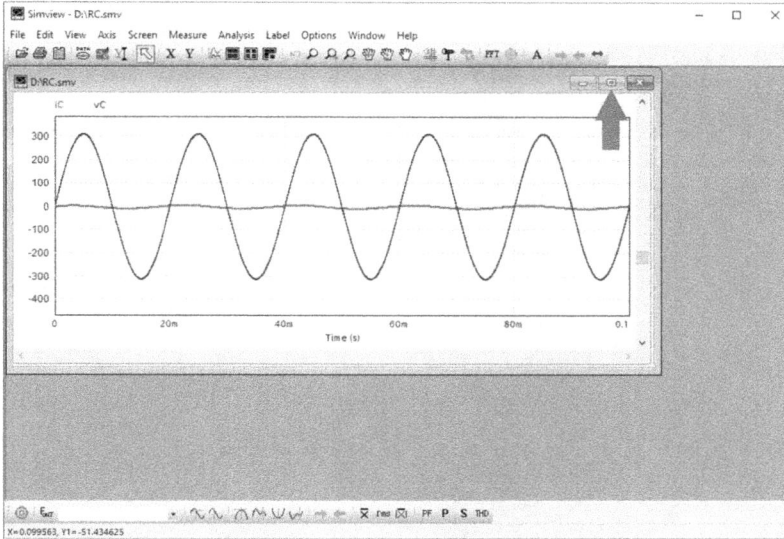

Fig. 3.12.

Click the maximize button to have a better view (Fig. 3.13).

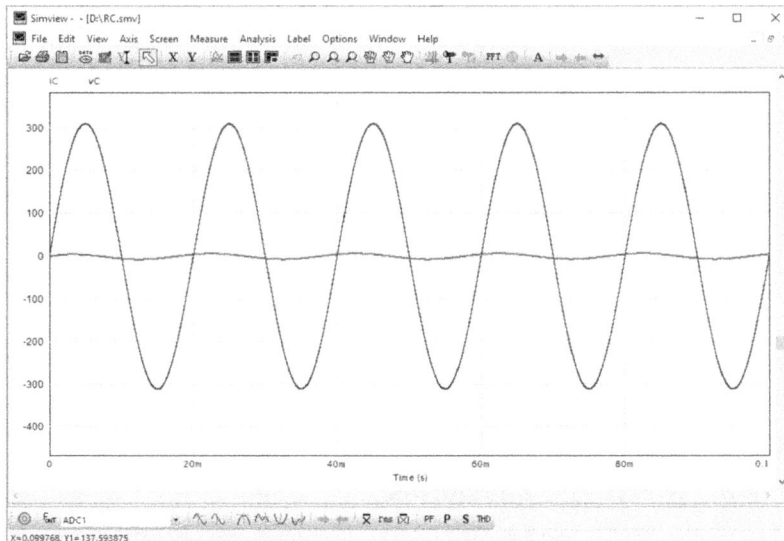

Fig. 3.13.

If you double click on the graph or click the icon shown in Fig. 3.14, then the Properties window appears again (Fig. 3.15). So, you can add/remove some waveforms.

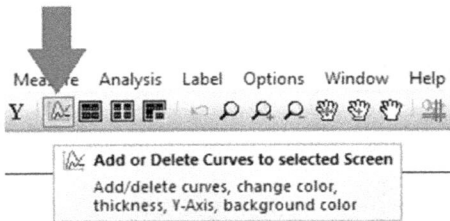

Fig. 3.14.

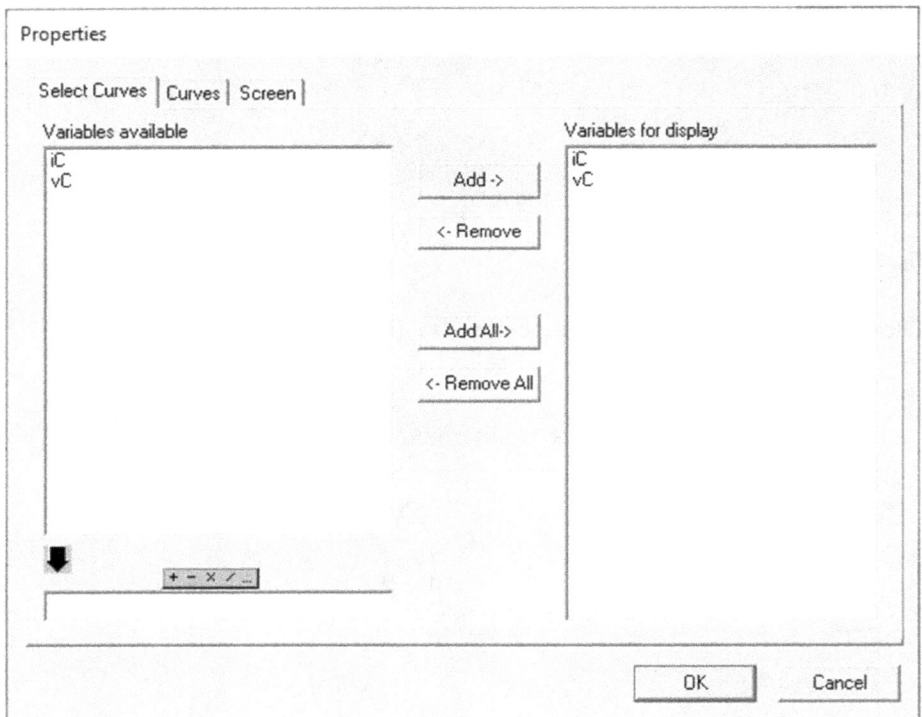

Fig. 3.15.

3.3 Addition of label to the waveforms

You can add text label to the waveforms with the aid of Label menu. In order to do this, click on the Label>Text (Fig. 3.16).

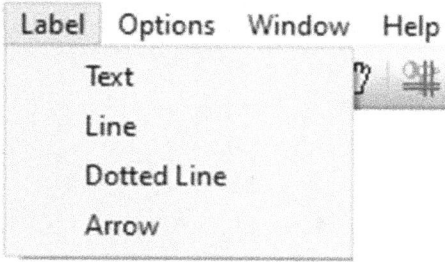

Fig. 3.16.

The window shown in Fig. 3.17 will appear. Click the Font button if you like to change the font size, font type, and so on. Type the text and then click the OK button.

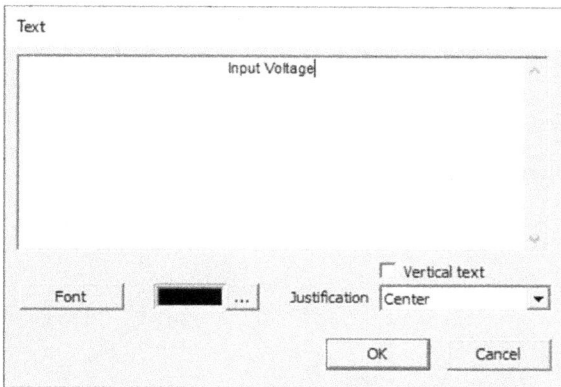

Fig. 3.17.

Now click desired location of graph and the text will be added to there (Fig. 3.18).

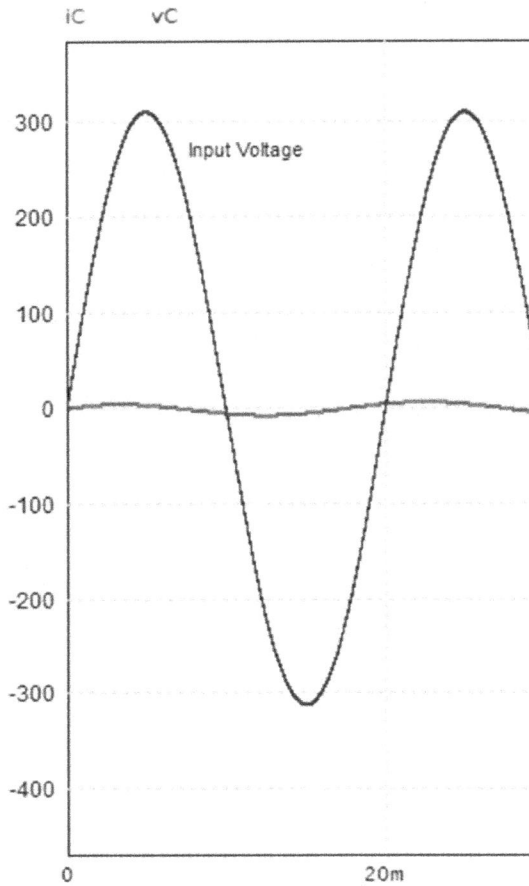

Fig. 3.18.

3.4 Changing the color of waveforms and addition of new axis to the graph

Name of waveforms shown in Simview is written in the upper left section of the screen (Fig. 3.19).

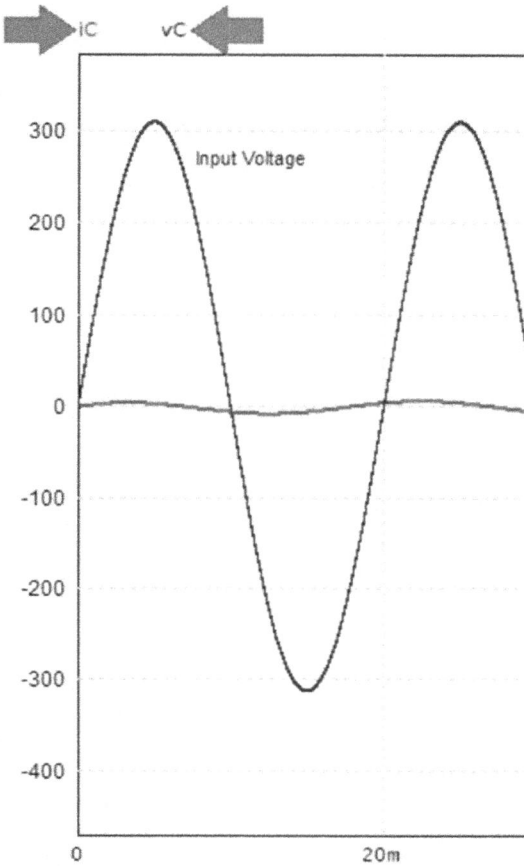

Fig. 3.19.

If you double click on the waveform names, the Properties window shown in Fig. 3.20 will appear. You can change the color of waveforms if you like. In order to change the color of waveforms, click on the color that is shown in front of the waveform. After clicking, the color palette shown in Fig. 3.21 will appear and you can select the desired color from it.

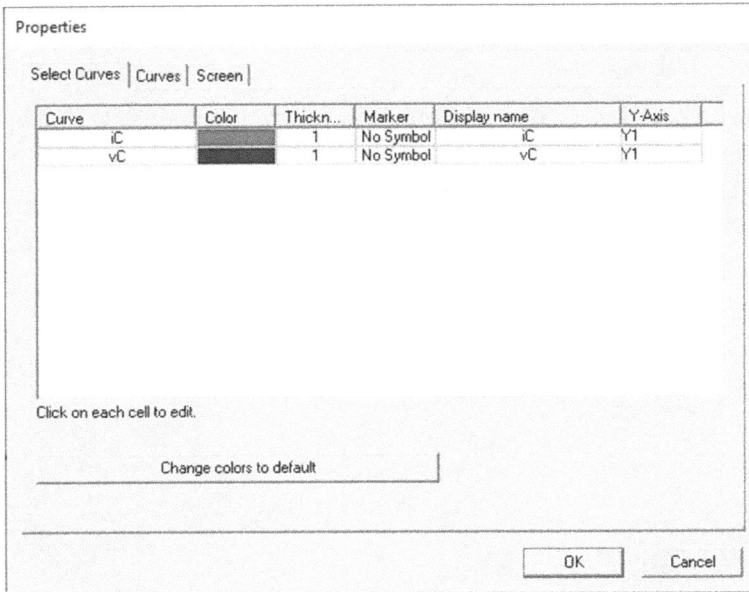

Fig. 3.20.

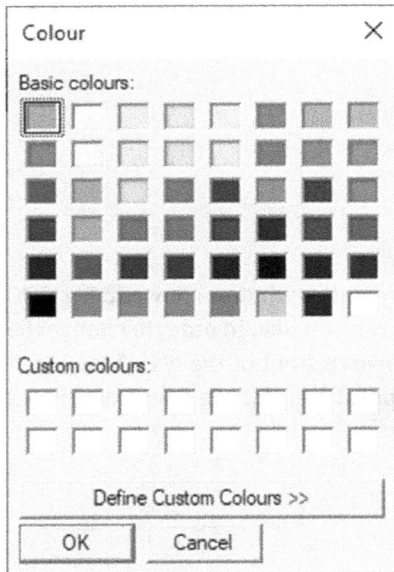

Fig. 3.21.

If you take a look at Fig. 3.13, you understand that values of iC waveform are smaller than vC. Using the same axis for both of these two waveforms caused the smaller waveform (iC) to not being shown well. We can use different axes for these two waveforms. In order to add a new axis, click the Y1 in the Y-Axis column. Then click the New Y-Axis from the appeared menu (Fig. 3.22). After clicking the New Y-Axis, a new axis will be added to the Y-Axis column.

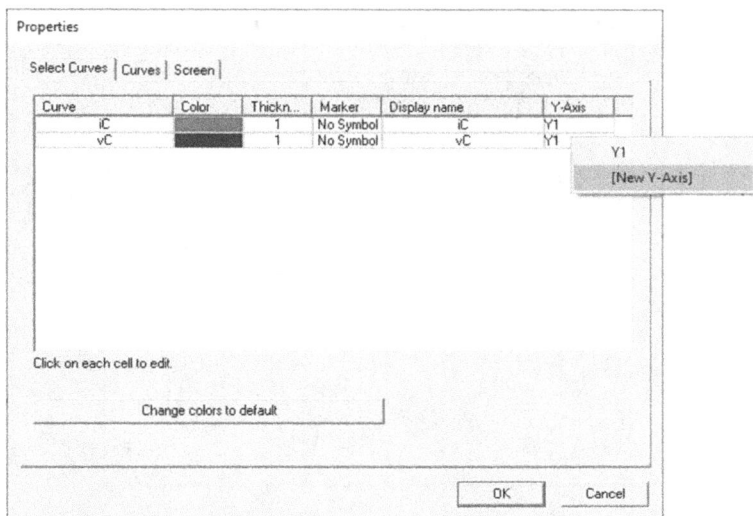

Fig. 3.22.

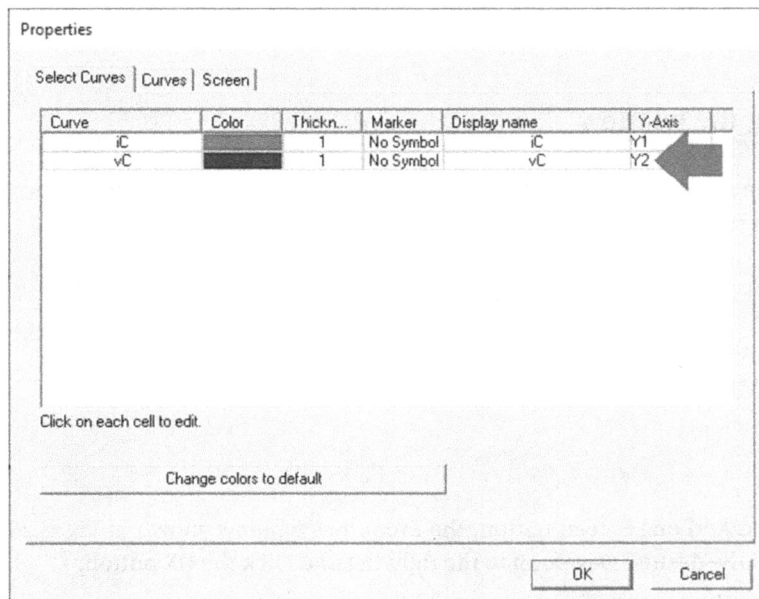

Fig. 3.23.

Click the OK button in Fig. 3.23. Now, the graph has two axes: One for iC and another for vC (Fig. 3.24).

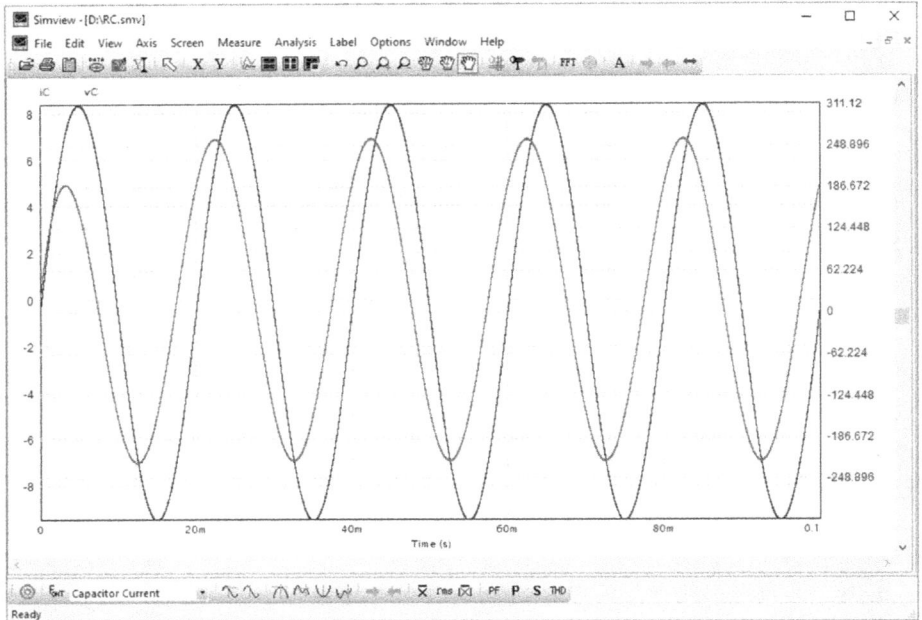

Fig. 3.24.

3.5 Splitting the window

You can split the screen into some sections and show the desired waveform in each section. In order to do this, click the Add one Screen icon (Fig. 3.25).

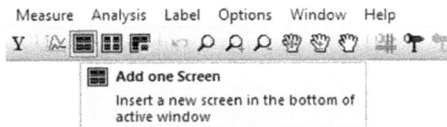

Fig. 3.25.

After clicking the Add one Screen button, the Properties window shown in Fig. 3.26 will appear. Add the desired waveform to the right list and click the OK button.

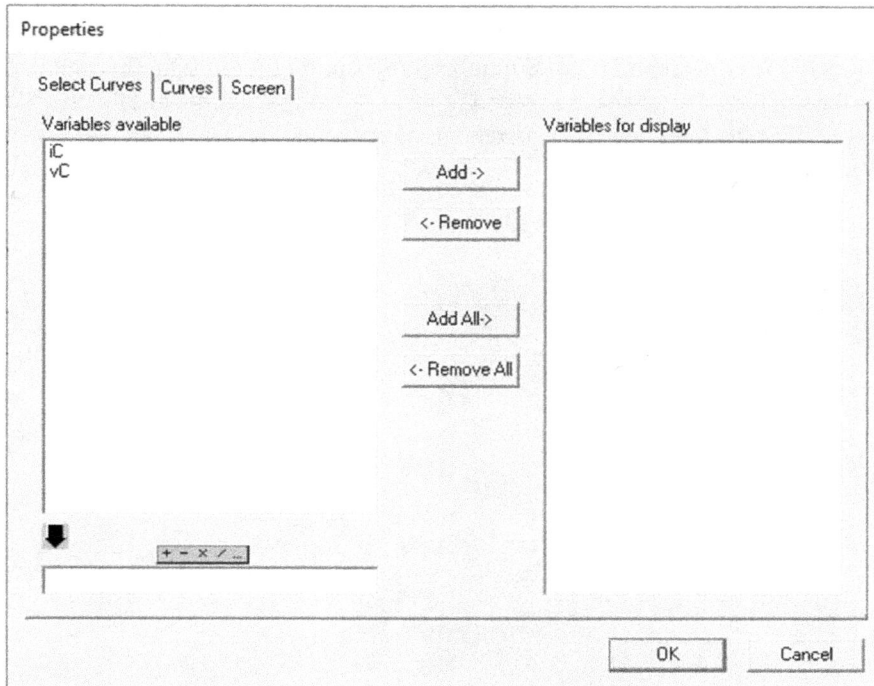

Fig. 3.26.

After clicking the OK button, a new section will be added to the screen (Fig. 3.27).

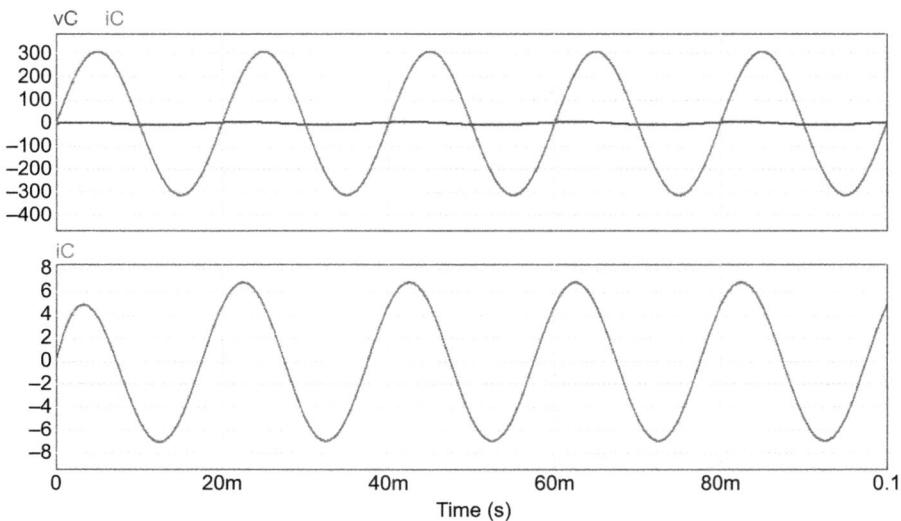

Fig. 3.27.

You can split the window using the Split Window icon (Fig. 3.28) as well. After clicking the Split Window icon, a list of numbers appears (Fig. 3.29). The first number shows the number of rows and the second number shows the number of columns. For instance, 2*3 splits the screen into six regions as shown in Fig. 3.30. Each region can show the desired waveform(s).

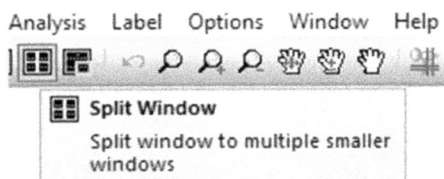

Fig. 3.28.

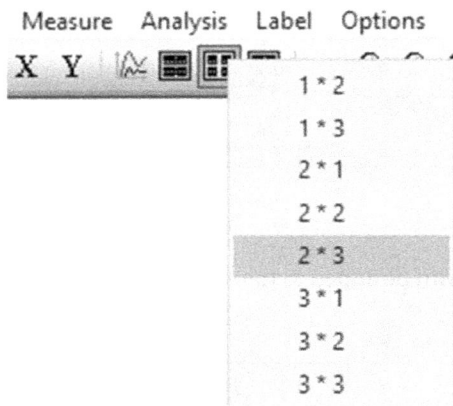

Fig. 3.29.

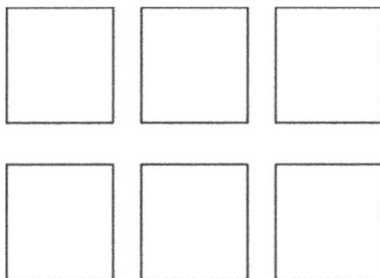

Fig. 3.30.

For instance, assume that we want to split the screen into two regions. After clicking the 2*1, the window changes to what shown in Fig. 3.31. Now you can double click on any region and add/remove the desired waveform(s).

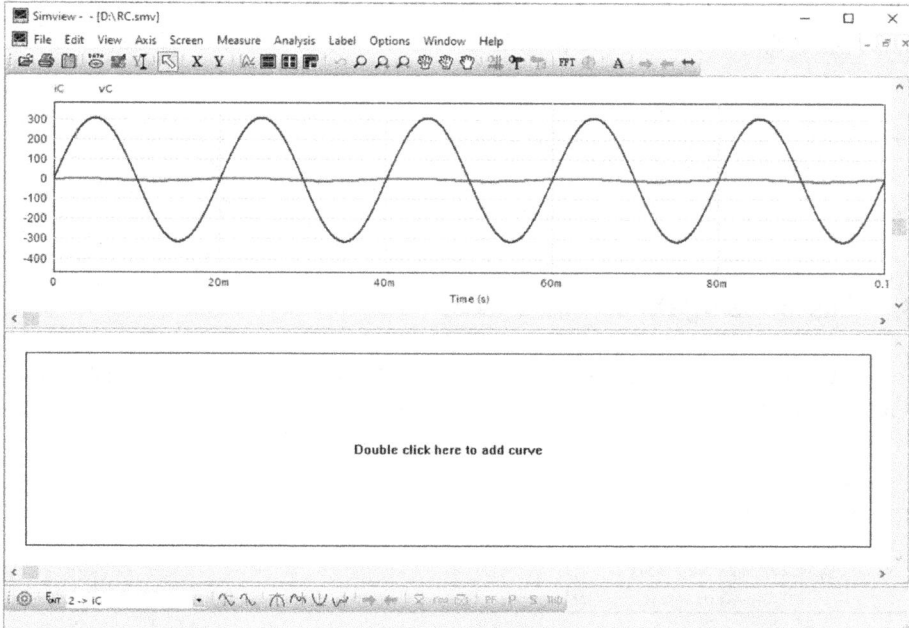

Fig. 3.31.

3.6 Zoom in, zoom out, and move

You can do the zoom in, zoom out, and move with the aid of icons shown in Fig. 3.32. You can do the zoom in/out with holding down the Ctrl key and then rotating the mouse wheel, as well.

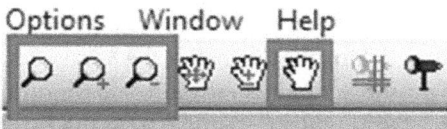

Fig. 3.32.

You can obtain the original unzoomed views by clicking the Re-draw icon (Fig. 3.33).

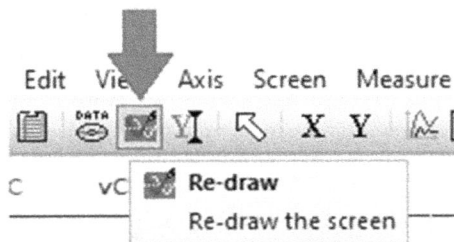

Fig. 3.33.

3.7 Customizing the X- and Y-axes

You can use the Customize the X-Axis icon (Fig. 3.34) and Customize the Y-Axis icon (Fig. 3.35) in order to customize the X- and Y-axes, respectively.

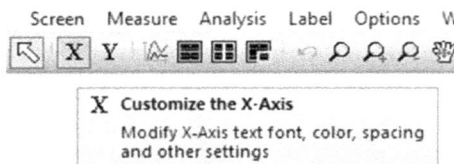

Fig. 3.34.

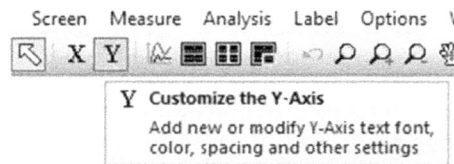

Fig. 3.35.

The window shown in Fig. 3.36 will appear after clicking the Customize the X-Axis. You can determine the type of scale (linear or logarithmic) and its range in this window. For instance, in order to see the [10 ms, 20 ms] range, enter 0.01 and 0.02 to the From and To boxes, respectively.

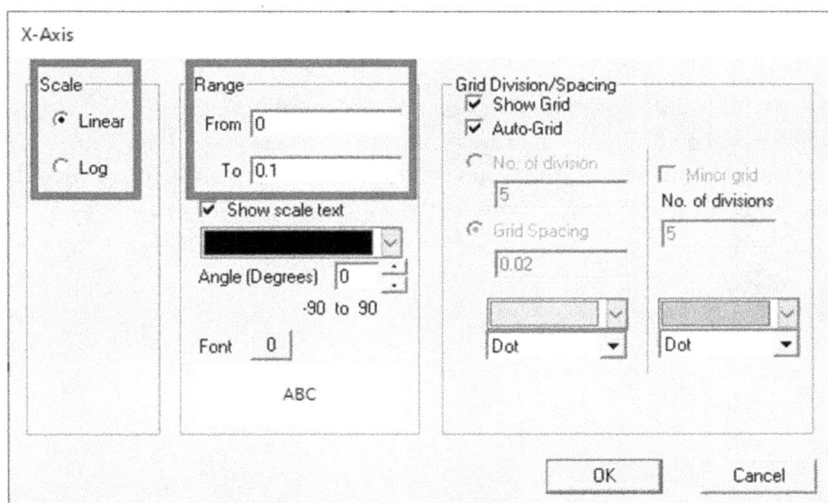

Fig. 3.36.

The window shown in Fig. 3.37 will appear after clicking the Customize the Y-Axis. You can determine the type of scale (linear, logarithmic or db µv) and its range in this window. db µv (decibels-microvolt) is used in electromagnetic interference and electromagnetic compatibility designs.

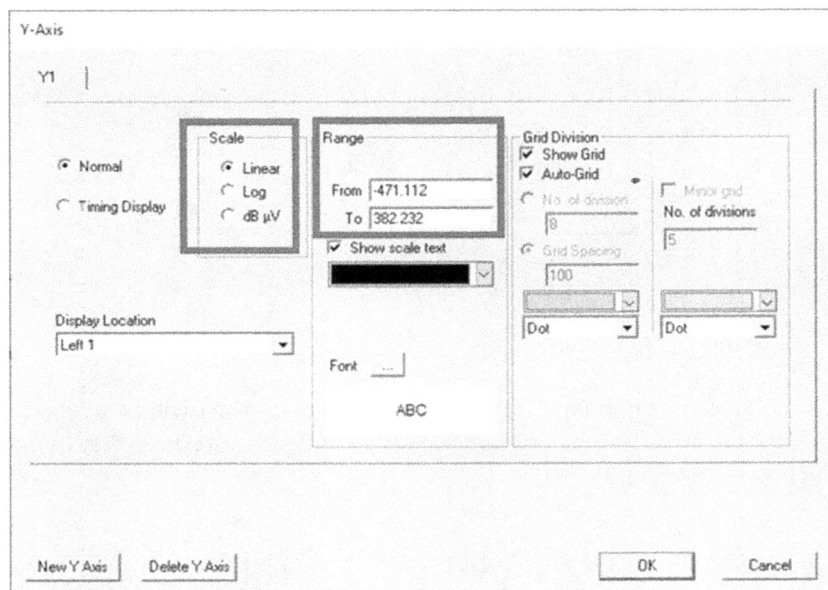

Fig. 3.37.

3.8 Cursors

You can see the coordinate of different points of the graph with the aid of cursors. Click the telescope icon (Fig. 3.38) in order to add cursor to the graph (Fig. 3.39). Now move the cursor(s) to the point(s) that you want to measure its coordinates. The coordinates will be shown in the Measure window.

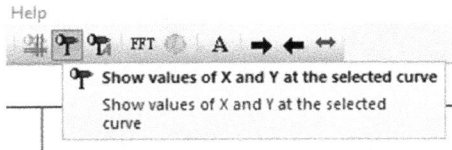

Fig. 3.38.

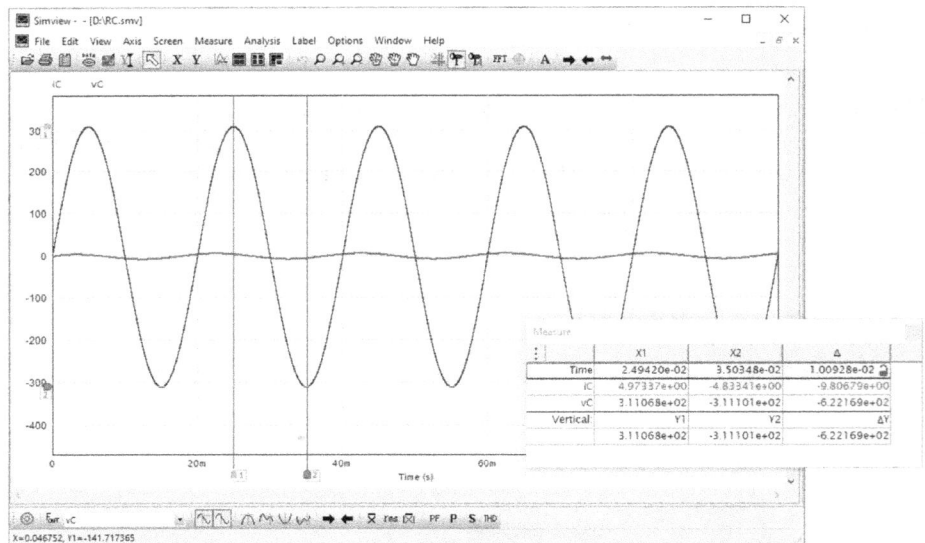

Fig. 3.39.

If you click the icon shown in Fig. 3.40, the coordinate of cursor 1 will be added to the graph (Fig. 3.41). You can click on the added label and press the delete key of the keyboard in order to remove it.

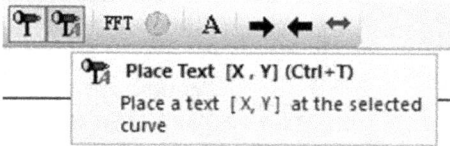

Fig. 3.40.

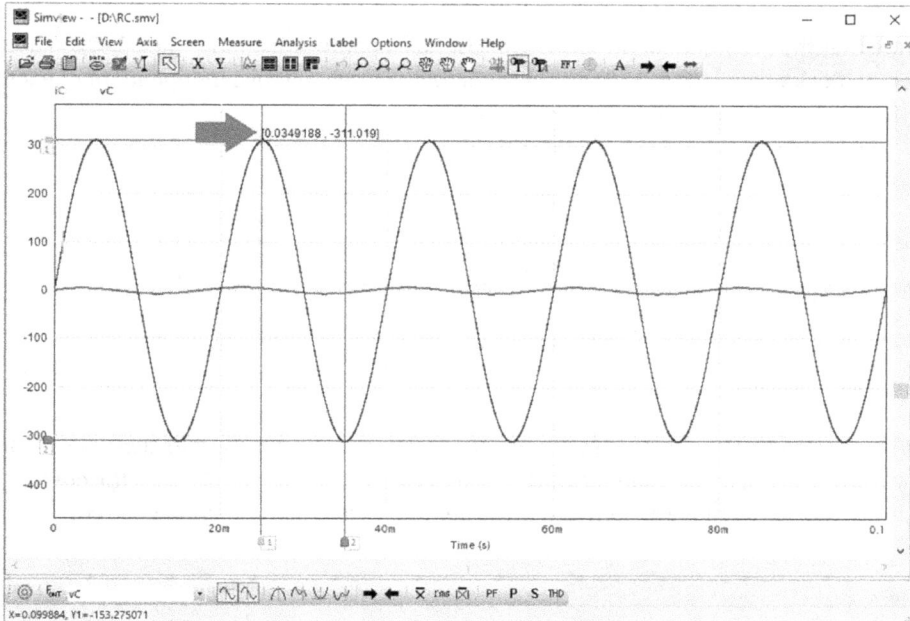

Fig. 3.41.

Sometimes, it is difficult to use the mouse in order to bring the cursor to the desired location. For instance, assume that you want to read the value of voltage at $t = 10$ ms. When you move the mouse using your hand, you may put the cursor at close point to 10 ms (i.e., 10.05 or 9.97 ms), but not exactly at the 10 ms point.

In such cases, you can double click the Time box for X1 row and enter 10 m or 0.01 or 10e-3. The cells of Measure window which have yellow background (Fig. 3.42) accept values from keyboard as well.

Measure				x
⋮		X1	X2	Δ
Time		2.50000e-02	7.50000e-02	5.00000e-02 🔒
iC		4.88517e+00	-4.88707e+00	-9.77223e+00
vC		3.11120e+02	-3.11120e+02	-6.22240e+02

Fig. 3.42.

Note that the Δ column of Measure window has a padlock icon in it (Fig. 3.42). When the padlock is open (Fig. 3.43), the two cursors are free to move independent of each other and the distance between them may be any number.

Measure				x
⋮		X1	X2	Δ
Time		4.03712e-02	5.04640e-02	1.00928e-02 🔓
iC		5.42267e+00	-5.54529e+00	-1.09680e+01
vC		3.62029e+01	-4.51950e+01	-8.13980e+01

Fig. 3.43.

When the padlock is locked (Fig. 3.44), the distance between the two cursors is constant and equals to the value behind the padlock icon. If you move one of the cursors, the other one will move as well and the difference between the cursors keep constant.

Measure				x
⋮		X1	X2	Δ
Time		3.00464e-02	8.00464e-02	5.00000e-02 🔒
iC		-4.95831e+00	4.95793e+00	9.91624e+00
vC		-4.53502e+00	4.53539e+00	9.07041e+00

Fig. 3.44.

Assume that you want to measure the frequency of waveform shown in Fig. 3.45. In order to do this, you need to put one of the cursors at the beginning of the period and the other one at the end of period.

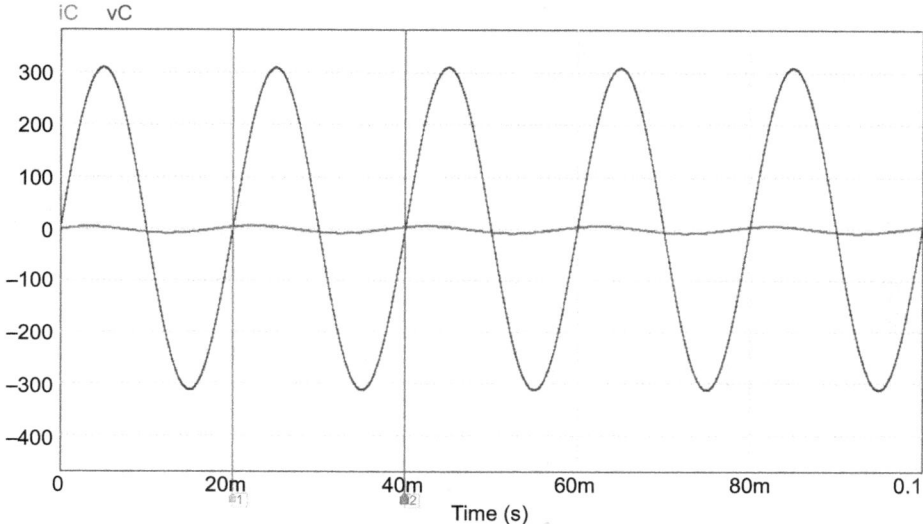

Fig. 3.45.

The measure window shows the time difference between the two cursors. This value gives the period of signal to us. We need to reverse this number in order to obtain the frequency.

Measure				☒
⋮		X1	X2	Δ
Time		2.00008e-02	3.99072e-02	1.99064e-02
iC		4.88397e+00	4.74726e+00	-1.36716e-01
vC		7.81881e-02	-9.06983e+00	-9.14802e+00

Fig. 3.46.

You can use the Simview calculator (Figs. 3.47 and 3.48) to calculate the reverse of period. However, there is another easy way to calculate the reverse.

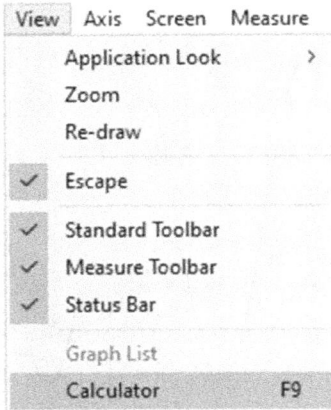

Fig. 3.47.

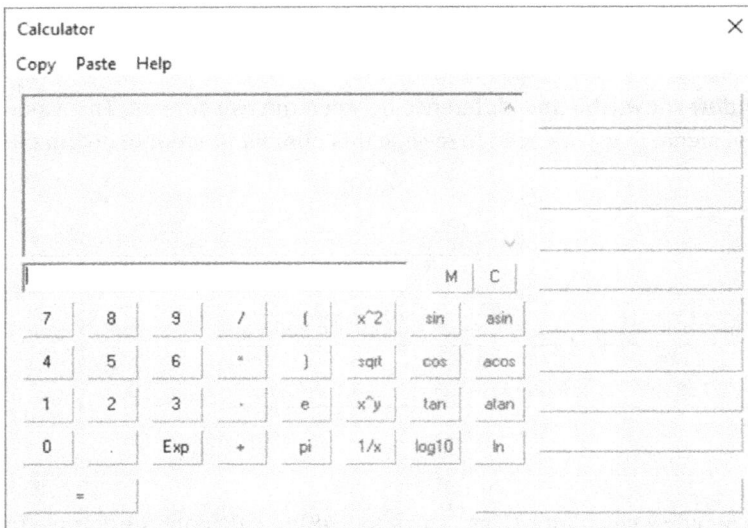

Fig. 3.48.

Click the three vertical dots (Fig. 3.49) in the Measure window and click the from $\frac{1}{\Delta}$ appeared list in order to calculate the reverse without being involved with any calculations.

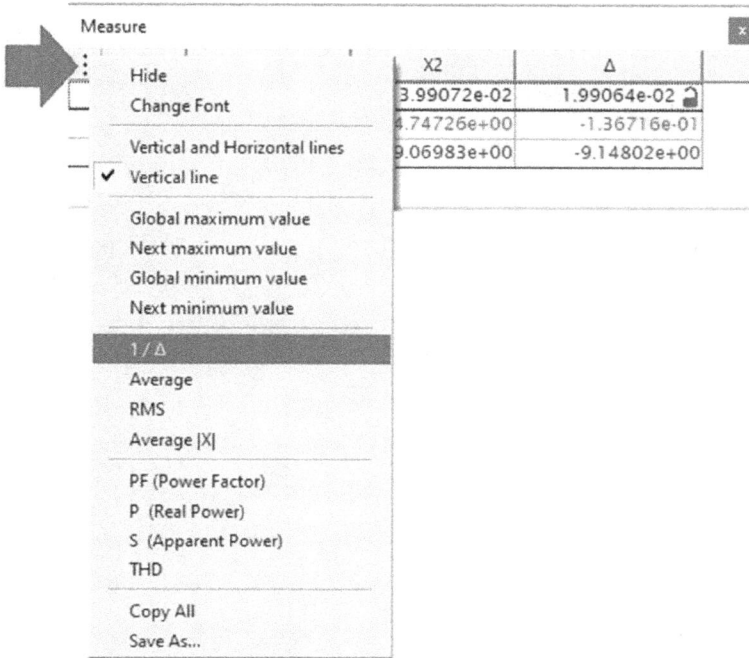

Fig. 3.49.

Now value of frequency can be read from the $\dfrac{1}{\Delta}$ column (Fig. 3.50).

	X1	X2	Δ	1/Δ
Time	2.00008e-02	3.99072e-02	1.99064e-02 🔒	5.02351e+01
iC	4.88397e+00	4.74726e+00	-1.36716e-01	
vC	7.81881e-02	-9.06983e+00	-9.14802e+00	

Fig. 3.50.

3.9 Calculation of average and root mean square

You can calculate the average and root mean square (RMS) values of signals easily in the Simview environment. Click the average icon (Fig. 3.51) in order to calculate the average value of signal. You can click the Avg section from Analysis menu as well (Fig. 3.52).

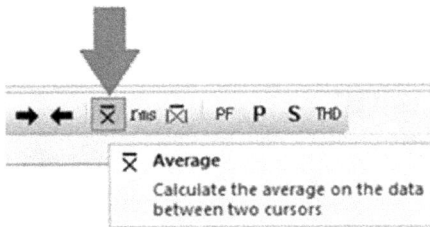

Fig. 3.51.

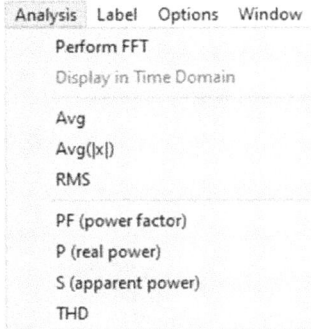

Fig. 3.52.

Two cursors will be added to the graph after clicking the average icon (Fig. 3.53). Ensure that they surround an integer number of cycles.

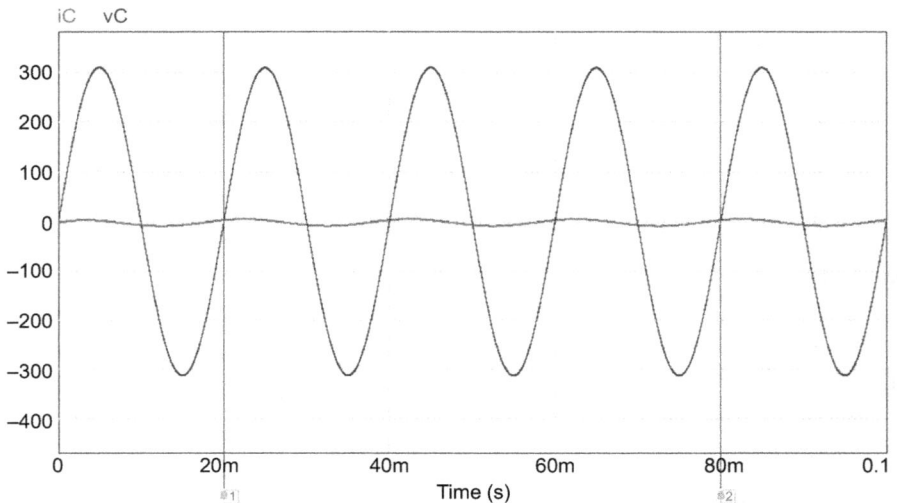

Fig. 3.53.

Simview calculates the average value and show it in the Average column of Measure window (Fig. 3.54).

	X1	X2	Δ	Average
Time	2.00080e-02	8.00020e-02	5.99940e-02	
iC	4.89037e+00	4.89028e+00	-9.55045e-05	3.31686e-04
vC	7.81924e-01	1.95477e-01	-5.86446e-01	8.14370e-05

Fig. 3.54.

You can calculate the RMS value with the aid of RMS icon (or RMS section of Analysis menu) (Fig. 3.55).

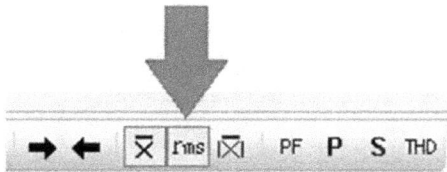

Fig. 3.55.

Two cursors will be added to the graph after clicking the RMS icon. Ensure that they surround an integer number of cycles. Simview calculates the RMS value and show it in the RMS column of Measure window.

	X1	X2	Δ	Average	RMS
Time	2.00080e-02	8.00020e-02	5.99940e-02		
iC	4.89037e+00	4.89028e+00	-9.55045e-05	3.31686e-04	4.88666e+00
vC	7.81924e-01	1.95477e-01	-5.86446e-01	8.14370e-05	2.19977e+02

Fig. 3.56.

3.10 Calculation of power factor, active power, and apparent power

You can calculate the power factor, average power, and apparent power factor in Simview. In order to do these calculations, you need to have both current and voltage waveforms on the screen. In order to calculate the power factor, click the PF icon.

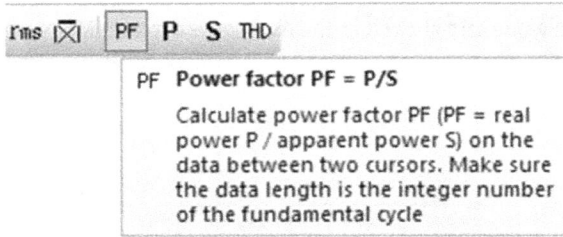

rms ⊠ | PF | P S THD

> PF **Power factor PF = P/S**
>
> Calculate power factor PF (PF = real power P / apparent power S) on the data between two cursors. Make sure the data length is the integer number of the fundamental cycle

Fig. 3.57.

After clicking the PF icon, two cursors will be added to the graph. Make sure the distance between the two cursors is an integer multiple of signals period. The number shown in the PF column of Measure window shows the power factor (Fig. 3.58).

Measure				☒
	X1	X2	Δ	PF
Time	2.00080e-02	8.00020e-02	5.99940e-02	
iC	4.89037e+00	4.89028e+00	-9.55045e-05	7.06919e-01
vC	7.81924e-01	1.95477e-01	-5.86446e-01	

Fig. 3.58.

In order to calculate the average power, click the P icon (Fig. 3.59).

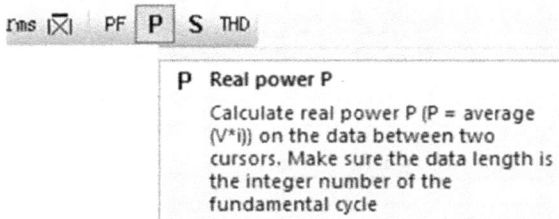

rms ⊠ PF | P | S THD

> P **Real power P**
>
> Calculate real power P (P = average (V*i)) on the data between two cursors. Make sure the data length is the integer number of the fundamental cycle

Fig. 3.59.

After clicking the P icon, two cursors will be added to the graph. Make sure the distance between the two cursors is an integer multiple of signals period. The number shown in the P column of Measure window shows the average power (Fig. 3.60).

Measure					x
⋮		X1	X2	Δ	P
Time		2.00080e-02	8.00020e-02	5.99940e-02 🔒	
iC		4.89037e+00	4.89028e+00	-9.55045e-05	7.59903e+02
vC		7.81924e-01	1.95477e-01	-5.86446e-01	

Fig. 3.60.

In order to calculate the apparent power, click the S icon (Fig. 3.61).

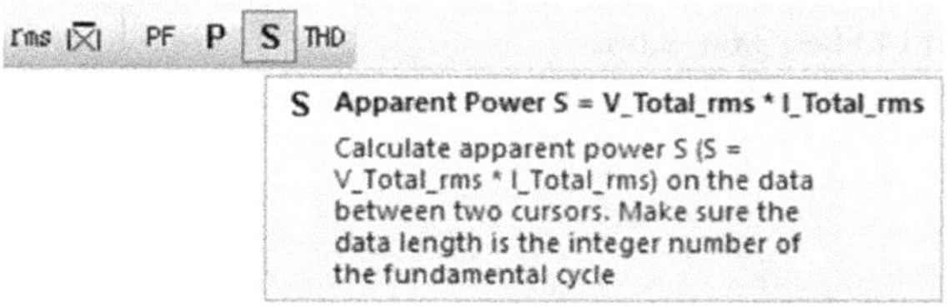

rms ⊠ PF **P S** THD

S Apparent Power S = V_Total_rms * I_Total_rms
Calculate apparent power S (S = V_Total_rms * I_Total_rms) on the data between two cursors. Make sure the data length is the integer number of the fundamental cycle

Fig. 3.61.

After clicking the S icon, two cursors will be added to the graph. Make sure the distance between the two cursors is an integer multiple of signals period. The number shown in the S column of Measure window shows the apparent power (Fig. 3.62).

Measure					x
⋮		X1	X2	Δ	S
Time		2.00080e-02	8.00020e-02	5.99940e-02 🔒	
iC		4.89037e+00	4.89028e+00	-9.55045e-05	1.07495e+03
vC		7.81924e-01	1.95477e-01	-5.86446e-01	

Fig. 3.62.

3.11 Calculation of total harmonic distortion

You can measure total harmonic distortion (THD) in Simview easily. Assume a simple circuit as shown in Fig. 3.63. The input voltage source is a square wave with amplitude of 5 V.

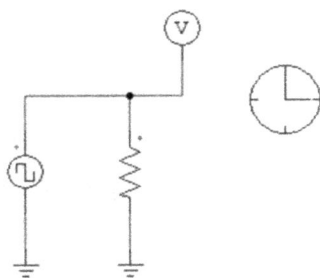

Fig. 3.63.

Setting of used elements is shown in Figs. 3.64–3.66.

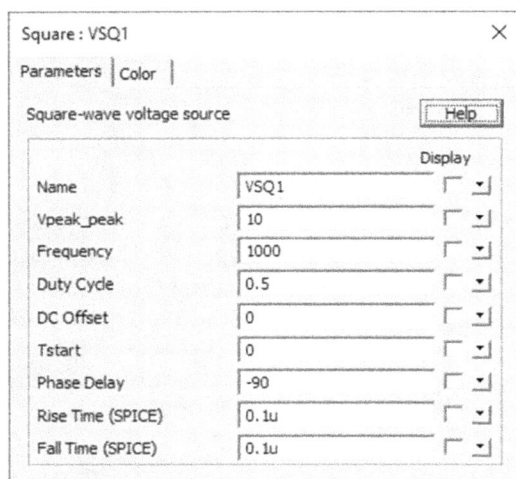

Fig. 3.64.

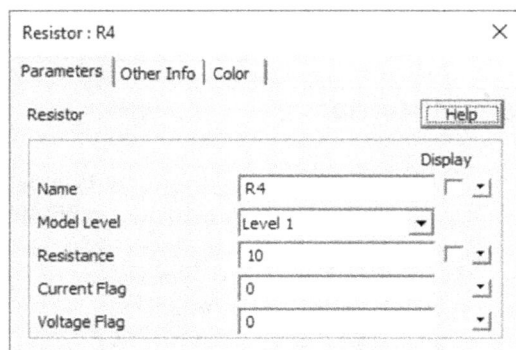

Fig. 3.65.

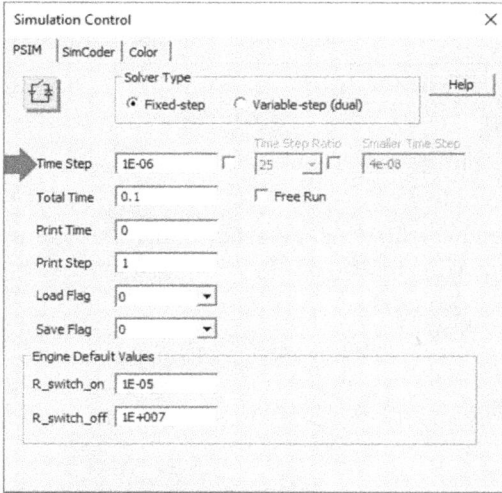

Fig. 3.66.

Run the simulation. Click the THD button (Fig. 3.67).

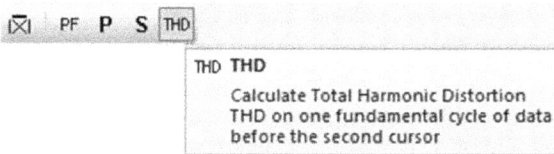

Fig. 3.67.

After clicking the THD icon, the window shown in Fig. 3.68 will appear. Enter the Fundamental frequency of the signal here. After clicking the OK button, the THD will be shown in the THD column of Measure window (Fig. 3.69).

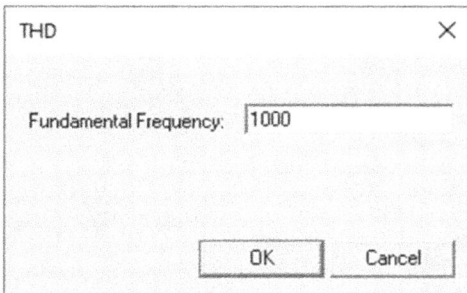

Fig. 3.68.

	X1	X2	Δ	THD
Time	8.07398e-02	8.17532e-02	1.01346e-03	freq=1000
Vout	0.00000e+00	1.00000e+01	1.00000e+01	1.21238e+00

Measure

Fig. 3.69.

Let's check the Simview result with hand calculations. Remember that the Fourier series representation of square wave shown in Fig. 3.70 is

$$x(t) = \frac{V_m}{2} + \sum_{n=0}^{\infty} \frac{2V_m}{(2n+1)\pi} \sin\left(2n+1 \times \frac{2\pi}{T} \times t\right).$$

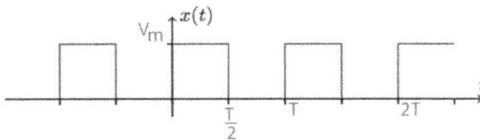

Fig. 3.70.

When

$$V_m = 10, \quad x(t) = 5 + \sum_{n=0}^{\infty} \frac{20}{(2n+1)\pi} \sin\left(2n+1 \times \frac{2\pi}{T} \times t\right). \quad V_1 = \frac{20}{\pi} = 6.366V$$

or

$$V_{1,RMS} = \frac{\frac{20}{\pi}}{\sqrt{2}} = \frac{6.366}{\sqrt{2}} = 4.502V_{RMS}.$$

RMS value of signal $x(t)$ is

$$10\sqrt{0.5} = 7.07 \text{ V}.$$

The THD can be calculated using the

$$\sqrt{\frac{V_{RMS}^2 - V_{1,RMS}^2}{V_{1,RMS}^2}}$$

formula. The result (Fig. 3.71) is the same as that obtained by Simview.

```
Command Window                    ⊙
    >> vrms=10*sqrt(.5);
    >> vlrms=2*10/pi/1.4142;
    >> sqrt(vrms^2-vlrms^2)/vlrms

    ans =

        1.2113

fx >>
```

Fig. 3.71.

3.12 Fast Fourier transform analysis

With the aid fast Fourier transform (FFT) analysis, you can see the frequency content of a signal.

The resistor (Fig. 3.63) voltage waveform is shown in Fig. 3.72.

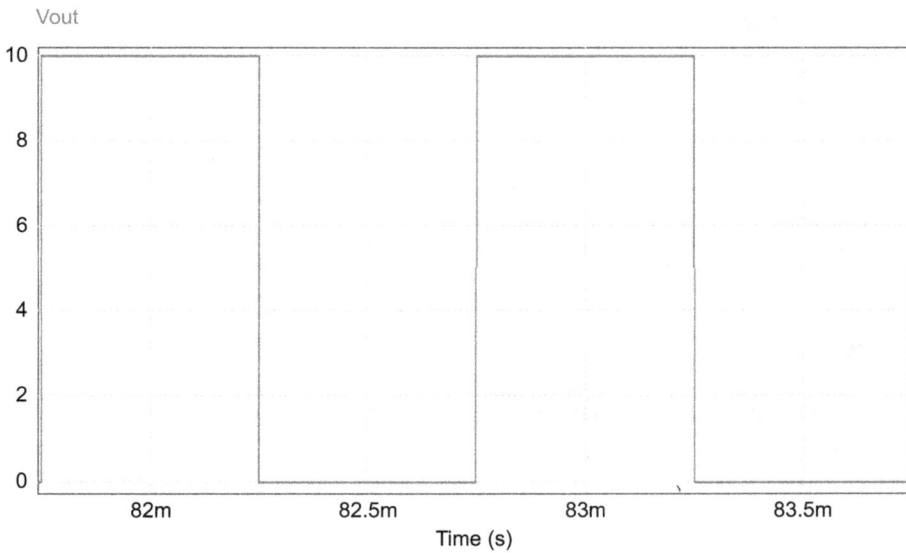

Fig. 3.72.

Click the FFT icon in order to see the frequency content of signal.

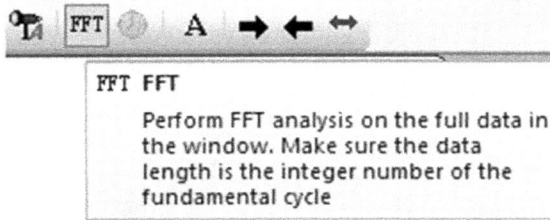

FFT **FFT**

Perform FFT analysis on the full data in the window. Make sure the data length is the integer number of the fundamental cycle

Fig. 3.73.

After clicking the FFT icon, the graph shown in Fig. 3.74 will appear.

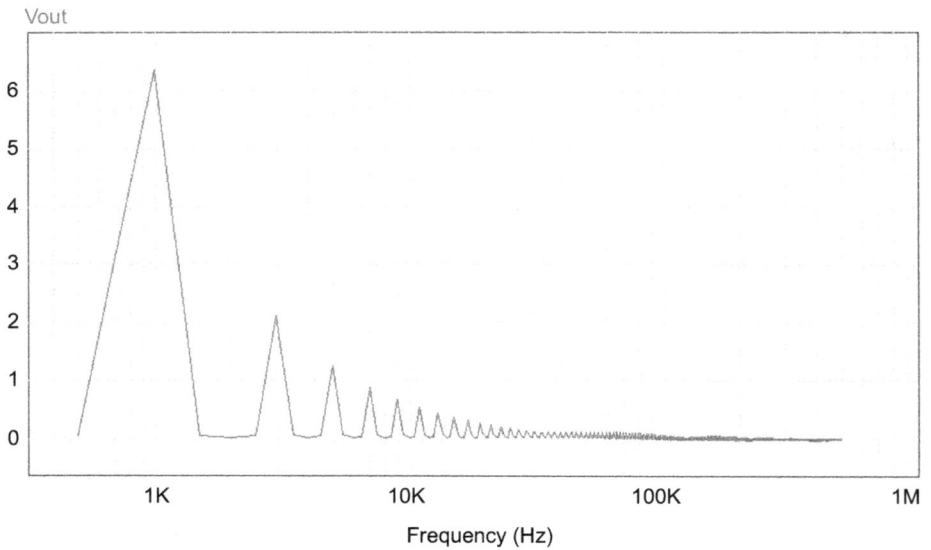

Fig. 3.74.

Double click on the X-axis (Frequency (HZ) axis) in order to determine the desired frequency range (Fig. 3.75).

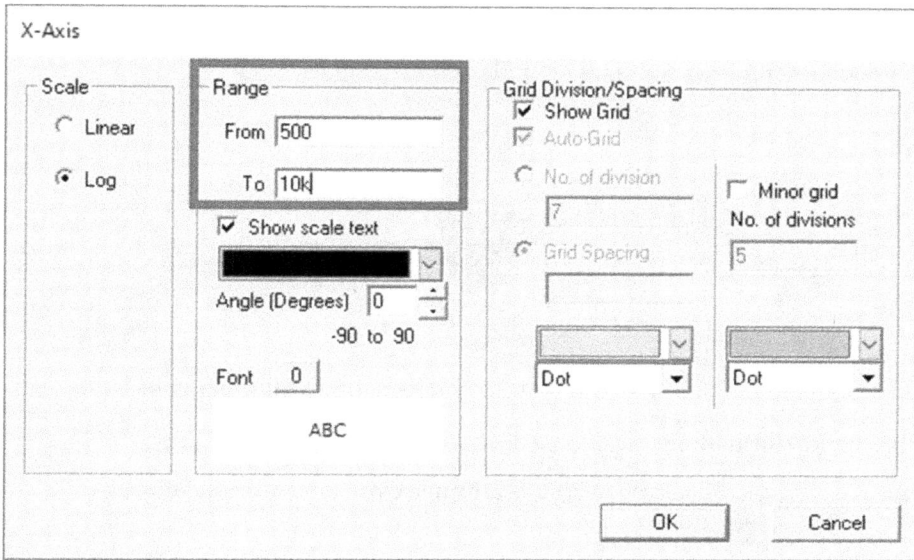

Fig. 3.75.

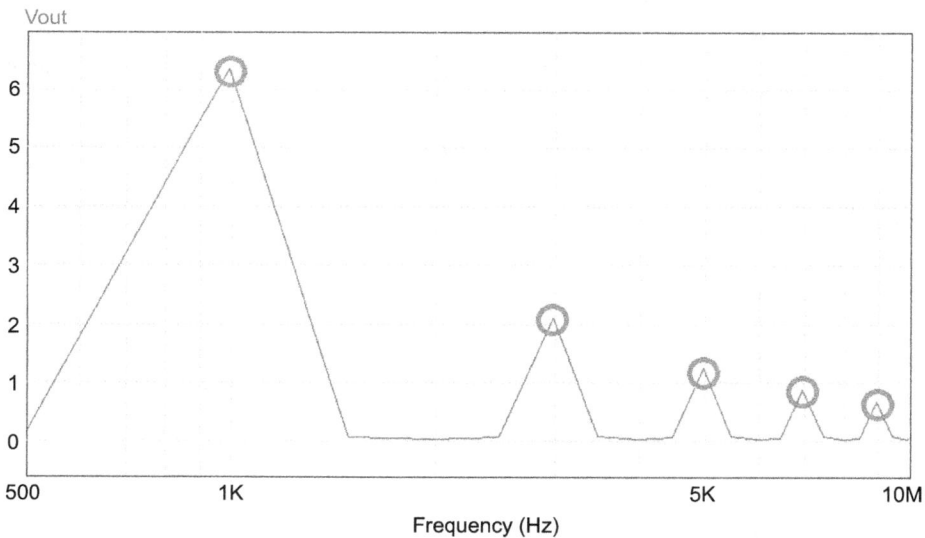

Fig. 3.76.

You can use the cursors in order to read the amplitude of each harmonic (Fig. 3.77).
Remember that frequency domain spectrum of periodic signals is composed of integer
multiple of fundamental frequency. So, values of this graph at non-integer multiples
of fundamental frequency has have no importance. The odd multiples of fundamental

frequency (in our example, 1 kHz) are shown with circles in Fig. 3.76. The square wave does not contain even multiples of fundamental frequency.

Measure				x
⋮	X1	X2	Δ	
Frequency	9.89612e+02	2.96248e+03	1.97287e+03	
Vout	6.33703e+00	2.05715e+00	-4.27988e+00	

Fig. 3.77.

Let's compare the Simview results for fundamental and third harmonics (Fig. 3.77) with hand calculations. According to $x(t)=5+\sum_{n=0}^{\infty}\frac{20}{(2n+1)\pi}\sin\left(2n+1\times\frac{2\pi}{T}\times t\right)$, $V_1=\frac{20}{\pi}=6.366\,V$ and $V_3=\frac{20}{3\pi}=2.122\,V$. The Simview results are quite close to hand calculations.

3.13 Exporting the waveforms to Excel® and MATLAB®

You can export the Simview waveforms as Comma Separated Values (CSV) format which is readable for Excel and MATLAB.

Assume that we want to export the vC and iC waveforms shown in Fig. 3.78 as a CSV file.

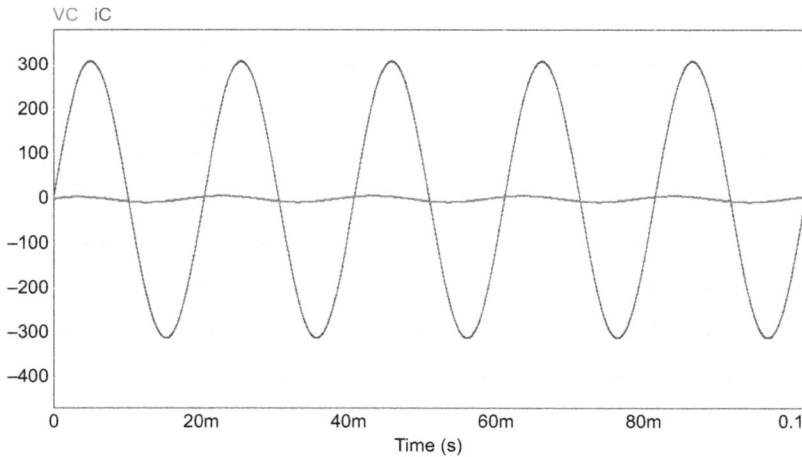

Fig. 3.78.

Click the Save As (Fig. 3.79) in order to do this.

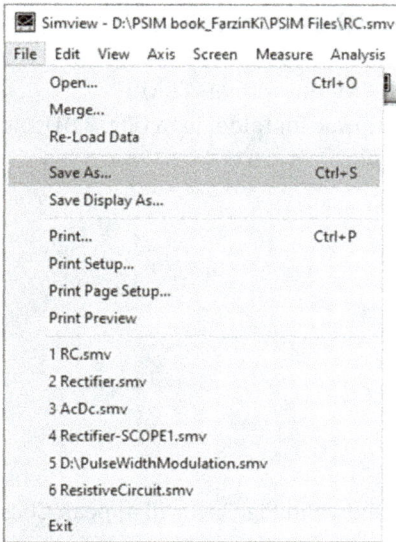

Fig. 3.79.

Enter desired name for the file and select the Comma separated file for Save as type box (Fig. 3.80).

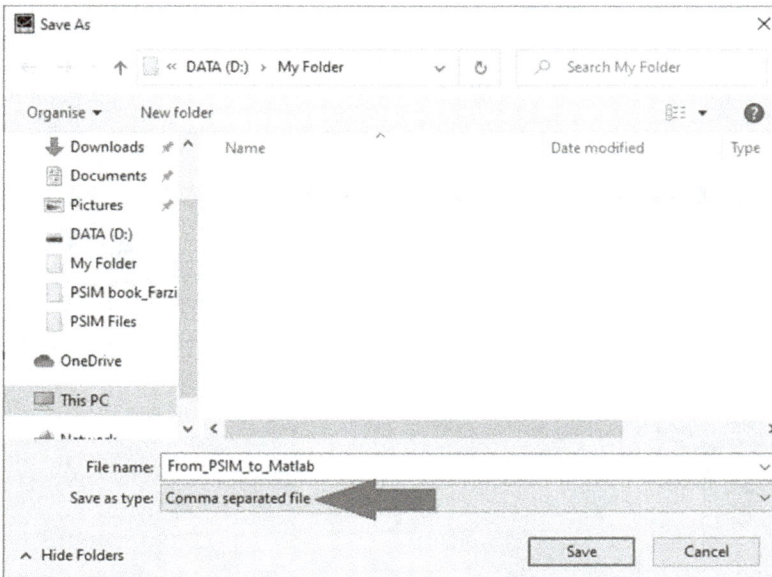

Fig. 3.80.

3.14 Reading the exported file in MATLAB environment

In this section, we show how to read the exported CSV file with MATLAB.

First of all, open MATLAB. Then click on the Browse for folder icon (Fig. 3.81) and open the folder which contains the exported CSV file.

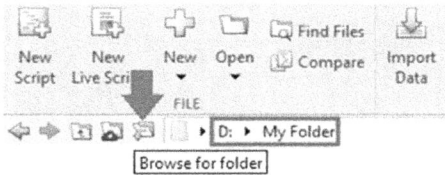

Fig. 3.81.

Double click the CSV file (Fig. 3.82). The window shown in Fig. 3.83 will appear. Click the Import Selection icon.

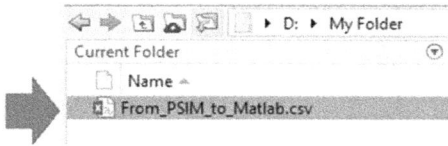

Fig. 3.82.

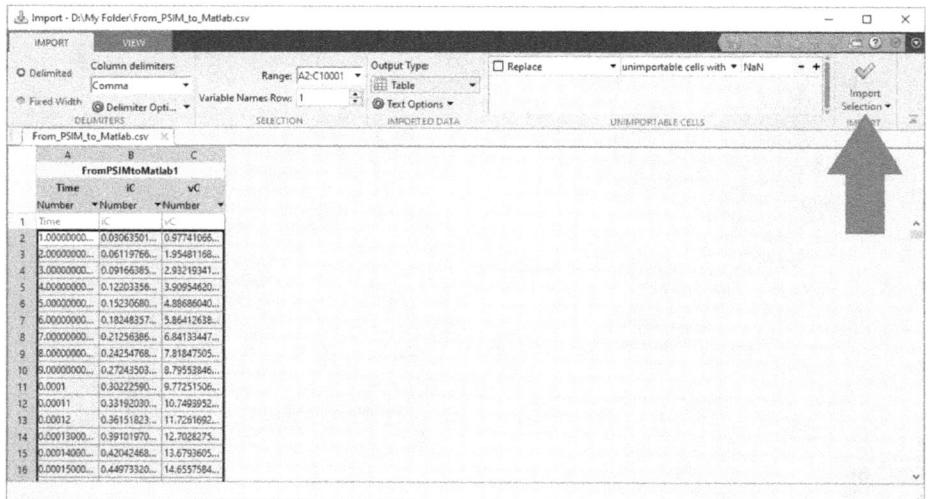

Fig. 3.83.

You can draw the plot of imported data with the aid commands shown in Fig. 3.84.
Results of these commands are shown in Fig. 3.85.

```
Command Window
>> plot(table2array(FromPSIMtoMatlab(:,1)),table2array(FromPSIMtoMatlab(:,3)))
>> hold on
>> plot(table2array(FromPSIMtoMatlab(:,1)),table2array(FromPSIMtoMatlab(:,2)))
fx >>
```

Fig. 3.84.

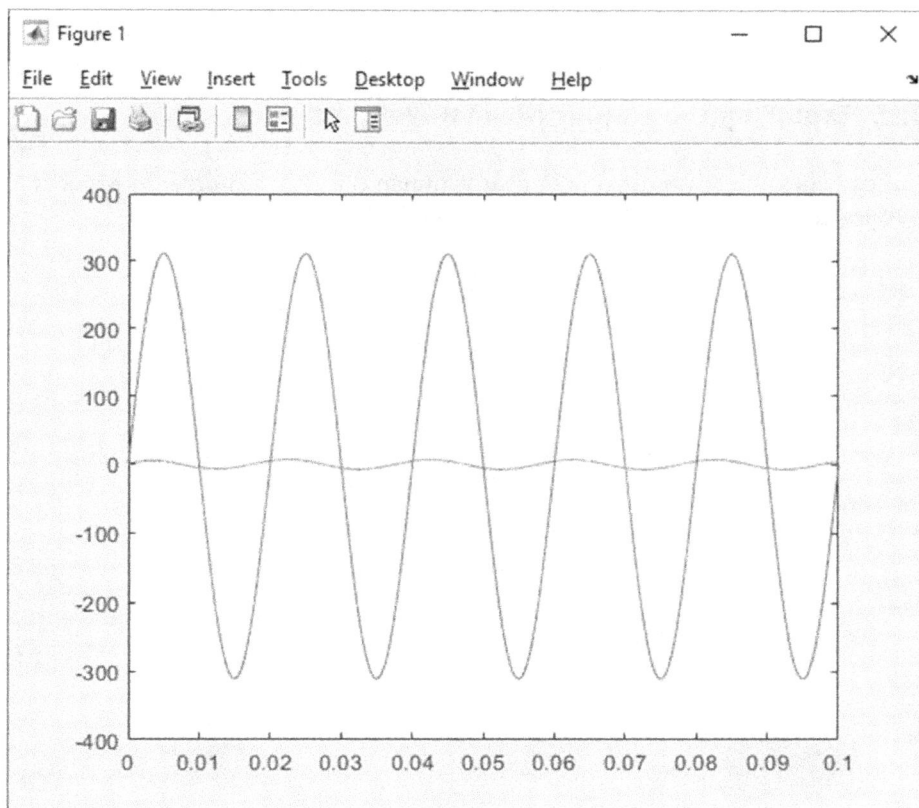

Fig. 3.85.

You can analyze the imported data with MATLAB commands.

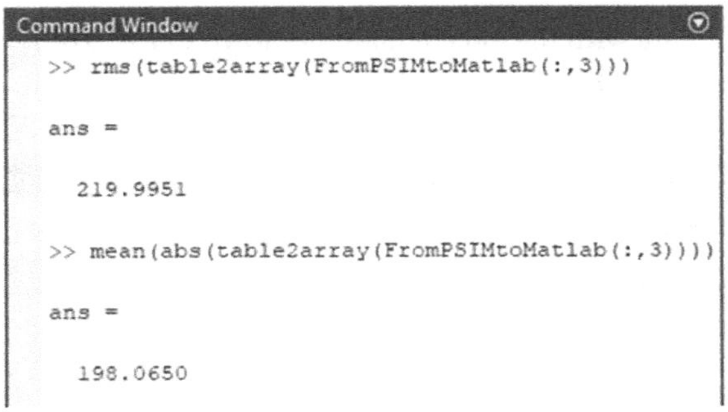

Command Window
>> rms(table2array(FromPSIMtoMatlab(:,3)))
ans =
219.9951
>> mean(abs(table2array(FromPSIMtoMatlab(:,3))))
ans =
198.0650

Fig. 3.86.

3.15 Exporting the graphic file of waveforms

You can click the Edit>Copy to Clipboard>Bitmap (Fig. 3.87) in order to export the waveforms that you see on the screen as graphical files.

Simview - D:\RC.smv
File
Edit
View
Axis
Screen
Measure
Analysis
Label
Options
Undo
Copy to Clipboard
View Data Points
Metafile format
Bitmap

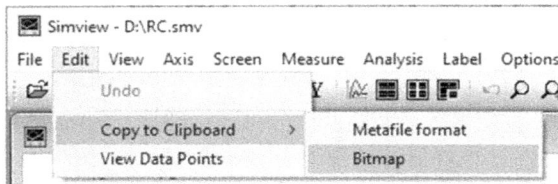

Fig. 3.87.

Chapter 4
PSIM's elements

4.1 Introduction

PSIM has many elements which are accessible in the Elements menu. This chapter introduces the commonly used ones.

4.2 Resistor, inductor, and capacitor

Resistor (R), inductor (L), and capacitor (C) are building blocks of any electric circuit. Resistor, inductor, and capacitors can be found in the Elements> Power> RLC Branches section.

https://doi.org/10.1515/9783110740653-004

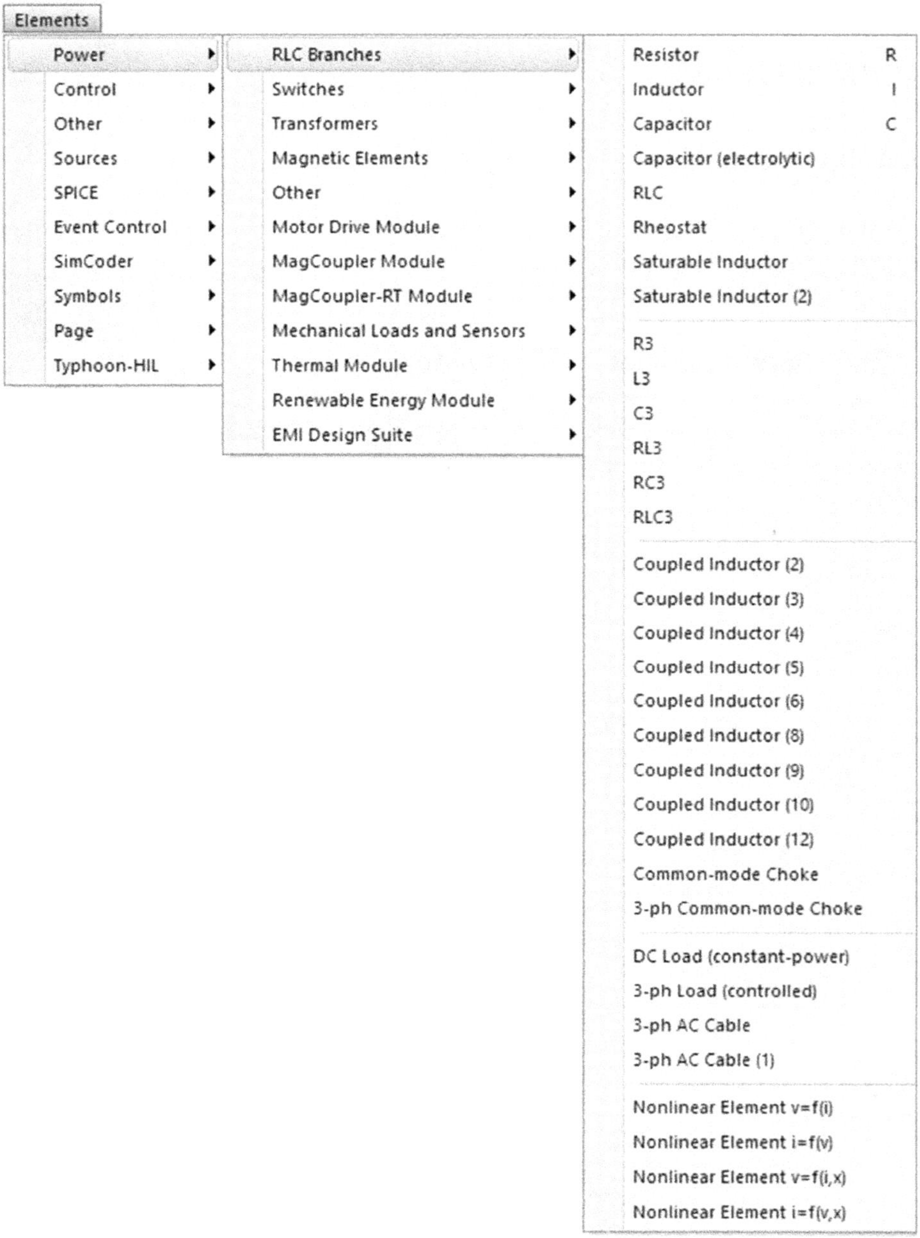

Elements		
Power ▶	RLC Branches ▶	Resistor R
Control ▶	Switches ▶	Inductor I
Other ▶	Transformers ▶	Capacitor C
Sources ▶	Magnetic Elements ▶	Capacitor (electrolytic)
SPICE ▶	Other ▶	RLC
Event Control ▶	Motor Drive Module ▶	Rheostat
SimCoder ▶	MagCoupler Module ▶	Saturable Inductor
Symbols ▶	MagCoupler-RT Module ▶	Saturable Inductor (2)
Page ▶	Mechanical Loads and Sensors ▶	R3
Typhoon-HIL ▶	Thermal Module ▶	L3
	Renewable Energy Module ▶	C3
	EMI Design Suite ▶	RL3
		RC3
		RLC3
		Coupled Inductor (2)
		Coupled Inductor (3)
		Coupled Inductor (4)
		Coupled Inductor (5)
		Coupled Inductor (6)
		Coupled Inductor (8)
		Coupled Inductor (9)
		Coupled Inductor (10)
		Coupled Inductor (12)
		Common-mode Choke
		3-ph Common-mode Choke
		DC Load (constant-power)
		3-ph Load (controlled)
		3-ph AC Cable
		3-ph AC Cable (1)
		Nonlinear Element v=f(i)
		Nonlinear Element i=f(v)
		Nonlinear Element v=f(i,x)
		Nonlinear Element i=f(v,x)

Fig. 4.1.

Different types of R, L, and C's, which are available in PSIM, are shown in Fig. 4.2. Note that PSIM has two types of capacitors: capacitor and capacitor (electrolytic).

Both are the same from PSIM view point, that is, both of them obey the $i(t)=C\frac{dv(t)}{dt}$ relationship. The only difference is the symbol that is used to show them on the schematic. Such a difference in the capacitor symbols permits the designer to show the correct polarity for electrolytic capacitors. Components that have the number 3 behind them (R3, L3, C3, RL3, RC3, and RLC3) are used in three-phase systems.

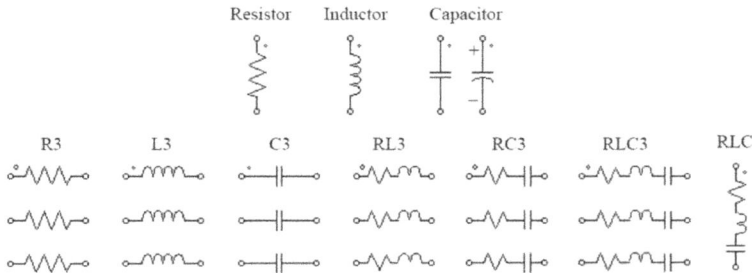

Fig. 4.2.

If you double click a resistor in the schematic, the window shown in Fig. 4.3 will appear. The value of resistor must be entered to the Resistance box. You can use the prefixes shown in Tab. 4.1 as well. For instance, in order to define a ten-kilo Ohm resistor, you can type 10k, 10,000, or 1e4 in the Resistance box.

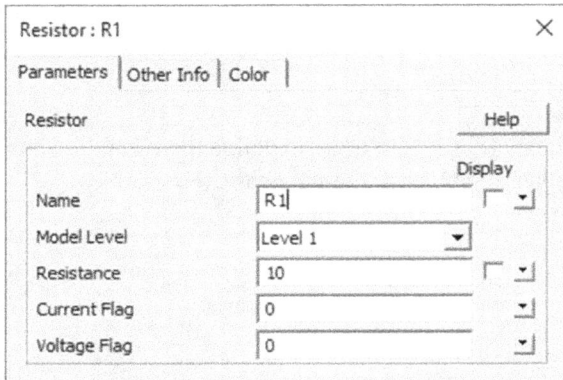

Fig. 4.3.

Tab. 4.1: Available prefixes in PSIM.

G	10^9
M	10^6
k or K	10^3
m	10^{-3}
u	10^{-6}
n	10^{-9}
p	10^{-12}

You can enter a variable name in the Resistance box as well. For instance, you enter R1 in resistance box. However, you need to define the value of a resistor, otherwise you will face an error message.

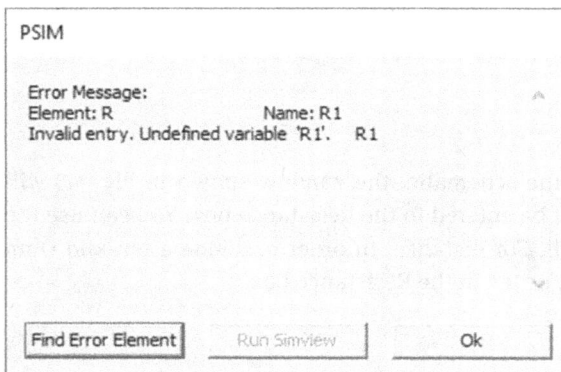

PSIM

Error Message:
Element: R   Name: R1
Invalid entry. Undefined variable 'R1'.   R1

Find Error Element   Run Simview   Ok

Fig. 4.4.

You need to use a Parameter File block (Fig. 4.5) in order to define the variable. Go to Elements> Other and add a Parameter File block to the schematic (Fig. 4.6).

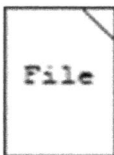

File

Fig. 4.5.

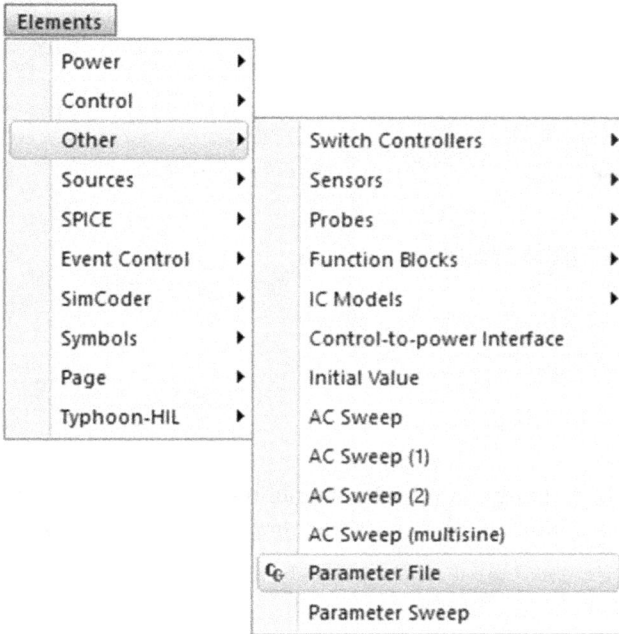

Fig. 4.6.

Double click the added Parameter File and define the variable as shown in Fig. 4.7.

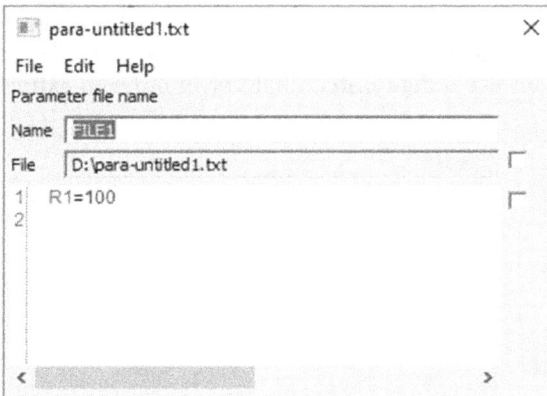

Fig. 4.7.

Model Level drop down list has two choices: Level 1 and Level 2.

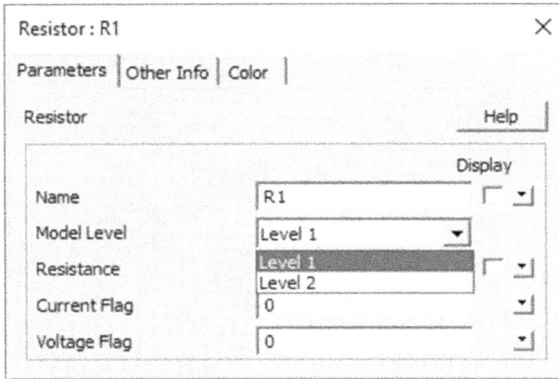

Fig. 4.8.

When you select the Level 1, you want to model a simple resistor which is shown in Fig. 4.9. The only parameter which such a model requires is the resistance of the resistor.

Fig. 4.9.

When you select Model 2, you want to use the model shown in Fig. 4.10. Such a model considered the equivalent series inductance (ESL) and parallel capacitance (Cp) of the resistor as well. Value of ESL and Cp must be entered to the Inductance ESL and Parallel Capacitance boxes. You can use a digital RLC analyzer in order to extract these values.

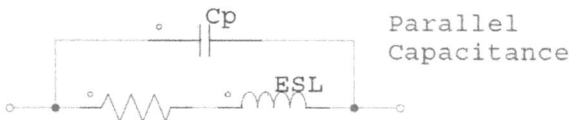

Fig. 4.10.

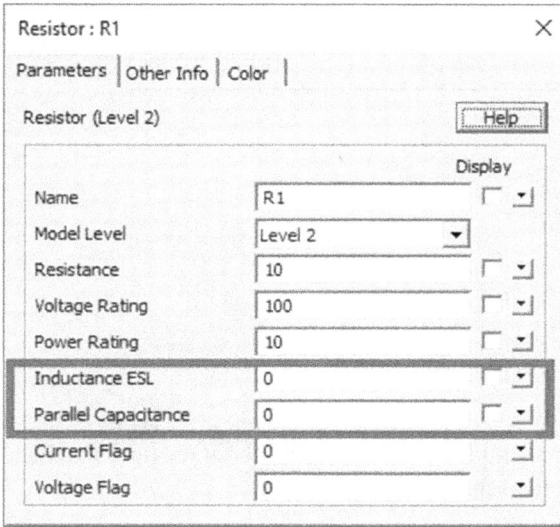

Fig. 4.11.

Level 1 and Level 2 models of inductors are shown in Figs. 4.12 and 4.13, respectively. In Fig. 4.13, Rp, Cp and ESL show the parallel resistance, parallel capacitance, and equivalent series inductance of the inductor, respectively.

Level 1 Model

Fig. 4.12.

Level 2 Model

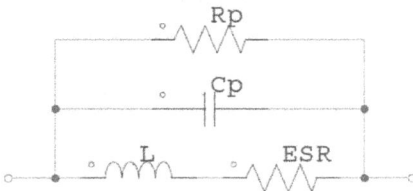

Fig. 4.13.

Level 1 and Level 2 models of capacitors are shown in Figs. 4.14 and 4.15, respectively. In Fig. 4.15, the Rp shows the parallel resistance of the capacitor. Rp shows the capacitor dielectric losses.

Level 1 Model

Fig. 4.14.

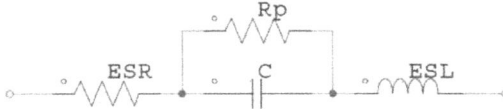

Fig. 4.15.

With the aid of Other Info tab, you can add some information about the used part. The information of this tab has no effect on the simulation process.

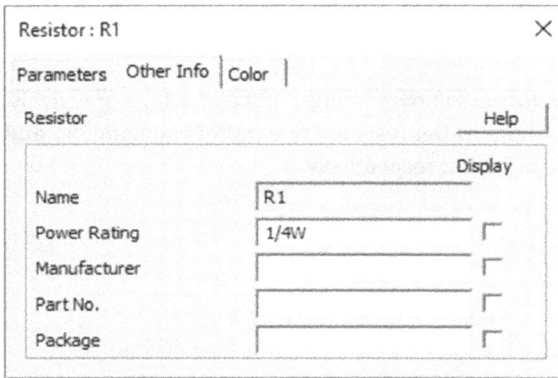

Fig. 4.16.

With the aid of Color tab, you can change the components' color and symbol.

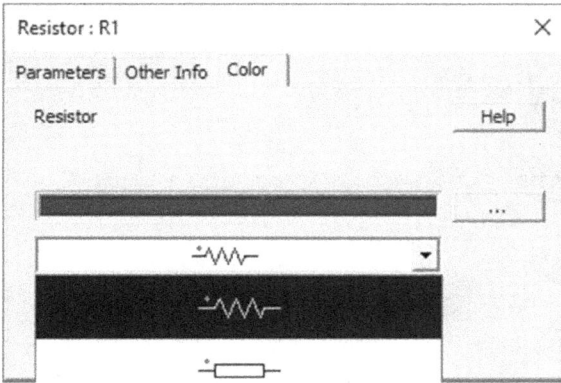

Fig. 4.17.

You can measure the current and voltage of element with the aid of Current Flag and Voltage Flag.

For instance, assume the circuit shown in Fig. 4.19. In this circuit, an ammeter and a voltmeter measure the current and voltage of resistor R2, respectively. You can measure the voltage and current of an element without using any voltmeter and ammeter (Fig. 4.20). In order to measure current and voltage of resistor R2 in Fig. 4.20, just double click the resistor R2 and enter one to Current Flag and Voltage Flag boxes.

With the aid of Voltage Flag, you measure the $V_{terminal\ with\ dot} - V_{terminal\ without\ dot}$. Note that Current Flag measures the elements current. The current that entered into the dot is assumed to be positive.

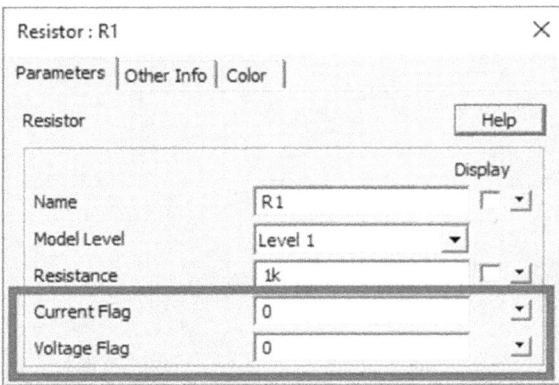

Fig. 4.18.

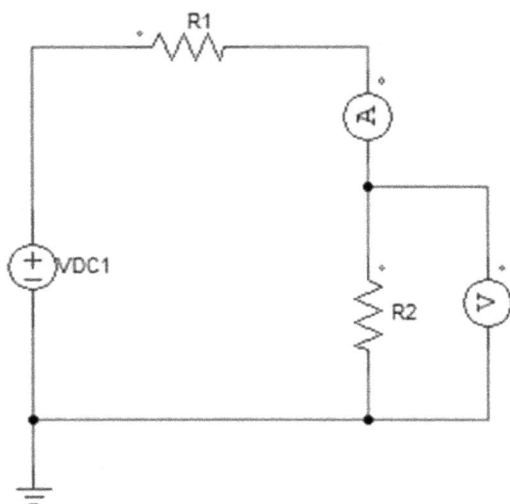

Fig. 4.19.

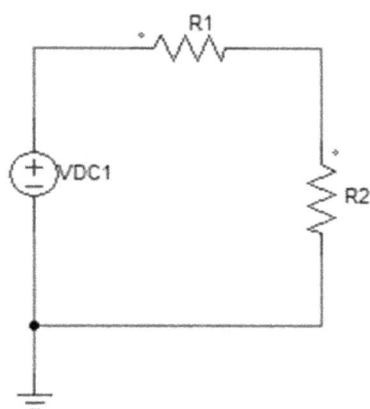

Fig. 4.20.

PSIM has rheostat block as well (Fig. 4.21).

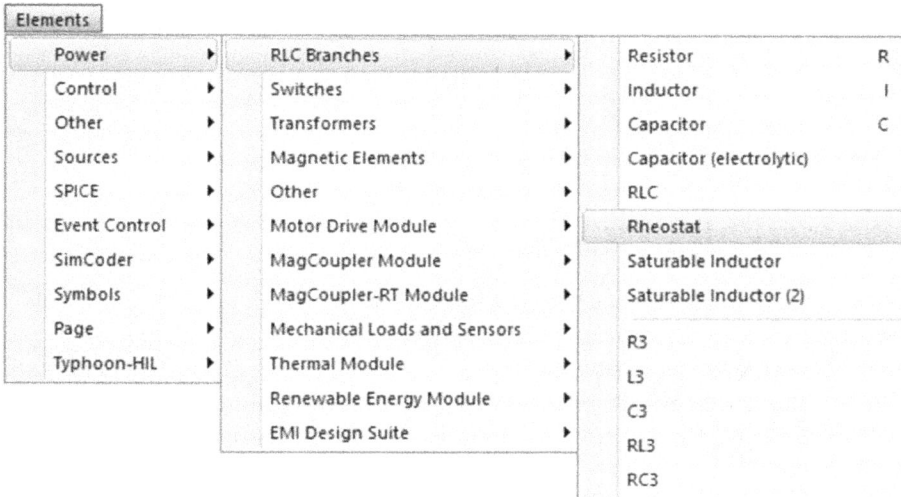

Elements					
Power ▶	RLC Branches ▶	Resistor	R
Control ▶	Switches ▶	Inductor	I
Other ▶	Transformers ▶	Capacitor	C
Sources ▶	Magnetic Elements ▶	Capacitor (electrolytic)	
SPICE ▶	Other ▶	RLC	
Event Control ▶	Motor Drive Module ▶	Rheostat	
SimCoder ▶	MagCoupler Module ▶	Saturable Inductor	
Symbols ▶	MagCoupler-RT Module ▶	Saturable Inductor (2)	
Page ▶	Mechanical Loads and Sensors ▶	R3	
Typhoon-HIL ▶	Thermal Module ▶	L3	
	Renewable Energy Module ▶	C3	
	EMI Design Suite ▶	RL3	
		RC3	

Fig. 4.21.

If you double click on a rheostat block on your schematic, window shown in Fig. 4.22 will appear.

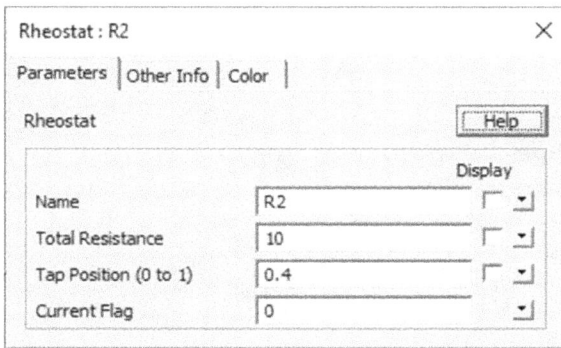

Rheostat : R2   ✕

Parameters | Other Info | Color |

Rheostat   [Help]

		Display
Name	R2	☐ ▾
Total Resistance	10	☐ ▾
Tap Position (0 to 1)	0.4	☐ ▾
Current Flag	0	▾

Fig. 4.22.

The rheostat block has three terminals: k, t, and m (Fig. 4.23). The value entered to the Total Resistance box determines the resistance between the k and m terminals. The value entered to the Tap Position (0 to 1) box determines the resistance between k and t terminals.

Fig. 4.23.

4.3 Saturable inductor block

Saturation is the state reached when an increase in applied external magnetic field H cannot increase the magnetization of the material further, so the total magnetic flux density B more or less levels off. Saturation is a characteristic of ferromagnetic materials. Different ferromagnetic materials have different saturation levels.

You can consider the saturation effect of inductors with the aid of saturable inductor block.

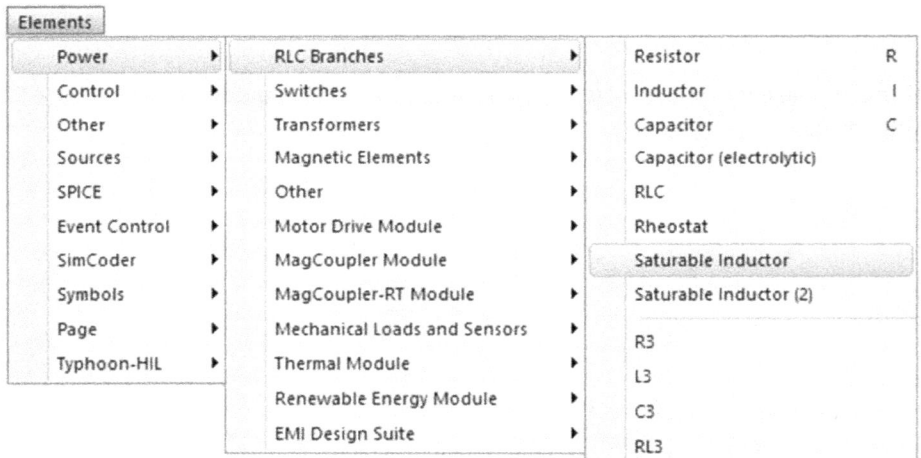

Fig. 4.24.

The change of inductance versus current for the nonlinear inductor must be entered to the Current versus Inductance box.

For instance, assume the flux versus current shown in Fig. 4.25. In order to model such a flux versus current behavior, Current versus Inductance box must be filled as shown in Fig. 4.26 (remember that $\lambda = L \times i$).

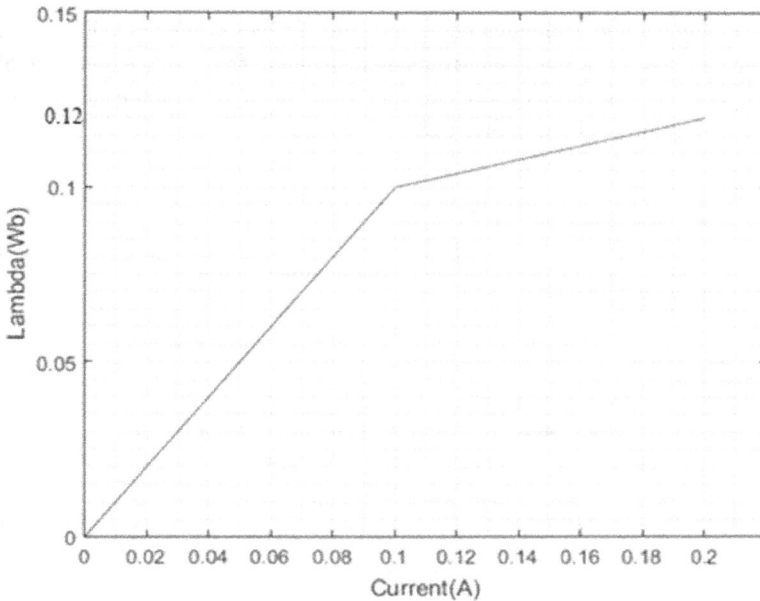

Fig. 4.25.

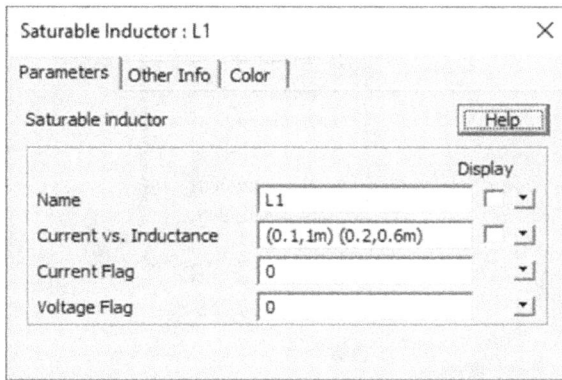

Fig. 4.26.

4.4 Coupled inductor block

A coupled inductor has two or more windings on a common core. Coupled inductors function in dc–dc converters by transferring energy from one winding to the other through the common core. They are available in many sizes, inductance values, and current ratings and most are magnetically shielded for low electromagnetic interference.

You can model the coupled inductors using the Coupled Inductor blocks (Fig. 4.27). The Coupled Inductor (n) block has n windings. For instance, the circuit schematic of coupled inductor (2), coupled inductor (3), coupled inductor (4), coupled inductor (5), and coupled inductor (6) are shown in Fig. 4.28.

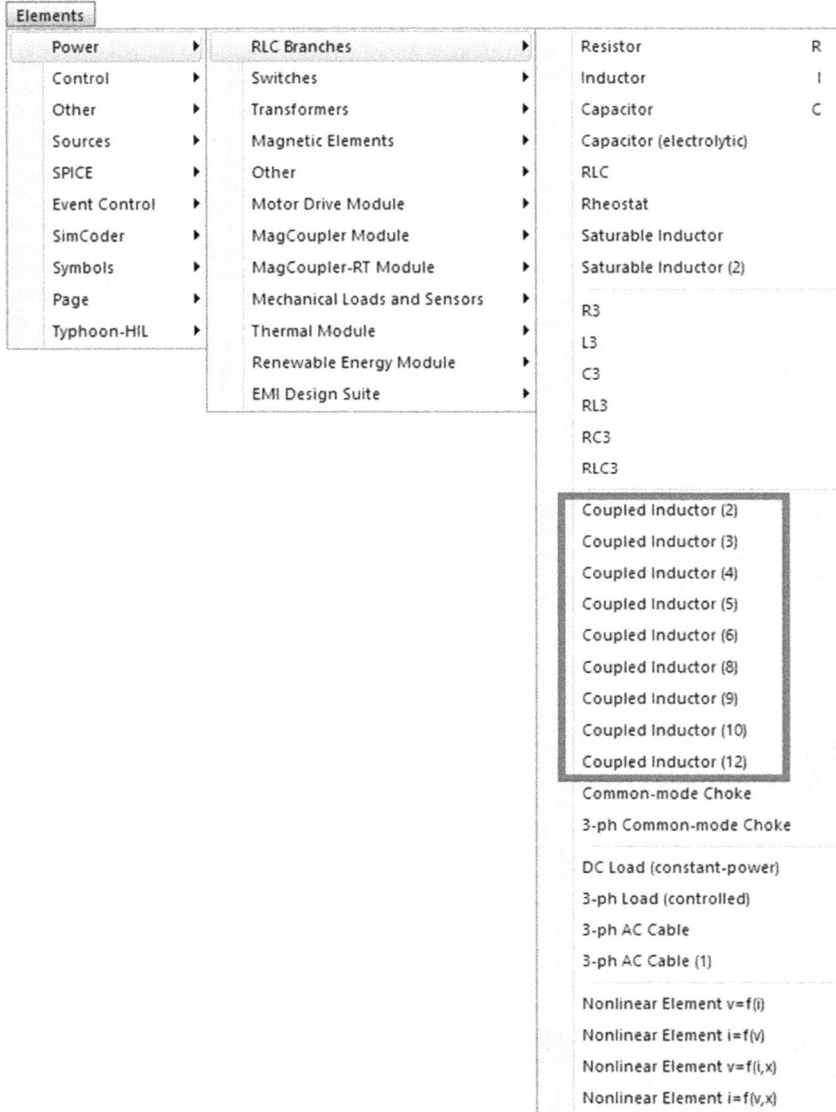

Elements		
Power ▸	RLC Branches ▸	Resistor R
Control ▸	Switches ▸	Inductor I
Other ▸	Transformers ▸	Capacitor C
Sources ▸	Magnetic Elements ▸	Capacitor (electrolytic)
SPICE ▸	Other ▸	RLC
Event Control ▸	Motor Drive Module ▸	Rheostat
SimCoder ▸	MagCoupler Module ▸	Saturable Inductor
Symbols ▸	MagCoupler-RT Module ▸	Saturable Inductor (2)
Page ▸	Mechanical Loads and Sensors ▸	R3
Typhoon-HIL ▸	Thermal Module ▸	L3
	Renewable Energy Module ▸	C3
	EMI Design Suite ▸	RL3
		RC3
		RLC3
		Coupled Inductor (2)
		Coupled Inductor (3)
		Coupled Inductor (4)
		Coupled Inductor (5)
		Coupled Inductor (6)
		Coupled Inductor (8)
		Coupled Inductor (9)
		Coupled Inductor (10)
		Coupled Inductor (12)
		Common-mode Choke
		3-ph Common-mode Choke
		DC Load (constant-power)
		3-ph Load (controlled)
		3-ph AC Cable
		3-ph AC Cable (1)
		Nonlinear Element v=f(i)
		Nonlinear Element i=f(v)
		Nonlinear Element v=f(i,x)
		Nonlinear Element i=f(v,x)

Fig. 4.27.

| 2-branch | 3-branch | 4-branch | 5-branch | 6-branch |

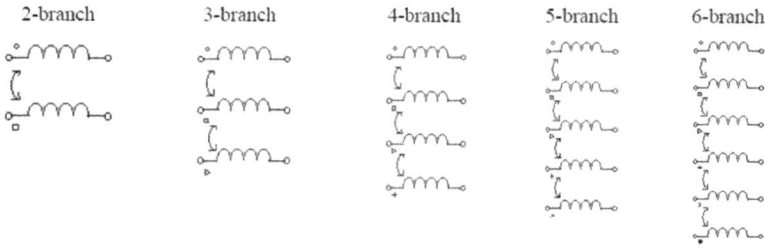

Fig. 4.28.

Assume that we want to analyze the circuit shown in Fig. 4.29 (V_{in} is the unit step function).

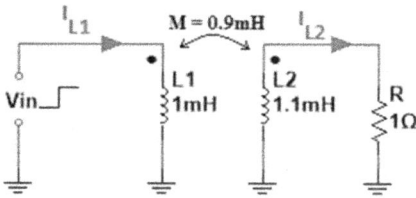

Fig. 4.29.

From basic circuit theory,

$$\begin{cases} L_1\dfrac{di_{L_1}}{dt} - M\dfrac{di_{L_2}}{dt} = V_{in}(t) \\ Ri_{L_2} + L_2\dfrac{di_{L_2}}{dt} - M\dfrac{di_{L_1}}{dt} = 0 \end{cases}$$

Take the Laplace transform of both sides:

$$\begin{bmatrix} L_1s & -Ms \\ -Ms & R+L_2s \end{bmatrix} \times \begin{bmatrix} I_{L1}(s) \\ I_{L2}(s) \end{bmatrix} = \begin{bmatrix} V_{in}(s) \\ 0 \end{bmatrix}$$

So,

$$\begin{bmatrix} I_{L1}(s) \\ I_{L2}(s) \end{bmatrix} = \begin{bmatrix} L_1s & -Ms \\ -Ms & R+L_2s \end{bmatrix}^{-1} \times \begin{bmatrix} V_{in}(s) \\ 0 \end{bmatrix}$$

$$V_{in}(s)=\frac{1}{s}, \text{ so}$$

$$\begin{bmatrix} I_{L1}(s) \\ I_{L2}(s) \end{bmatrix} = \begin{bmatrix} \dfrac{(11s+10{,}000)\times10{,}000}{s^2\times(29s+100{,}000)} \\ \dfrac{90000}{s(29s+100{,}000)} \end{bmatrix}$$

You can use the commands shown in Fig. 4.30 in order to see the time domain graph of I_{L1} and I_{L2}. Graph of I_{L1} and I_{L2} are shown in Figs. 4.31 and 4.32, respectively.

```
Command Window

>> I1=tf([11e4 1e8],[29 1e5 0 0]);
>> I2=tf(9e4,[29 1e5 0]);
>> figure(1);impulse(I1), grid minor
>> figure(2);impulse(I2), grid minor
fx >>
```

Fig. 4.30.

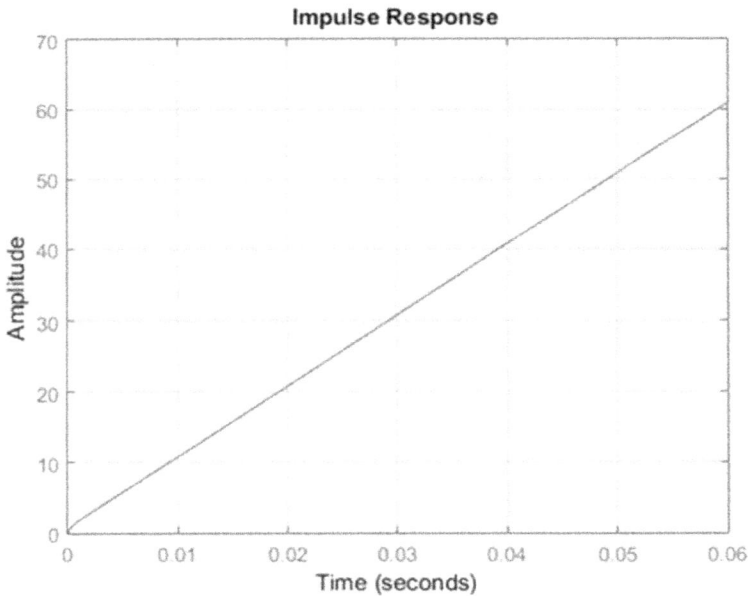

Fig. 4.31.

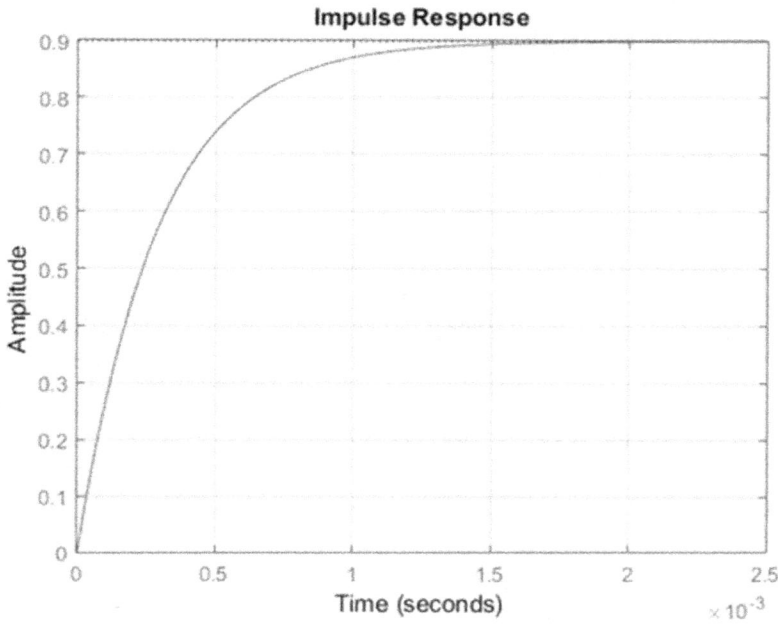

Fig. 4.32.

Let's check our calculations with PSIM. Assume the schematic shown in Fig. 4.33.

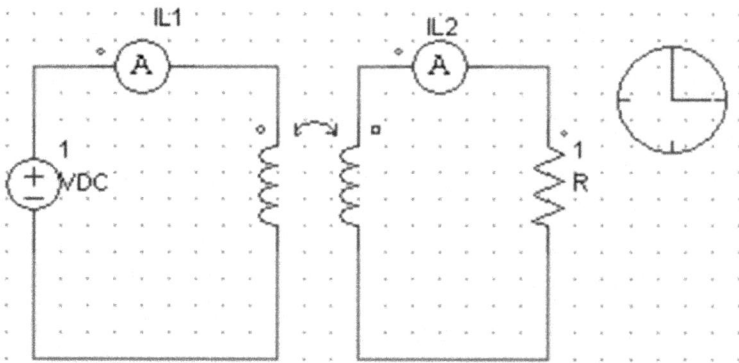

Fig. 4.33.

Settings of used blocks are shown in Figs. 4.34–4.39.

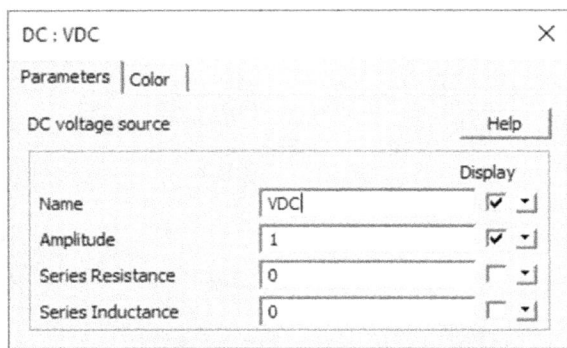

Fig. 4.34.

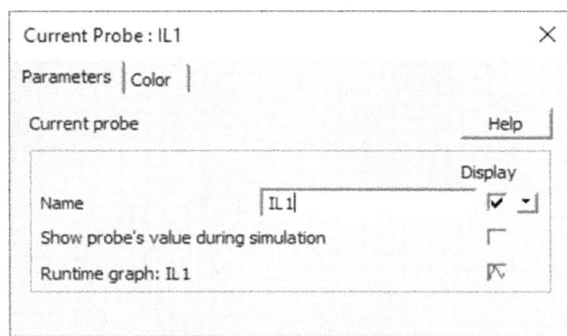

Fig. 4.35.

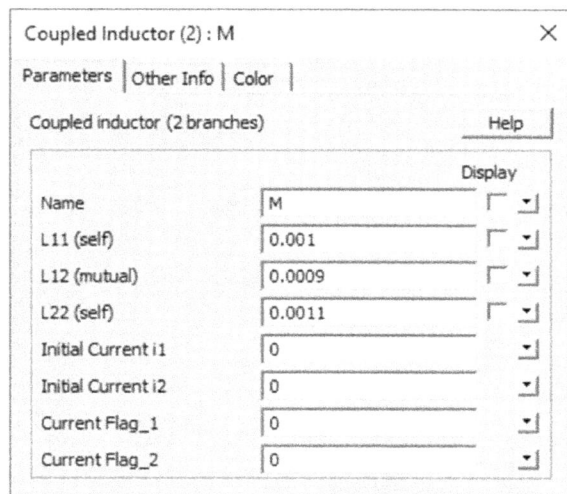

Fig. 4.36.

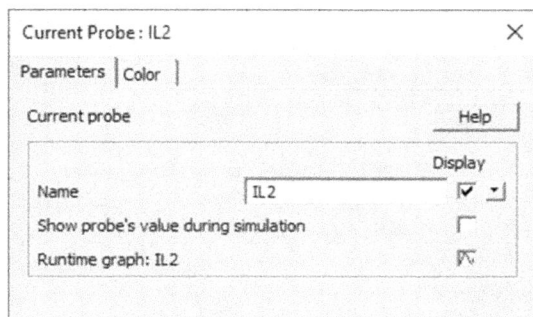

Fig. 4.37.

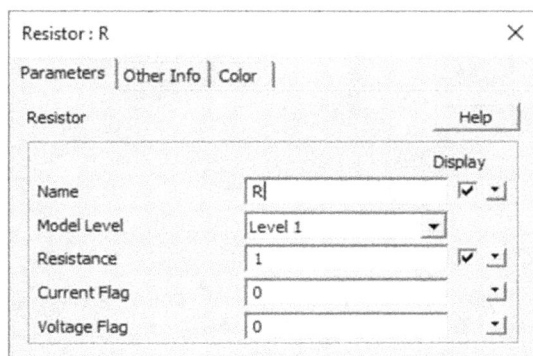

Fig. 4.38.

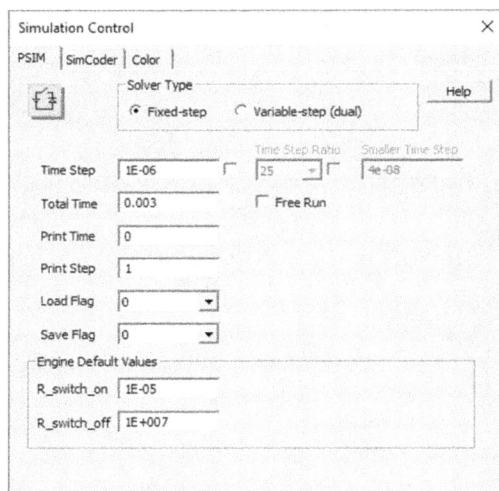

Fig. 4.39.

Run the simulation (press F8). The simulation results are shown in Figs. 4.40 and 4.41.

IL1

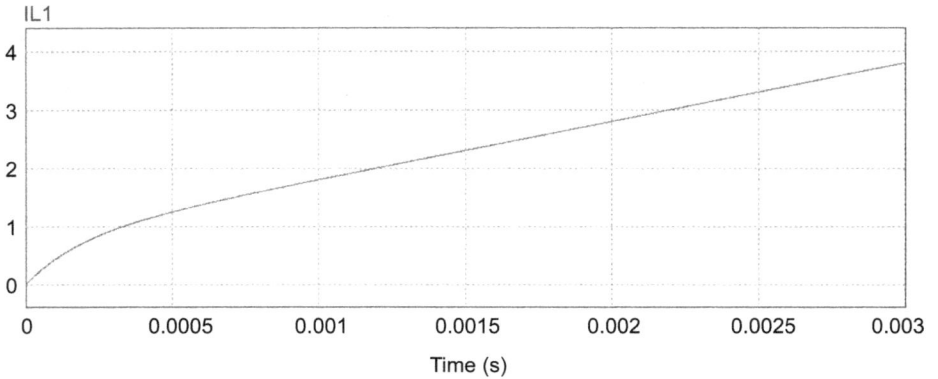

Fig. 4.40.

IL2

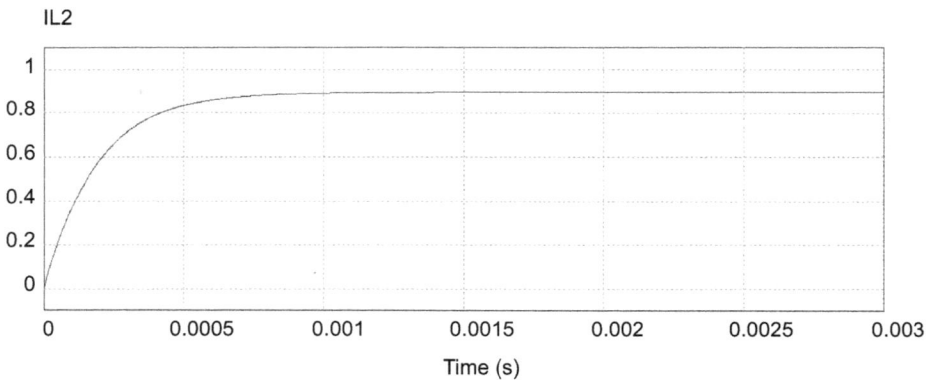

Fig. 4.41.

You can use the use cursors in order to read different points of the graph. For instance, for t = 0.201 s, i_{L1} = 0.606 A and i_{L2} = 0.45 A.

The values of i_{L1} and i_{L1} for are shown in Figs. 4.42 and 4.43. Hand analysis and PSIM analysis leaded to the same result.

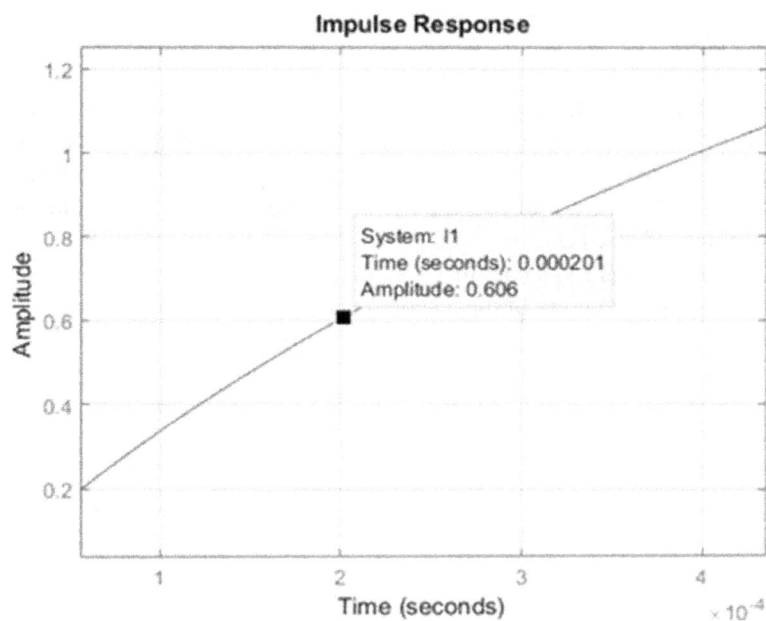

Fig. 4.42.

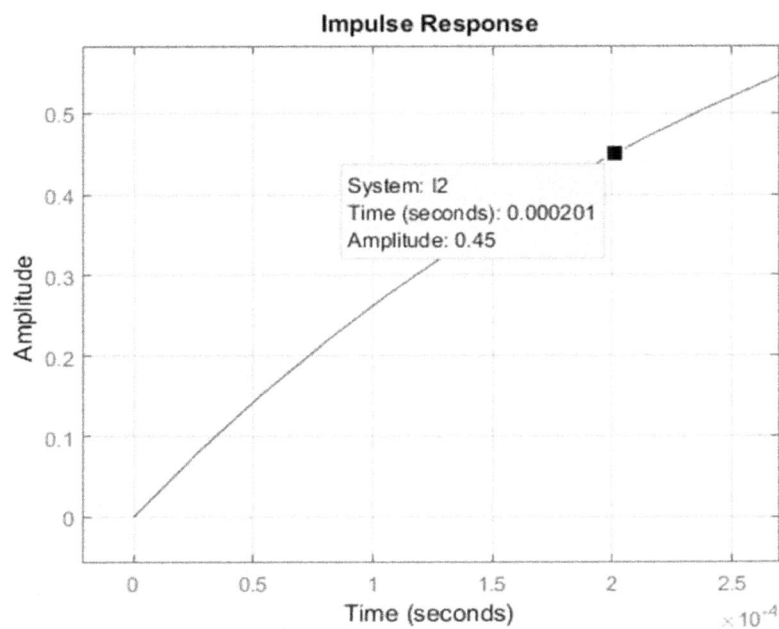

Fig. 4.43.

4.5 DC Load block

When you use a resistor as a load, the consumed power is a function of applied voltage or current: $p(t)=R{\times}i(t)^2=\dfrac{v(t)^2}{R}$. For instance, for a 10 Ω resistor: When the applied voltage is 10 V, the consumed power is 10 W and when the applied voltage is 20 V, the consumed power is 40 W. If you want to consume 10 W for the 20 V, then the value of resistor must be increased to 40 Ω.

PSIM has a block called DC Load (constant-power) which consume defined amount of power despite of changes in the voltage/current.

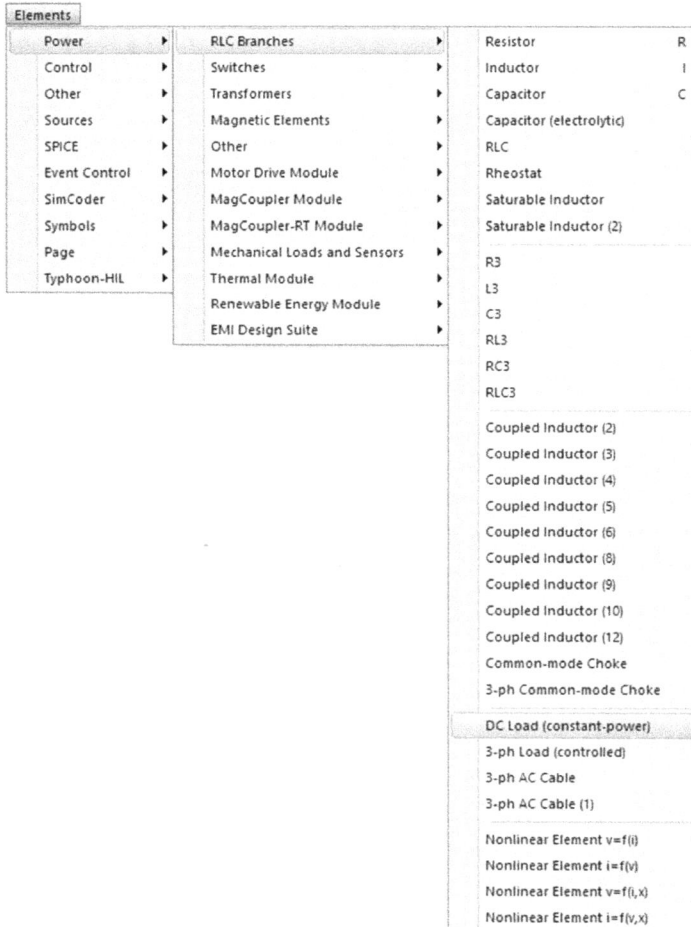

Elements		
Power ▸	RLC Branches ▸	Resistor R
Control ▸	Switches ▸	Inductor I
Other ▸	Transformers ▸	Capacitor C
Sources ▸	Magnetic Elements ▸	Capacitor (electrolytic)
SPICE ▸	Other ▸	RLC
Event Control ▸	Motor Drive Module ▸	Rheostat
SimCoder ▸	MagCoupler Module ▸	Saturable Inductor
Symbols ▸	MagCoupler-RT Module ▸	Saturable Inductor (2)
Page ▸	Mechanical Loads and Sensors ▸	R3
Typhoon-HIL ▸	Thermal Module ▸	L3
	Renewable Energy Module ▸	C3
	EMI Design Suite ▸	RL3
		RC3
		RLC3
		Coupled Inductor (2)
		Coupled Inductor (3)
		Coupled Inductor (4)
		Coupled Inductor (5)
		Coupled Inductor (6)
		Coupled Inductor (8)
		Coupled Inductor (9)
		Coupled Inductor (10)
		Coupled Inductor (12)
		Common-mode Choke
		3-ph Common-mode Choke
		DC Load (constant-power)
		3-ph Load (controlled)
		3-ph AC Cable
		3-ph AC Cable (1)
		Nonlinear Element v=f(i)
		Nonlinear Element i=f(v)
		Nonlinear Element v=f(i,x)
		Nonlinear Element i=f(v,x)

Fig. 4.44.

The DC Load (constant-power) block is shown in Fig. 4.45. The amount of consumed power is determined by the P terminal at the bottom of the block.

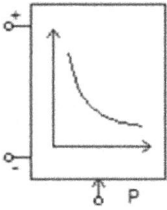

Fig. 4.45.

When you double click the block, the window shown in Fig. 4.46 is appeared.

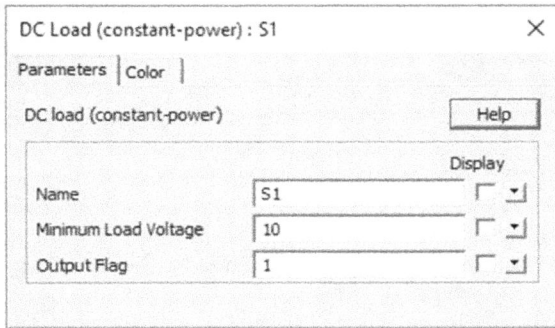

Fig. 4.46.

Click the help button to see the description for Minimum Load Voltage and Output Flag boxes.

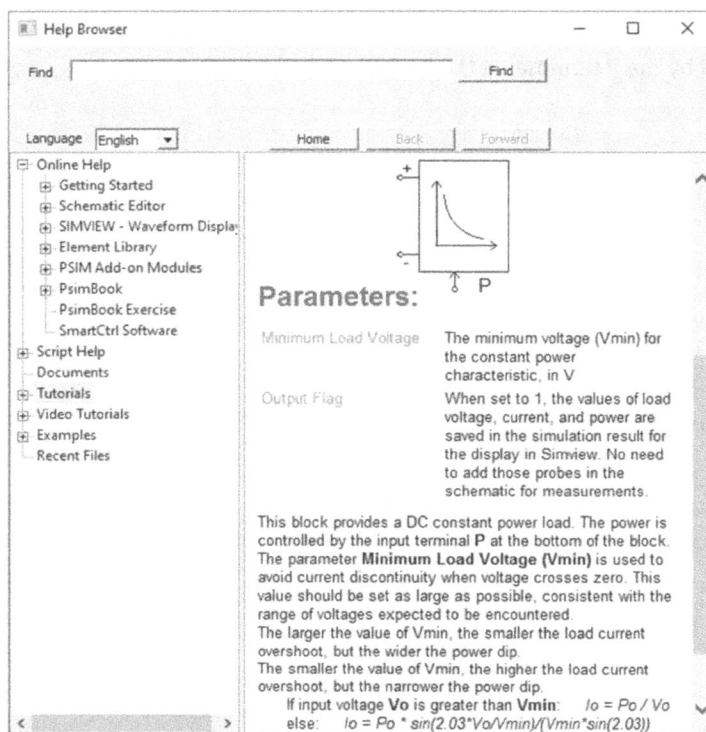

Fig. 4.47.

Let's use an example to see how DC Load (constant-power) block works. Assume the schematic shown in Fig. 4.48. Block C1 is a constant block (Fig. 4.49).

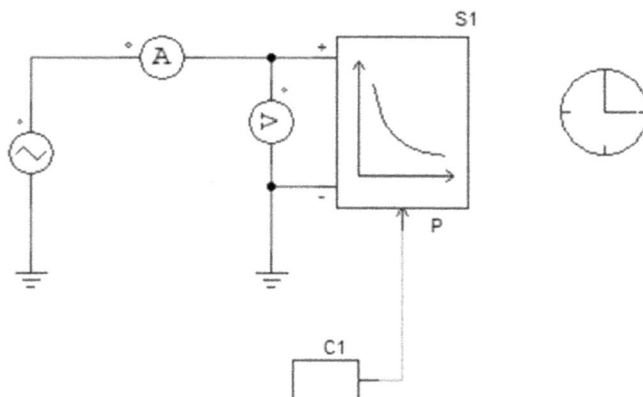

Fig. 4.48.

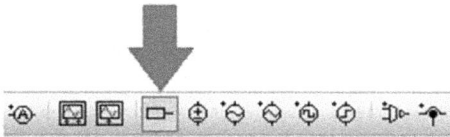

Fig. 4.49.

The settings of the blocks are shown in Figs. 4.50–4.55.

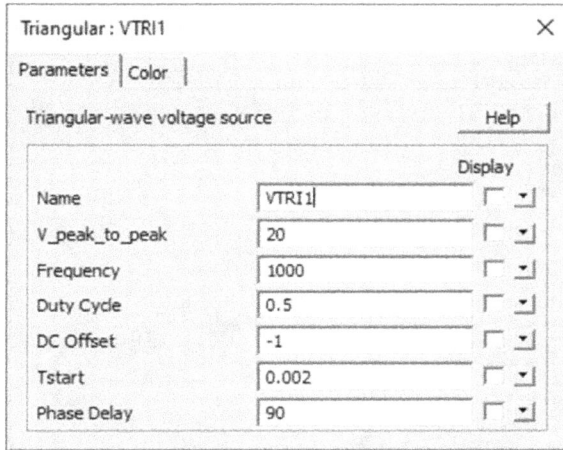

Fig. 4.50.

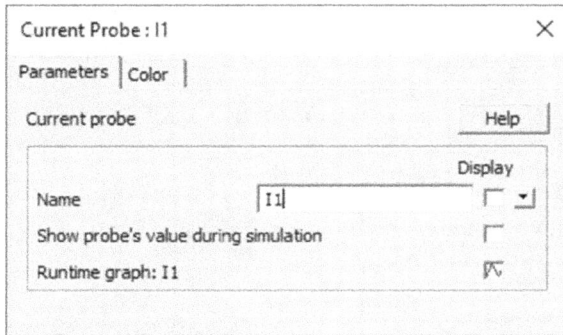

Fig. 4.51.

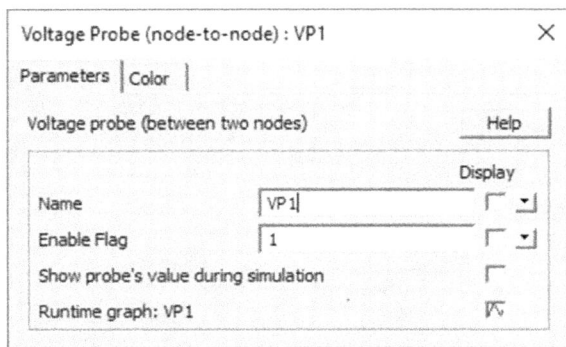

Fig. 4.52.

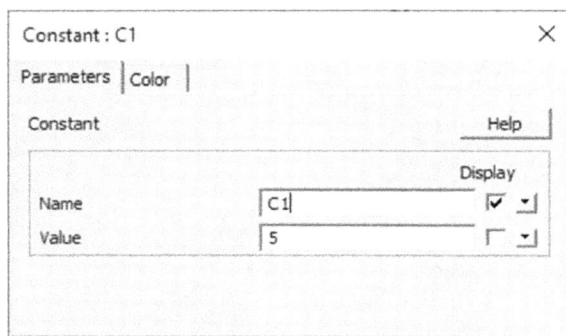

Fig. 4.53.

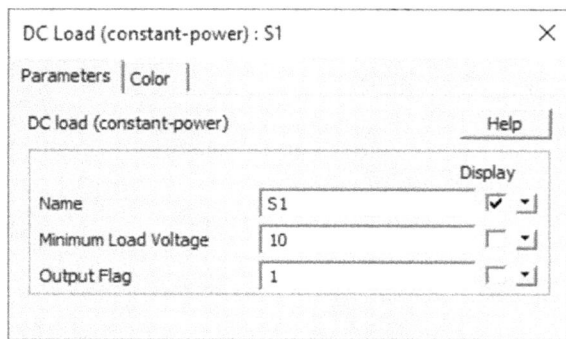

Fig. 4.54.

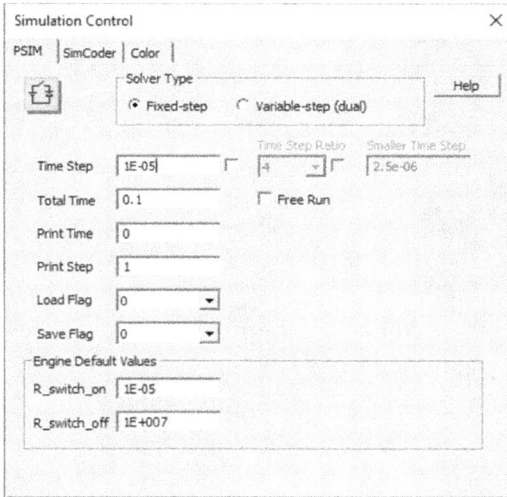

Fig. 4.55.

Run the simulation. After running the simulation, the window shown in Fig. 4.56 will appear. Click the I1 from left hand side list, then click the add button in order to add it to the right hand side list. Then click the OK button.

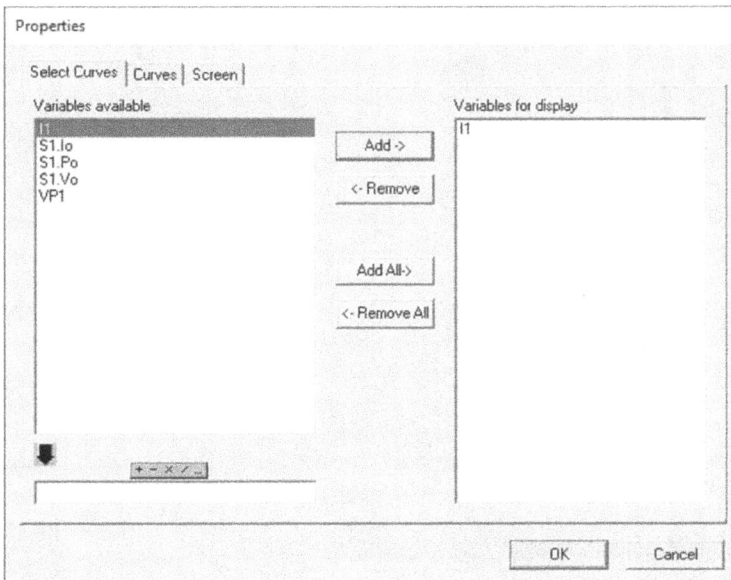

Fig. 4.56.

Now the Simview shows the graph of current I1.

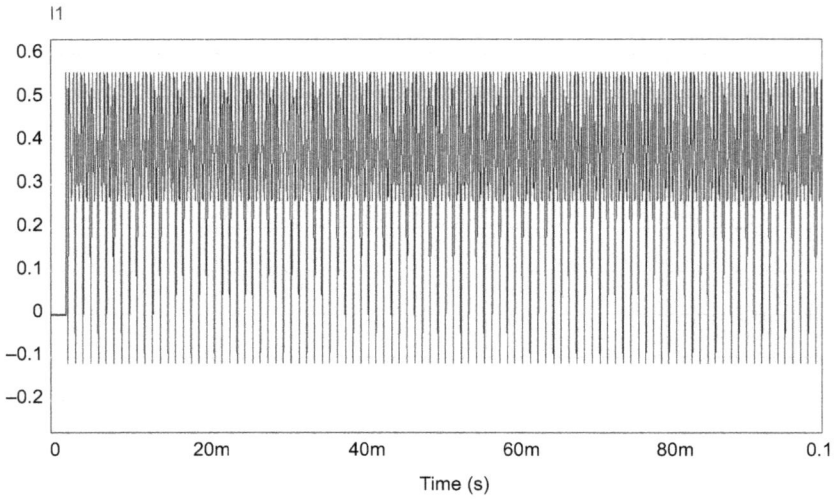

Fig. 4.57.

Click the Axis> Choose X-Axis Variable.

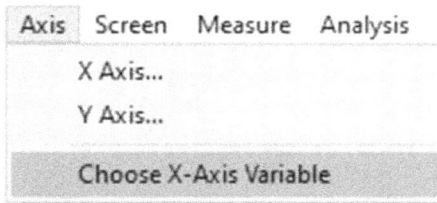

Fig. 4.58.

Select the VP1 from the list.

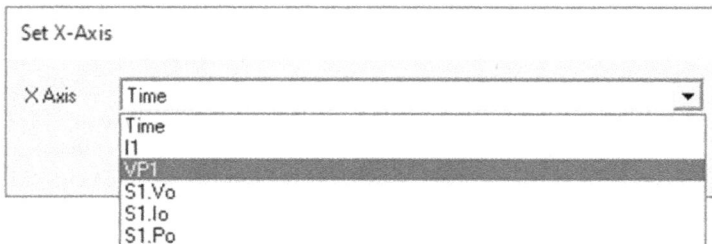

Fig. 4.59.

Now, the current I1 is drawn as a function of VP1 (Fig. 4.60). In other words, we have *I–V* characteristic for the DC Load (constant-power) block used in the schematic.

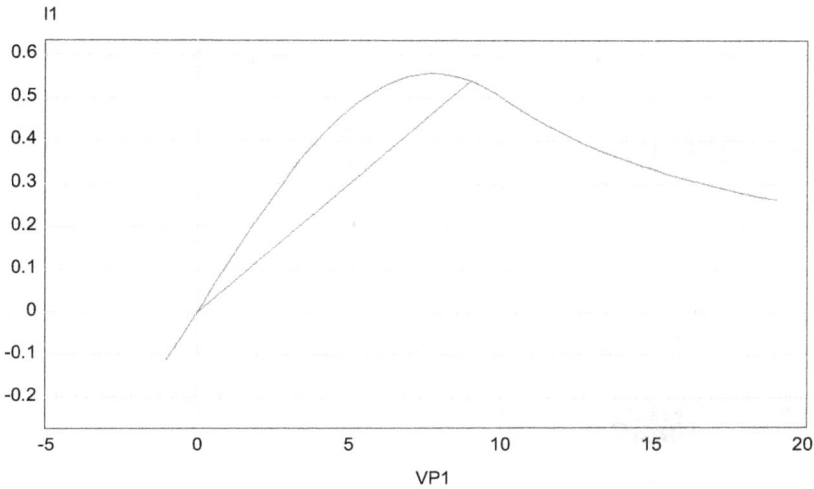

Fig. 4.60.

According to Fig. 4.54, when voltage is bigger than 10 V, the consumed power is constant and equals to 5 W. The consumed power is determined by the constant block connected to the bottom port of DC Load (constant-power) block. The *V* > 10 region of Fig. 4.60 is shown in Fig. 4.61.

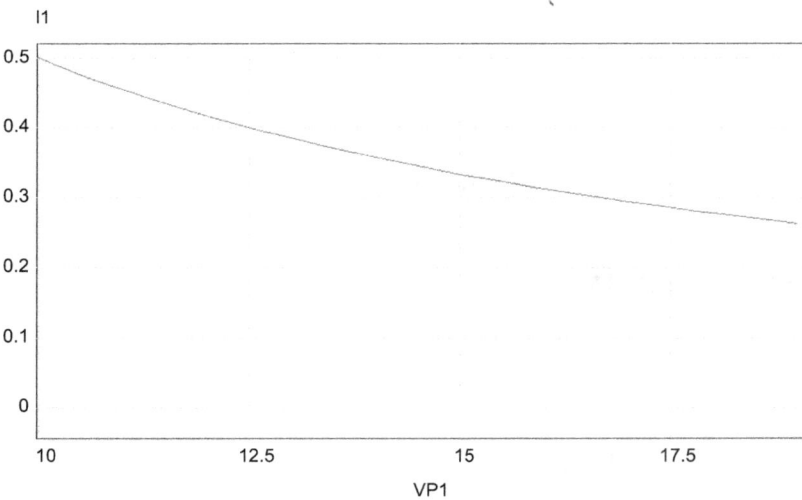

Fig. 4.61.

You can use cursors to read different points of this graph. If you calculate the product of current into voltage for different points of Fig. 4.61, you see that the product is constant despite of change in current/voltage.

Measure				
⋮		X1	X2	Δ
VP1		1.60739e+01	1.24076e+01	-3.66628e+00
I1		3.11105e-01	4.03083e-01	9.19776e-02

Fig. 4.62.

Figure 4.63 shows the product of current into voltage for coordinates shown in Fig. 4.62. You can see the result is 5 W despite of changes in the current/voltage.

```
Command Window

>> 16.0739*0.311105

ans =

     5.0007

>> 12.4076*0.403083

ans =

     5.0013

fx >>
```

Fig. 4.63.

4.6 Nonlinear elements

Simulation of nonlinear elements are very easy in PSIM.

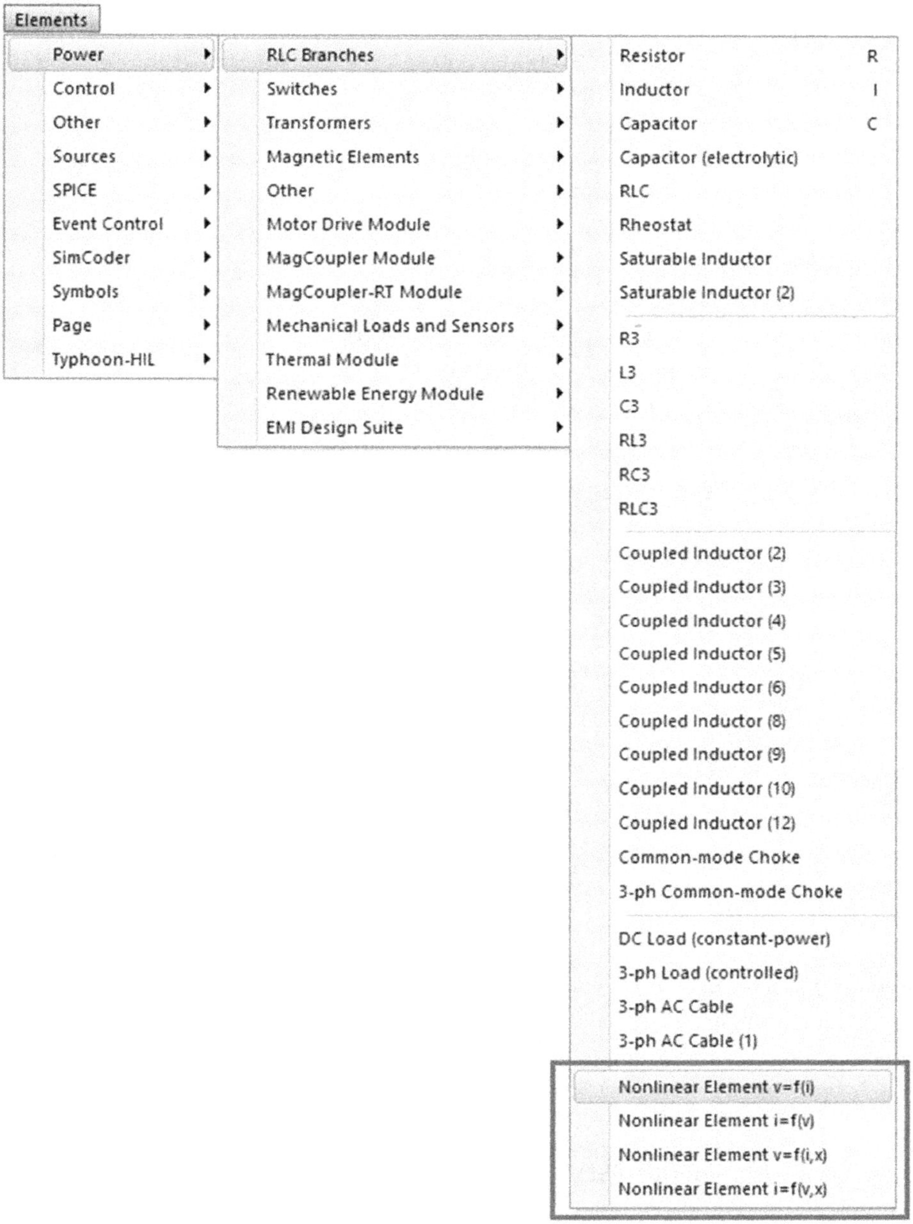

Elements			
Power ▶	RLC Branches ▶	Resistor	R
Control ▶	Switches ▶	Inductor	I
Other ▶	Transformers ▶	Capacitor	C
Sources ▶	Magnetic Elements ▶	Capacitor (electrolytic)	
SPICE ▶	Other ▶	RLC	
Event Control ▶	Motor Drive Module ▶	Rheostat	
SimCoder ▶	MagCoupler Module ▶	Saturable Inductor	
Symbols ▶	MagCoupler-RT Module ▶	Saturable Inductor (2)	
Page ▶	Mechanical Loads and Sensors ▶	R3	
Typhoon-HIL ▶	Thermal Module ▶	L3	
	Renewable Energy Module ▶	C3	
	EMI Design Suite ▶	RL3	
		RC3	
		RLC3	
		Coupled Inductor (2)	
		Coupled Inductor (3)	
		Coupled Inductor (4)	
		Coupled Inductor (5)	
		Coupled Inductor (6)	
		Coupled Inductor (8)	
		Coupled Inductor (9)	
		Coupled Inductor (10)	
		Coupled Inductor (12)	
		Common-mode Choke	
		3-ph Common-mode Choke	
		DC Load (constant-power)	
		3-ph Load (controlled)	
		3-ph AC Cable	
		3-ph AC Cable (1)	
		Nonlinear Element v=f(i)	
		Nonlinear Element i=f(v)	
		Nonlinear Element v=f(i,x)	
		Nonlinear Element i=f(v,x)	

Fig. 4.64.

Let's study an example. Assume the circuit shown in Fig. 4.65. The I–V characteristic of the R_nonlinear is $V = 5.I^2$. The initial voltage of capacitor at $t = 0$ is zero, that is, $V_{0c} = 0$.

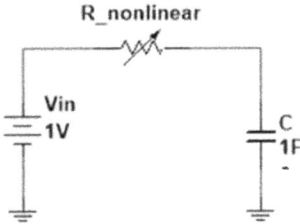

Fig. 4.65.

From basic circuit theory: Let's take the derivative of both sides:

$$V_{R\_nonlinear} + V_C = V_{in}$$

$$5 \times i^2 + \frac{1}{C} \int_0^t i(\tau) + V_{o,c} = V_{in}$$

$$10 \times i \times \frac{d}{dt}(i) + \frac{1}{1} \times i + 0 = \frac{d}{dt}(V_{in})$$

$$\left(10 \times \frac{d}{dt}(i) + 1\right) \times i = 0$$

$$i(t) = -0.1 \times t + i_0$$

Voltage of R_nonlinear at t = 0 equals to $V_{in} - V_{0,C} = 1 - 0 = 1$ V. According to the $V = 5.I^2$ the initial current at $t = 0$ equals to $\frac{1}{\sqrt{5}} = 0.4472$ A. So

$$i(t) = -0.1 \times t + 0.4472$$

Let's check this result with PSIM. Assume the schematic shown in Fig. 4.66.

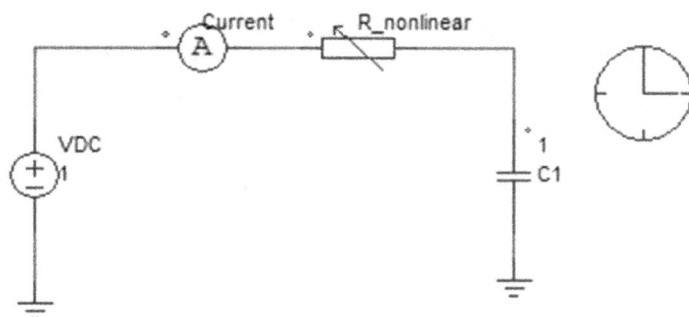

Fig. 4.66.

Settings of the blocks are shown in Figs. 4.67–4.71.

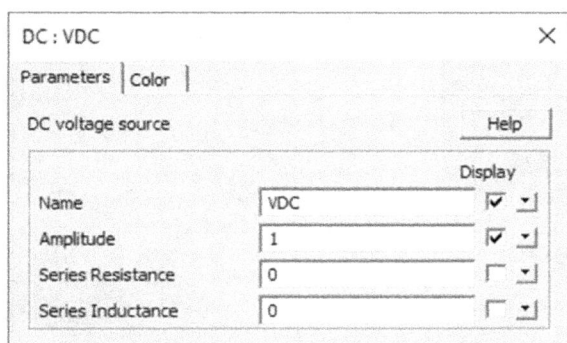

Fig. 4.67.

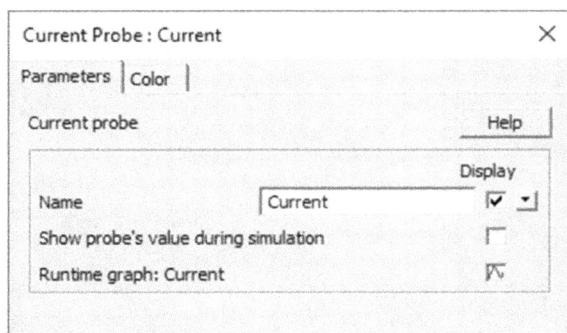

Fig. 4.68.

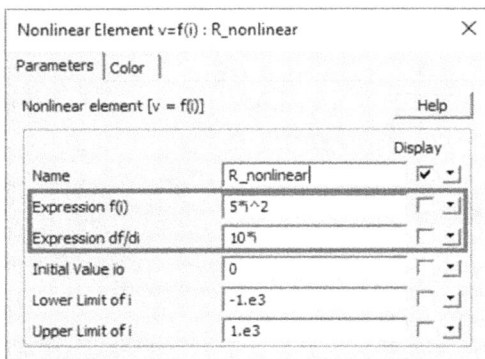

Fig. 4.69.

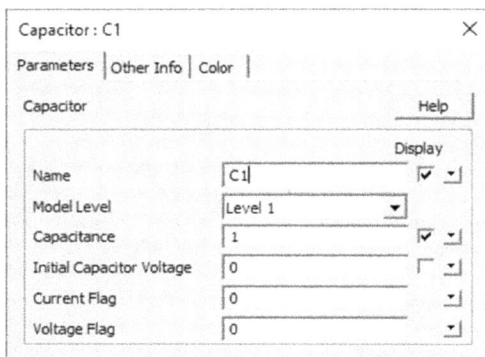

Fig. 4.70.

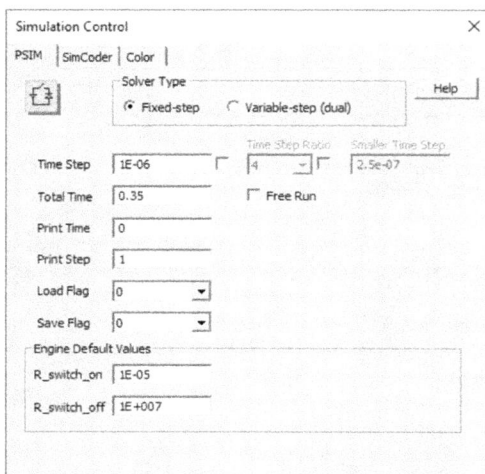

Fig. 4.71.

Run the simulation. The result shown in Fig. 4.72 is obtained.

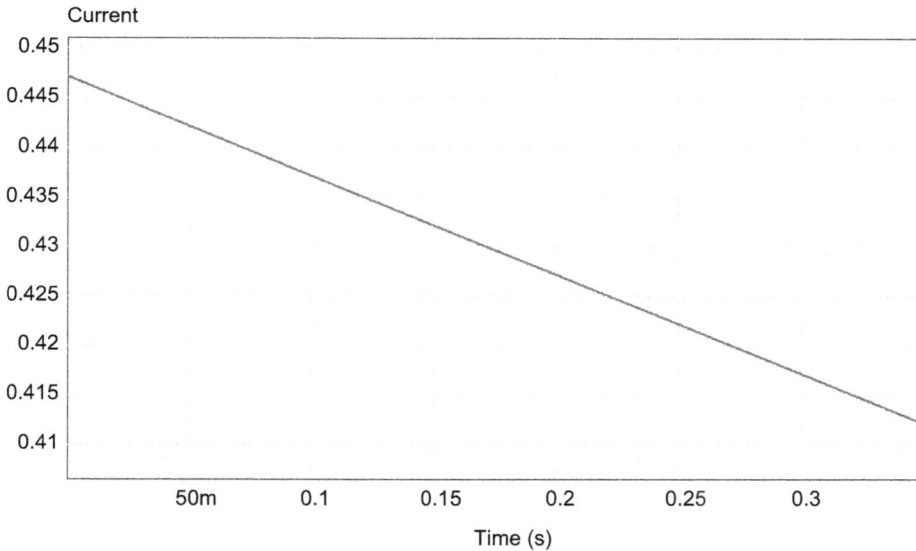

Fig. 4.72.

You can use the cursors in order to read different points of the graph. For instance, according to Fig. 4.73, the current equals to 0.432 A at $t = 150.34$ ms. If we calculate $i(t) = -0.1 \times t + 0.4472$ for $t = 150.34$ ms, we obtain the same result.

Measure				
		X1	X2	Δ
	Time	1.50340e-01	2.80000e-01	1.29660e-01
	Current	4.32179e-01	4.19213e-01	-1.29660e-02

Fig. 4.73.

4.7 Filters

PSIM has ready to use first and second filter blocks.

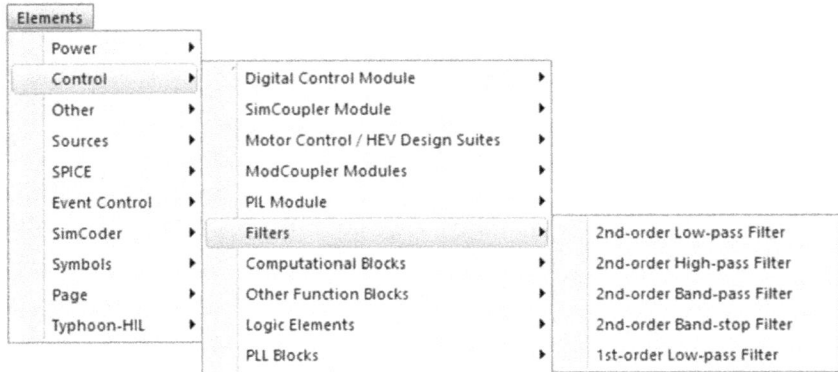

Elements		
Power ▶		
Control ▶	Digital Control Module ▶	
Other ▶	SimCoupler Module ▶	
Sources ▶	Motor Control / HEV Design Suites ▶	
SPICE ▶	ModCoupler Modules ▶	
Event Control ▶	PIL Module ▶	
SimCoder ▶	Filters ▶	2nd-order Low-pass Filter
Symbols ▶	Computational Blocks ▶	2nd-order High-pass Filter
Page ▶	Other Function Blocks ▶	2nd-order Band-pass Filter
Typhoon-HIL ▶	Logic Elements ▶	2nd-order Band-stop Filter
	PLL Blocks ▶	1st-order Low-pass Filter

Fig. 4.74.

Transfer functions of filters are given in Tab. 4.2. ω_c and ξ in the low-pass- and high-pass filters show the cut-off frequency and damping ratio, respectively. B and ω_0 in the band-pass- and band-stop filters show the band width and center frequency, respectively. Unit of B, ω_0 and ω_c is $\dfrac{\text{Rad}}{\text{s}}$. ξ is unitless.

Tab. 4.2: Transfer functions of filters.

Filter name	Transfer function
Second-order low-pass filter	$G(s) = k.\dfrac{\omega_c^2}{s^2 + 2\xi\omega_c s + \omega_c^2}$
Second-order high-pass filter	$G(s) = k.\dfrac{s^2}{s^2 + 2\xi\omega_c s + \omega_c^2}$
Second-order band-pass filter	$G(s) = k.\dfrac{B.s}{s^2 + B.s + \omega_o^2}$
Second-order band-stop filter	$G(s) = k.\dfrac{s^2 + \omega_o^2}{s^2 + B.s + \omega_o^2}$
First-order low-pass filter	$G(s) = k.\dfrac{\omega_c}{s + \omega_c}$

The input signal of filter blocks must be a control signal. In the schematic shown in Fig. 4.75, a part of wire that connects the filter block to resistor is green and the other part is red. This shows that an inconvenient signal entered the block. Sometimes, the simulation can be run despite of such a warning.

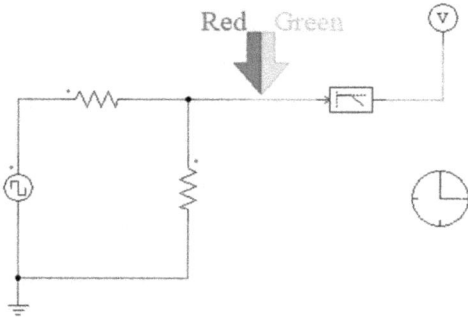

Fig. 4.75.

In order to solve this problem, put a voltage sensor block before the filter block. The output signal of voltage and current sensor is control signals. Now the wire that connects the voltage sensor block to the filter is completely green.

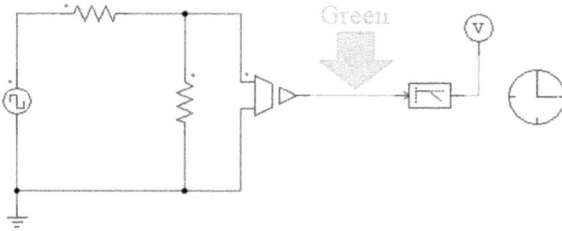

Fig. 4.76.

Let's see a simple simulation. Assume the schematic shown in Fig. 4.77. This input source is a step voltage source (Fig. 4.78). The band pass filter has $B=2\pi\times20=125.66\frac{Rad}{s}$, $\omega_0=2\pi\times60=377\frac{Rad}{s}$ and . $k=1$. So, according to Tab. 4.2, the transfer function of filter is

$$G(s)=\frac{125.66s}{s^2+125.66s+1.421\times10^5}.$$

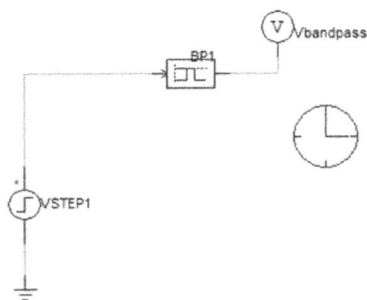

Fig. 4.77.

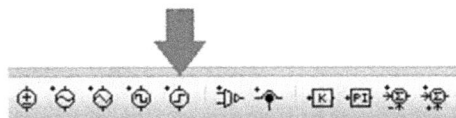

Fig. 4.78.

Settings of used elements are shown in Figs. 4.79–4.82.

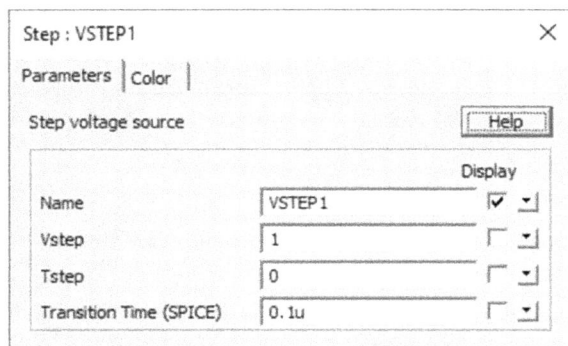

Fig. 4.79.

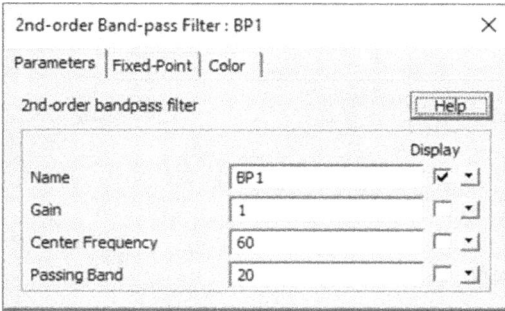

Fig. 4.80.

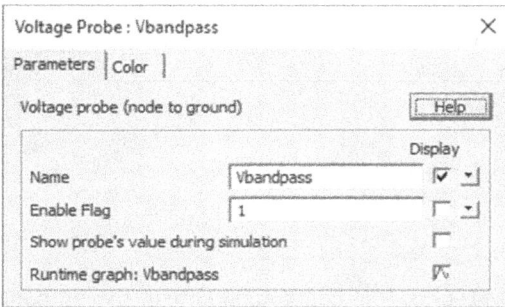

Fig. 4.81.

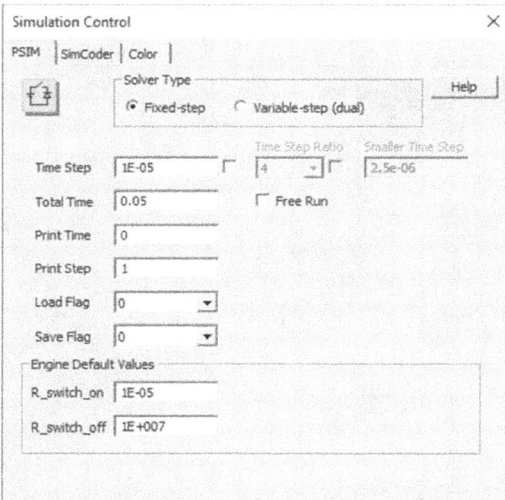

Fig. 4.82.

Run the simulation. The result shown in Fig. 4.83 will be obtained.

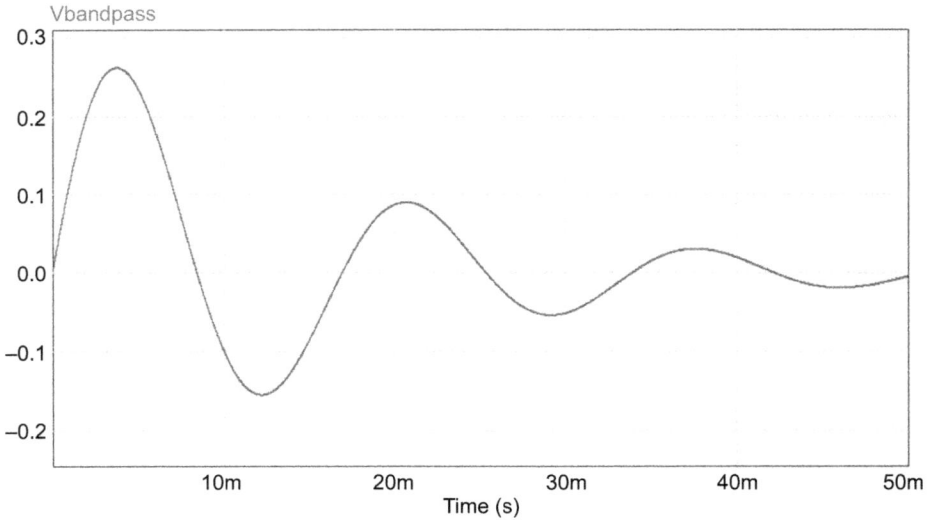

Fig. 4.83.

Let's check the obtained result with MATLAB. The commands shown in Fig. 4.84 draw the step response of $G(s)=\dfrac{125.66s}{s^2+125.66s+1.421\times10^5}$. Result is shown in Fig. 4.85. You can use cursors in order to compare the two graphs and ensure that both of them are the same.

```
Command Window                            ⊙
   >> G=tf([125.66 0],[1 125.66 1.421e5]);
   >> step(G)
   >> grid on
fx >>
```

Fig. 4.84.

Step Response

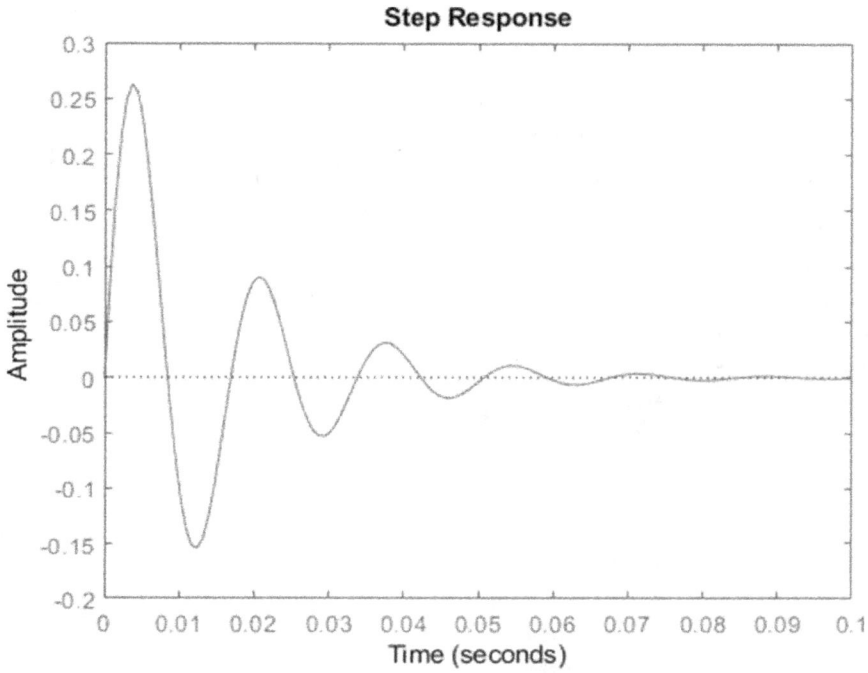

Fig. 4.85.

4.8 Computational blocks

PSIM has many ready to use computational blocks. These blocks help you to do required computations easily.

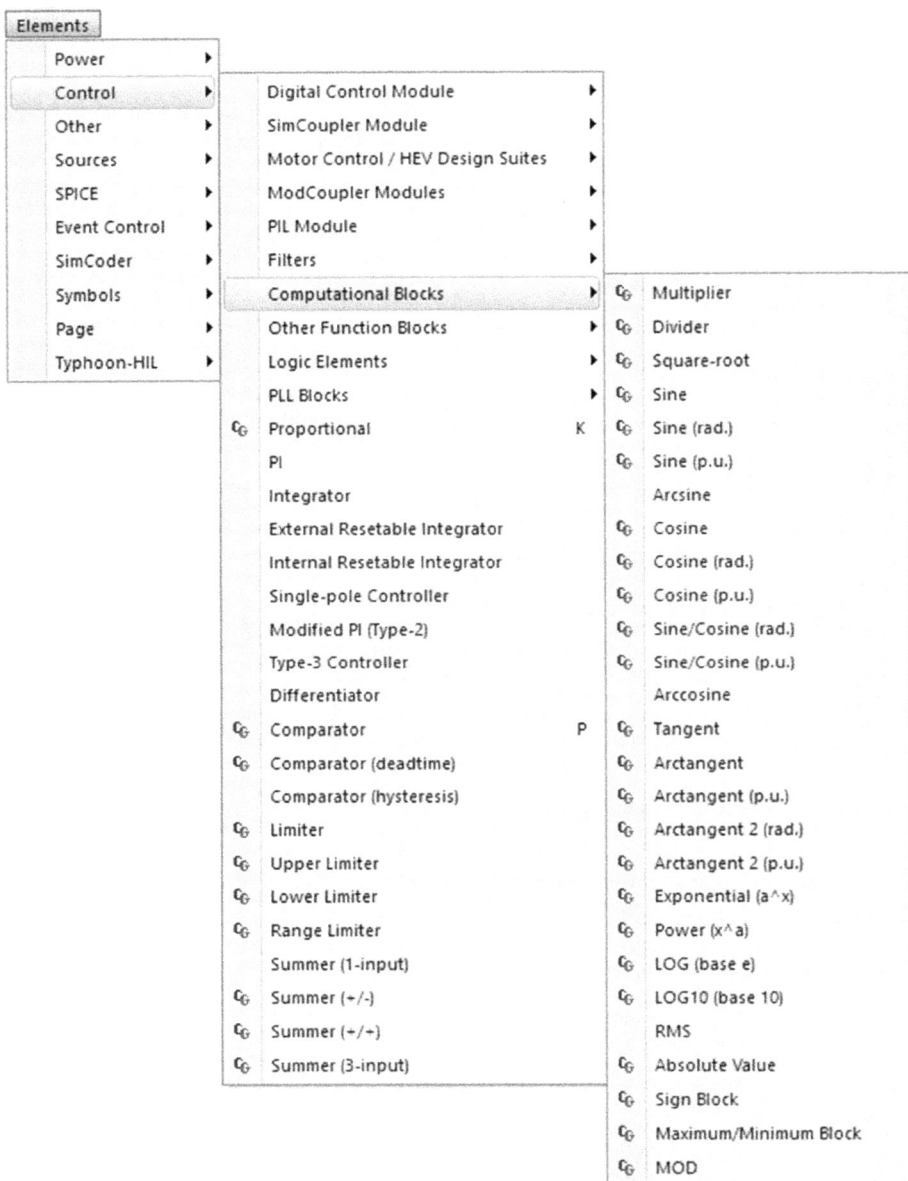

Fig. 4.86.

Let's study a simple example. Assume the schematic shown in Fig. 4.87. We want to calculate the RMS of capacitor voltage.

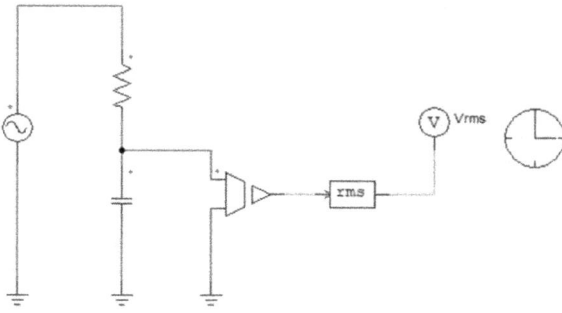

Fig. 4.87.

Settings of used components are shown in Figs. 4.88–4.94.

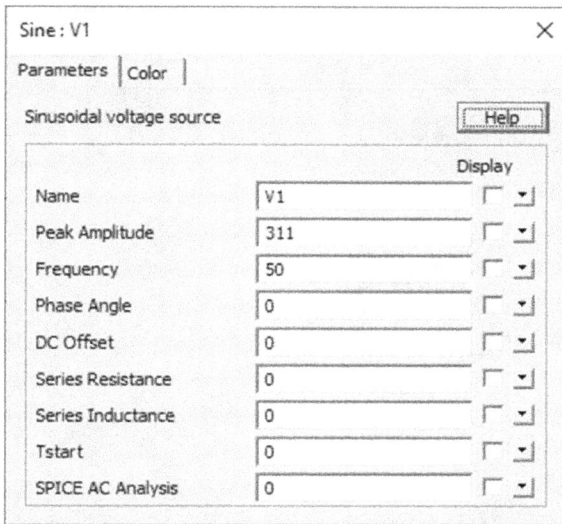

Fig. 4.88.

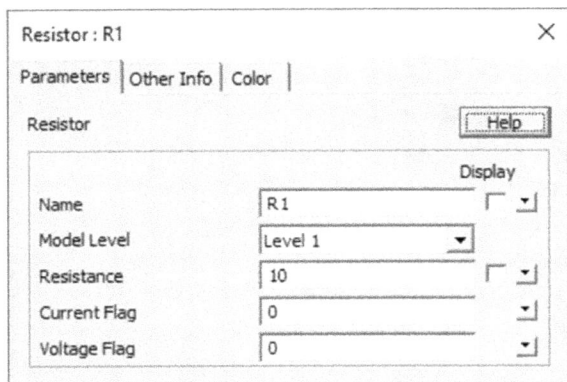

Fig. 4.89.

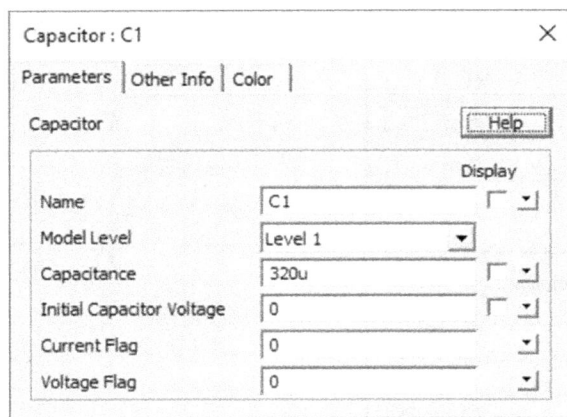

Fig. 4.90.

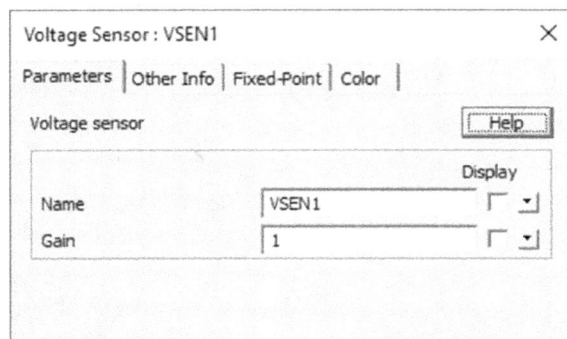

Fig. 4.91.

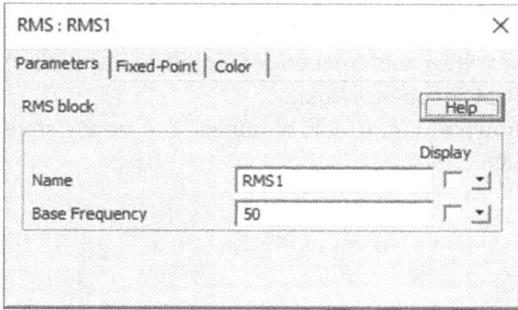

Fig. 4.92.

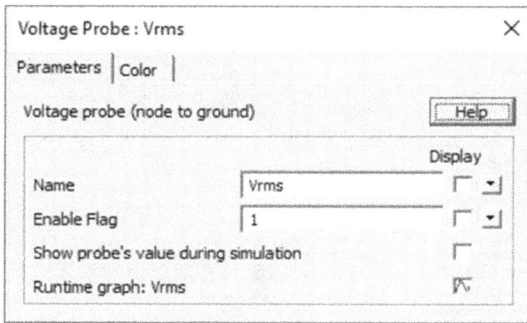

Fig. 4.93.

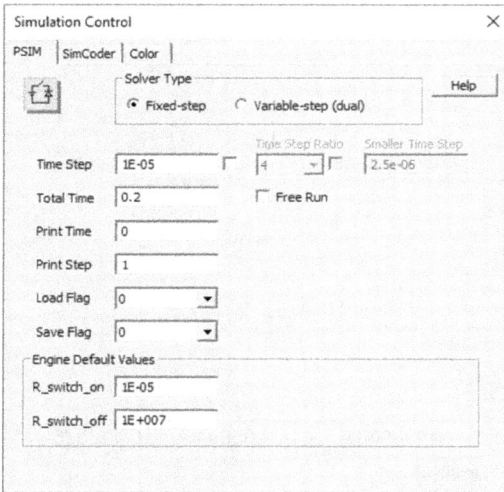

Fig. 4.94.

Note that the frequency of input signal must be entered to the RMS block (Fig. 4.92). In this example, frequency of capacitor voltage and frequency of input source (50 Hz) are the same. So, 50 is entered to the Base Frequency box.

Run the simulation. The result shown in Fig. 4.95 is obtained. The steady-state section of the graph is shown in Fig. 4.96.

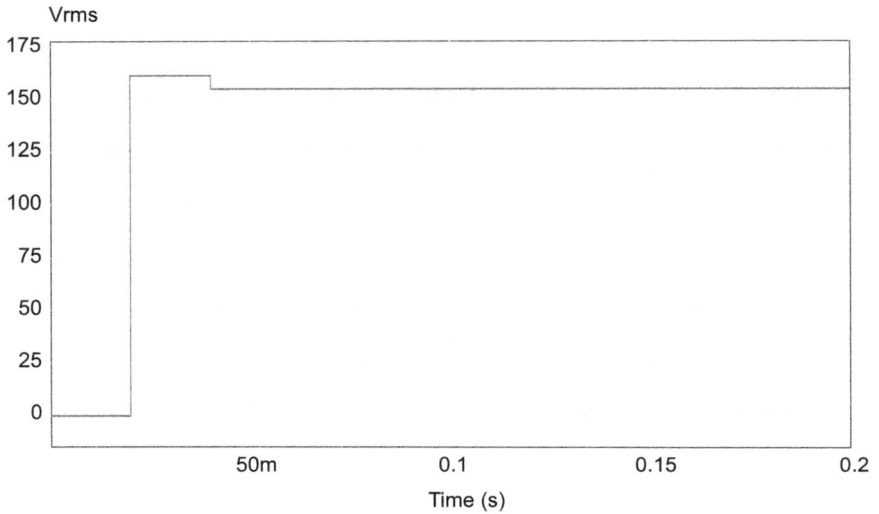

Fig. 4.95.

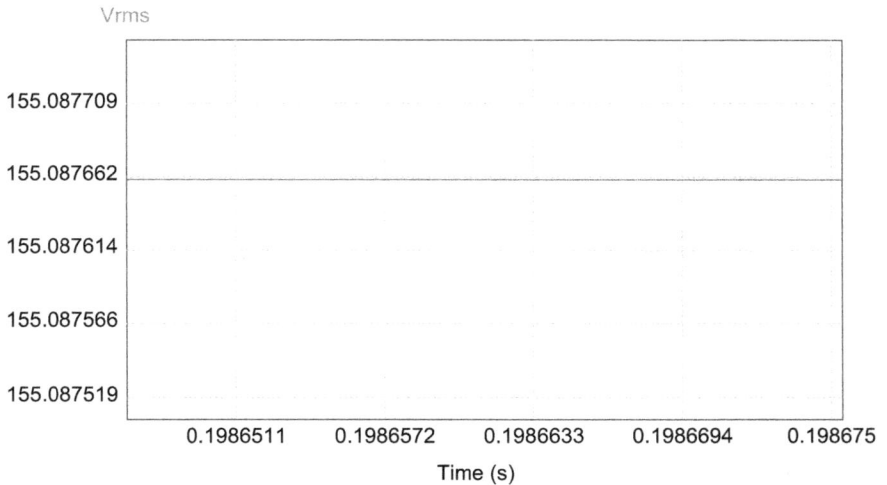

Fig.4.96.

Let's check the obtained result. The MATLAB code shown in Fig. 4.97 shows that PSIM result is correct.

Let's see one more example. Assume that you have a 220 Vrms, 50 Hz source (Fig. 4.98). Your load is 10 Ω and it requires about 37 Vrms.

```
Command Window                    ⊙
>> Xc=-i/2/pi/50/320e-6;
>> R=10;
>> abs(Xc/(Xc+R)*311)/sqrt(2)

ans =

    155.0877

fx >>
```

Fig. 4.97.

Fig. 4.98.

Using a step down transformer is the best solution from energy efficiency viewpoint. Let's assume that we have no access to transformer and we want to use a simple voltage divider shown in Fig. 4.99. We want to measure the efficiency ($\eta=\frac{P_{out}}{P_{in}}\times100\%$) of this circuit. The schematic shown in Fig. 4.100 does this job. This schematic uses multiplication and division blocks (Fig. 4.101) in order to calculate the input/output power and efficiency.

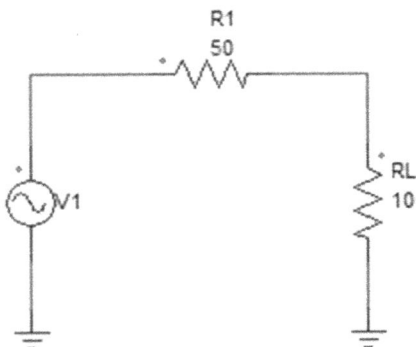

Fig. 4.99.

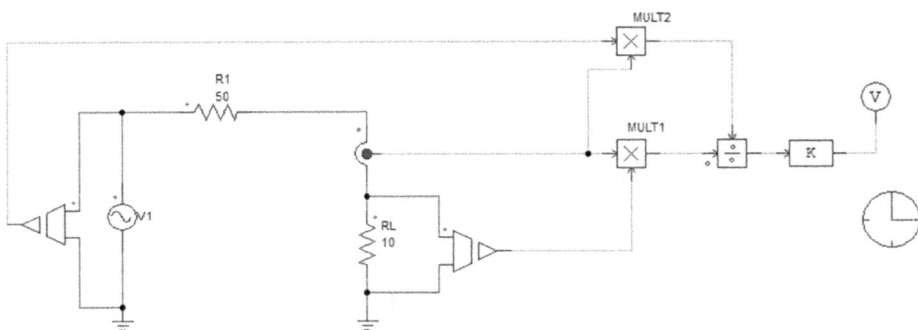

Fig. 4.100.

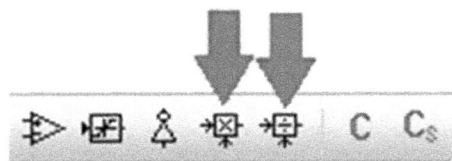

Fig. 4.101.

Settings of components used in the schematic are shown in Figs. 4.102–4.113.

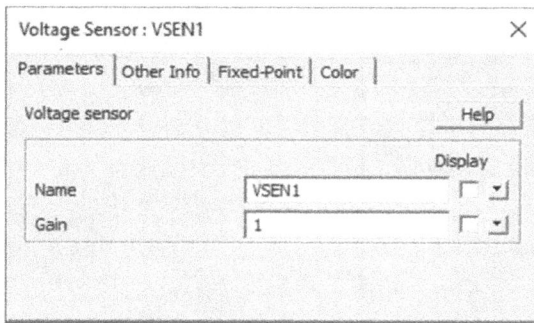

Fig. 4.102.

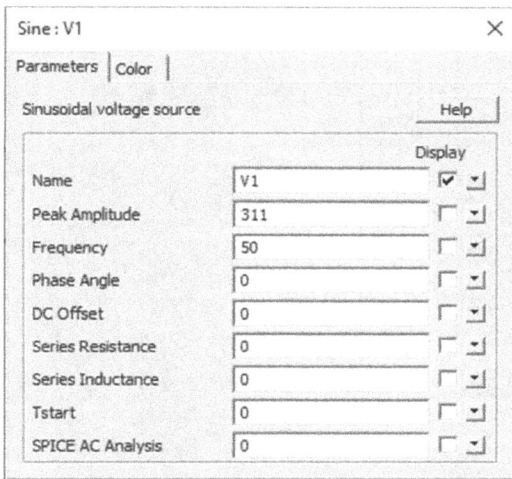

Fig. 4.103.

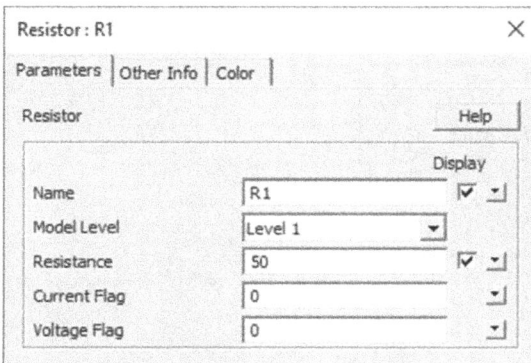

Fig. 4.104.

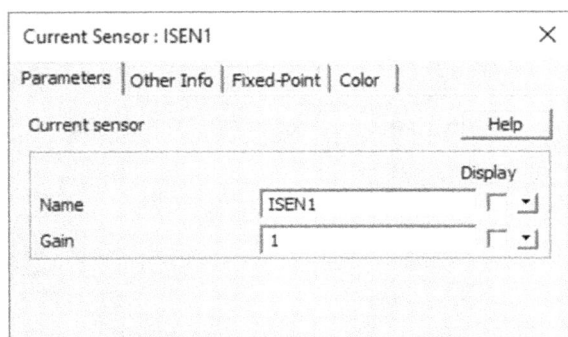

Fig. 4.105.

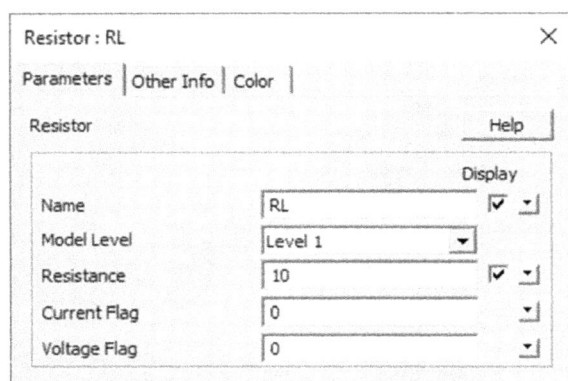

Fig. 4.106.

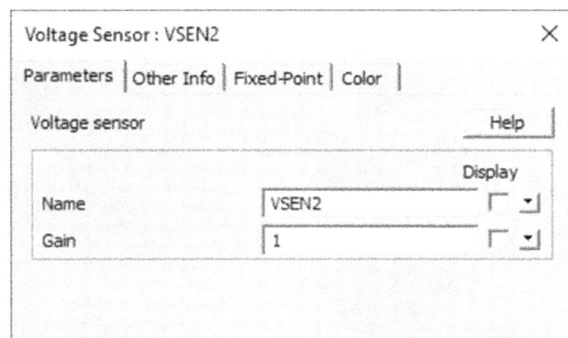

Fig. 4.107.

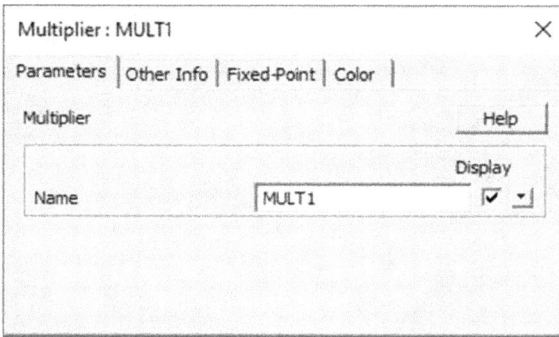

Fig. 4.108.

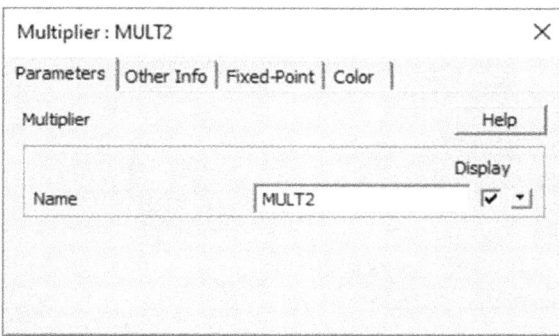

Fig. 4.109.

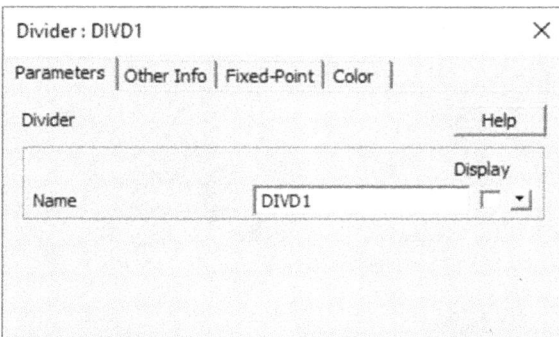

Fig. 4.110.

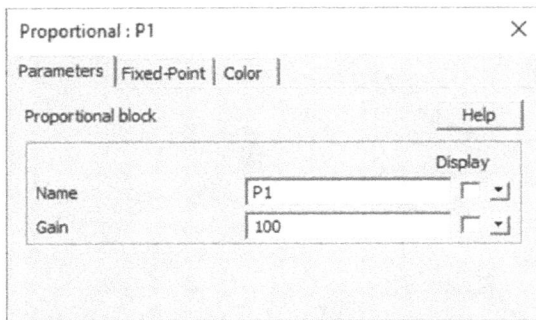

Fig. 4.111.

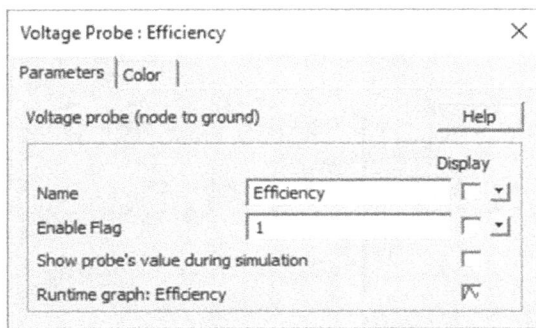

Fig. 4.112.

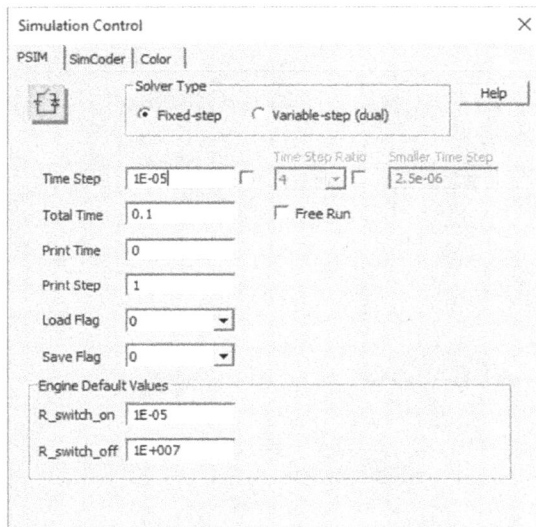

Fig. 4.113.

Run the simulation. The result shown in Fig. 4.114 is obtained. Figure 4.115 shows the steady-state value of this graph. So, the efficiency is only 16.6%. This value is expected from the basic circuit theory since $\eta=\frac{P_{out}}{P_{in}}\times100\%=\frac{R_L I^2}{R_L I^2+R_1 I^2}\times100\%$
$=\frac{10}{10+50}\times100\%=16.6\%$.

Efficiency

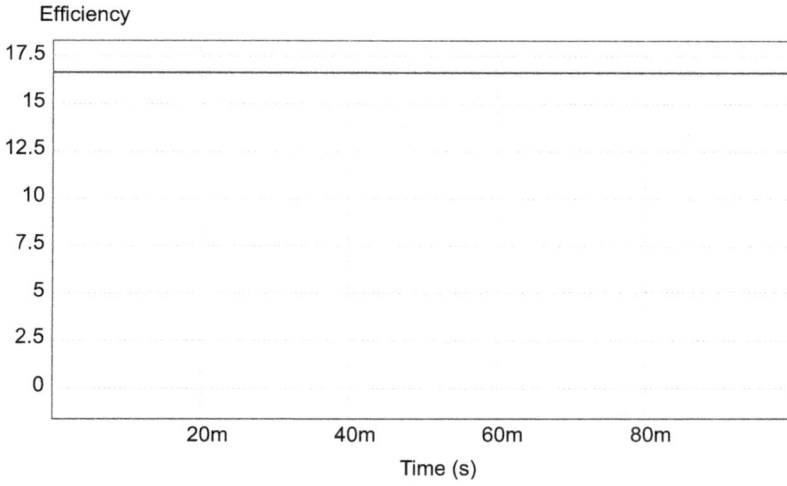

Fig. 4.114.

Efficiency

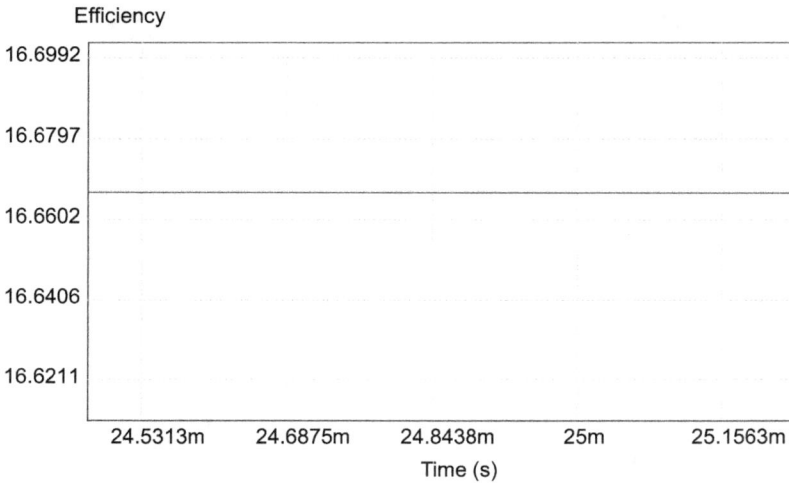

Fig. 4.115.

4.9 Transfer function block

If you want to use analog control techniques for control of the converter, then you need to use the s-domain Transfer Function block (Fig. 4.116) in order to model the designed controller.

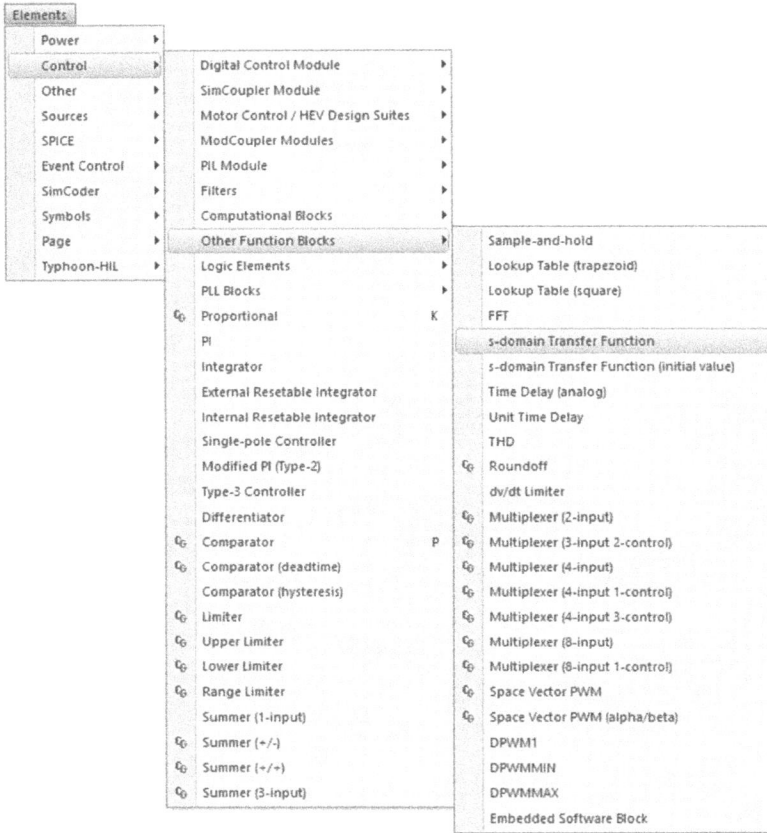

Fig. 4.116.

The s-domain Transfer Function block can be used to model any transfer function $H(s)=k\dfrac{b_n s^n + b_{n-1} s^{n-1} + \cdots + b_0}{a_n s^n + a_{n-1} s^{n-1} + \cdots + a_0}$. You need to enter the coefficients and degrees of the transfer function to the block. Figure 4.117 shows how the boxes must be filled for $H(s)=1.5\dfrac{400}{s^2 + 10s + 400}$.

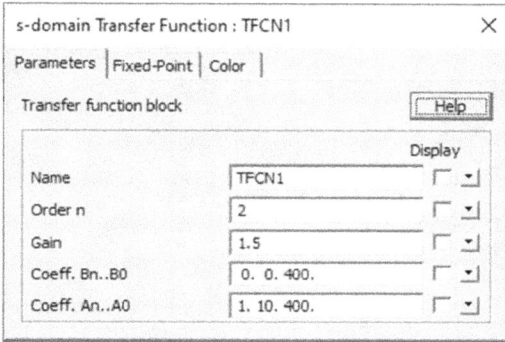

Fig. 4.117.

Commonly used controllers have ready to use blocks (Fig. 4.118). So, there is no need to use the s-domain Transfer Function block when you want to use a well-defined controller.

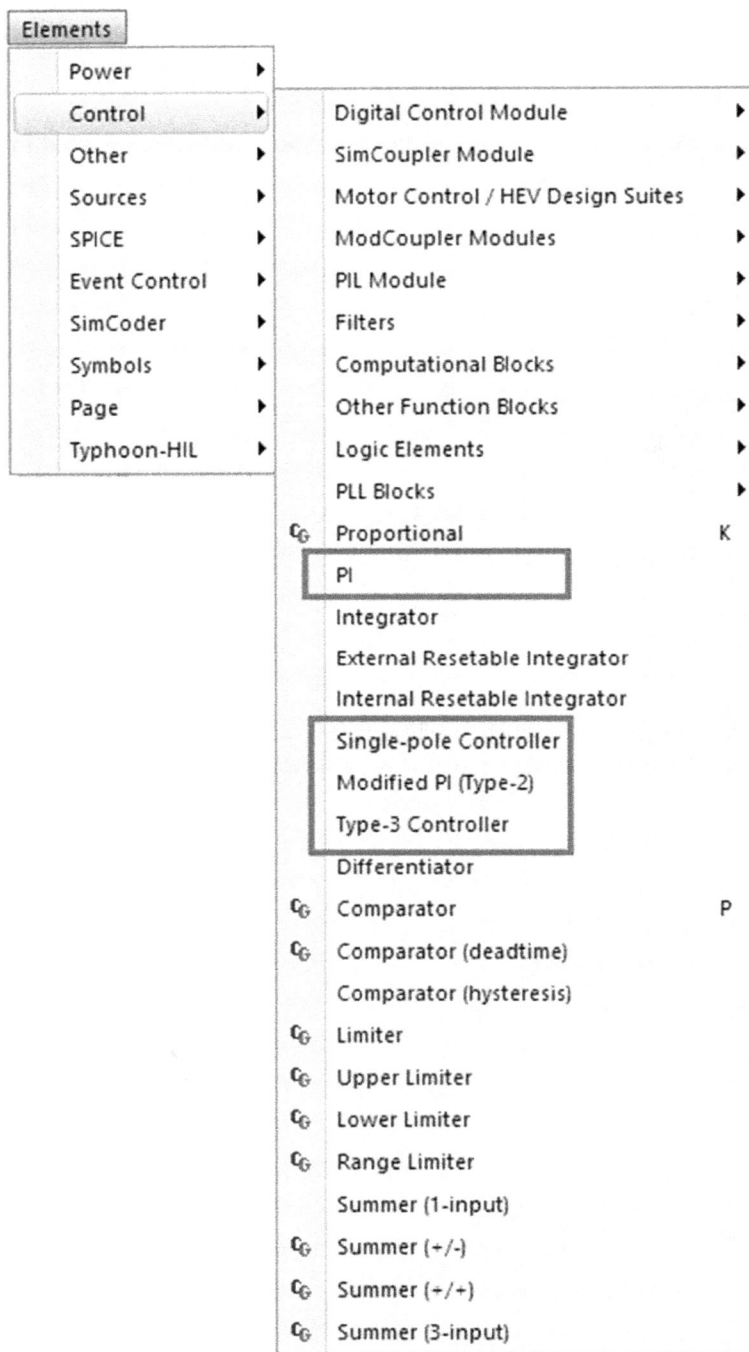

Elements		
Power ▶		
Control ▶	Digital Control Module ▶	
Other ▶	SimCoupler Module ▶	
Sources ▶	Motor Control / HEV Design Suites ▶	
SPICE ▶	ModCoupler Modules ▶	
Event Control ▶	PIL Module ▶	
SimCoder ▶	Filters ▶	
Symbols ▶	Computational Blocks ▶	
Page ▶	Other Function Blocks ▶	
Typhoon-HIL ▶	Logic Elements ▶	
	PLL Blocks ▶	
	𝒞ɢ Proportional	K
	PI	
	Integrator	
	External Resetable Integrator	
	Internal Resetable Integrator	
	Single-pole Controller	
	Modified PI (Type-2)	
	Type-3 Controller	
	Differentiator	
	𝒞ɢ Comparator	P
	𝒞ɢ Comparator (deadtime)	
	Comparator (hysteresis)	
	𝒞ɢ Limiter	
	𝒞ɢ Upper Limiter	
	𝒞ɢ Lower Limiter	
	𝒞ɢ Range Limiter	
	Summer (1-input)	
	𝒞ɢ Summer (+/-)	
	𝒞ɢ Summer (+/+)	
	𝒞ɢ Summer (3-input)	

Fig. 4.118.

PSIM has plenty of blocks for digital controllers as well. Elements which are required to model a digital control system are available in the Elements> Control> Digital Control Module section (Fig. 4.119).

The z-domain Transfer Function block is one of the important blocks of this section. This block permits you to model any transfer function $H(z)=\dfrac{b_0 z^N + b_1 z^{N-1}+\cdots+b_N}{a_0 z^N + a_1 z^{N-1}+\cdots+a_N}$.

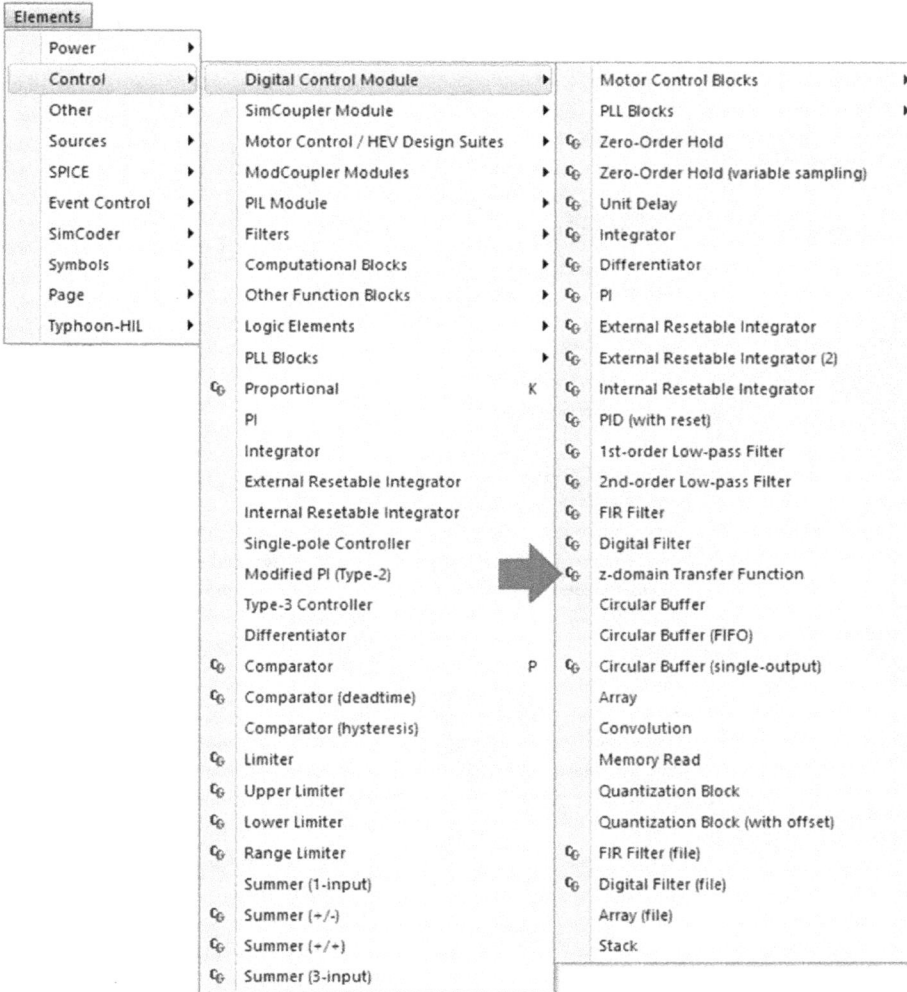

Fig. 4.119.

4.10 Logic elements

PSIM has many logic elements that can help you to model desired controller and modulator (Fig. 4.120).

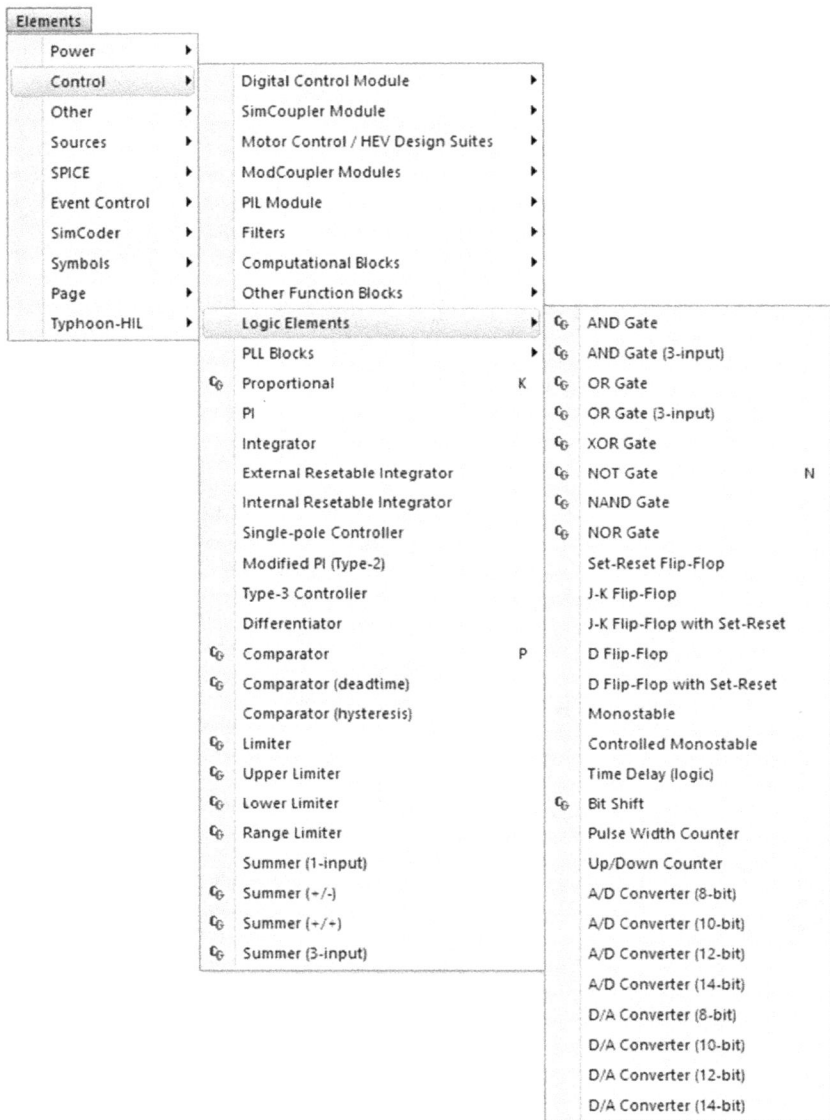

Fig. 4.120.

Click the File> Search Examples in order to see a control system which uses these elements. When Help Browser window is appeared, type sync rectifier (Fig. 4.121) and click the Find button. Then click the "sync rectifier.sch" in order to open it (Fig. 4.122).

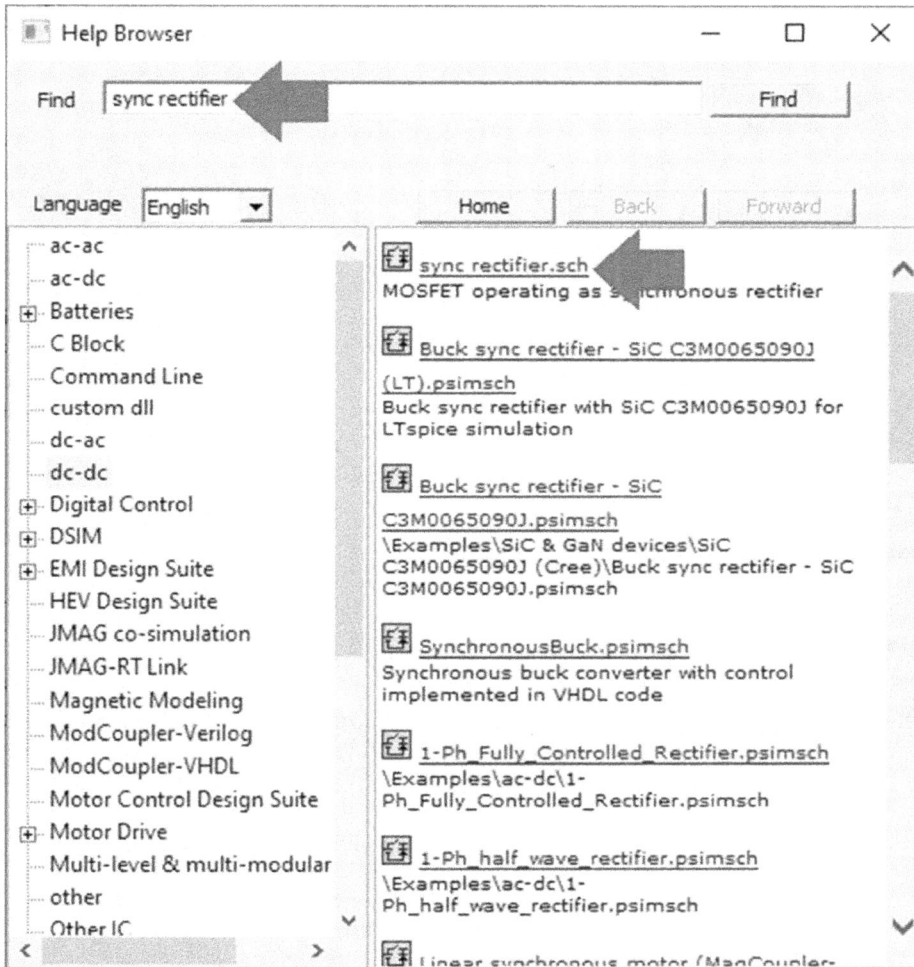

Fig. 4.121.

Buck Converter with Synchronous Rectifier

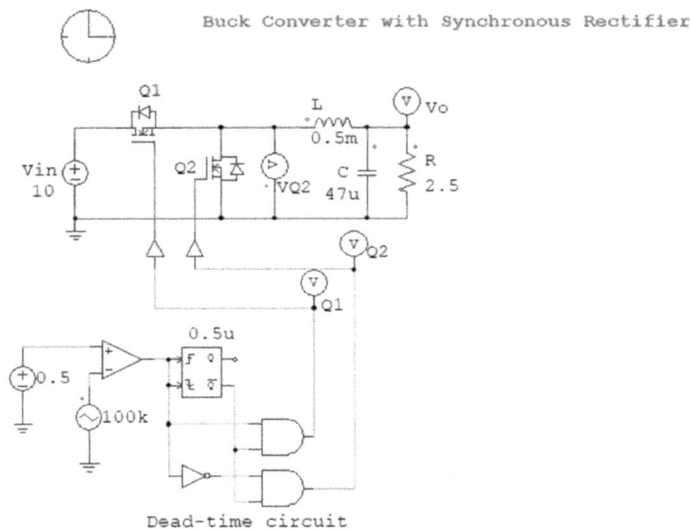

Dead-time circuit

This circuit illustrates how a MOSFET can be used as a synchronous rectifier.
The waveform VQ2 shows that, during the dead time, the diode of Q2 conducts.
When Q2 is on, the active switch conducts, reducing the conduction voltage dr
    from Vdiode to IL*Rdon.

Fig. 4.122.

4.11 Math function

The Math function block (Fig. 4.123) is a simple way to apply the desired mathematical
function to a control signal.

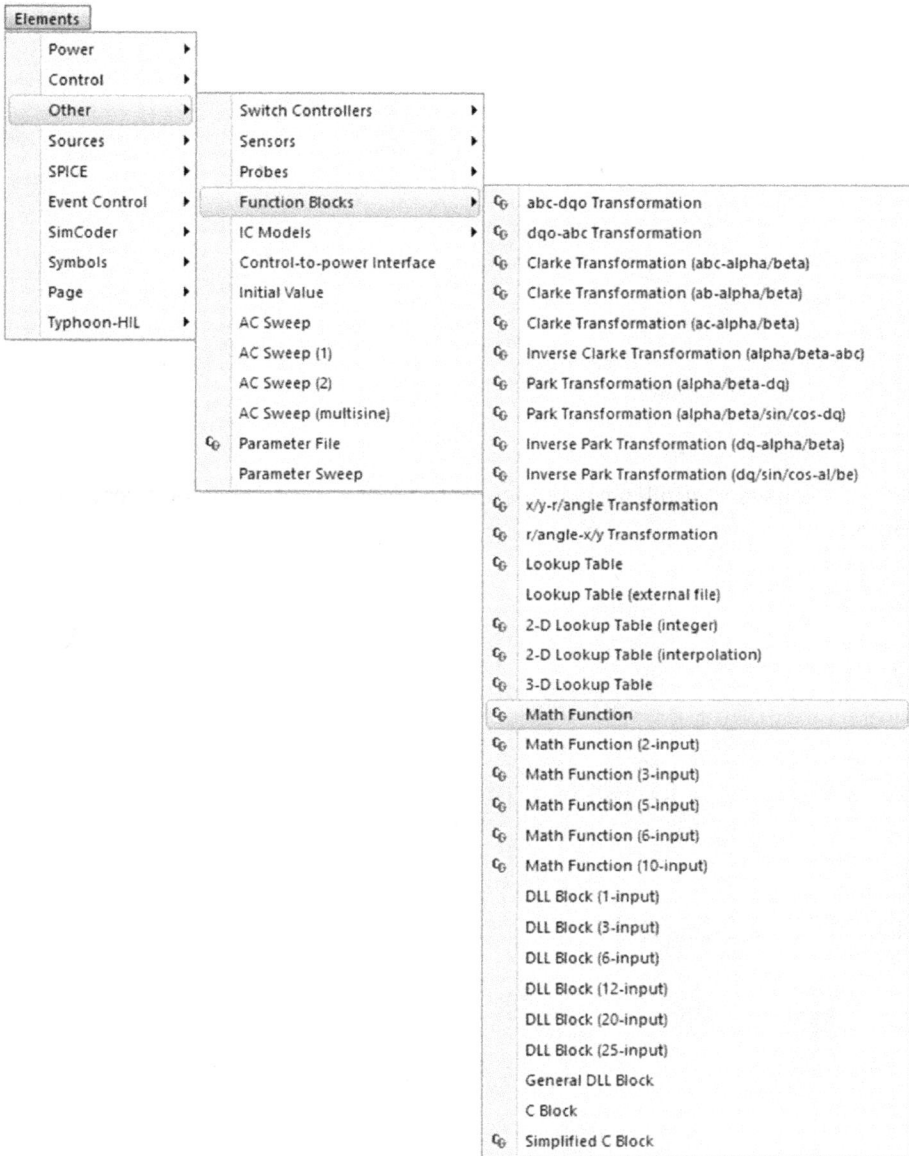

Fig. 4.123.

Let's study an example. Assume that we want to model a non-linear resistor with $I-V$ characteristic given by $V = \sqrt{I}$. PSIM has different types of (dependent and independent) sources. The sources can be found in the Elements> Sources section (Fig. 4.124).

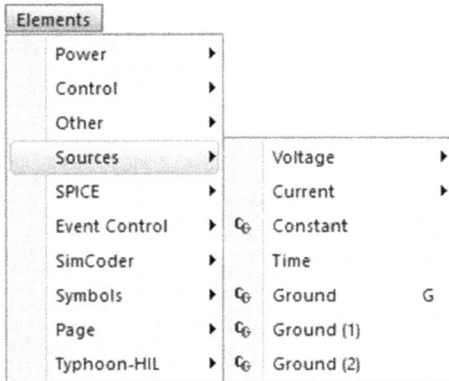

Fig. 4.124.

We use a voltage-controlled current source in order to model the non-linear resistor (Fig. 4.125).

Fig. 4.125.

The schematic shown in Fig. 4.126, reads the voltage of terminals and set the value of current source to the square of terminal voltage, that is, $I = V^2$ or $V = \sqrt{I}$.

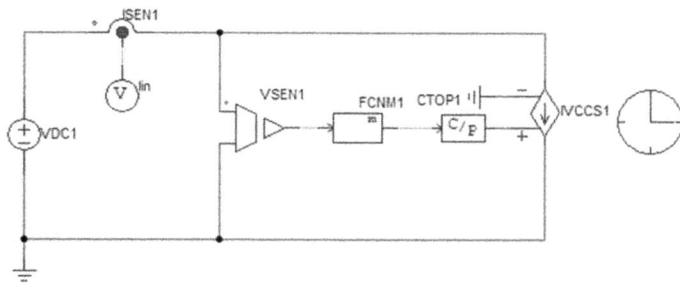

Fig. 4.126.

Settings of used elements are shown in Figs. 4.127–4.134.

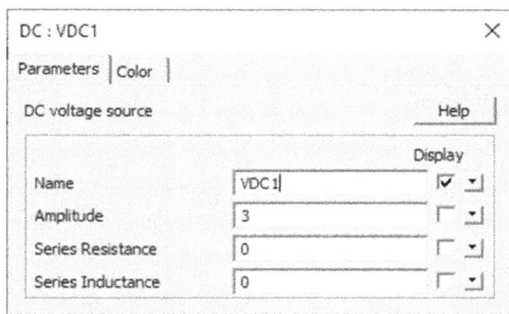

Fig. 4.127.

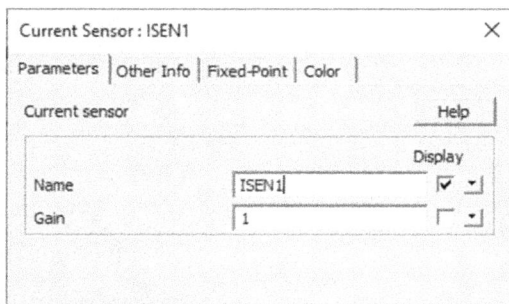

Fig. 4.128.

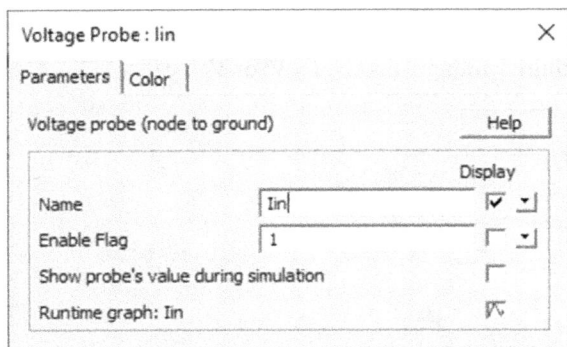

Fig. 4.129.

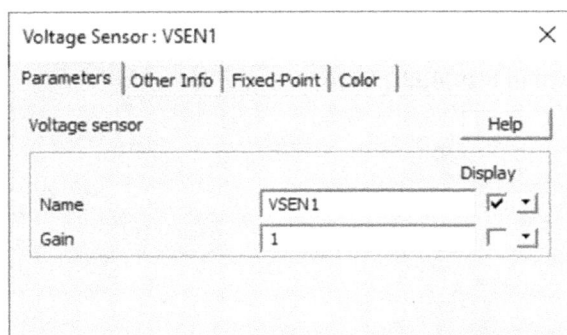

Fig. 4.130.

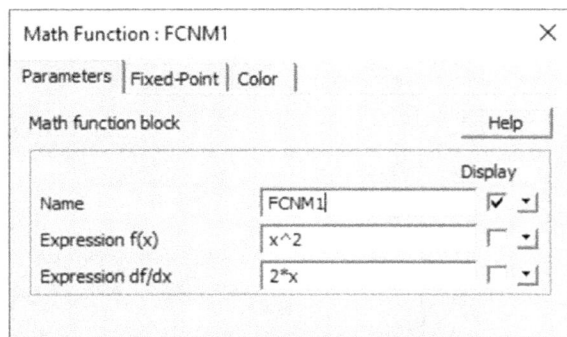

Fig. 4.131.

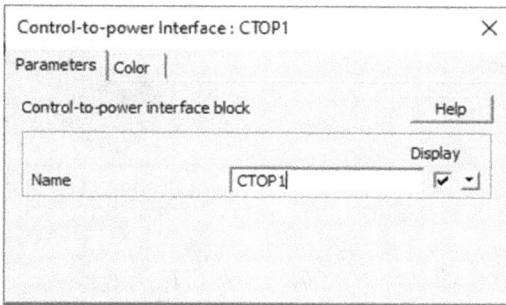

Fig. 4.132.

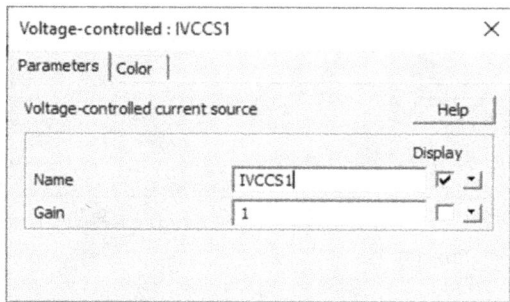

Fig. 4.133.

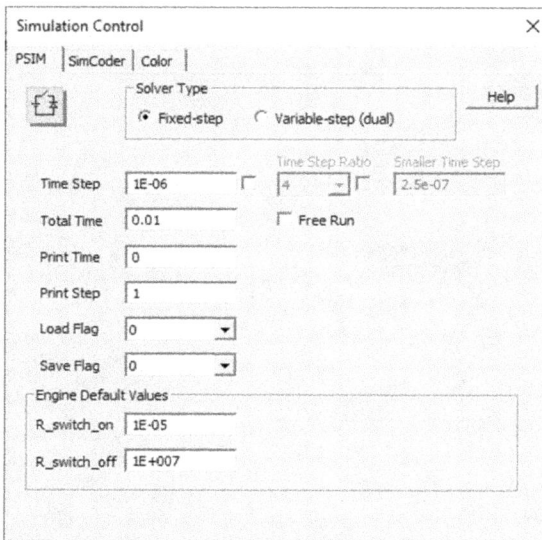

Fig. 4.134.

Run the simulation. The result shown in Fig. 4.135 will be obtained. The current is the square of input voltage (3 V) as expected.

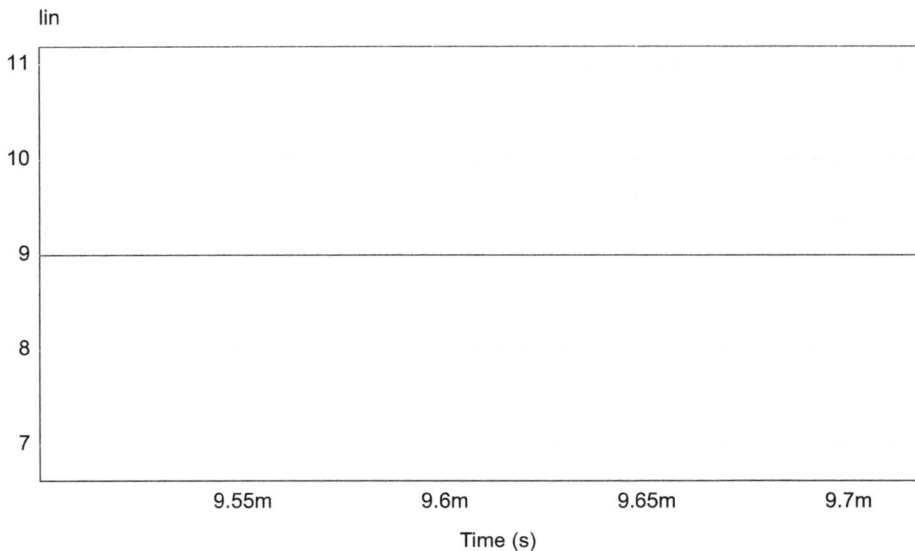

Fig. 4.135.

Now add a resistor in series to the voltage source with a value of 0.2 Ω.

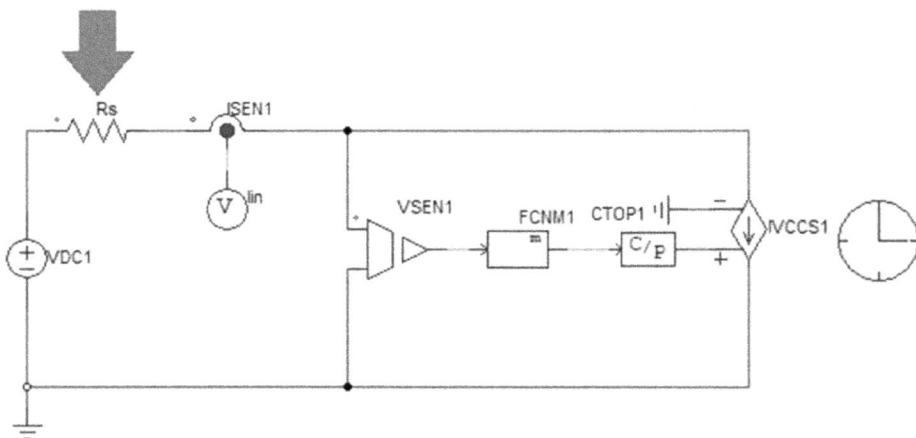

Fig. 4.136.

Run the simulation again. The current waveform is shown in Fig. 4.137. The circuit current is 4.451 A.

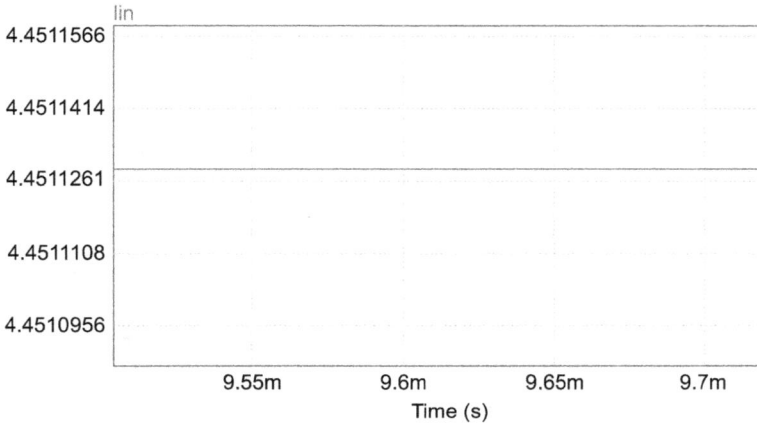

Fig. 4.137.

Let's check the PSIM result using hand analysis. According to Kirchhoff's voltage law $0.2I + \sqrt{I} = 3$. So, the intersection of $y_1 = 0.2x + \sqrt{x}$ and $y_2 = 3$ gives the circuit current. According to Fig. 4.138, the circuit current is 4.451 A.

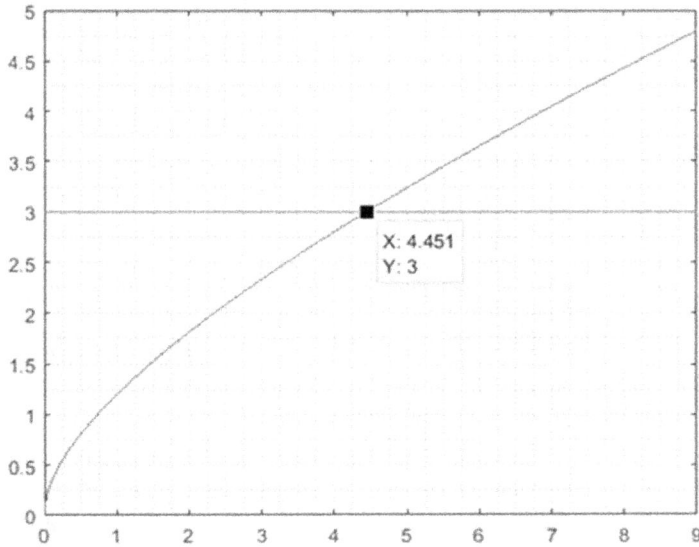

Fig. 4.138.

4.12 Switches

PSIM has plenty of switches which can be found in the Elements> Power> Switches section (Fig. 4.139).

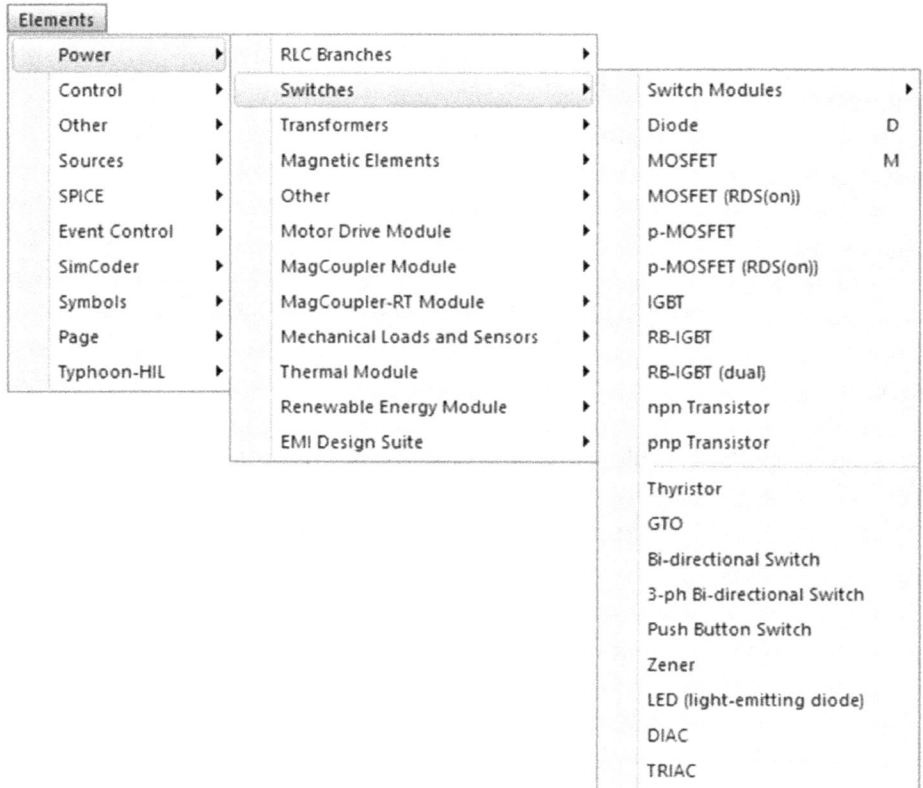

Fig. 4.139.

Switches can have different models (Ideal, Level 1, Level 2) during the simulation. Level 1 has more detail in comparison to Ideal model. Level 2 has more detail in comparison to Level 1, as well.

For instance, Ideal model settings and Level 2 settings for diode are shown in Figs. 4.140 and 4.141, respectively. As you see, Ideal model (Fig. 4.140) assumes a simple model (a DC voltage source in series with a resistor) for a forward-biased diode. Level 1 model considers the parasitic elements and reverse recovery time as well. You can click the Help button in Figs. 4.140 and 4.141 in order to see the description of each parameter.

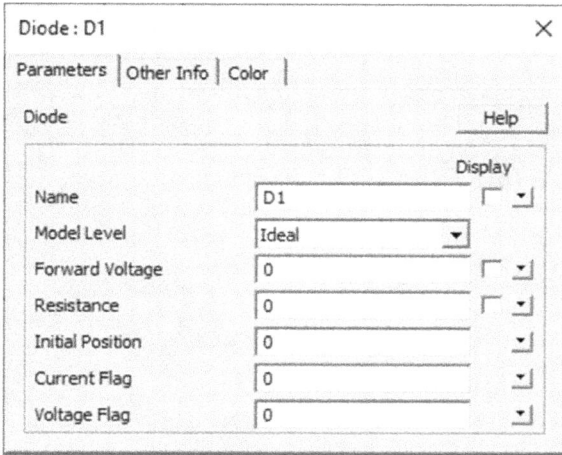

Fig. 4.140.

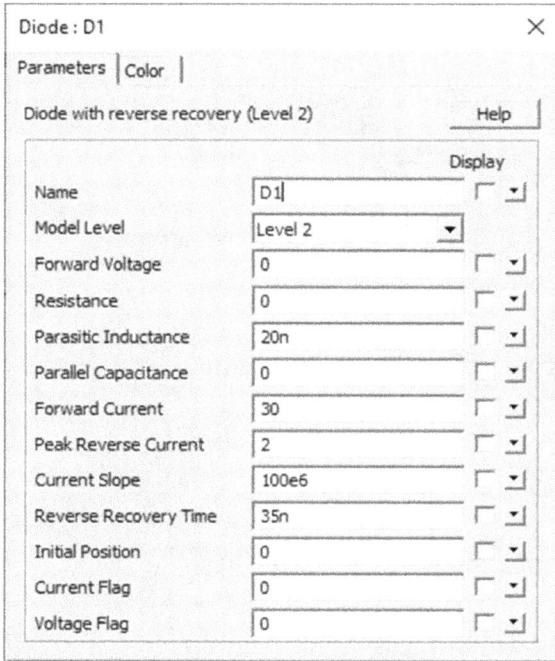

Fig. 4.141.

Let's study an example. Assume the circuit shown in Fig. 4.142. Input voltage is a square waveform which alternates between +10 V and –10 V with frequency of 1 kHz.

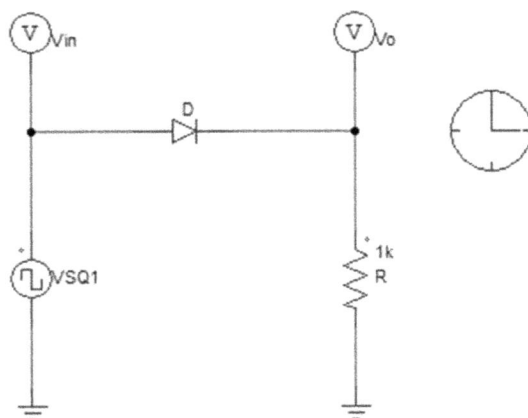

Fig. 4.142.

Settings of the blocks are shown in Figs. 4.143–4.148.

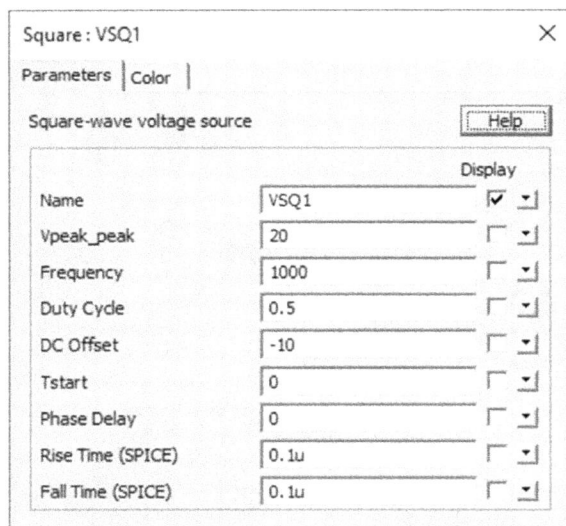

Square : VSQ1   ✕

Parameters | Color |

Square-wave voltage source   Help

Display

Name	VSQ1	☑ ▾
Vpeak_peak	20	▾
Frequency	1000	▾
Duty Cycle	0.5	▾
DC Offset	-10	▾
Tstart	0	▾
Phase Delay	0	▾
Rise Time (SPICE)	0.1u	▾
Fall Time (SPICE)	0.1u	▾

Fig. 4.143.

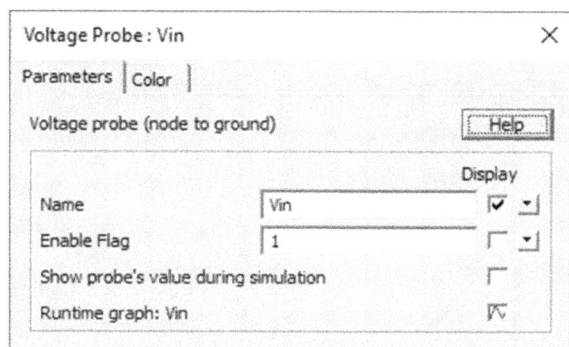

Fig. 4.144.

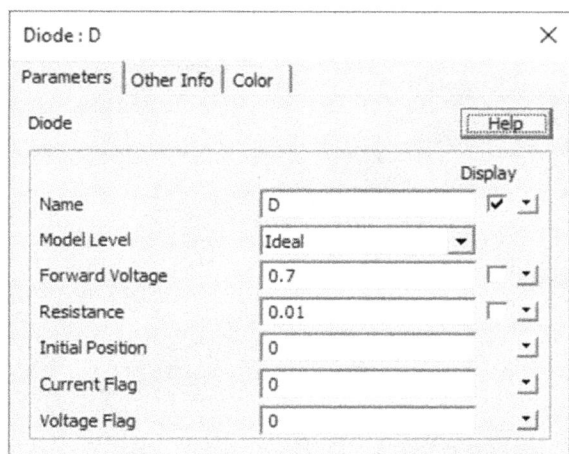

Fig. 4.145.

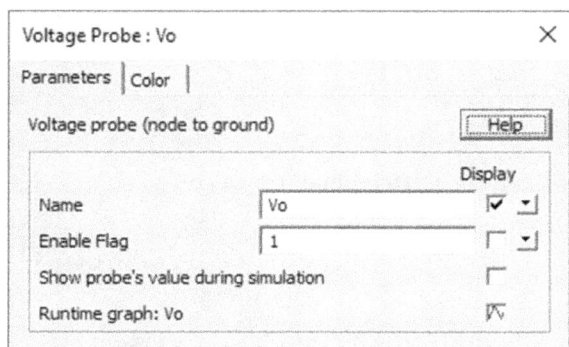

Fig. 4.146.

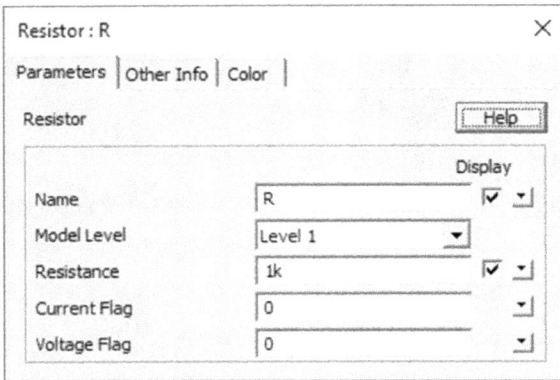

Fig. 4.147.

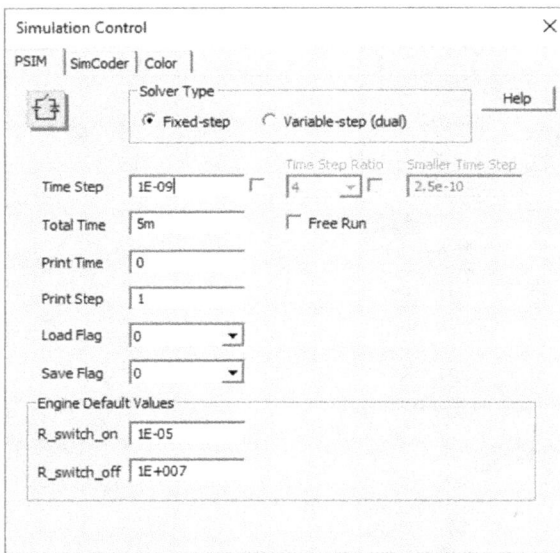

Fig. 4.148.

Run the simulation. The result shown in Fig. 4.149 is obtained.

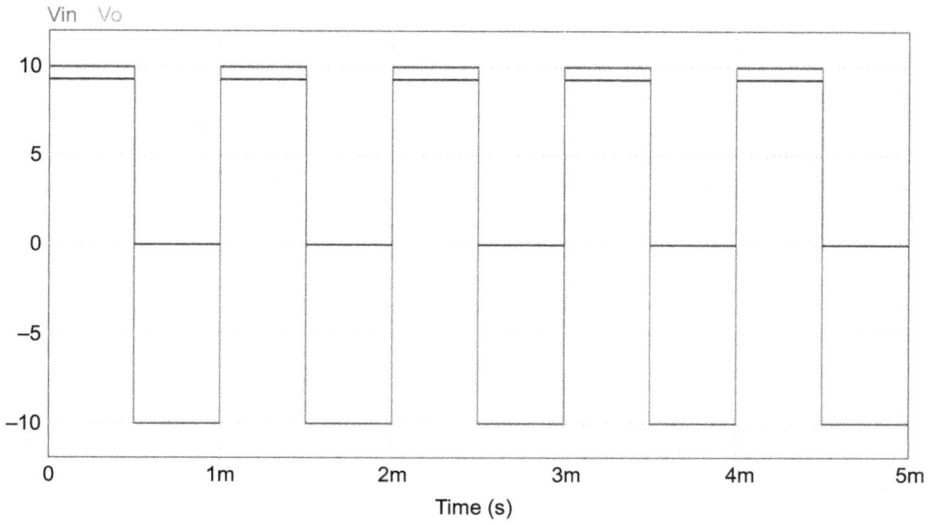

Fig. 4.149.

The falling and rising edges are shown in Figs. 4.150 and 4.151. The output voltage is positive in these two figures.

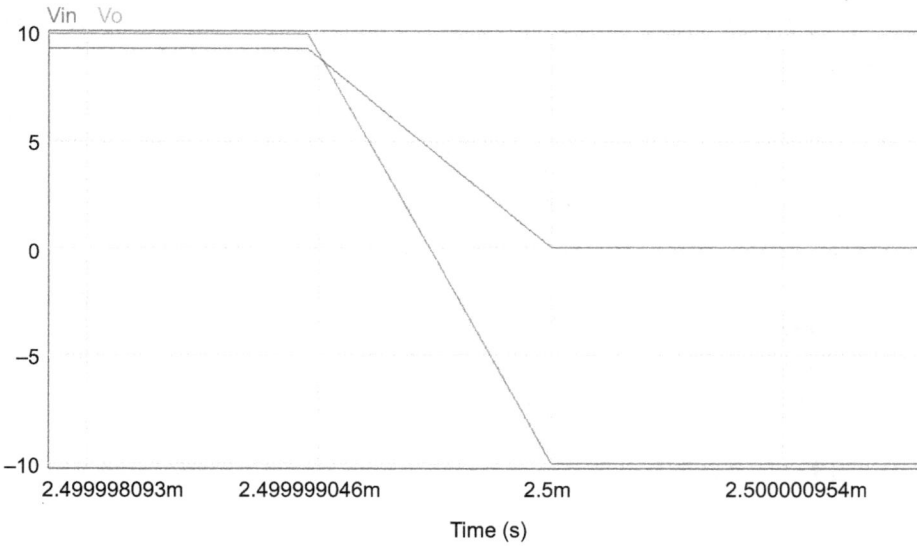

Fig. 4.150.

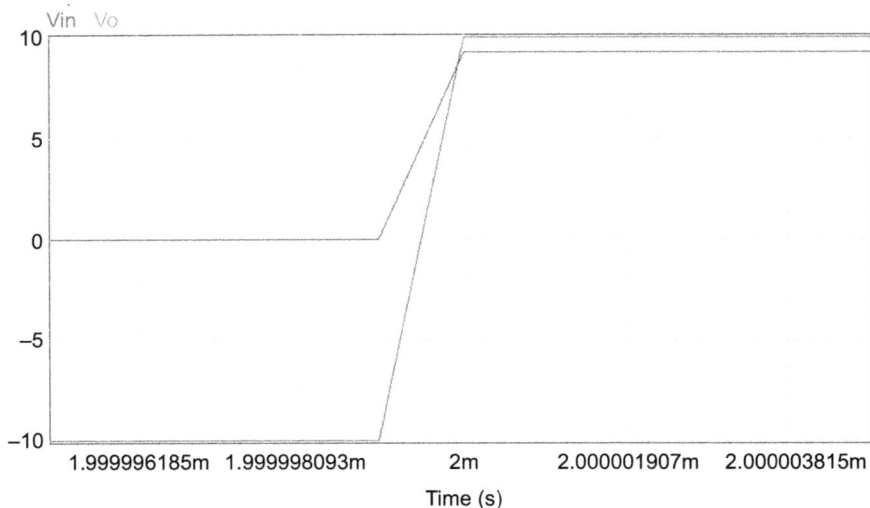

Fig. 4.151.

Now double click on the diode block and change the model to Level 2.

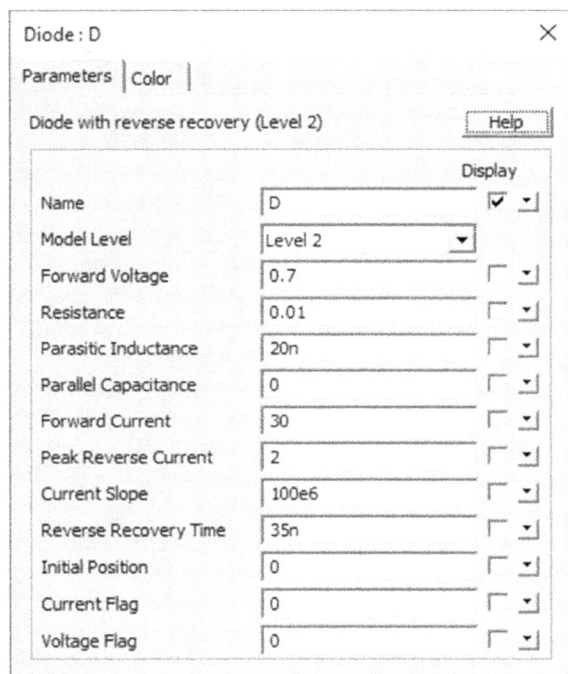

Fig. 4.152.

Rerun the simulation. The waveform shown in Fig. 4.153 is obtained.

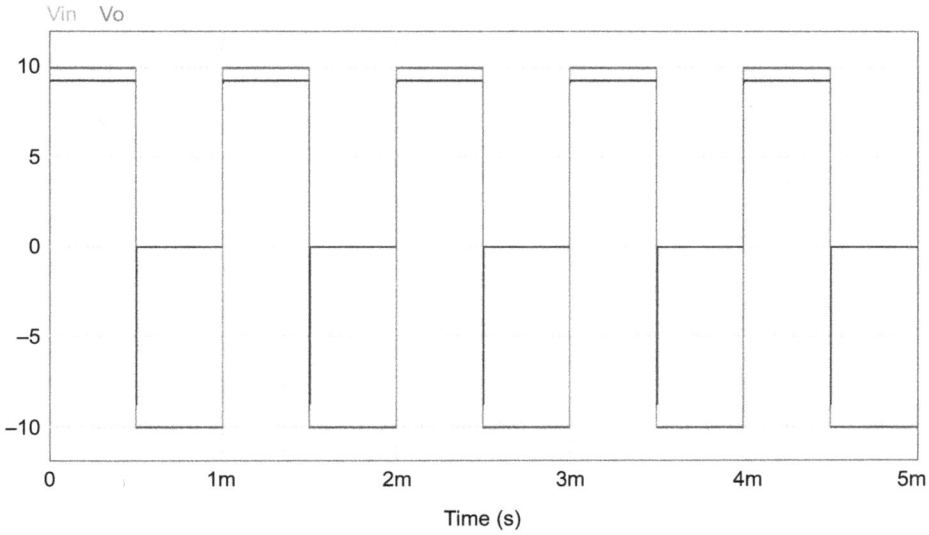

Fig. 4.153.

Figs. 4.154 and 4.155 show the falling and rising edges. Note that output voltage became negative in Fig. 4.154.

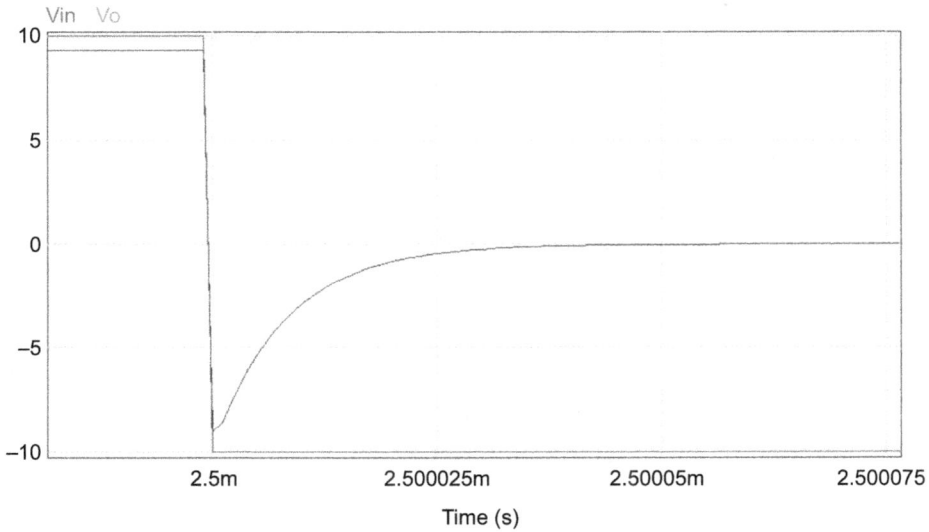

Fig. 4.154.

Vin Vo

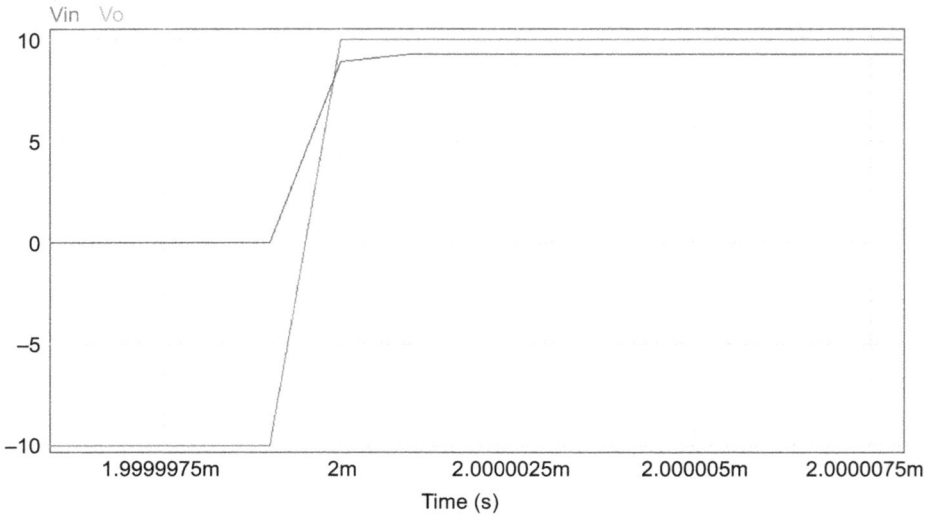

Fig. 4.155.

Figs. 4.156 and 4.157 compare falling and rising edges of Figs. 4.149 and 4.153. The behavior of two models in the rising edge is similar. However, the Ideal model ignores the reverse recovery effect. So, we see different behaviors in the falling edge.

Vo Vo_(IdealModeL)

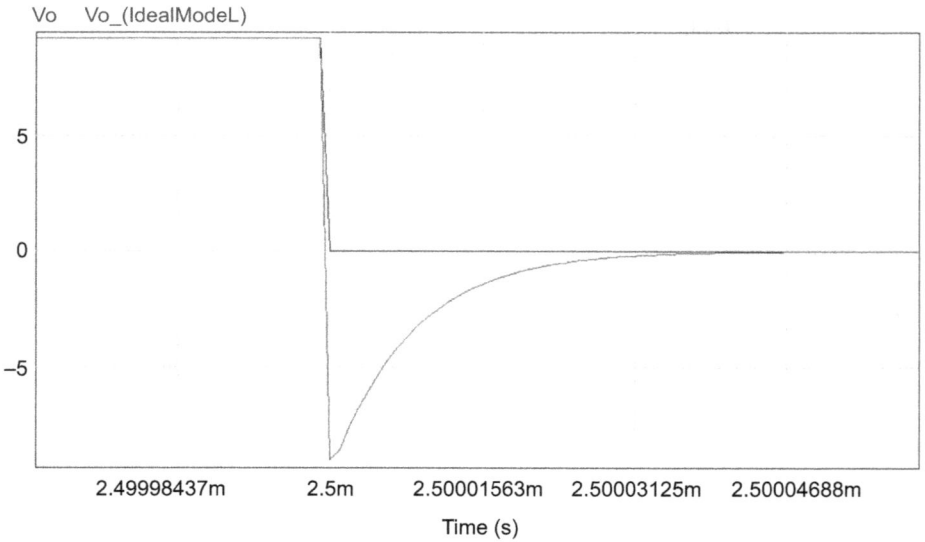

Fig. 4.156.

Vo   Vo_(IdealModeL)

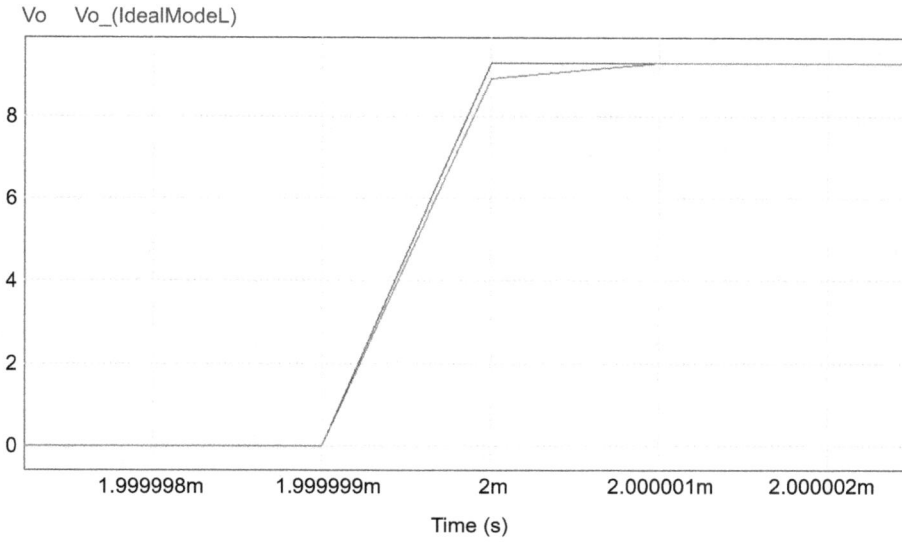

Fig. 4.157.

4.13 Transformers

Transformers are one of important elements that are used in circuits. Simulation of a circuit which contains a transformer is easy since PSIM has model of different type of transformers (Fig. 4.158).

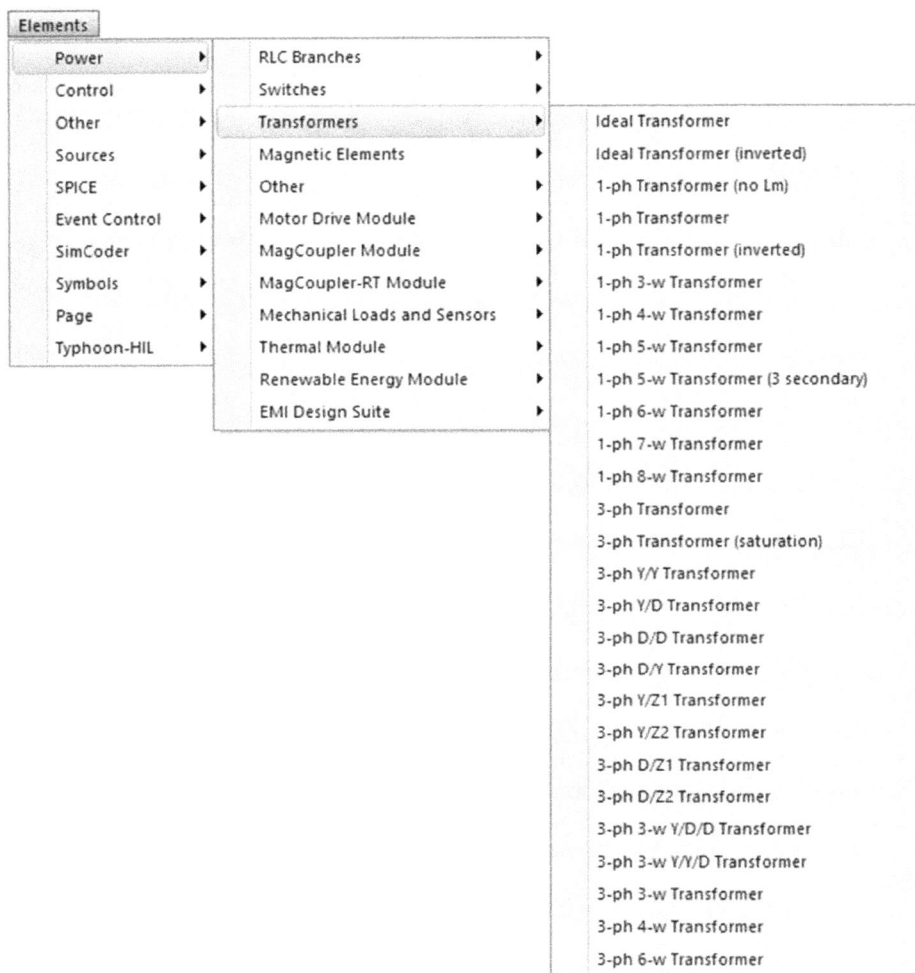

Elements			
Power ▶	RLC Branches ▶	Ideal Transformer
Control ▶	Switches ▶	Ideal Transformer (inverted)
Other ▶	Transformers ▶	1-ph Transformer (no Lm)
Sources ▶	Magnetic Elements ▶	1-ph Transformer
SPICE ▶	Other ▶	1-ph Transformer (inverted)
Event Control ▶	Motor Drive Module ▶	1-ph 3-w Transformer
SimCoder ▶	MagCoupler Module ▶	1-ph 4-w Transformer
Symbols ▶	MagCoupler-RT Module ▶	1-ph 5-w Transformer
Page ▶	Mechanical Loads and Sensors ▶	1-ph 5-w Transformer (3 secondary)
Typhoon-HIL ▶	Thermal Module ▶	1-ph 6-w Transformer
	Renewable Energy Module ▶	1-ph 7-w Transformer
	EMI Design Suite ▶	1-ph 8-w Transformer
		3-ph Transformer
		3-ph Transformer (saturation)
		3-ph Y/Y Transformer
		3-ph Y/D Transformer
		3-ph D/D Transformer
		3-ph D/Y Transformer
		3-ph Y/Z1 Transformer
		3-ph Y/Z2 Transformer
		3-ph D/Z1 Transformer
		3-ph D/Z2 Transformer
		3-ph 3-w Y/D/D Transformer
		3-ph 3-w Y/Y/D Transformer
		3-ph 3-w Transformer
		3-ph 4-w Transformer
		3-ph 6-w Transformer

Fig. 4.158.

Most of circuits can be simulated with the aid of single-phase transformer blocks shown in Fig. 4.159.

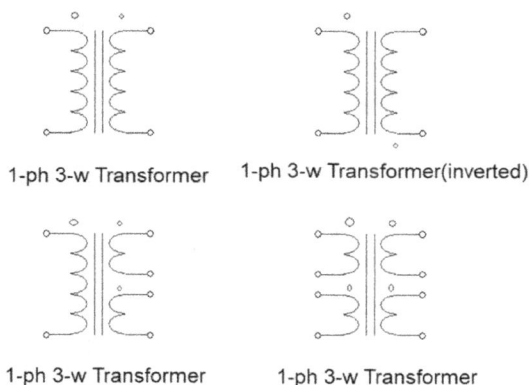

1-ph 3-w Transformer   1-ph 3-w Transformer(inverted)

1-ph 3-w Transformer   1-ph 3-w Transformer

Fig. 4.159.

If you double click on any transformer block in the schematic, the transformer parameter window is opened. If you click the Help button in this window, you can see the description of each parameter. For instance, description of parameters shown in Fig. 4.160 is shown in Fig. 4.161.

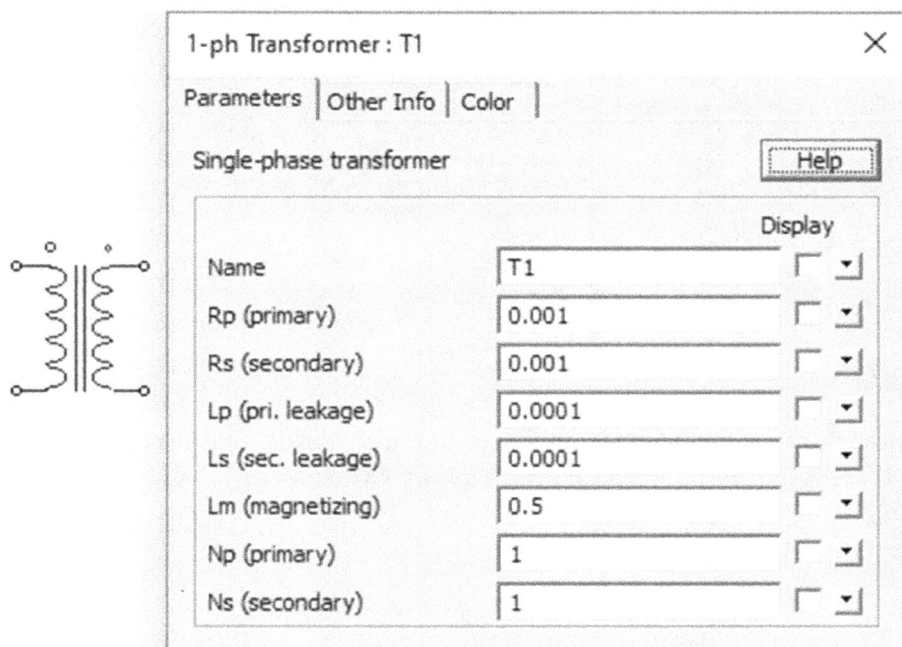

Fig. 4.160.

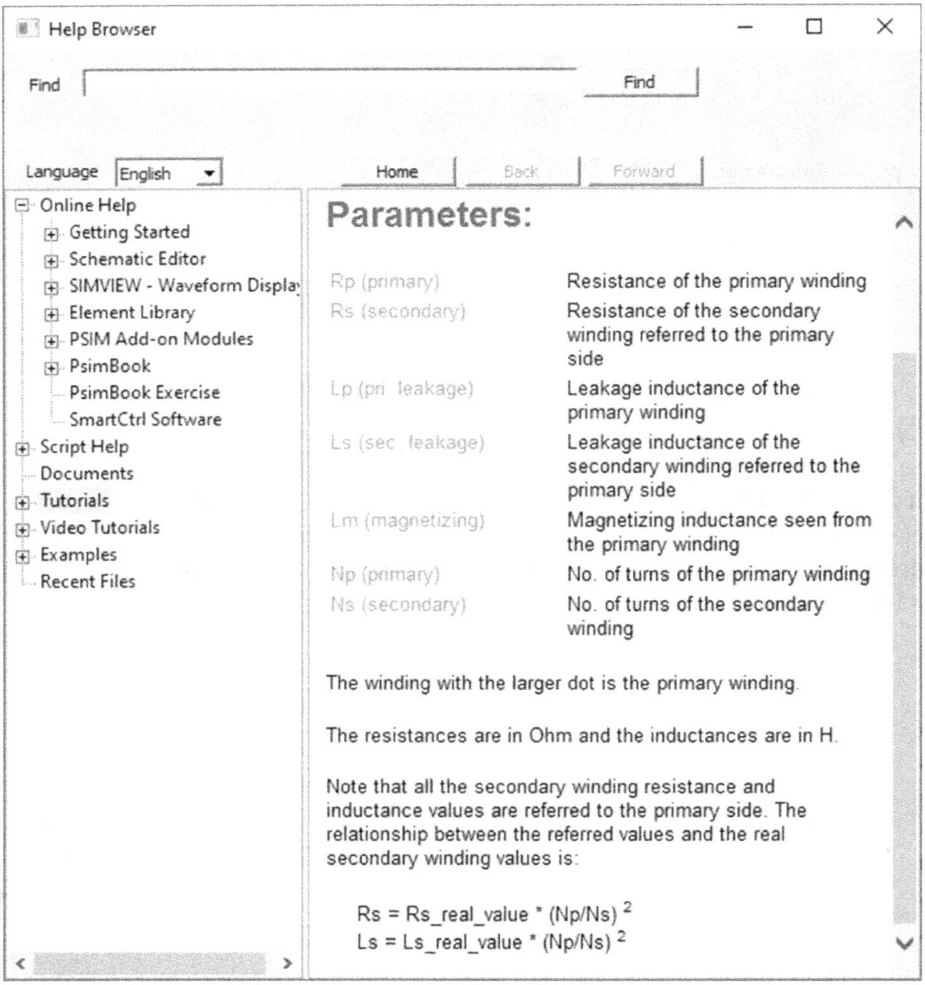

Help Browser — □ ×

Find | Find

Language | English ▼ | Home | Back | Forward

- Online Help
  - Getting Started
  - Schematic Editor
  - SIMVIEW - Waveform Display
  - Element Library
  - PSIM Add-on Modules
  - PsimBook
    - PsimBook Exercise
    - SmartCtrl Software
  - Script Help
  - Documents
  - Tutorials
  - Video Tutorials
  - Examples
  - Recent Files

Parameters:

Rp (primary)	Resistance of the primary winding
Rs (secondary)	Resistance of the secondary winding referred to the primary side
Lp (pri. leakage)	Leakage inductance of the primary winding
Ls (sec. leakage)	Leakage inductance of the secondary winding referred to the primary side
Lm (magnetizing)	Magnetizing inductance seen from the primary winding
Np (primary)	No. of turns of the primary winding
Ns (secondary)	No. of turns of the secondary winding

The winding with the larger dot is the primary winding.

The resistances are in Ohm and the inductances are in H.

Note that all the secondary winding resistance and inductance values are referred to the primary side. The relationship between the referred values and the real secondary winding values is:

$$Rs = Rs\_real\_value * (Np/Ns)^2$$
$$Ls = Ls\_real\_value * (Np/Ns)^2$$

Fig. 4.161.

Click File > Search Example and then type transformer in the Find box in order to see some sample simulations which use the transformer block (Fig. 4.162).

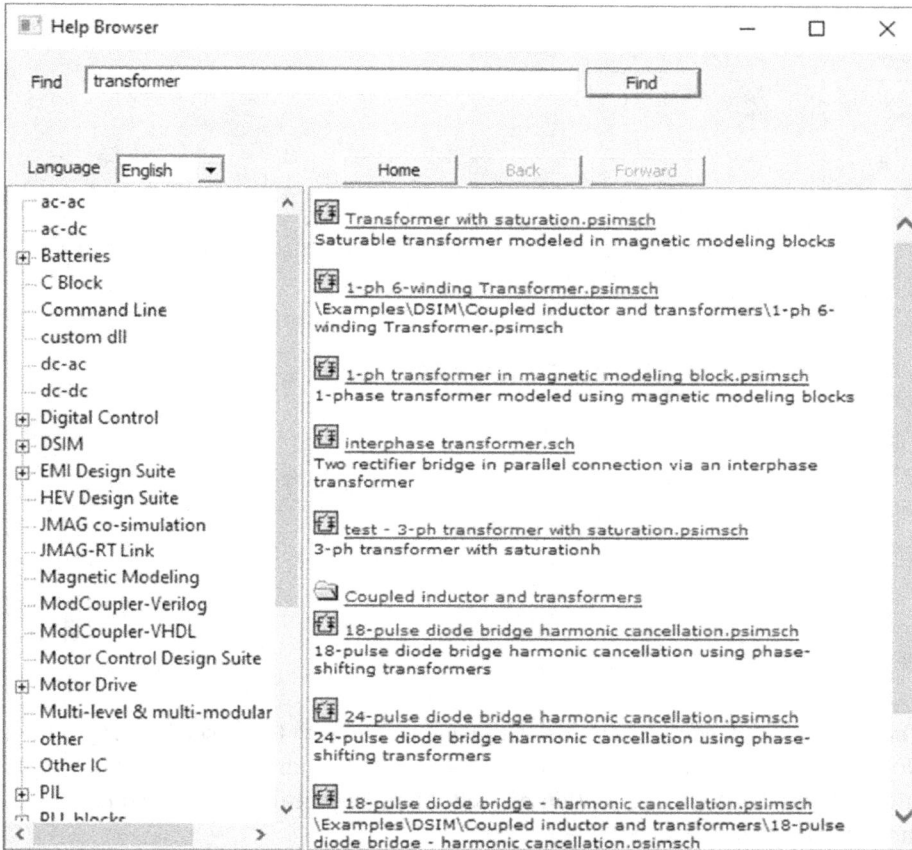

Fig. 4.162.

4.14 Op amp

Op amps are the heart of analog controllers. For instance, in Fig. 4.163, a buck converter with an analog controller is shown.

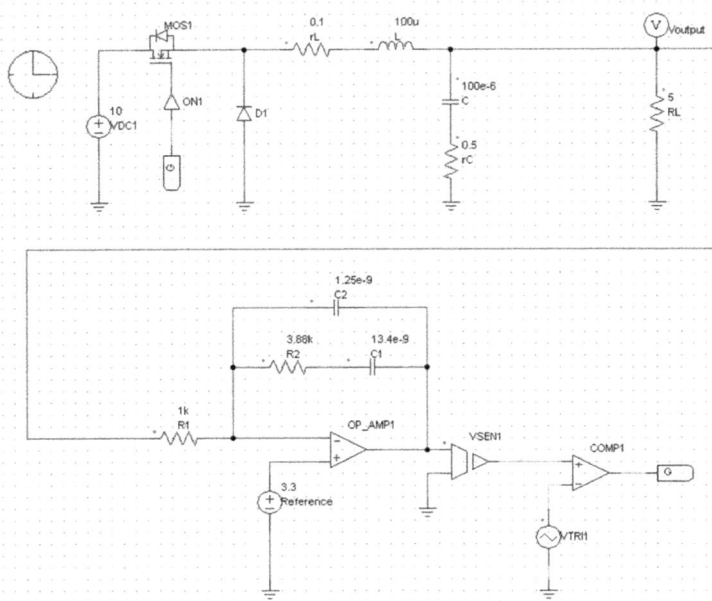

Fig. 4.163.

The op amp block can be found in the Elements> Power> Other section (Fig. 4.164). Note that Optocoupler and TL431 adjustable precision shunt voltage regulator blocks can be found in the Other section as well. Many practical control circuits contain these two blocks.

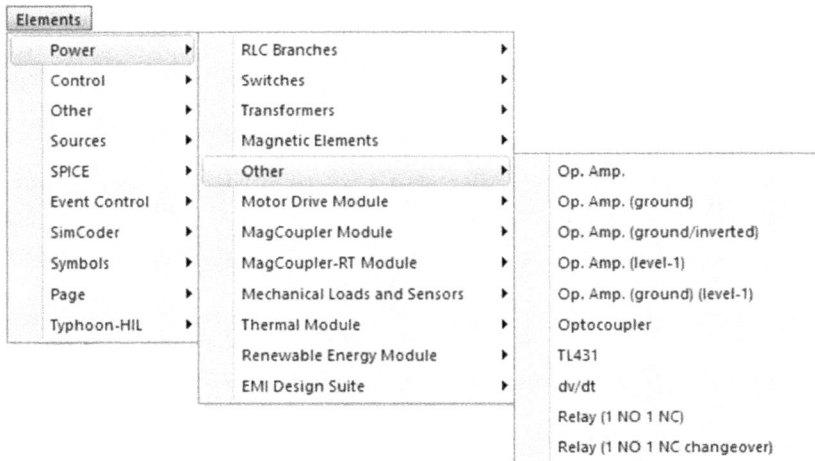

Fig. 4.164.

Let's study a simple filter circuit. Assume the circuit shown in Fig. 4.165.

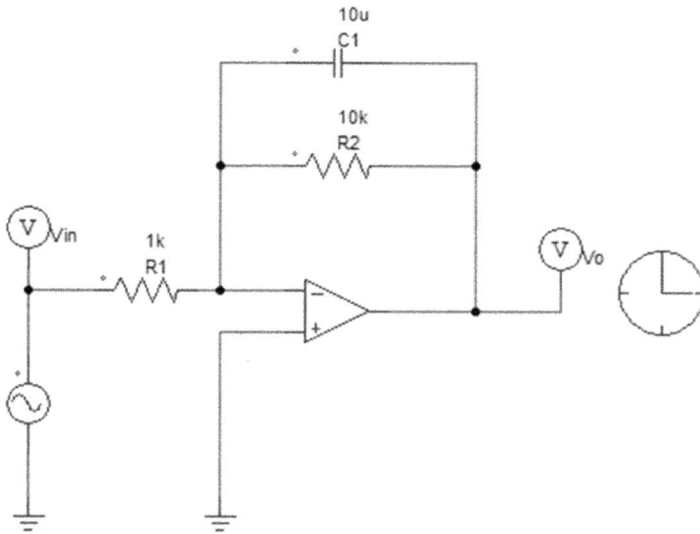

Fig. 4.165.

Settings of the blocks are shown in Figs. 4.166–4.168.

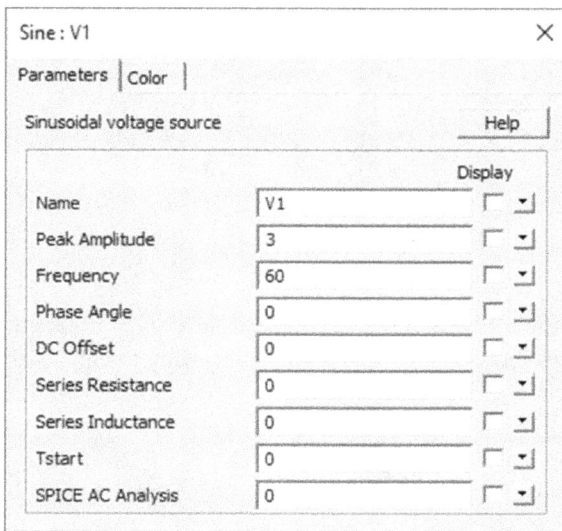

Fig. 4.166.

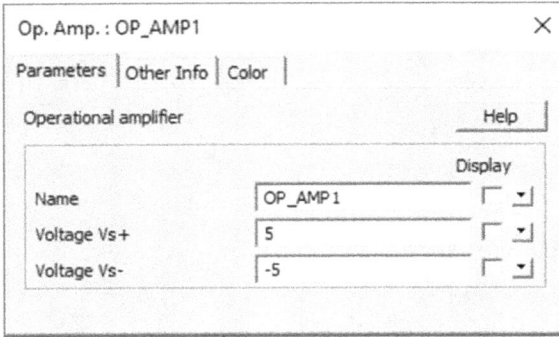

Fig. 4.167.

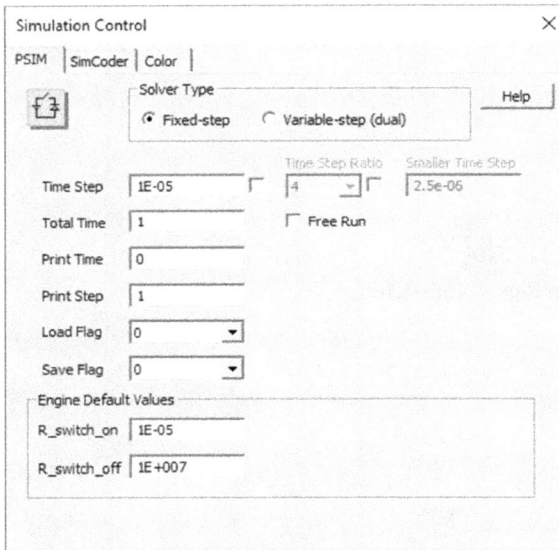

Fig. 4.168.

Run the simulation. The result shown in Fig. 4.169 is obtained. Add two cursors in order to measure the time difference between the two waves. According to Fig. 4.170, the time difference between the two waves is about 4.18 ms.

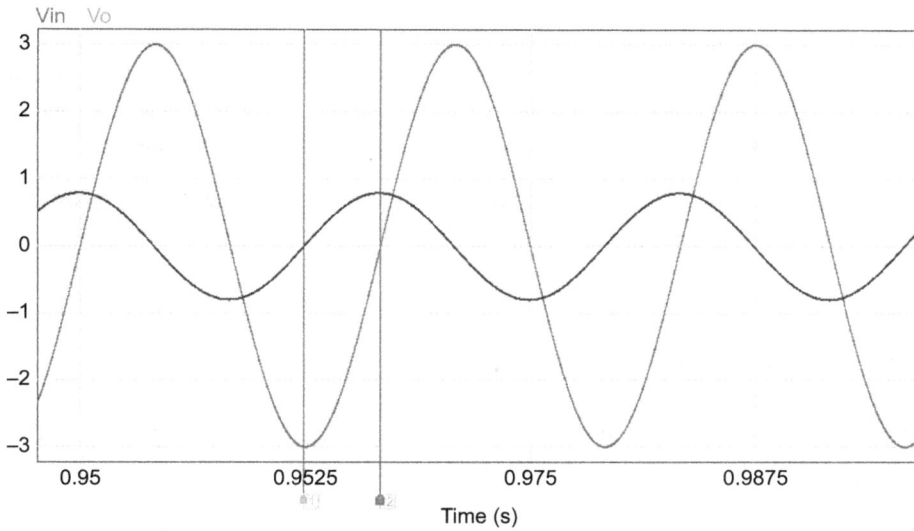

Fig. 4.169.

	X1	X2	Δ
Time	9.62446e-01	9.66629e-01	4.18351e-03
Vin	-2.99937e+00	-4.23492e-02	2.95702e+00
Vo	4.76252e-03	7.95373e-01	7.90611e-01

Measure

Fig. 4.170.

Now we can measure the phase difference between the two waves:

$$\frac{\Delta T}{\frac{T}{2}} \times 180^\circ = \frac{4.1835m}{\frac{1}{2} \times \frac{1}{60}} \times 180^\circ = 90.3636^\circ$$

Let's use hand calculations in order to check the result. For the schematic of Fig. 4.165, the transfer function equals to $H(s) = \frac{V_O(s)}{V_I(s)} = -\frac{Z_2(s)}{Z_1(s)}$, where $Z_1(s) = R_1$ and

$$Z_2(s)=\frac{\frac{1}{C_1 s}.R_2}{\frac{1}{C_1 s}+R_2}=\frac{R_2}{R_2 C_1 s+1}$$. So, $H(s)=-\frac{\frac{R_2}{R_1}}{R_2 C_1 s+1}=-\frac{100}{s+10}$. At $s=2\pi\times60i=120\pi i$, the value of

transfer function is $\frac{100}{120\pi i+10}=0.2652e^{i1.5973}$. Note that 1.5973 Rad = 91.5186°.

Obtained values are quite close. Since the cursors did not place exactly at starting points (zero) of the waves, the phase difference calculated with PSIM is a little bit smaller.

4.15 IC models

Some of the most commonly used IC's of power electronics are modeled in PSIM. The model for these IC's can be found in the Elements> Other> IC Models section.

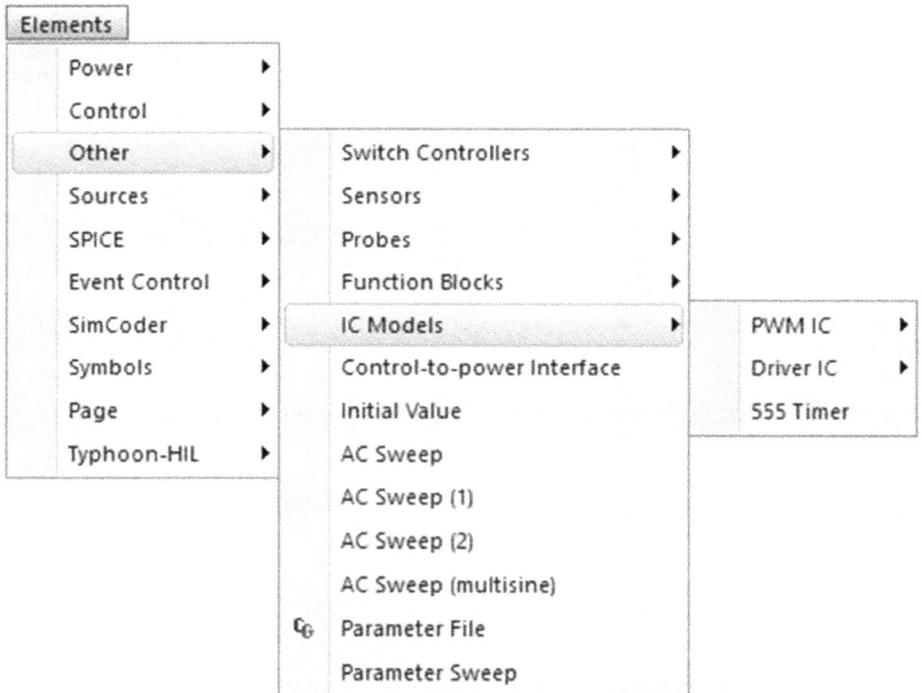

Fig. 4.171.

Click File> Search Example and then type ic in the Find box in order to see some sample simulations with theses block (Fig. 4.172).

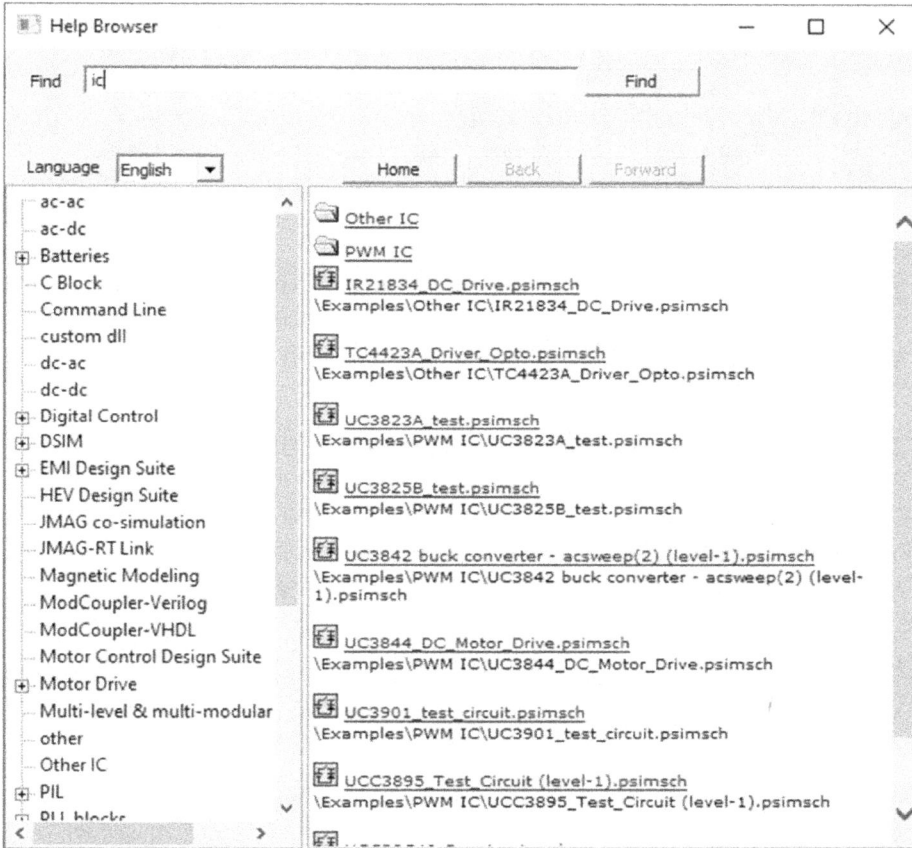

Fig. 4.172.

Chapter 5
Simulation of power electronic converters

5.1 Introduction

This chapter shows how PSIM can be used for analysis of power electronic converters. It is highly recommended to do some pencil-and-paper analysis for each circuit and compare the results with the one produced by PSIM. Try to find the source of discrepancy if hand analysis and simulation results are not the same. This helps you to learn the concepts deeply.

For instance, the voltage drop of diodes is neglected in hand analysis. So, the hand analysis result (which ignores the voltage drop of diodes) and simulation result (which considers the voltage drop of diodes) are not the same exactly.

5.2 Example 1: simulation of a simple RC circuit

In this example, we want to simulate a simple RC circuit. Consider the schematic shown in Fig. 5.1. The settings of the blocks are shown in Figs. 5.2–5.6. In this simulation, we assumed that initial voltage of capacitor is 0 V.

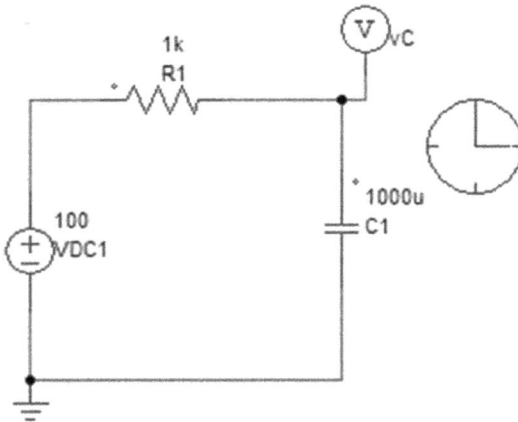

Fig. 5.1.

https://doi.org/10.1515/9783110740653-005

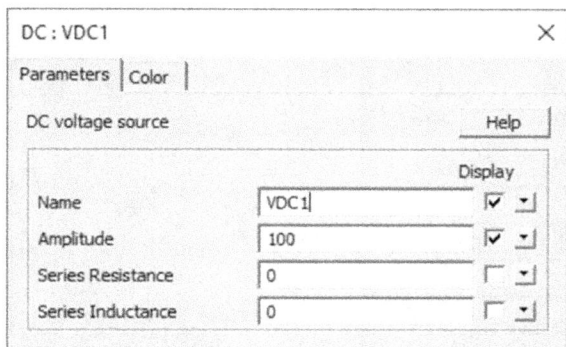

Fig. 5.2.

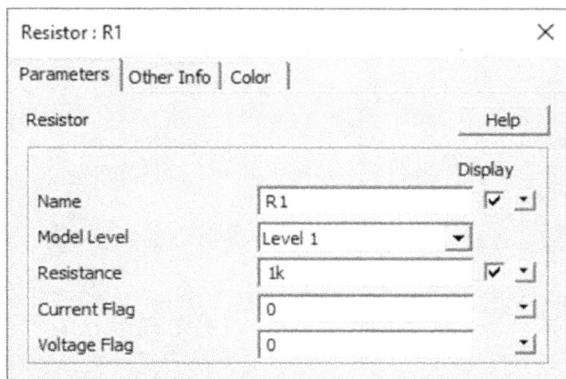

Fig. 5.3.

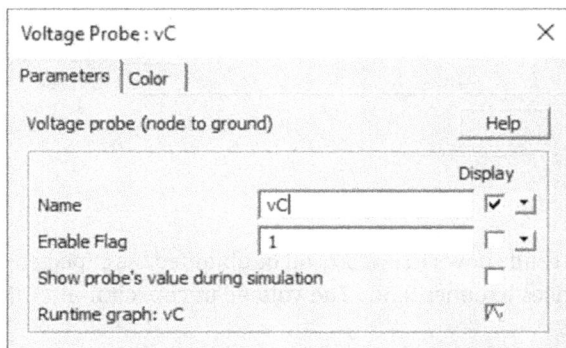

Fig. 5.4.

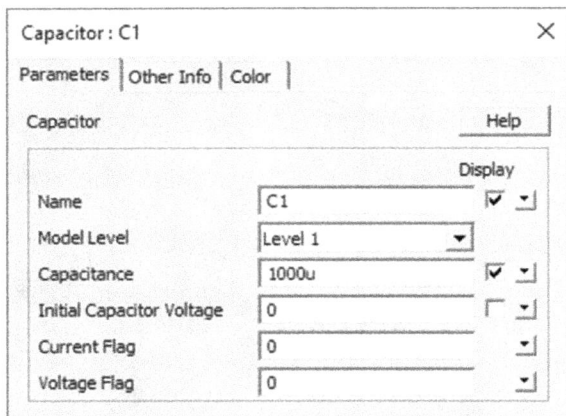

Fig. 5.5.

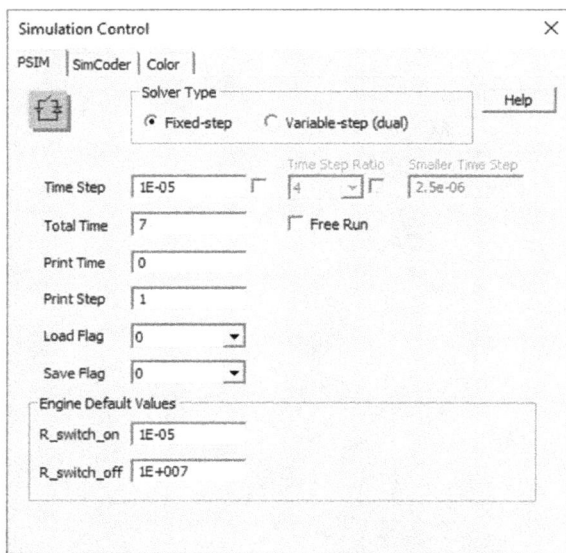

Fig. 5.6.

After running the simulation, the result shown in Fig. 5.7 will be obtained. As expected, the voltage across the capacitor rises exponentially. The voltage of capacitor at $t = 0$ equals to zero.

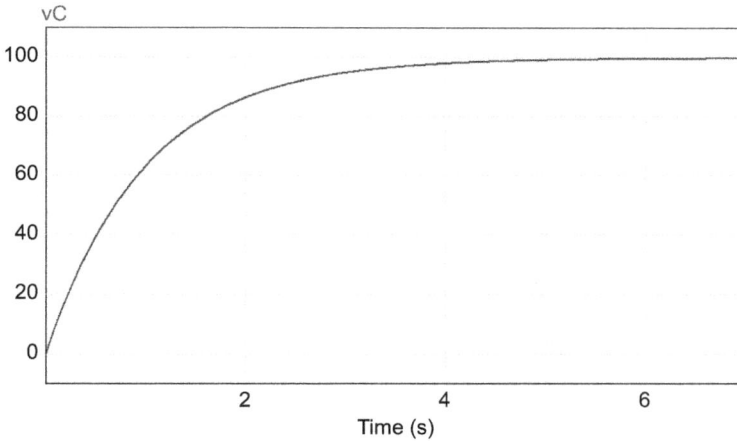

Fig. 5.7.

Now we want to simulate the circuit with nonzero initial conditions. Assume that the initial capacitor voltage is 10 V. In order to set the initial capacitor voltage, double click the capacitor and enter the desired initial voltage in the "Initial Capacitor Voltage" box (Fig. 5.8). Note that one of the capacitor terminals has a dot behind it. The value entered to the "Initial Capacitor Voltage" shows the $V_{terminal\,with\,dot} - V_{terminal\,without\,dot}$. If you enter 10 in the "Initial Capacitor Voltage" box, it means that $V_{terminal\,with\,dot} - V_{terminal\,without\,dot} = 10\ V$. Similar to the capacitor block, the PSIM's inductor block has a small dot behind one of the terminals. The current that enters the dot is assumed to be positive.

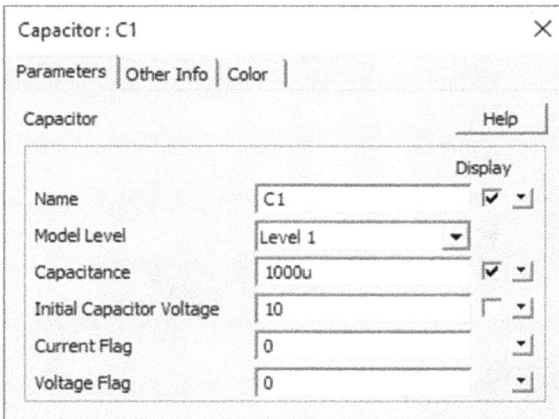

Fig. 5.8.

After running the simulation, the result shown in Fig. 5.9 will be obtained. The voltage waveform starts from 10 V.

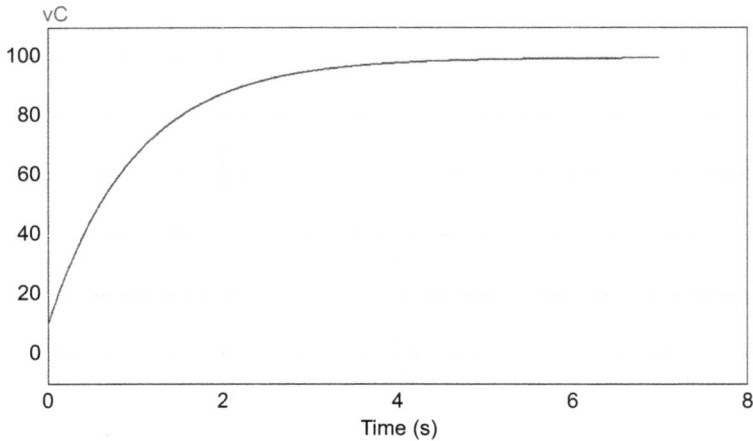

Fig. 5.9.

5.3 Example 2: effect of capacitor value on output waveform

Assume that you want to compare the output voltage of circuit shown in Fig. 5.1 for $C = 1,000\ \mu F$ and $C = 2,000\ \mu F$. If we draw the output voltages for these two values on the same axis, then comparison can be done easily.

In order to do this, double click the capacitor and enter 1,000 u in the capacitance box and simulate the circuit. Then click the File>Save as . . . (Fig. 5.10) and save the file with desired name (we used the C_1000_uf for File name) and "Save as type: Simview files" (Fig. 5.11).

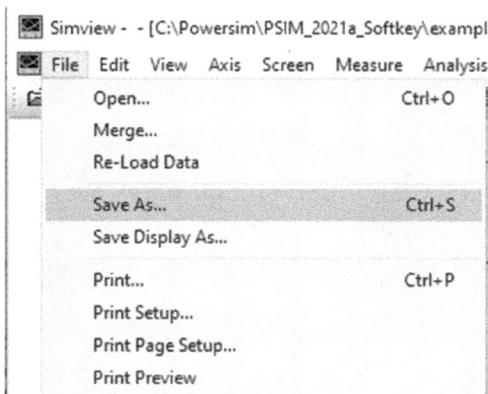

Fig. 5.10.

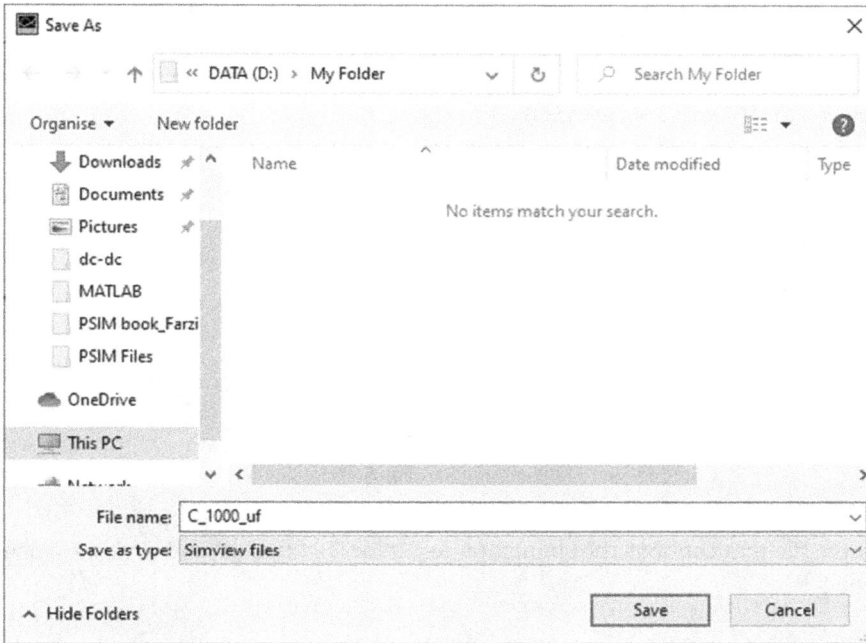

Fig. 5.11.

Now double click the capacitor and enter 2,000 u in the capacitance box. Simulate the circuit. The simulation result for C = 2,000 μF is shown in Fig. 5.12.

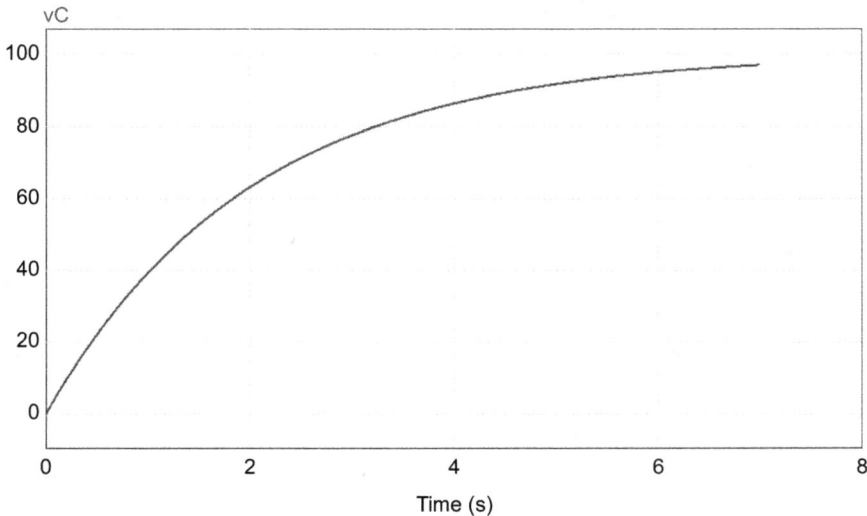

Fig. 5.12.

Click the File>Merge (Fig. 5.13).

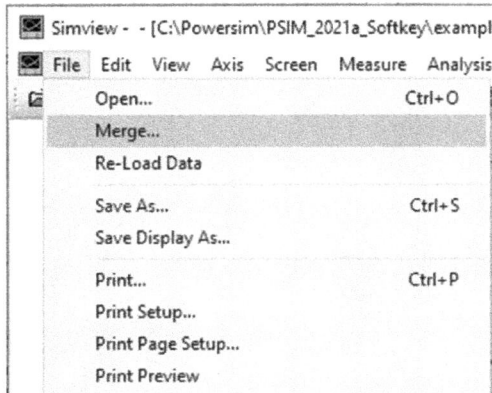

Fig. 5.13.

Select the file that contains the simulation result for C = 1,000 μF and click the Open button.

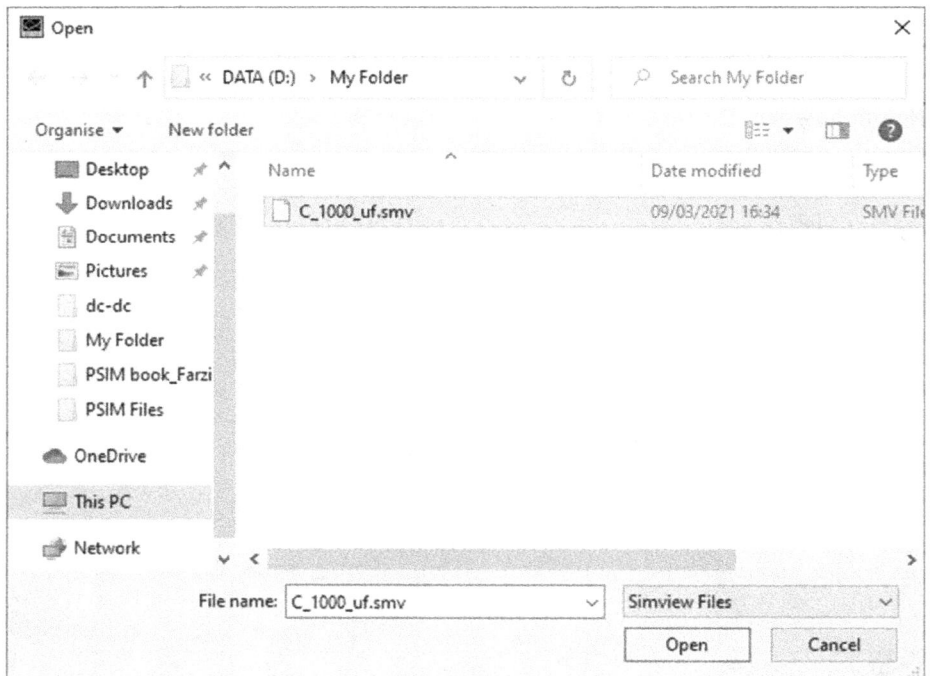

Fig. 5.14.

After clicking the Open button, the Properties window will appear (Fig. 5.15). Select the two analysis result from the left list (Variables available) and click the Add button in order to add the analysis results to the right list (Variables for display) and click the OK button.

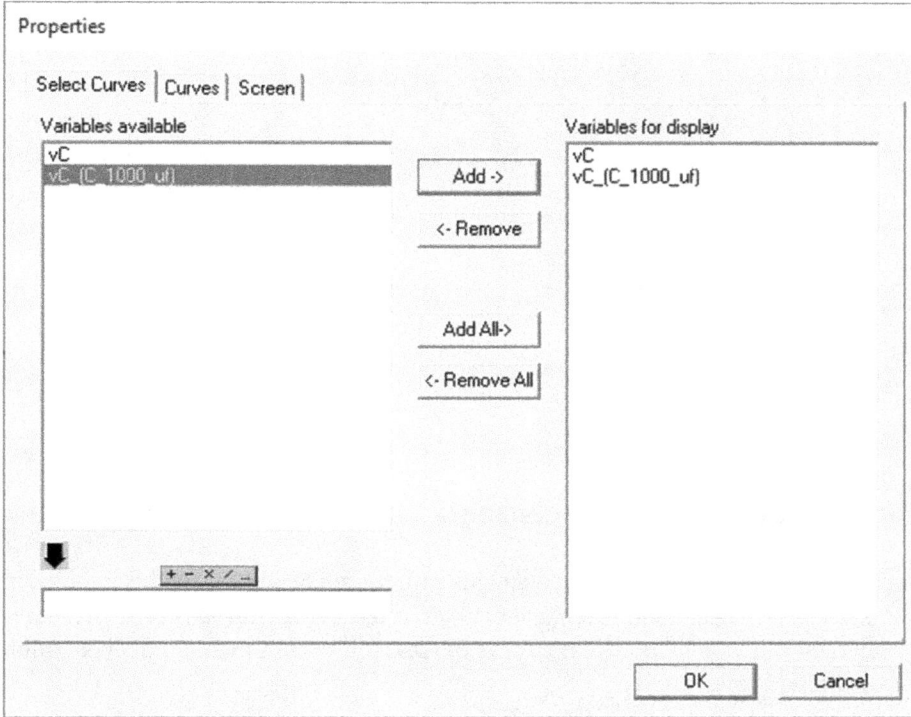

Fig. 5.15.

The two analysis results will be shown on the same axis. As expected, the voltage waveform for $C = 1,000\ \mu F$ reaches the steady state faster than the voltage waveform for $C = 2,000\ \mu F$.

vC   vC_(C_1000_uf)

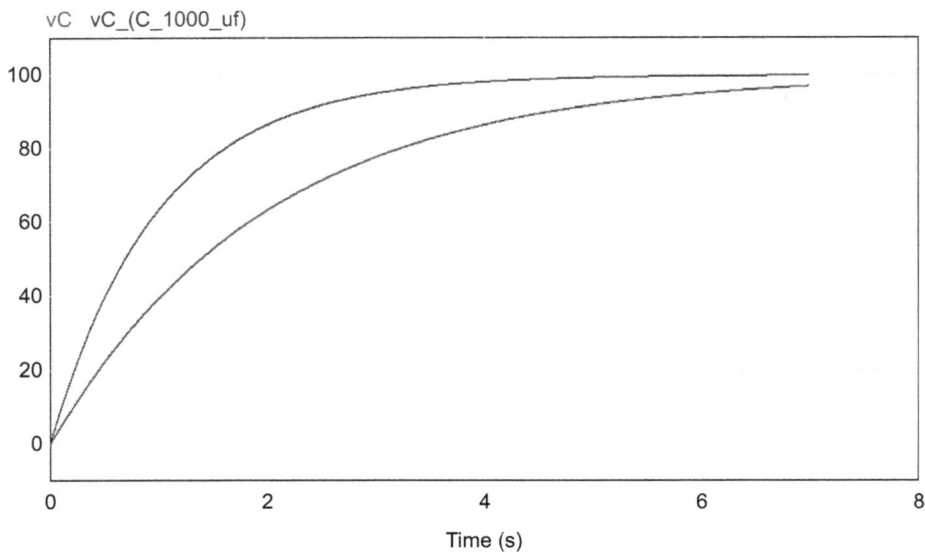

Fig. 5.16.

5.4 Example 3: nonlinear elements

You can model different nonlinear elements with the aid of nonlinear blocks shown in Fig. 5.17. For instance, nonlinear element $v = f(i)$ is used when the voltage of element is a nonlinear function of its current. These blocks require both the nonlinear equation and its derivative.

Elements		
Power ▶	RLC Branches ▶	Resistor R
Control ▶	Switches ▶	Inductor I
Other ▶	Transformers ▶	Capacitor C
Sources ▶	Magnetic Elements ▶	Capacitor (electrolytic)
SPICE ▶	Other ▶	RLC
Event Control ▶	Motor Drive Module ▶	Rheostat
SimCoder ▶	MagCoupler Module ▶	Saturable Inductor
Symbols ▶	MagCoupler-RT Module ▶	Saturable Inductor (2)
Page ▶	Mechanical Loads and Sensors ▶	R3
Typhoon-HIL ▶	Thermal Module ▶	L3
	Renewable Energy Module ▶	C3
	EMI Design Suite ▶	RL3
		RC3
		RLC3
		Coupled Inductor (2)
		Coupled Inductor (3)
		Coupled Inductor (4)
		Coupled Inductor (5)
		Coupled Inductor (6)
		Coupled Inductor (8)
		Coupled Inductor (9)
		Coupled Inductor (10)
		Coupled Inductor (12)
		Common-mode Choke
		3-ph Common-mode Choke
		DC Load (constant-power)
		3-ph Load (controlled)
		3-ph AC Cable
		3-ph AC Cable (1)
		Nonlinear Element v=f(i)
		Nonlinear Element i=f(v)
		Nonlinear Element v=f(i,x)
		Nonlinear Element i=f(v,x)

Fig. 5.17.

The nonlinear elements $v = f(i,x)$ or nonlinear elements $i = f(v,x)$ have an input which permits you to make more advanced nonlinear functions. This input is shown with an arrow in Fig. 5.18. The voltage (current) of the element is a function of current (voltage) of element and this input. In other words, you can make nonlinear functions which are functions of two variables.

Note that in Fig. 5.18, one of the terminals has a small dot behind it. The current that enters into the dot is assumed to be positive. The voltage is measured with respect to the terminal without dot, that is, $v_{terminal\ with\ dot} - v_{terminal\ without\ dot}$.

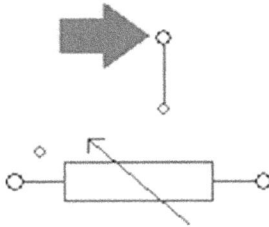

Fig. 5.18.

PSIM has a ready example for nonlinear examples. In order to see the example, click the File>Open Examples . . . and open the other folder. Then open the nonlinear diode file (Fig. 5.19).

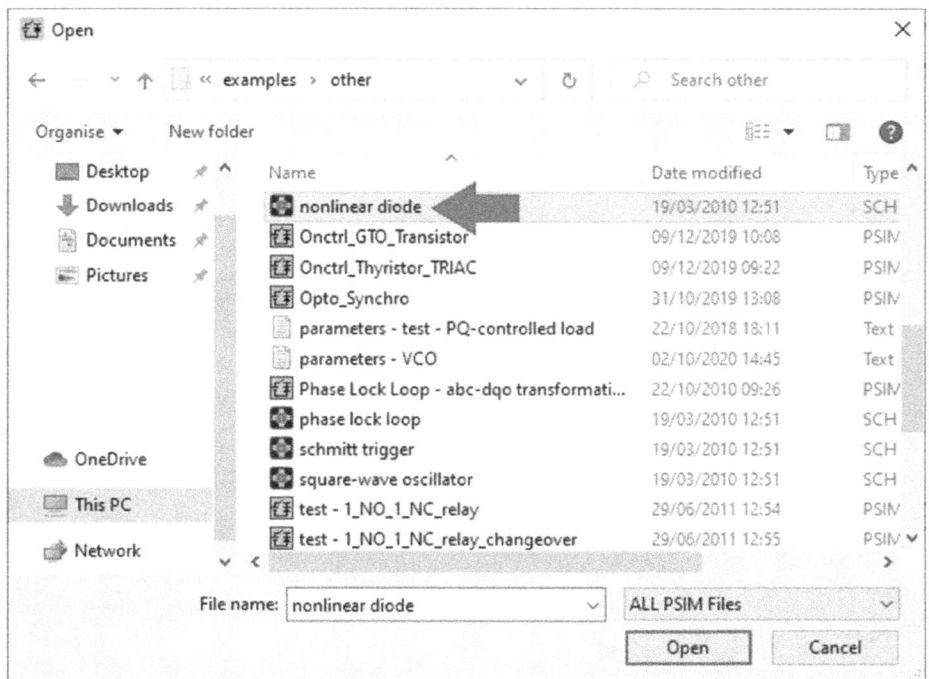

Fig. 5.19.

The simulation file of nonlinear "diode.sch" is shown in Fig. 5.20. This simulation models the I–V characteristic of a diode.

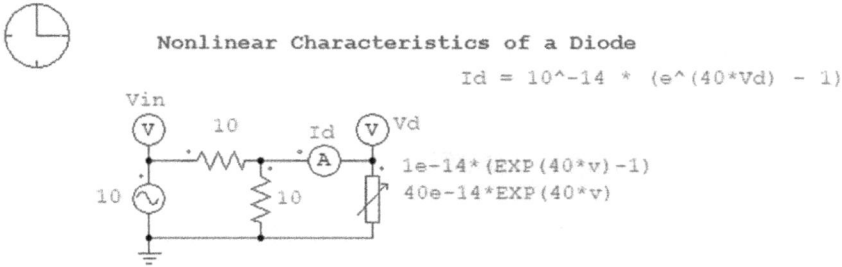

Fig. 5.20.

Run the simulation. The voltage V_d is shown in Fig. 5.21. This circuit of Fig. 5.20 acts like a half-wave rectifier (Fig. 5.22). As shown in Fig. 5.21, in negative half cycles, very small current passes form the nonlinear element (acts like a reverse-biased diode) and the voltage is divided between the resistors. In positive half cycles, the nonlinear element conducts current (acts like a forward-biased diode). The voltage drop is almost independent of current pass through the nonlinear element. It is about 0.75 V.

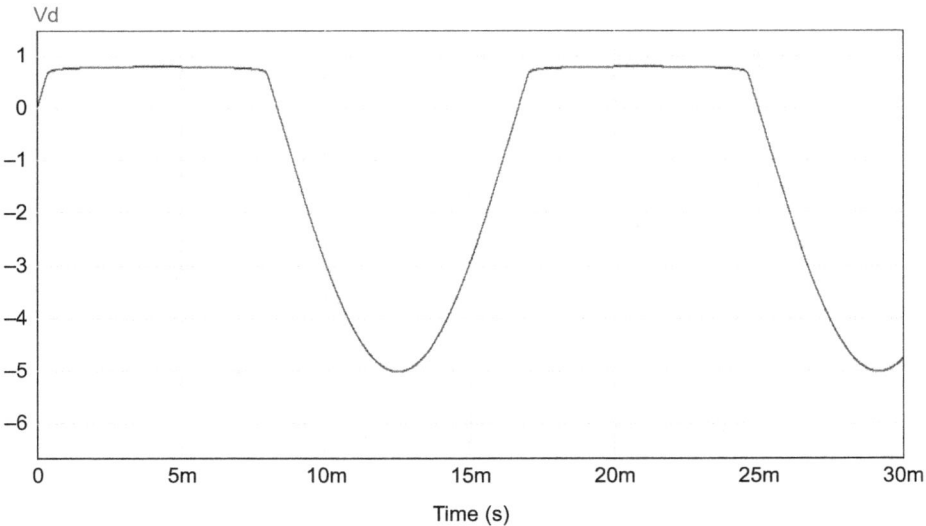

Fig. 5.21.

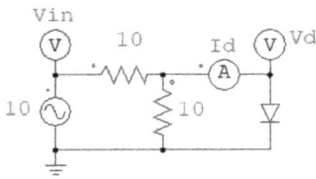

Fig. 5.22.

5.5 Example 4: switching the MOSFET

Assume the circuit shown in Fig. 5.23. In this circuit, a MOSFET acts as a switch and connects/disconnects a load current. The load is a simple resistive load. A gating block G1 (Fig. 5.24) is used to turn on/off the MOSFET.

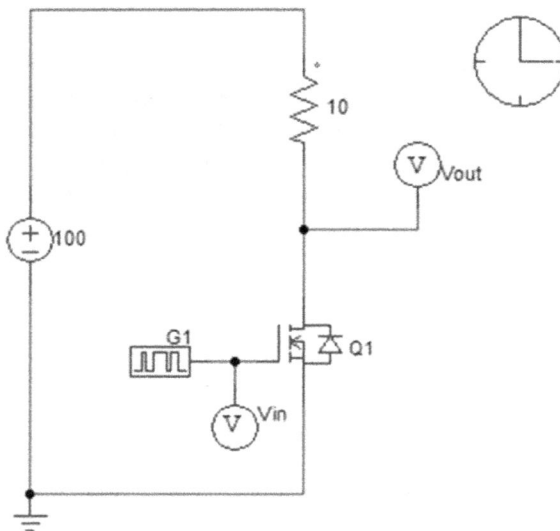

Fig. 5.23.

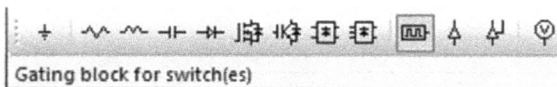

Gating block for switch(es)

Fig. 5.24.

Assume that you want to turn on/off a MOSFET similar to the pattern show in Fig. 5.25 with frequency of 1 kHz. Then the settings of the gating block must be similar to the one shown in Fig. 5.26.

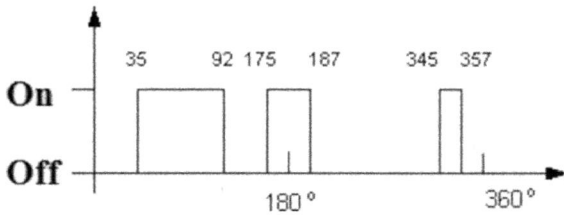

Fig. 5.25.

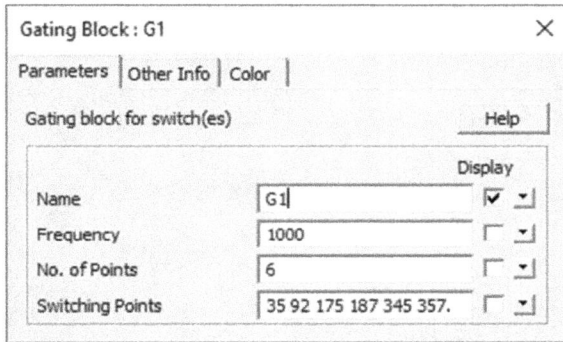

Fig. 5.26.

The settings for the elements of Fig. 5.23 are shown in Figs. 5.27–5.33. The turn on/off pattern of MOSFET in Fig. 5.23 is shown in Fig. 5.34. According to Fig. 5.34, the MOSFET is in the on state for half of the switching period and it is in the off state for the other half.

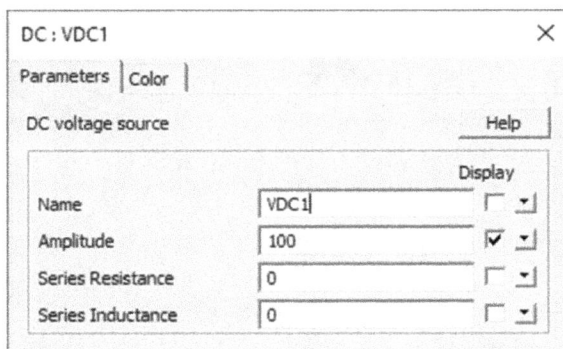

Fig. 5.27.

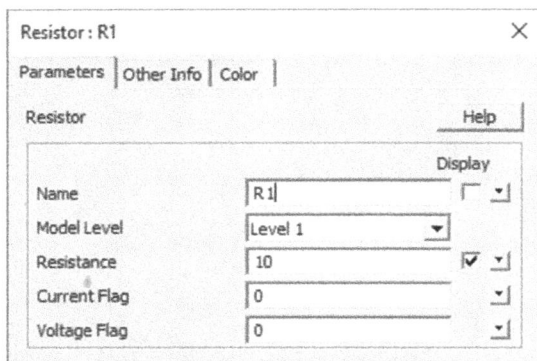

Fig. 5.28.

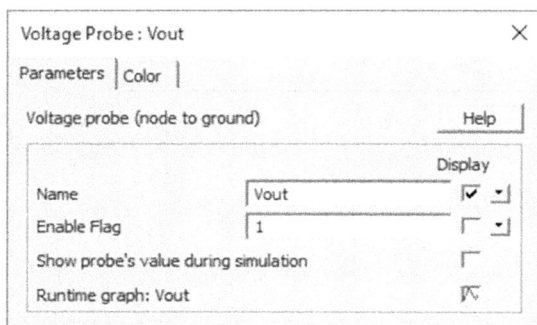

Fig. 5.29.

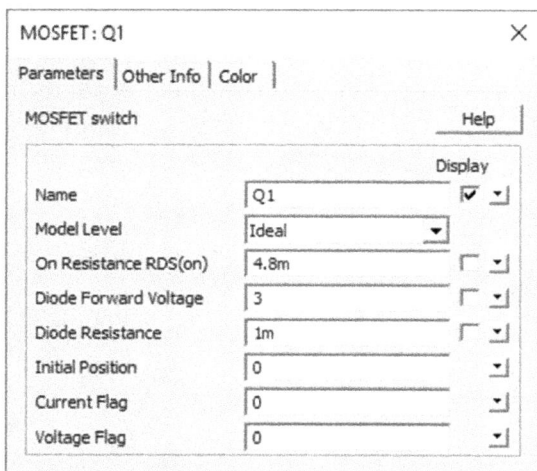

Fig. 5.30.

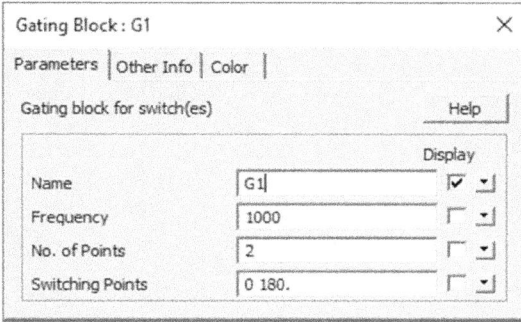

Fig. 5.31.

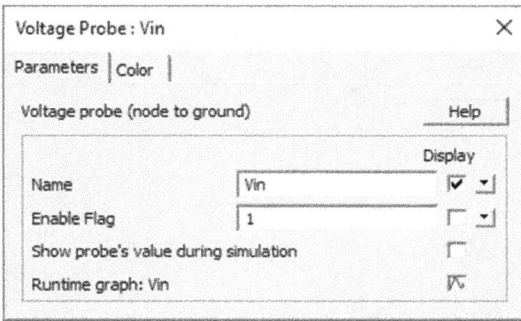

Fig. 5.32.

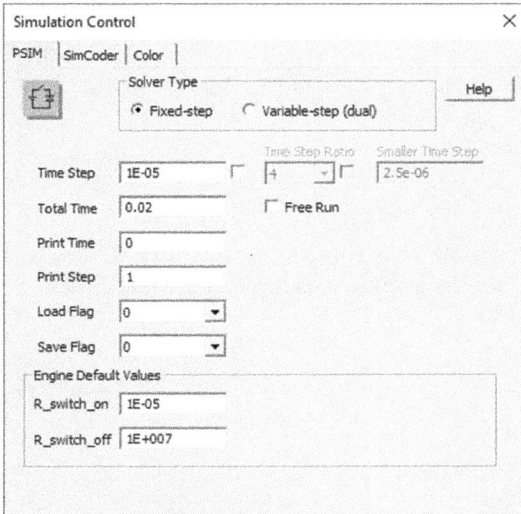

Fig. 5.33.

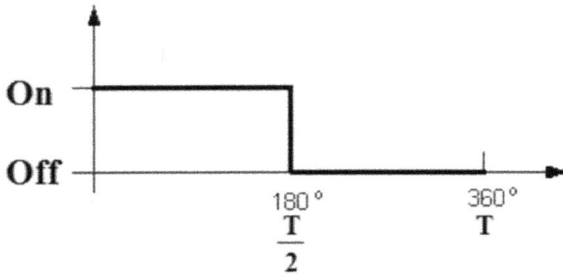

Fig. 5.34.

Run the simulation and see the output voltage (voltage of MOSFET drain) (Fig. 5.35).

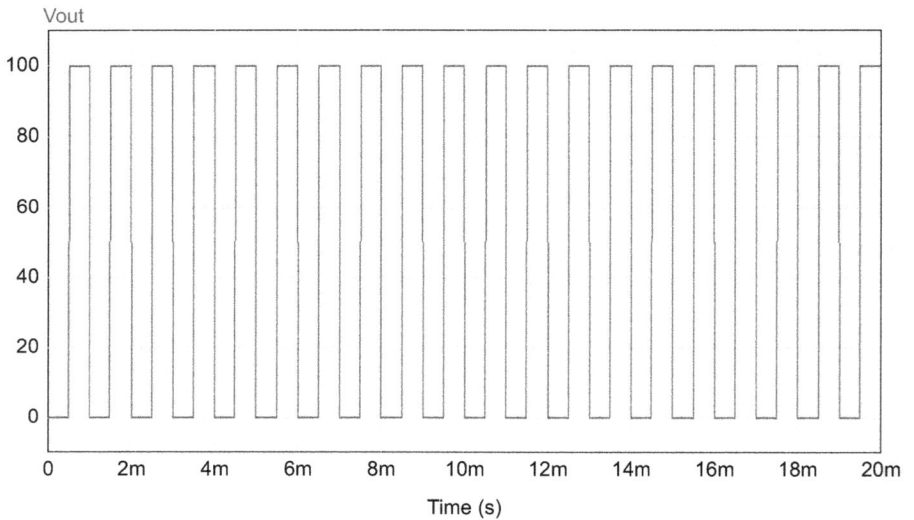

Fig. 5.35.

We want to measure the average and RMS value of this signal. Click the average and RMS value icons (Fig. 5.36) in order to see the average and RMS values (Fig. 5.37).

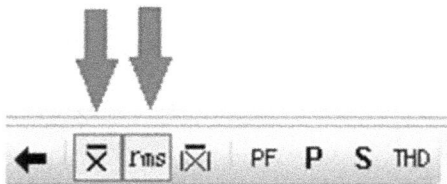

Fig. 5.36.

Measure						x
⋮		X1	X2	Δ	Average	RMS
	Time	4.00800e-03	1.60020e-02	1.19940e-02		
	Vout	2.00384e+01	8.00095e+01	5.99712e+01	5.00239e+01	7.07106e+01

Fig. 5.37.

Double click the Gating block G1 in the schematic and decrease the on time of MOSFET to $\frac{1}{4}$ of period (90°) (Fig. 5.38).

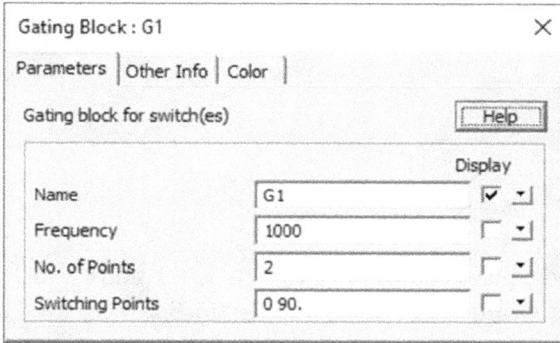

Gating Block : G1   ✕

Parameters | Other Info | Color

Gating block for switch(es)   [Help]

		Display
Name	G1	☑ ▾
Frequency	1000	☐ ▾
No. of Points	2	☐ ▾
Switching Points	0 90.	☐ ▾

Fig. 5.38.

Run the simulation. The result shown in Fig. 5.39 will be obtained.

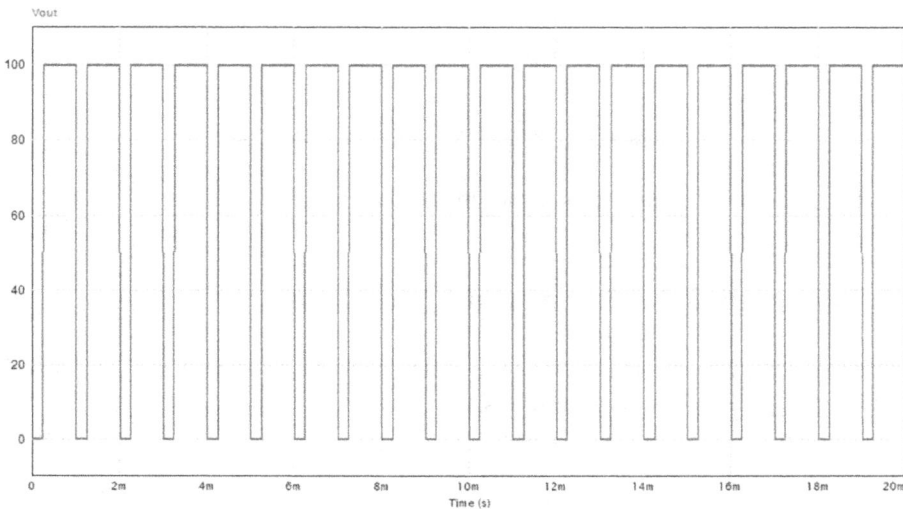

Fig. 5.39.

Use the average and RMS icons in order to calculate the average and RMS of waveform shown in Fig. 5.39. The average and RMS values are shown in Fig. 5.40.

Measure		X1	X2	Δ	Average	RMS
	Time	4.00800e-03	1.60020e-02	1.19940e-02		
	Vout	2.00384e+01	8.00095e+01	5.99712e+01	7.49911e+01	8.65904e+01

Fig. 5.40.

Let's check the software results with hand analysis. Assume the periodic pulse shown in Fig. 5.41.

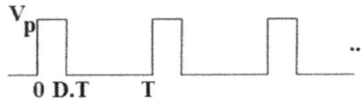

Fig. 5.41.

Average and RMS values of periodic pulse shown in Fig. 5.41 is $V_p \times D$ and $V_p \times \sqrt{D}$, respectively. In the first part of this example $V_p = 100\ V$, and $D = 0.5$. So,

$$V_{Average} = V_p \times D = 100 \times 0.5 = 50V,$$

$$V_{RMS} = V_p \times \sqrt{D} = 100 \times \sqrt{0.5} = 70.71V.$$

In the second part of this example, and 0.75. So,

$$V_{Average} = V_p \times D = 100 \times 0.75 = 75V$$

$$V_{RMS} = V_p \times \sqrt{D} = 100 \times \sqrt{0.75} = 86.60V$$

which is quite close to the results shown in Figs. 5.37 and 5.40.

5.6 Example 5: R_switch_on and R_switch_off

In this example, we want to study the "R_switch_on" and "R_switch_off" sections of Simulation Control block. Assume the simple circuit shown in Fig. 5.42.

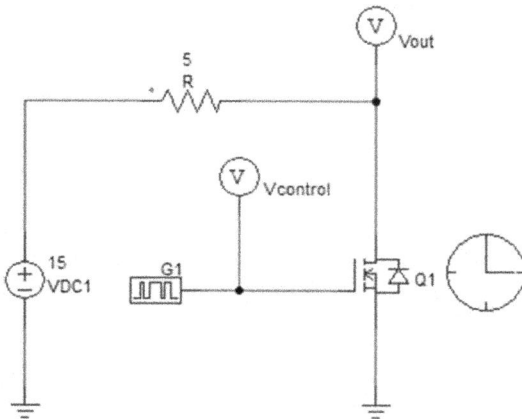

Fig. 5.42.

Settings of the blocks are shown in Figs. 5.43–5.48. Note that in this circuit we set the on-state resistance of MOSFET (On Resistance RDS(on) box in Fig. 5.47) to be 5 Ω.

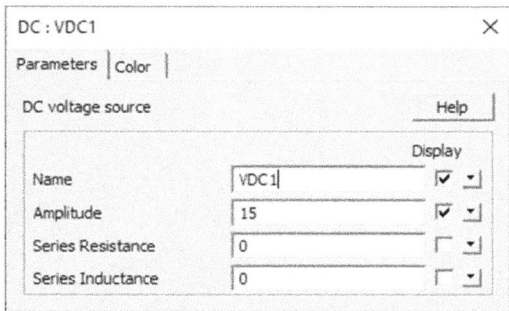

Fig. 5.43.

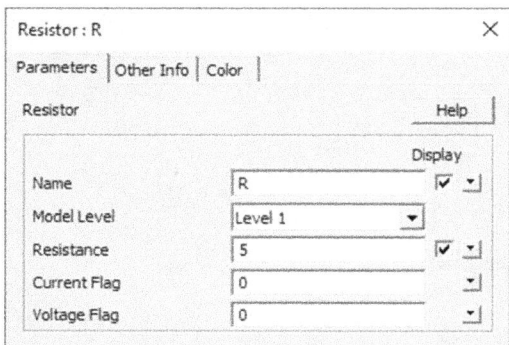

Fig. 5.44.

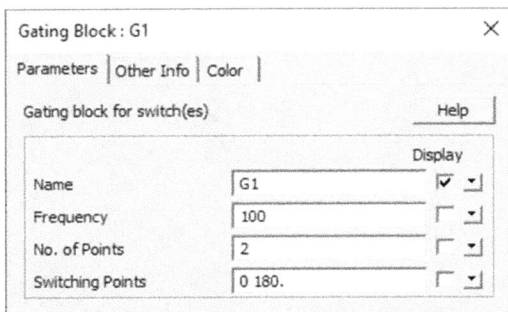

Fig. 5.45.

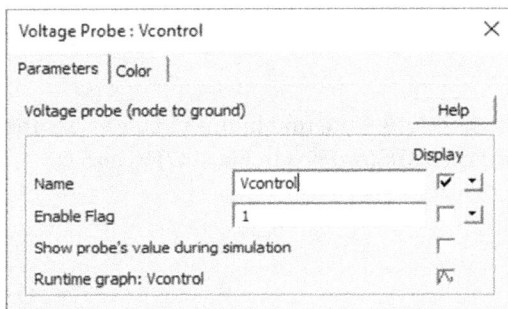

Fig. 5.46.

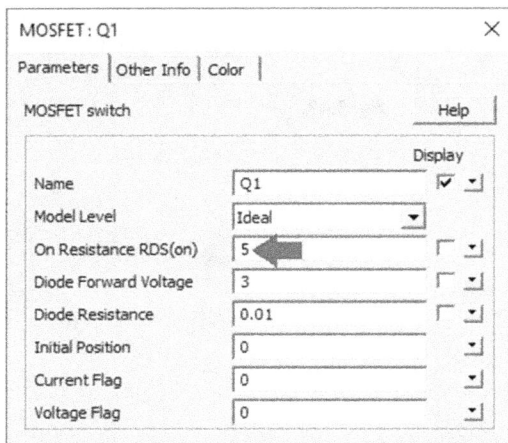

Fig. 5.47.

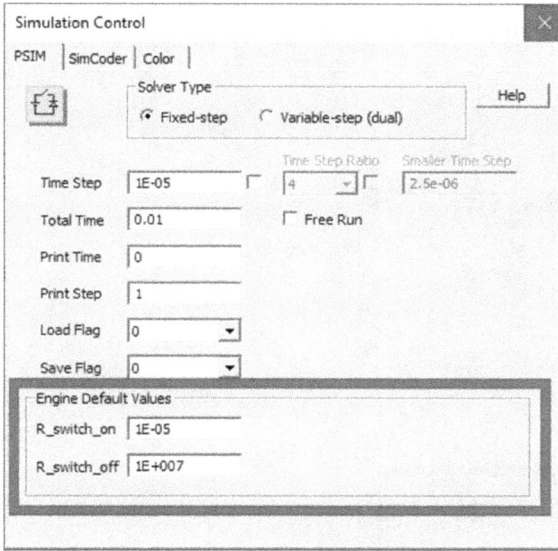

Fig. 5.48.

Run the simulation. The result is shown in Fig. 5.49. When the MOSFET is in the on state, the output voltage is $\frac{5}{5+5} \times 15 = 7.5V$ because we set the drain-source resistance of on state to be 5 Ω. The off-state resistance of drain source is very big so the output voltage is 15 V.

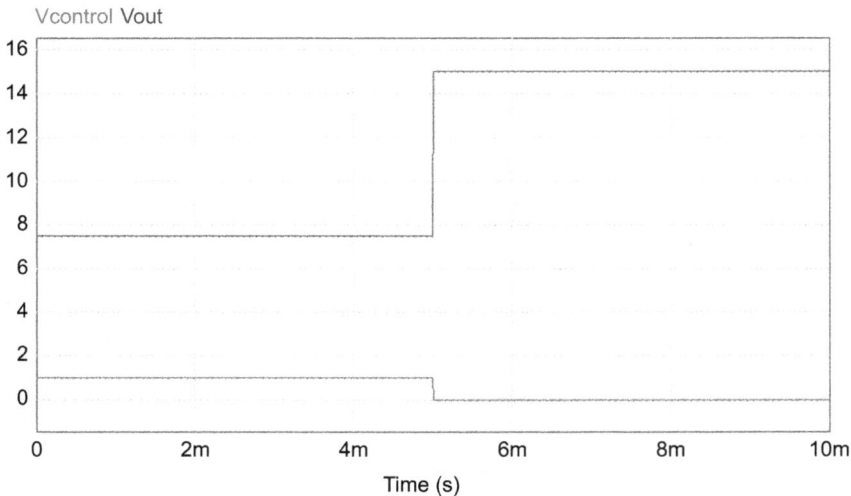

Fig. 5.49.

Now double click the Simulation Control block and set the R_switch_off to 15 Ω.

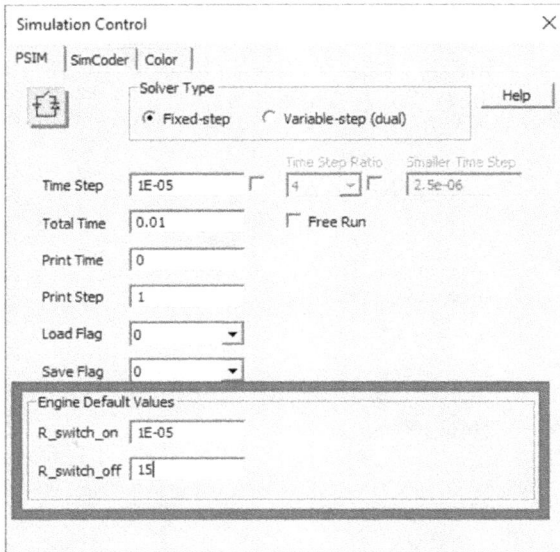

Fig. 5.50.

Run the simulation. The simulation result is shown in Fig. 5.51. Again, when the control signal equals to 1, that is, the MOSFET is on, the output voltage is 6.429 V; however, when the control signal is 0, the MOSFET is off, and the output voltage is 11.25 V.

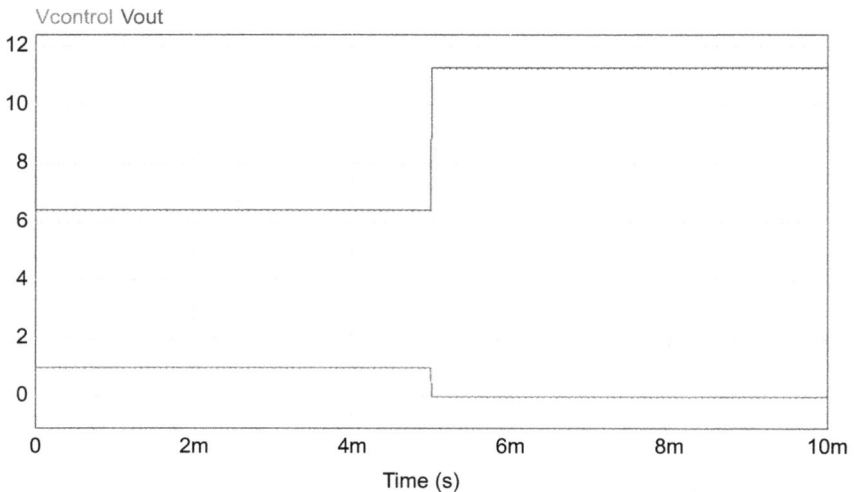

Fig. 5.51.

The equivalent circuit shown in Fig. 5.52 helps us understand the reason of changes in output voltage. Note that the value entered into the R_switch_on box has no effect on the MOSFET operation.

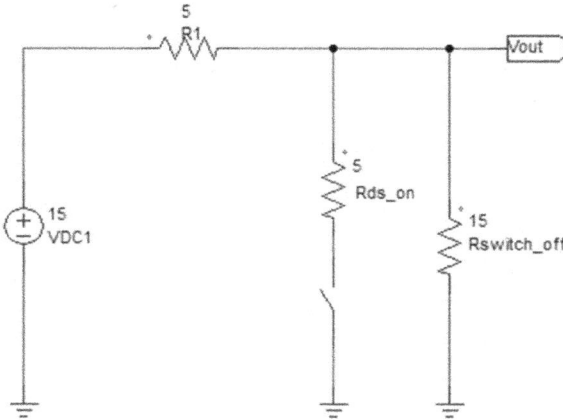

Fig. 5.52.

In Fig. 5.42, when the control signal is high, the switch is closed and the two resistors will be in parallel connection. In this case, the output voltage is

$$V_{out} = \frac{\dfrac{R_{switch\_off} \times R_{ds\_on}}{R_{switch\_off} + R_{ds\_on}}}{R_1 + \dfrac{R_{switch\_off} \times R_{ds\_on}}{R_{switch\_off} + R_{ds\_on}}} \times V_1 = \frac{3.75}{5 + 3.75} \times 15 = 6.428\,V$$

When the control signal is low, the switch is opened and the output voltage is

$$V_{out} = \frac{R_{switch\_off}}{R_{switch\_off} + R} \times V_1 = \frac{15}{15 + 5} \times 15 = \frac{225}{20} = 11.25\,V$$

Now, replace the MOSFET with an npn Transistor (Fig. 5.53). The npn Transistor block can be found in the Elements>Power>Switches menu (Fig. 5.54).

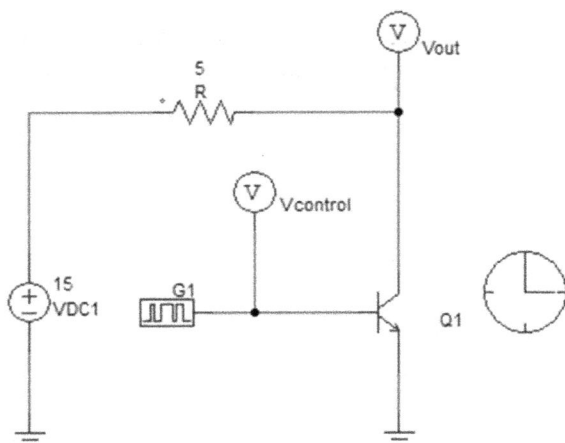

Fig. 5.53.

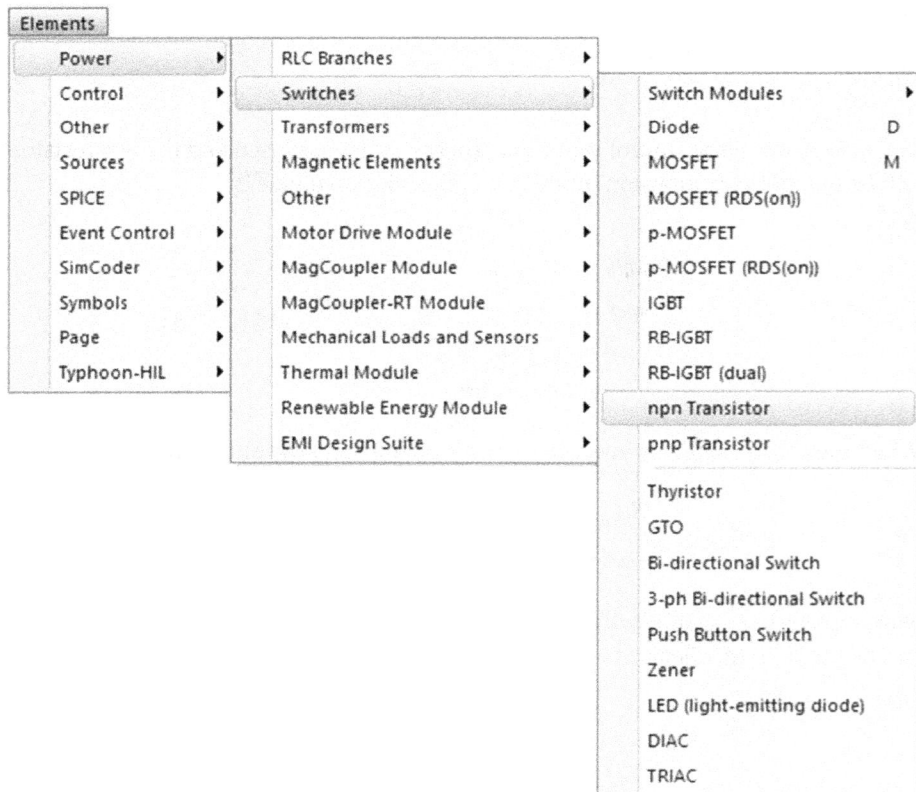

Fig. 5.54.

The settings of the npn Transistor in Fig. 5.53 are shown in Fig. 5.55. The collector emitter saturation voltage is assumed to be 0.1 V.

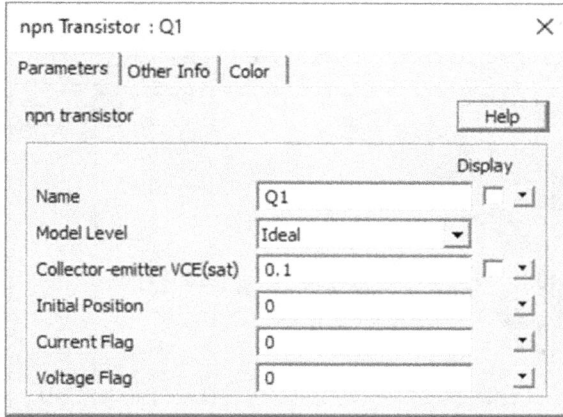

Fig. 5.55.

Now double click the Simulation Control block and set the R_switch_on and R_switch_off as shown in Fig. 5.56.

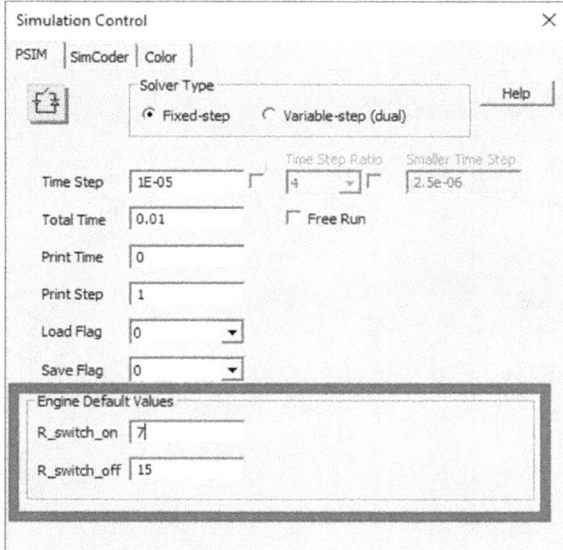

Fig. 5.56.

Run the simulation. The result shown in Fig. 5.57 will be obtained. When the control signal is high, the output voltage is 7.36 V and when the control signal is low, the output voltage is 11.25 V.

Vcontrol Vout

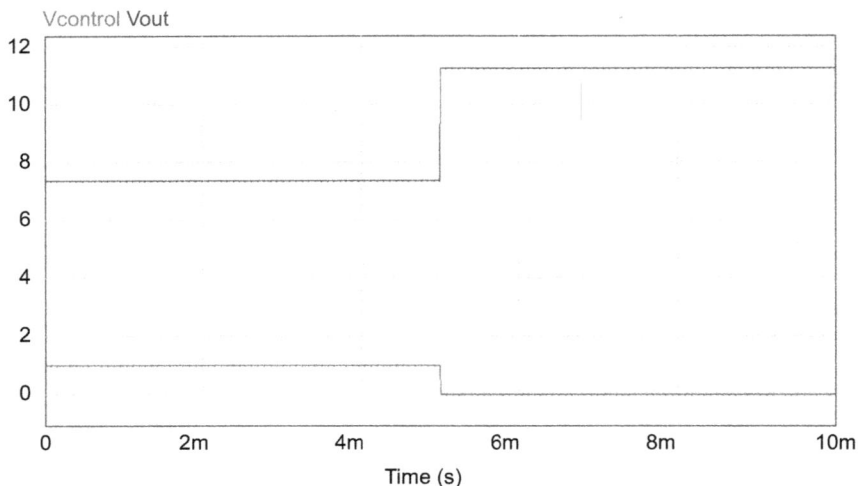

Fig. 5.57.

The circuit shown in Fig. 5.58 is the equivalent circuit of Fig. 5.53.

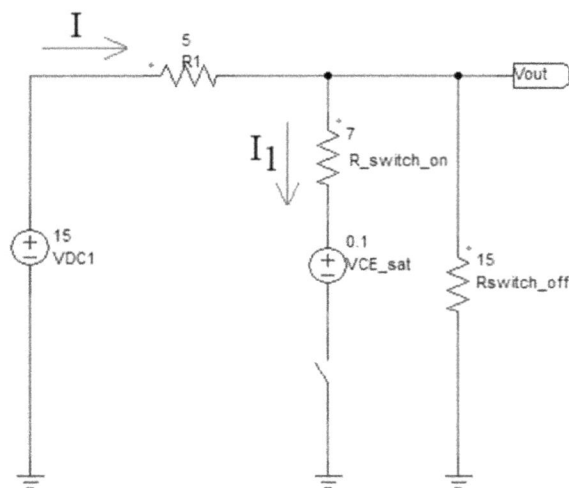

Fig. 5.58.

When the switch is opened, the output voltage is $\frac{15}{15+5}\times15=11.25\ V$. When the switch is closed, the output voltage is

$$15 = 5 \times I + 7 \times I_1 + 0.1$$

$$7 \times I_1 + 0.1 = 15 \times \left(I - I_1\right)$$

$$I = 1.528\ A \quad \text{and} \quad I_1 = 1.037\ A$$

$$V_{out} = 7 \times I_1 + 0.1 = 7.360\ V$$

Now, replace the npn Transistor with an IGBT switch (Fig. 5.59). The IGBT block can be found in Fig. 5.60 or Elements>Power>Switches menu (Fig. 5.61).

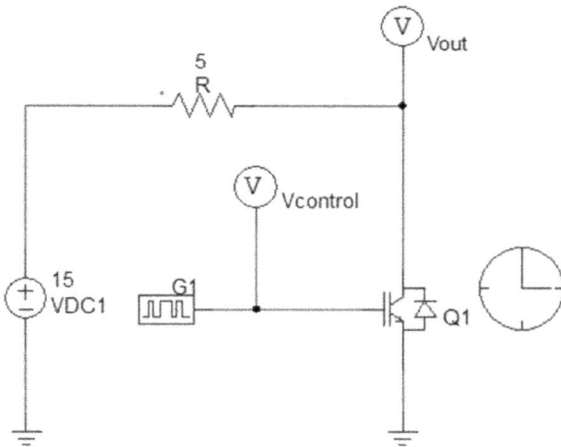

Fig. 5.59.

IGBT models

Fig. 5.60.

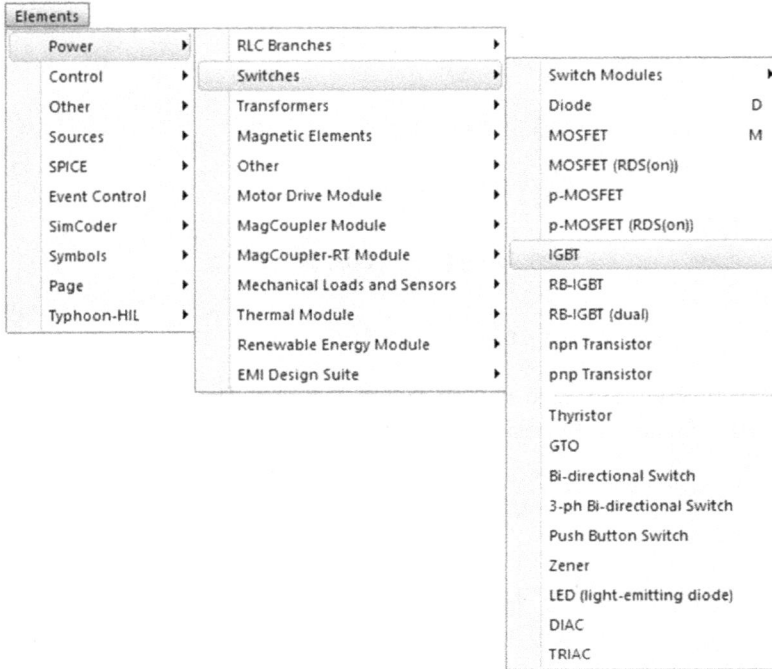

Fig. 5.61.

The settings of the IGBT block of Fig. 5.59 are shown in Fig. 5.62. The collector emitter saturation voltage is assumed to be 0.1 V and the on resistance of transistor (transistor resistance) is 1 Ω.

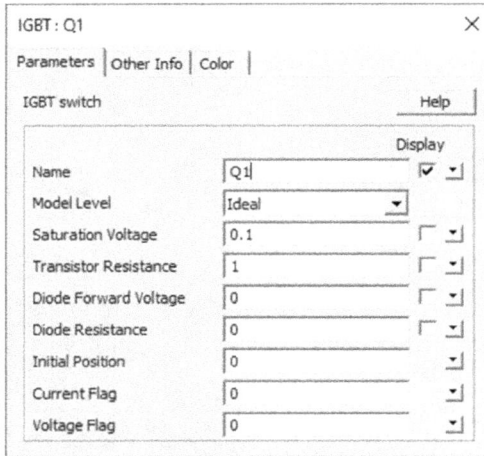

Fig. 5.62.

Run the simulation. The result shown in Fig. 5.63 will be obtained. When the control signal is high, the output voltage is 2.447 V and when the control signal is low, the output voltage is 11.25 V.

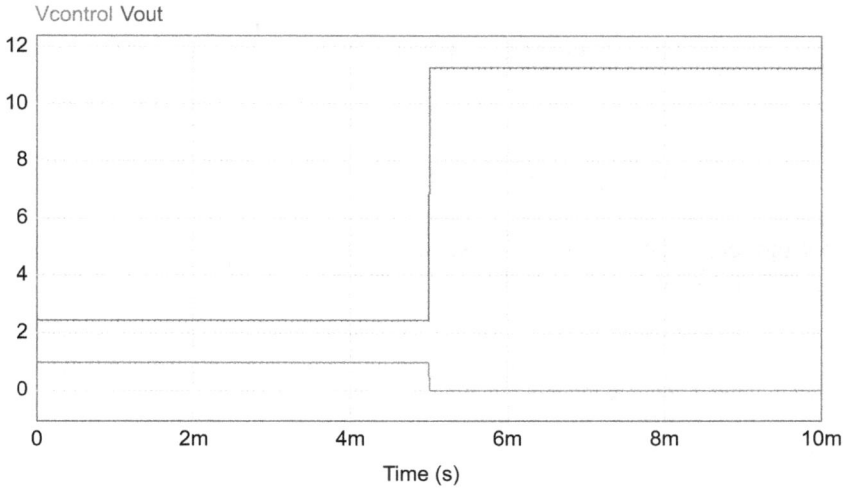

Fig. 5.63.

Equivalent circuit of Fig. 5.59 is shown in Fig. 5.64. Note that the value entered into the R_switch_on box has no effect on the IGBT operation.

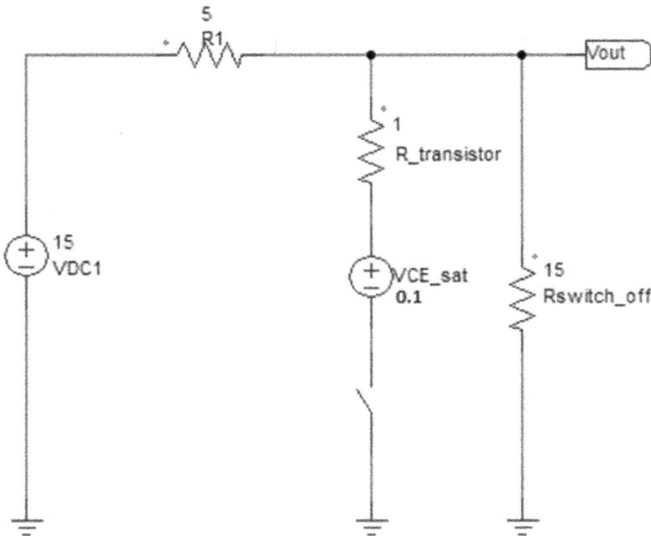

Fig. 5.64.

When the switch is opened, the output voltage is $\frac{15}{15+5}{\times}15{=}11.25\ V$. When the switch is closed, the output voltage is

$$15 = 5 \times I + 1 \times I_1 + 0.1$$

$$1 \times I_1 + 0.1 = 15 \times (I - I_1)$$

$$I = 2.510\ A \quad \text{and} \quad I_1 = 2.347\ A$$

$$V_{out} = 1 \times I_1 + 0.1 = 2.447\ V$$

The symbol and settings of Bi-directional Switch is shown in Fig. 5.65. The Bi-directional Switch is in the Elements>Power>Switches menu (Fig. 5.66).

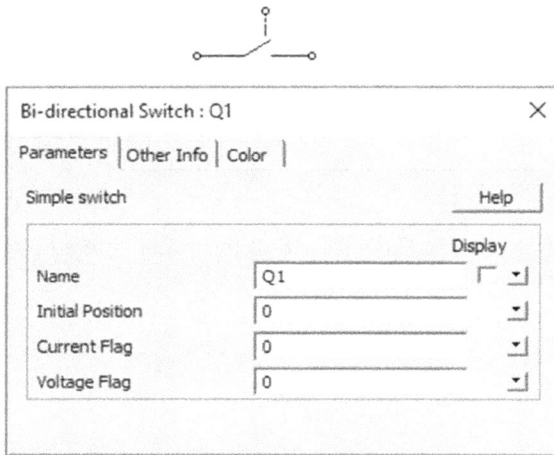

Fig. 5.65.

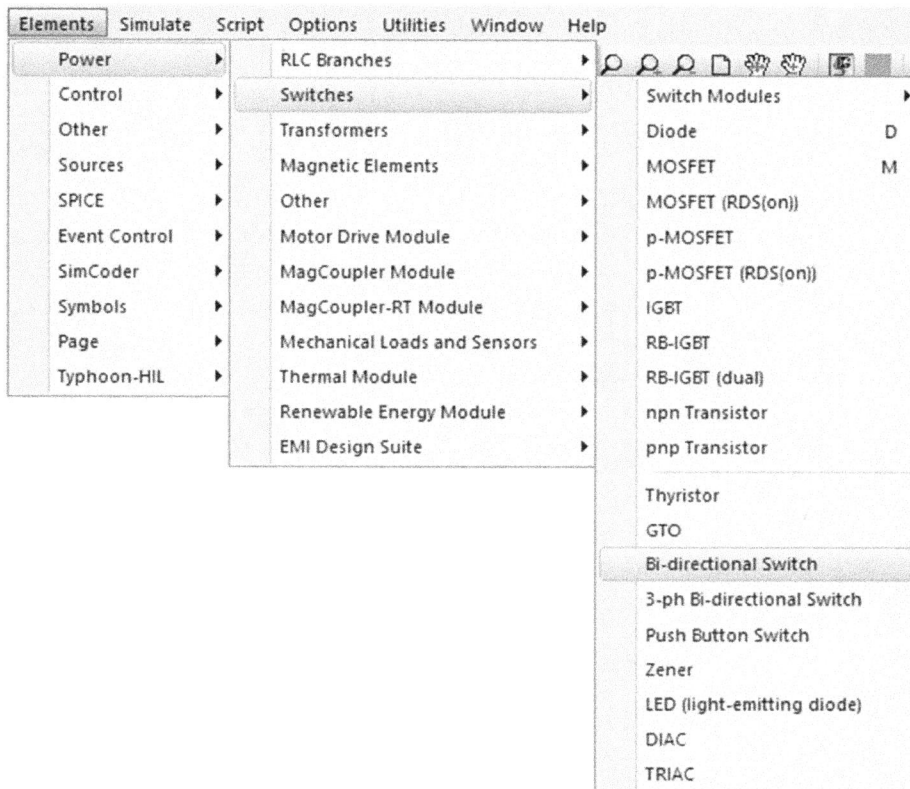

Fig. 5.66.

As shown in Fig. 5.65, the Bi-directional Switch dialog box has no box to get on-/off-state resistances. The on-state resistances of this device are set with the R_switch_on box of Simulation Control block. The off-state resistance of this switch is assumed to be infinity.

The Push Button Switch (Fig. 5.67) is similar to the Bi-directional Switch: The on-state resistances of this device are set with the R_switch_on box of Simulation Control block. The off-state resistance is assumed to be infinity.

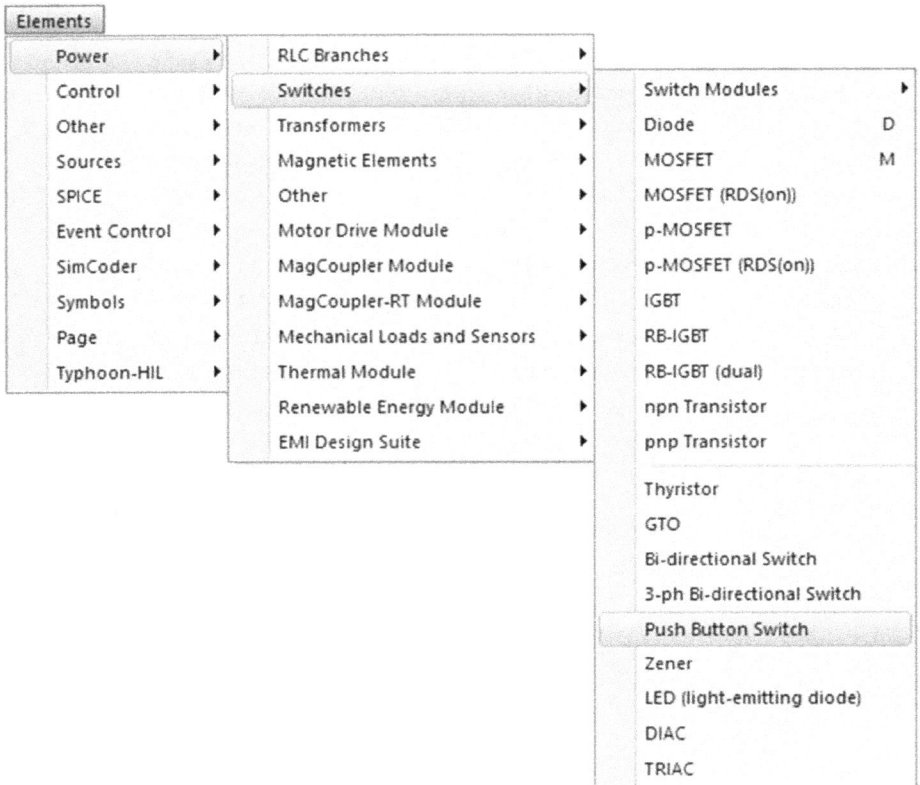

Elements						
Power ▶	RLC Branches ▶					
Control ▶	Switches ▶	Switch Modules ▶				
Other ▶	Transformers ▶	Diode D				
Sources ▶	Magnetic Elements ▶	MOSFET M				
SPICE ▶	Other ▶	MOSFET (RDS(on))				
Event Control ▶	Motor Drive Module ▶	p-MOSFET				
SimCoder ▶	MagCoupler Module ▶	p-MOSFET (RDS(on))				
Symbols ▶	MagCoupler-RT Module ▶	IGBT				
Page ▶	Mechanical Loads and Sensors ▶	RB-IGBT				
Typhoon-HIL ▶	Thermal Module ▶	RB-IGBT (dual)				
	Renewable Energy Module ▶	npn Transistor				
	EMI Design Suite ▶	pnp Transistor				
		Thyristor				
		GTO				
		Bi-directional Switch				
		3-ph Bi-directional Switch				
		Push Button Switch				
		Zener				
		LED (light-emitting diode)				
		DIAC				
		TRIAC				

Fig. 5.67.

5.7 Example 6: MOSFET conduction loss

In this example, we want to calculate the conduction losses of the MOSFET in Fig. 5.23. In order to calculate the conduction losses, we need to measure the drain-source voltage and the current of drain and multiply them together.

We can measure the drain-source voltage and drain current with the aid voltage flag and current flag, respectively. Double click the MOSFET and set the Current Flag and Voltage Flag to 1.

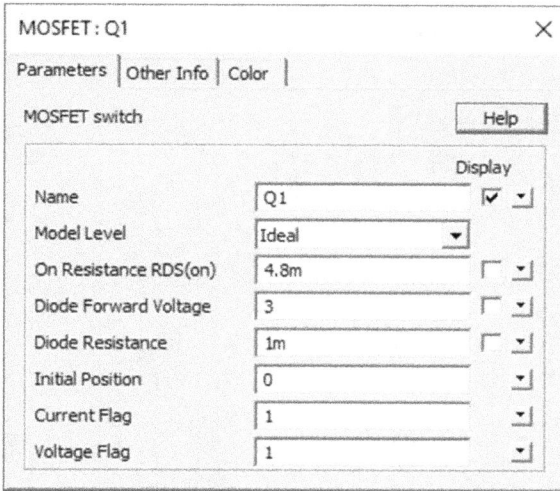

Fig. 5.68.

Now run the simulation. Add V(Q1)*I(Q1) expression to the Variables for display and click OK.

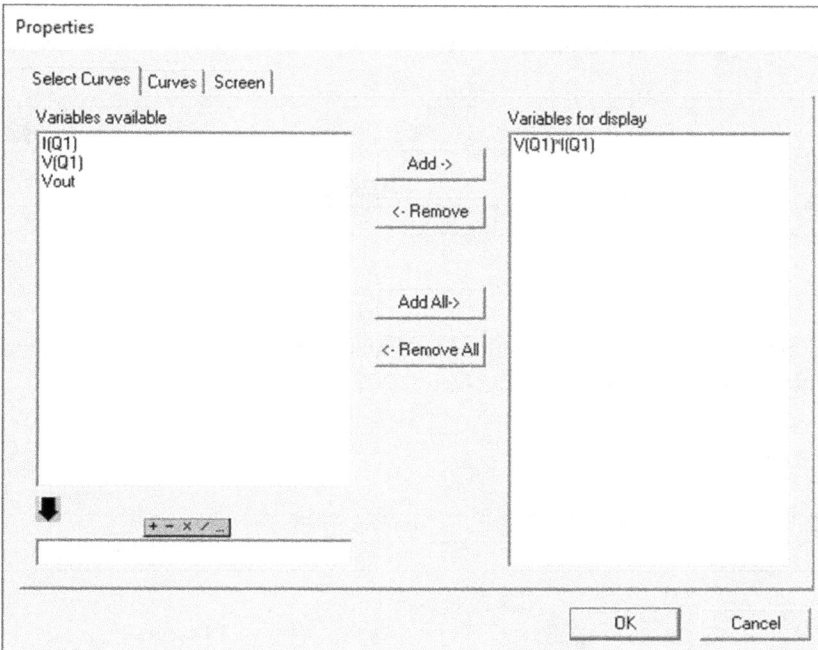

Fig. 5.69.

The dissipated power waveform appears on the screen.

V(Q1)*I(Q1)

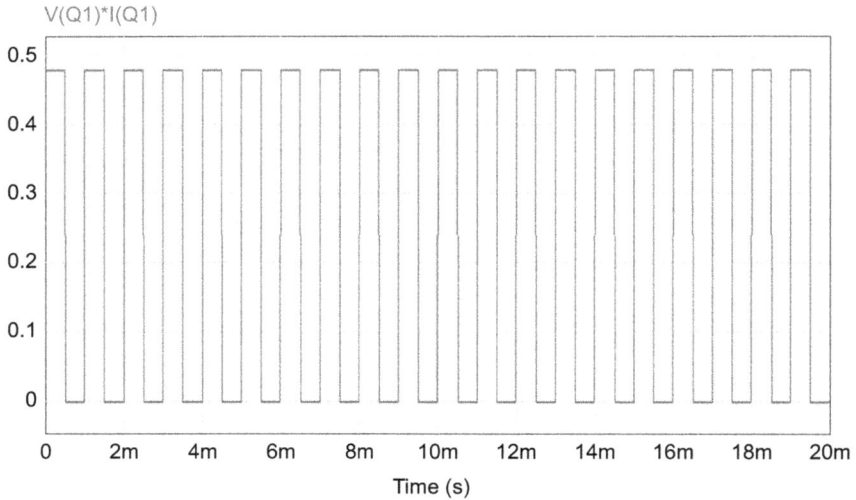

Fig. 5.70.

Use the average icon to measure the average dissipated power. According to Fig. 5.71, the average dissipated power is about 0.24 W.

Measure					
⋮	X1	X2	Δ	1/Δ	Average
Time	4.00800e-03	1.60020e-02	1.19940e-02	8.33750e+01	
V(Q1)*I(Q1)	3.83632e-01	9.59079e-02	-2.87724e-01		2.39770e-01

Fig. 5.71.

Let's check this result. Rerun the simulation and draw the drain current. Measure the RMS of drain current. According to Fig. 5.72, it is 7.07 A.

Measure					
⋮	X1	X2	Δ	1/Δ	RMS
Time	4.00800e-03	1.60020e-02	1.19940e-02	8.33750e+01	
I(Q1)	7.99616e+00	1.99904e+00	-5.99712e+00		7.06768e+00

Fig. 5.72.

The dissipated power can be calculated using the $P_{conduction\ loss} = R_{ds} \times I_{D,RMS}^2$. According to Fig. 5.73, this formula leads to the same value as the one shown in Fig. 5.71.

```
Command Window                ⊙

  >> 4.8e-3* (7.0676^2)


  ans =


      0.2398

fx >>
```

Fig. 5.73.

5.8 Example 7: calculation of total harmonic distortion

In this example we show how total harmonic distortion (THD) can be calculated with the THD button in Simview environment.

Consider the simple circuit shown in Fig. 5.74. The settings of elements are shown in Figs. 5.75–5.79.

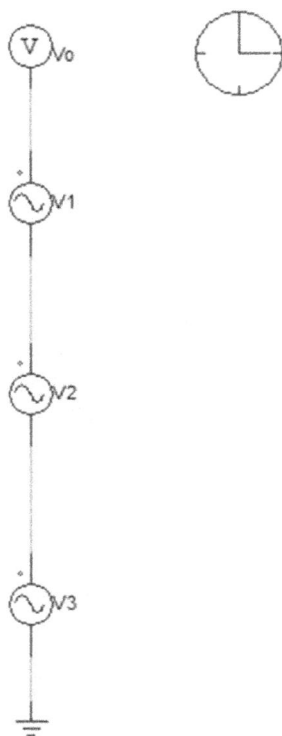

Fig. 5.74.

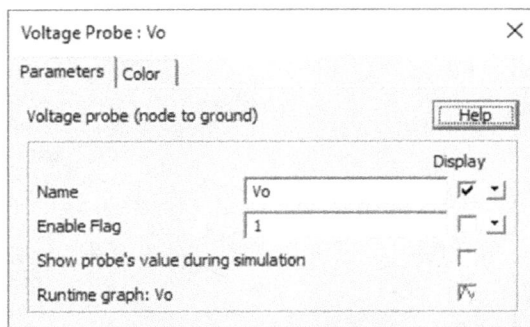

Fig. 5.75.

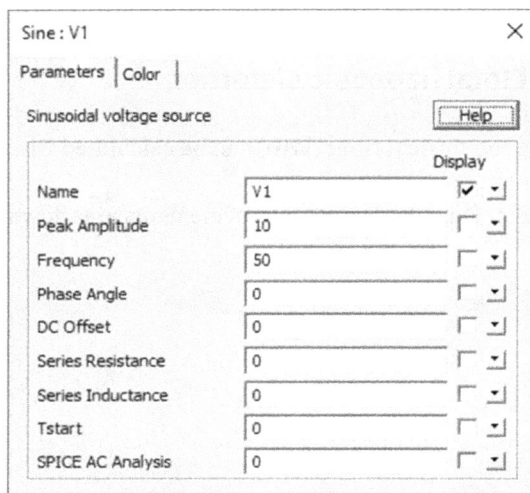

Fig. 5.76.

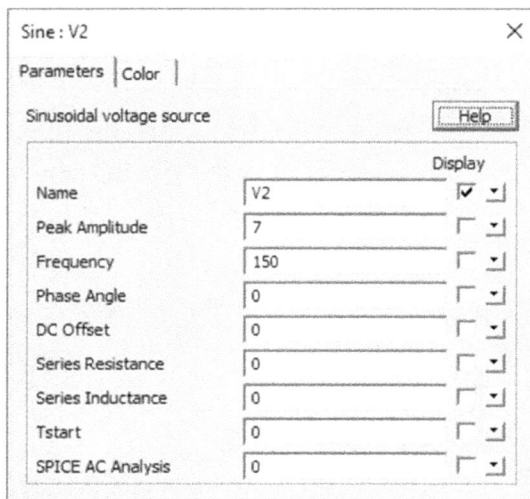

Fig. 5.77.

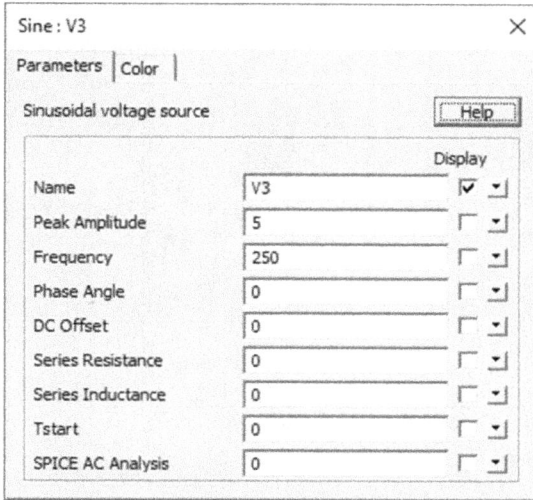

Fig. 5.78.

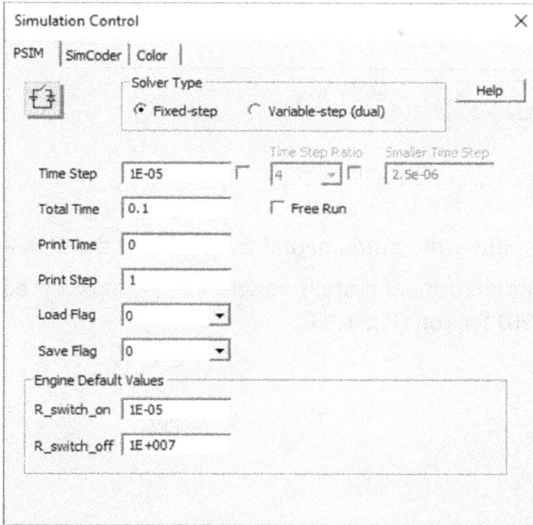

Fig. 5.79.

Run the simulation and see the output voltage V_o. The period of this signal is 20 ms.

Vo

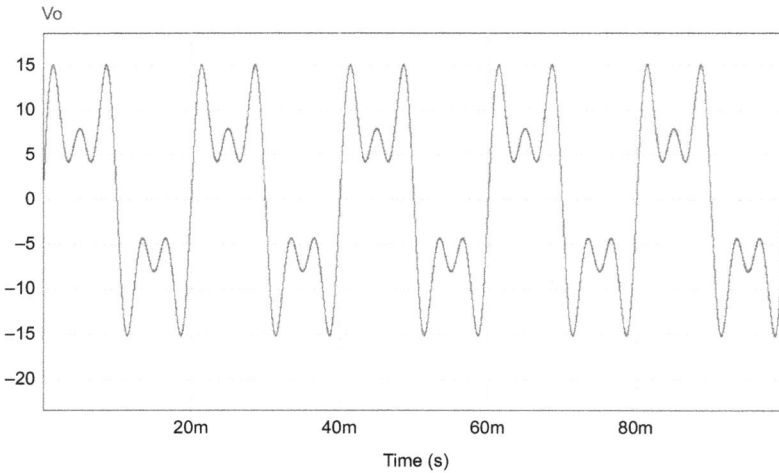

Fig. 5.80.

Click the THD icon.

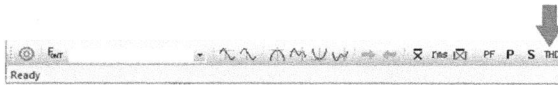

Fig. 5.81.

The THD window appears (Fig. 5.82). Enter the fundamental frequency of the signal and click the OK button. In this example, fundamental frequency is $\frac{1}{20m}$=50 Hz. The software calculates and shows the THD for you (Fig. 5.83).

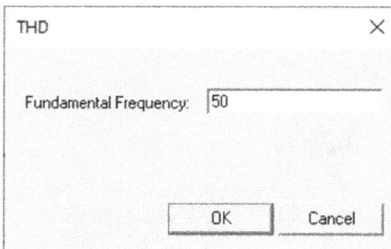

Fig. 5.82.

Measure					
	X1	X2	Δ	1/Δ	THD
Time	6.00020e-02	8.00020e-02	2.00000e-02	5.00000e+01	freq=50
Vo	3.51850e-02	3.51850e-02	2.32579e-11		8.59493e-01

Fig. 5.83.

Let's check the software results with hand analysis:

$$\text{THD} = \frac{\sqrt{\sum_{h=2}^{N} V_{h,\text{RMS}}^2}}{V_{1,\text{RMS}}} = \frac{\sqrt{\left(\frac{7}{\sqrt{2}}\right)^2 + \left(\frac{5}{\sqrt{2}}\right)^2}}{\frac{10}{\sqrt{2}}} = 0.8602$$

5.9 Example 8: THD block

You can use the THD block (Fig. 5.84) as another way for calculation of THD. The THD block measures the THD and filter out the fundamental component of the signal. It uses a second-order band pass filter (BPF) in order to obtain the fundamental component of the signal. The center frequency of the BPF is determined by the Fundamental Freq. box and the passing band of the filter is determined by the Passing Band Freq. box (Fig. 5.85).

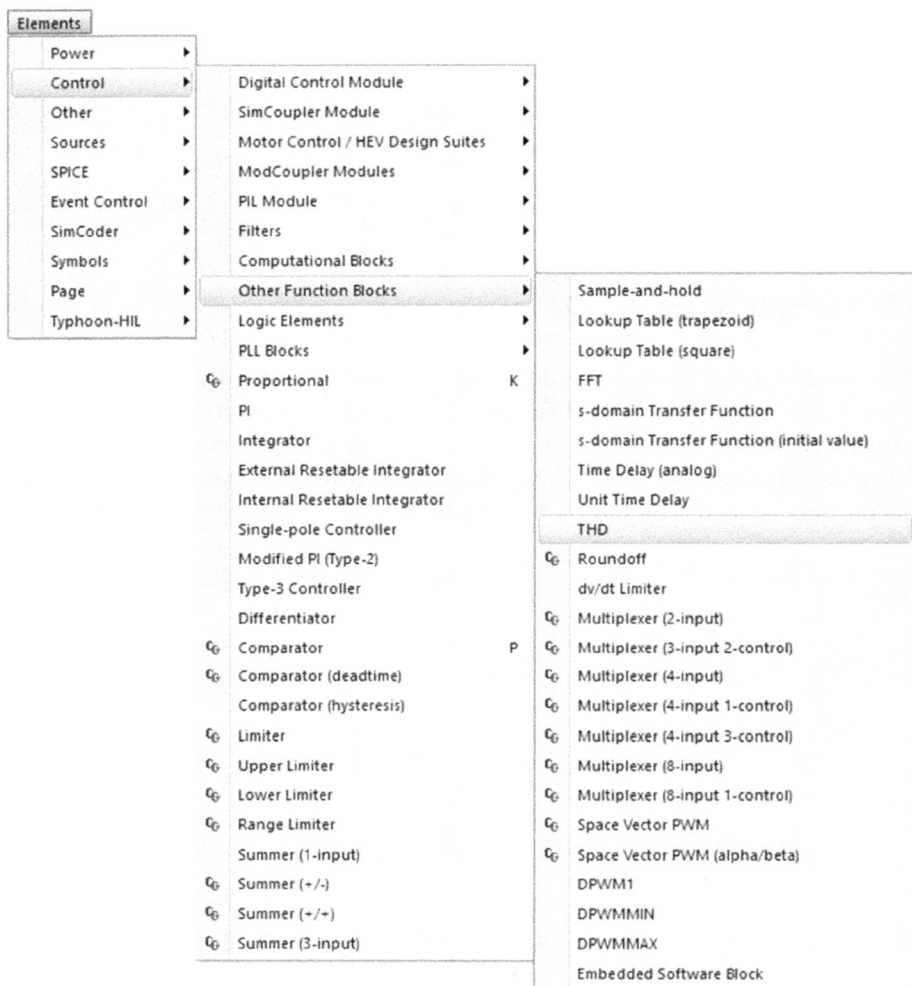

Fig. 5.84.

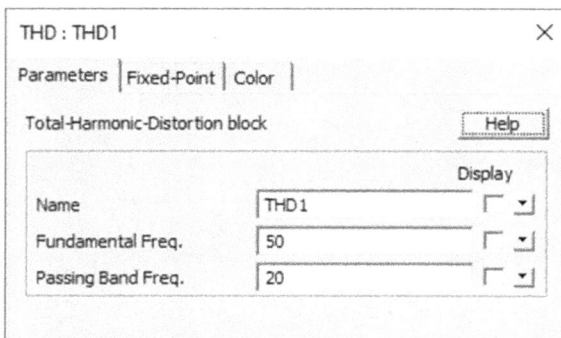

Fig. 5.85.

Change the schematic shown in Fig. 5.74 to the one shown in Fig. 5.86.

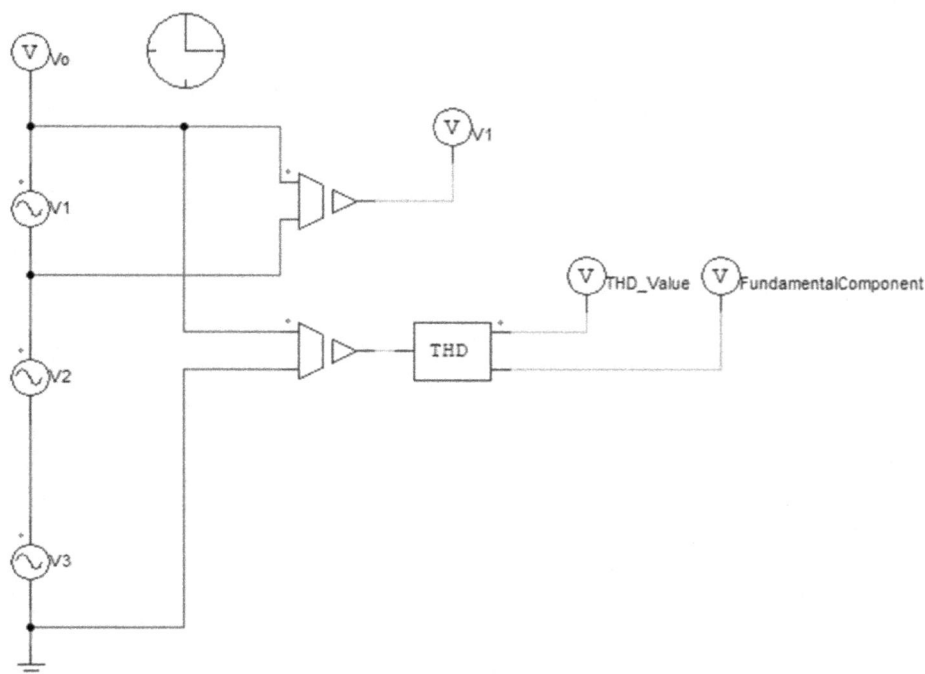

Fig. 5.86.

Settings of THD block and Simulation Control block are shown in Figs. 5.87 and 5.88.

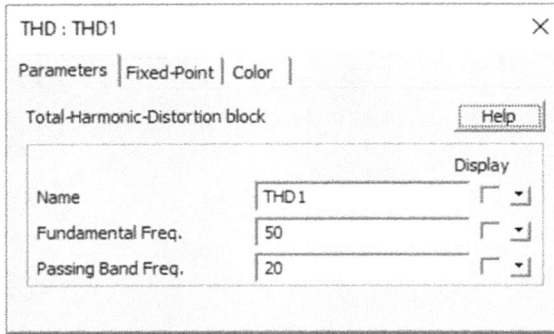

Fig. 5.87.

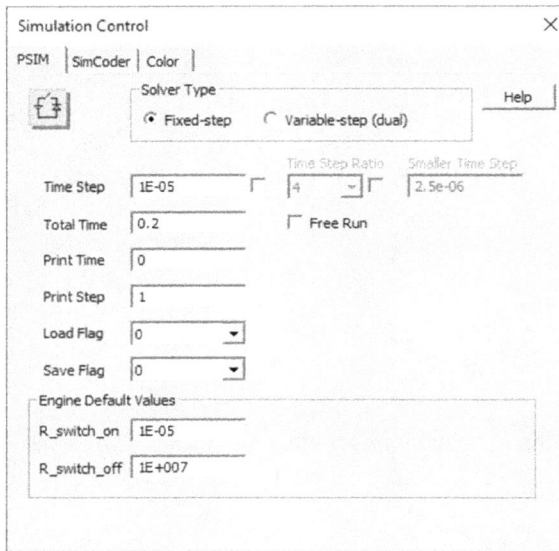

Fig. 5.88.

Run the simulation. The result shown in Fig. 5.89 will be obtained. The zoomed view of Fig. 5.89 is shown in Figs. 5.90 and 5.91. According to Fig. 5.91, the THD is 0.847. If you decrease the Passing Band Freq. value, then the THD value goes toward the correct value of 0.86. However, as you decrease the Passing Band Freq. value, you need to run the simulation for a longer interval in order to reach the steady-state value of THD. Only the steady-state section of THD waveform is important. The transient section has no importance.

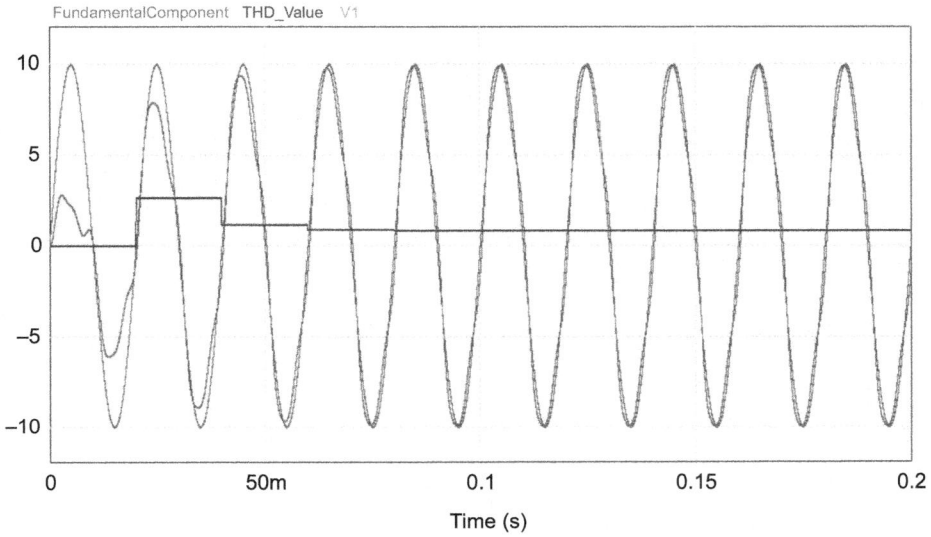

Fig. 5.89.

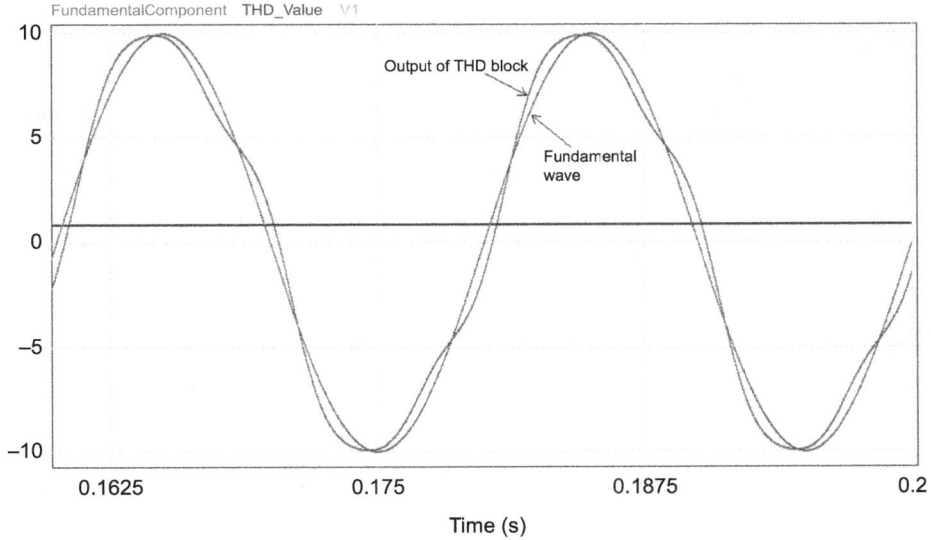

Fig. 5.90.

FundamentalComponent THD_Value V1

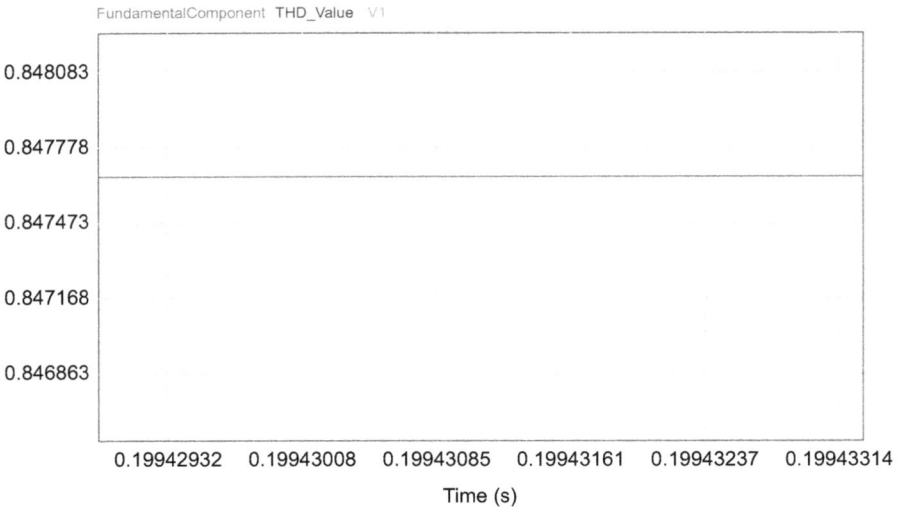

Fig. 5.91.

5.10 Example 9: Math Function voltage source block

In Example 7, we connected three sinusoidal AC sources together in order to generate the $10\sin(2\pi \times 50 \times t) + 7\sin(2\pi \times 150 \times t) + 5\sin(2\pi \times 250 \times t)$. Yet there is another way to obtain that voltage.

You can generate arbitrary voltage waveforms with the aid of Math Function block (Fig. 5.92). In order to do this, put the Math Function (voltage source) block onto the schematic, double click it and write the equation in the Expression box (Fig. 5.93).

For instance $v(t) = 10\sin(2\pi \times 50 \times t) + 7\sin(2\pi \times 150 \times t) + 5\sin(2\pi \times 250 \times t)$, in order to generate , you need to fill the Expression box of Math Function block with .

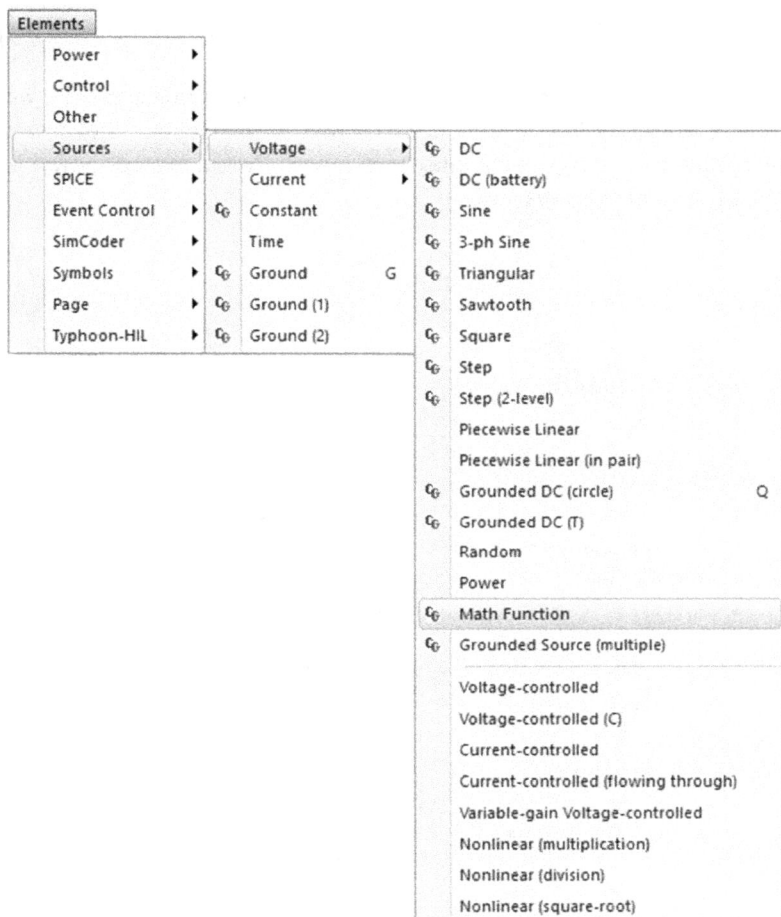

Fig. 5.92.

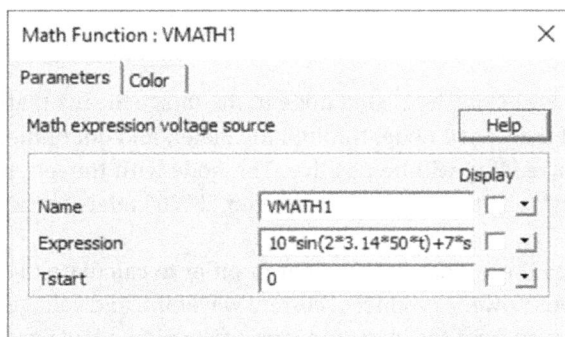

Fig. 5.93.

5.11 Example 10: Wattmeter block

PSIM has different blocks for power measurements (Fig. 5.94). In this example, we study the Wattmeter block. The Wattmeter block measures the real (or average) power and energy. Remember that energy is the integral of power. Unit of average power is Watts (W), while the unit of energy is Joules.

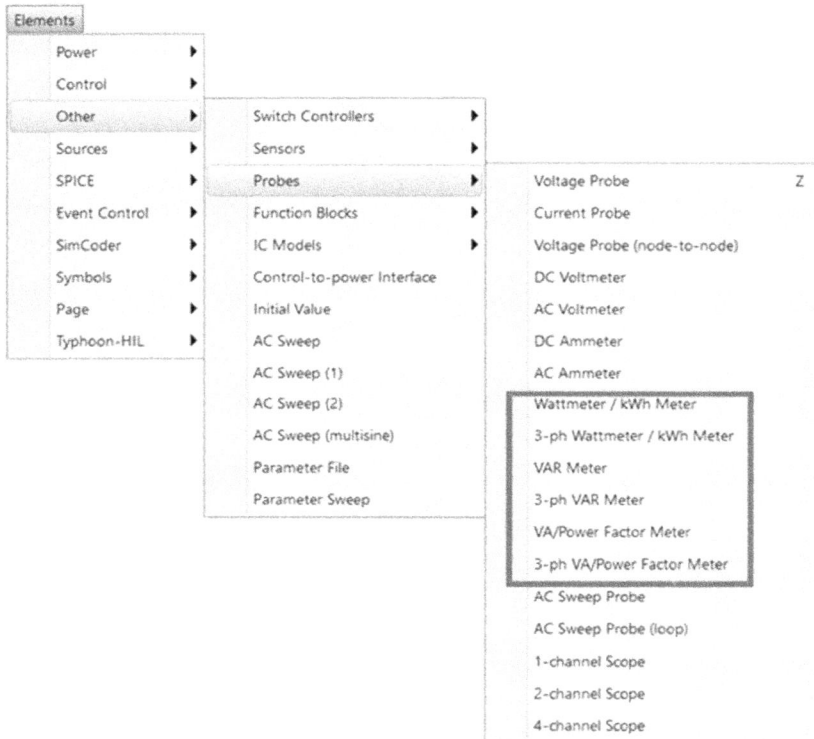

Fig. 5.94.

The Wattmeter block is shown in Fig. 5.95. The dotted node in the image means that when the power flows into side of the dotted node, through the meter, and out of the other side of the meter, the meter reading will be positive. The node with the letter "W" refers to the wattmeter output, and the node with the letter "kWh" refers to the kWh output.

The Wattmeter block measures the voltage and current in order to calculate the instantaneous power (instantaneous power = product of current waveform and voltage waveform). The real power is obtained from the instantaneous power after a low-pass filter (Fig. 5.96). The cut-off frequency of the low-pass filter needs to be specified.

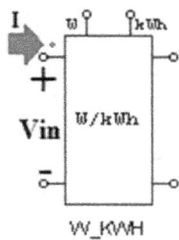

Fig. 5.95.

Fig. 5.96.

Assume the schematic shown in Fig. 5.97. We want to measure the average power drawn from the source. The settings of elements are shown in Figs. 5.98–5.103.

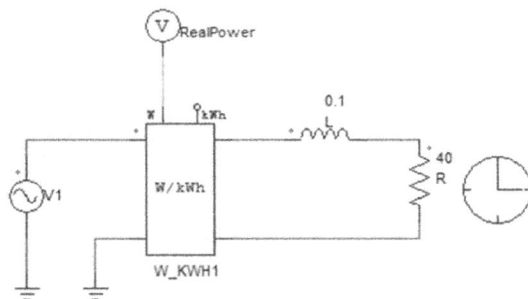

Fig. 5.97.

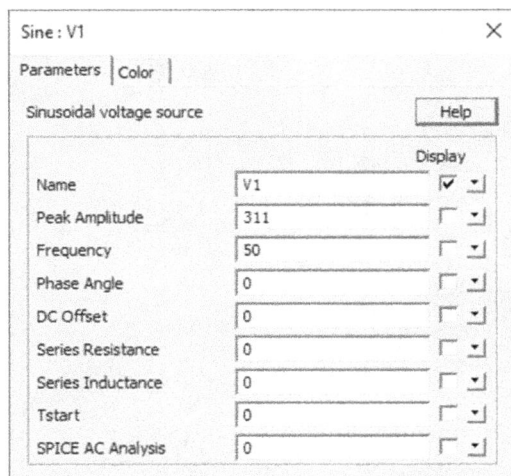

Fig. 5.98.

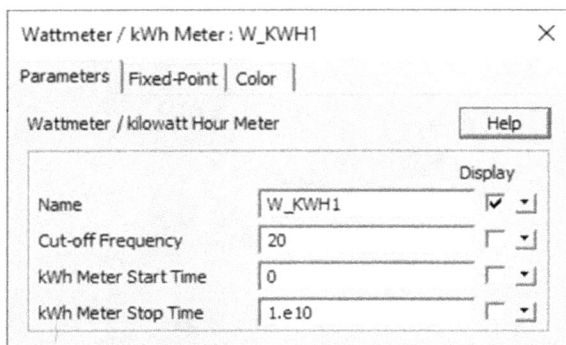

Fig. 5.99.

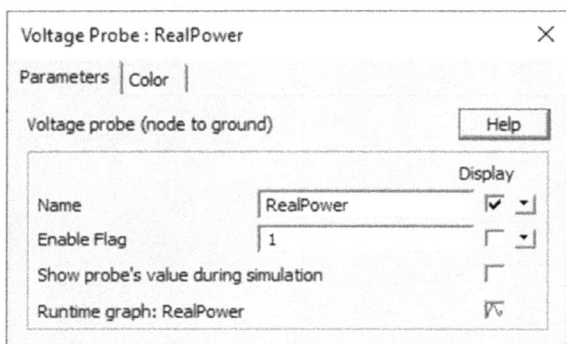

Fig. 5.100.

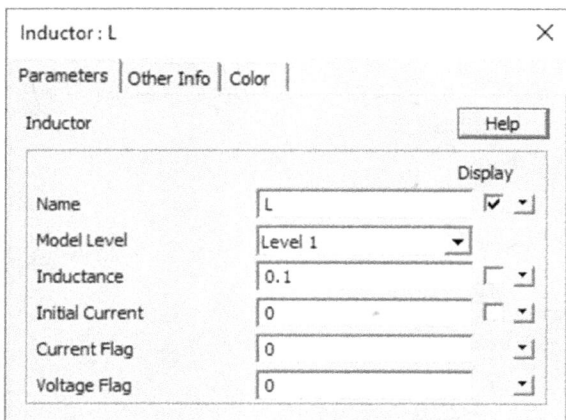

Fig. 5.101.

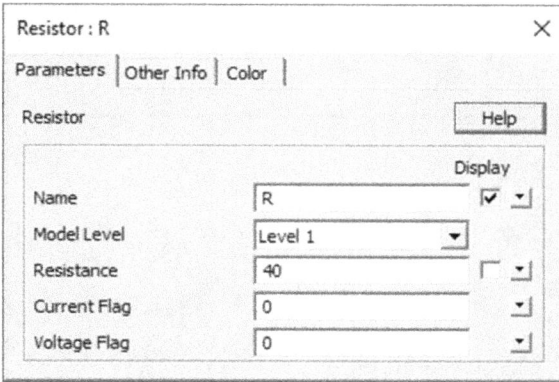

Fig. 5.102.

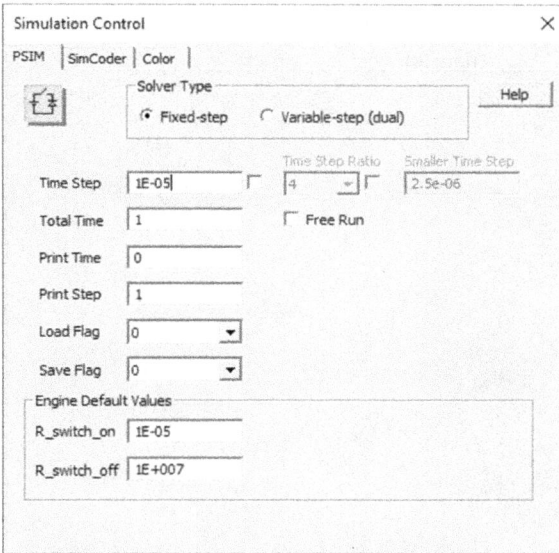

Fig. 5.103.

Run the simulation. The result shown in Fig. 5.104 will appear. Figure 5.105 is the zoomed view of Fig. 5.104.

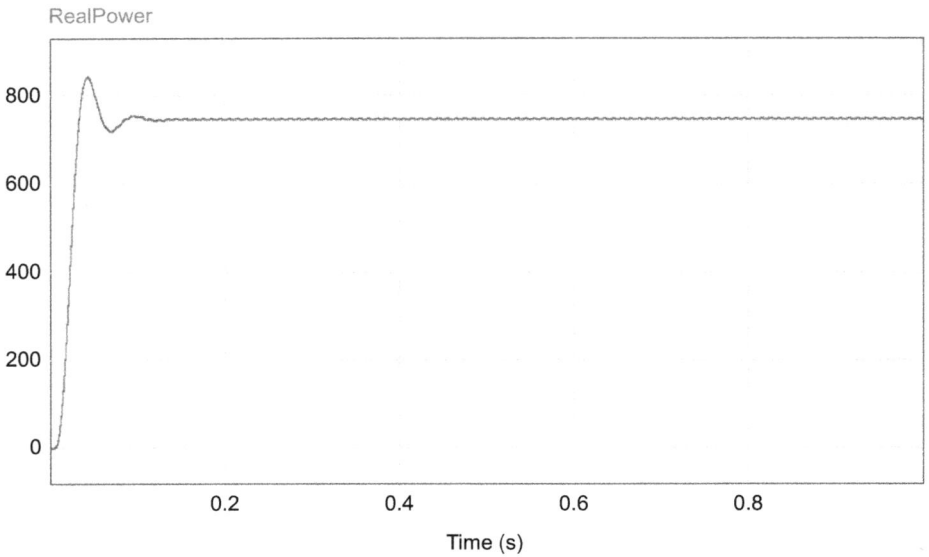

Fig. 5.104.

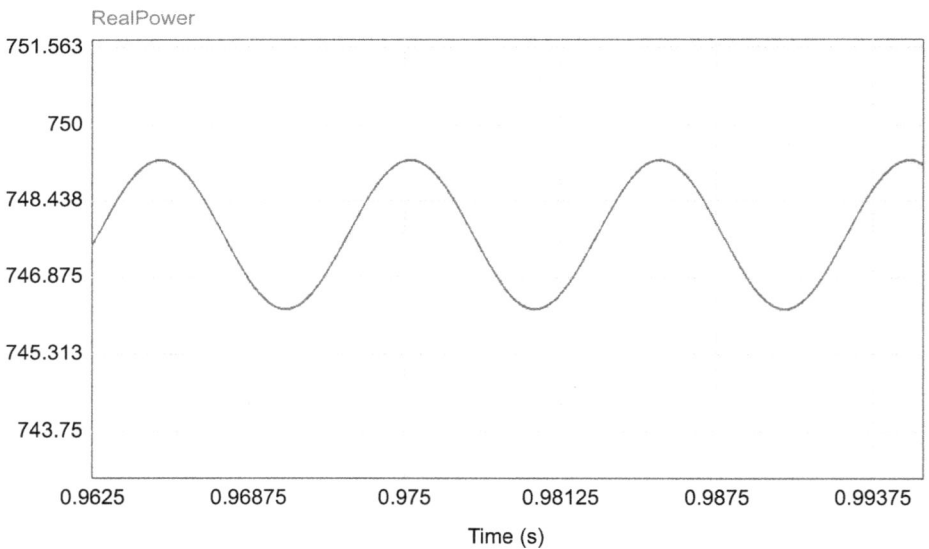

Fig. 5.105.

Figure 5.105 has a small ripple. The ripple decreases if we decrease the cut-off frequency of the low-pass filter. However, when you decrease the cut-off frequency of the low-pass filter, you need to simulate the circuit for a longer amount of time in order to reach to the steady state.

For instance, let's decrease the Cut-off Frequency to 2 Hz (Fig. 5.106). We change the Total Time box of Simulation Control to 10 s as well (Fig. 5.107). The simulation results are shown in Fig. 5.108. The zoomed view of Fig. 5.108 is shown in Fig. 5.109. As you see, the ripple decreased considerably. However, previous simulation reached steady-state value of average power in less than 0.2 s while the current simulation reaches the steady-state value of average power in more than 1 s.

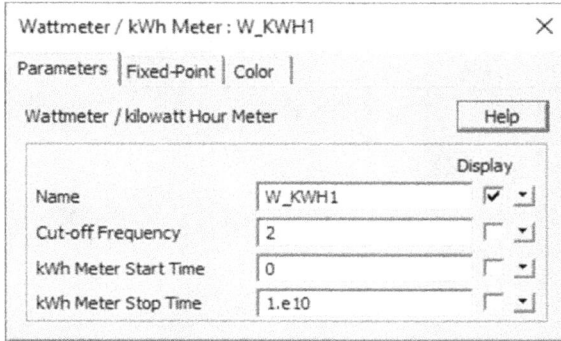

Fig. 5.106.

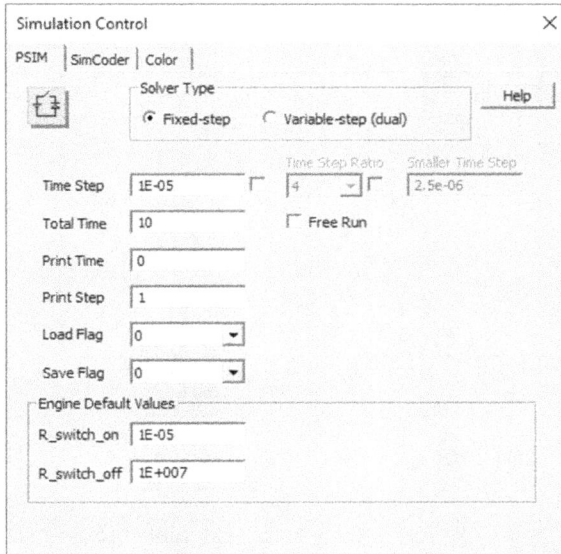

Fig. 5.107.

RealPower

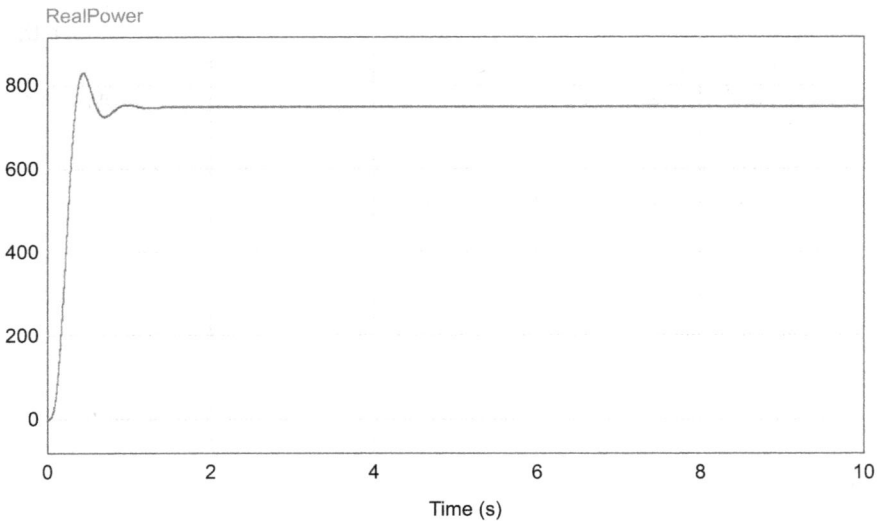

Fig. 5.108.

RealPower

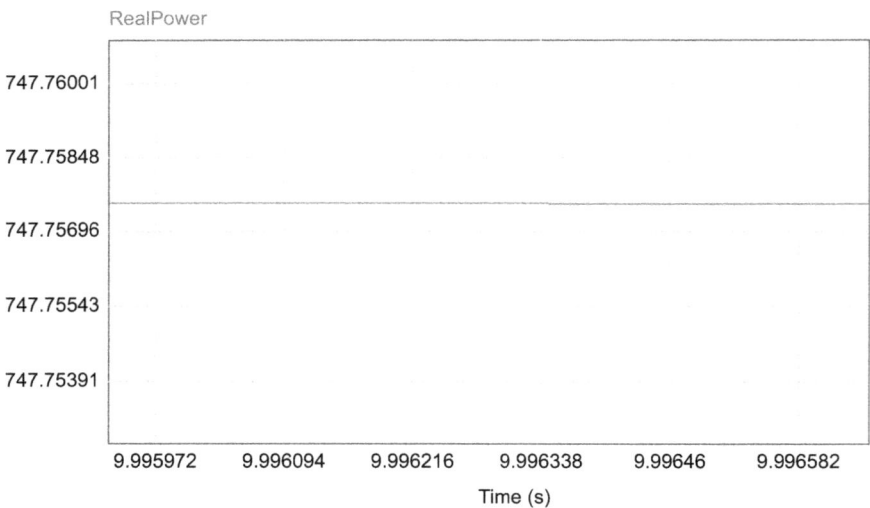

Fig. 5.109.

When you have a small amount of ripple like Fig. 5.105, you can calculate the average power by averaging the maximum value and minimum value of the graph.

For instance, according to Fig. 5.110, maximum value of Fig. 5.105 is 749.27 W and its minimum value is 746.237 W. Then, the average power could be approximated as (749.27 + 746.237)/2 = 747.75 W. Maximum and minimum of graph are measured with the aid of cursors.

Measure				x
	X1	X2	Δ	
Time	9.65205e-01	9.80189e-01	1.49842e-02	
RealPower	7.49278e+02	7.46237e+02	-3.04193e+00	

Fig. 5.110.

Let's use the MATLAB® in order to check the obtained results. The MATLAB code shown in Fig. 5.111 calculated the average power of circuit shown in Fig. 5.97. The obtained result is exactly the result obtained by PSIM in Fig. 5.109.

```
Command Window
>> syms i(t)
>> Di=diff(i,t);
>> cond=[i(0)==0, Di(0)==0];
>> current=dsolve(40*i+0.1*diff(i,1)==311*sin(314*t),cond)

current =

(244135*exp(-400*t))/64649 - (1555*64649^(1/2)*cos(314*t + atan(200/157)))/64649

>> Pin=current*311*sin(314*t);
>> T=0.01;
>> Pave=1/T*int(Pin,t,5,5+T)

Pave =

(3796299250*exp(-2000)*(157*cos(1570) + 200*sin(1570)))/4179493201 - (48360500*64649^(1/2)*((64649^(1/2)*(

>> eval(Pave)

ans =

   747.7405

fx >>
```

Fig. 5.111.

5.12 Example 11: average power measurement in presence of harmonics

In the previous example, the input voltage source had no harmonic contents. The Wattmeter block can measure the average power in presence of harmonics as well.

Consider the schematic shown in Fig. 5.112. In this example, we have an input voltage source which is composed of two different frequencies. The settings of the two voltage sources are shown in Figs. 5.113 and 5.114.

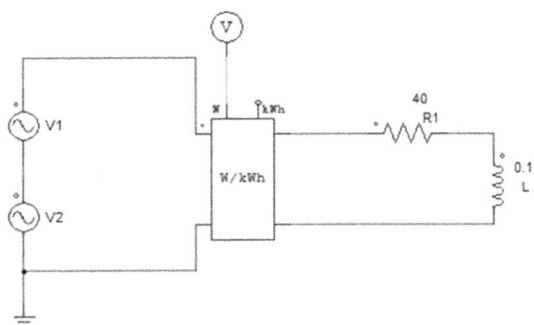

Fig. 5.112.

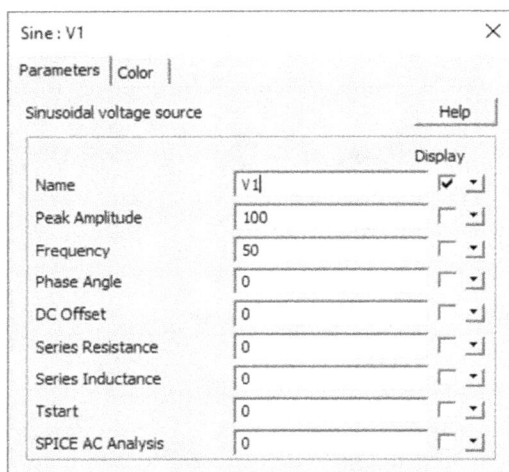

Fig. 5.113.

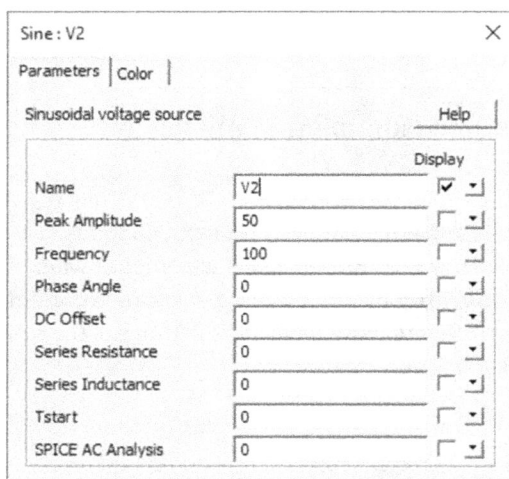

Fig. 5.114.

Run the simulation. The result shown in Fig. 5.115 will appear. Figure 5.116 is the zoomed view of Fig. 5.115.

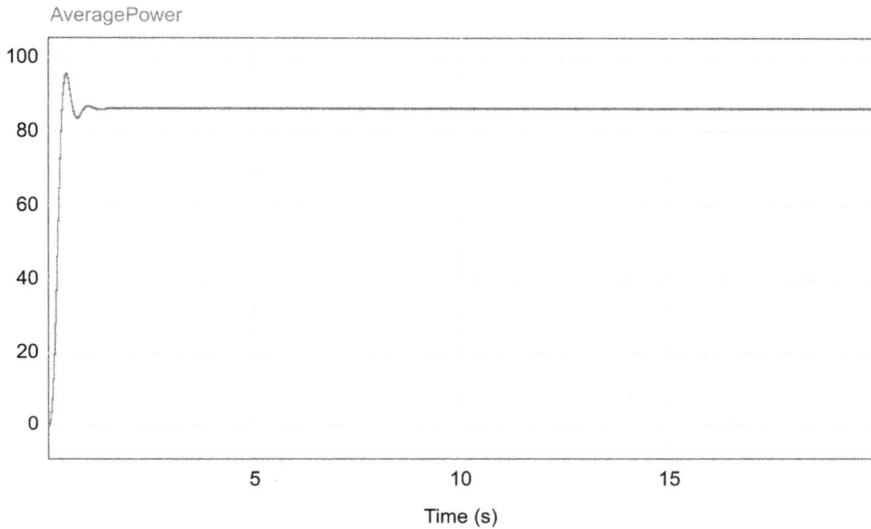

Fig. 5.115.

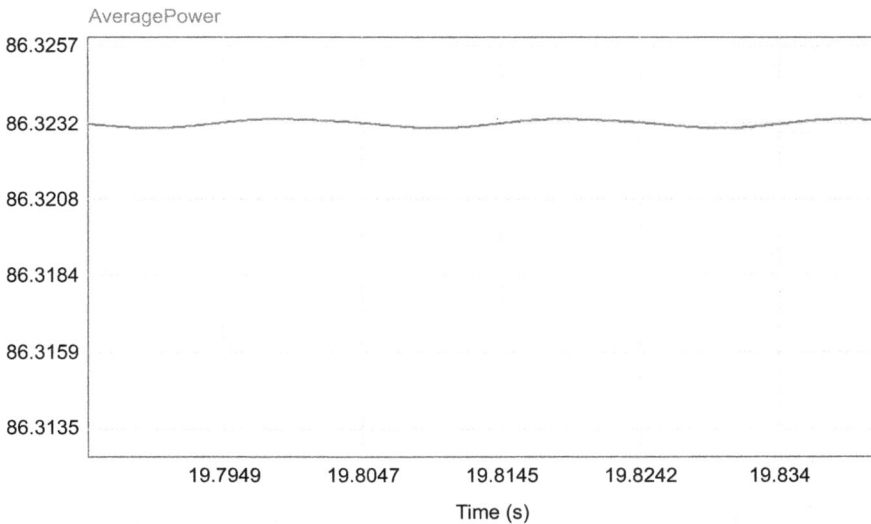

Fig. 5.116.

Let's check the results with MATLAB. The following MATLAB code calculates the average power for circuit shown in Fig. 5.112. After running the MATLAB code, the result shown in Fig. 5.117 will be obtained. This result is exactly the same as the result obtained in Fig. 5.116.

```
clc
V1=100;
V2=50;
f1=50;
T=1/f1;
w1=2*pi*f1;
f2=100;
w2=2*pi*f2;
L=0.1;
R=40;

Z1=40+i*2*pi*f1*L;
Z2=40+i*2*pi*f2*L;

I1=V1/Z1;
I2=V2/Z2;

syms t
I=abs(I1)*sin(w1*t+angle(I1))+abs(I2)*sin(w2*t+angle(I2));
V=V1*sin(w1*t)+V2*sin(w2*t);
Power=I*V;
Pave=1/T*int(Power,5,5+T);
disp('Avergae power =')
eval(Pave)
```

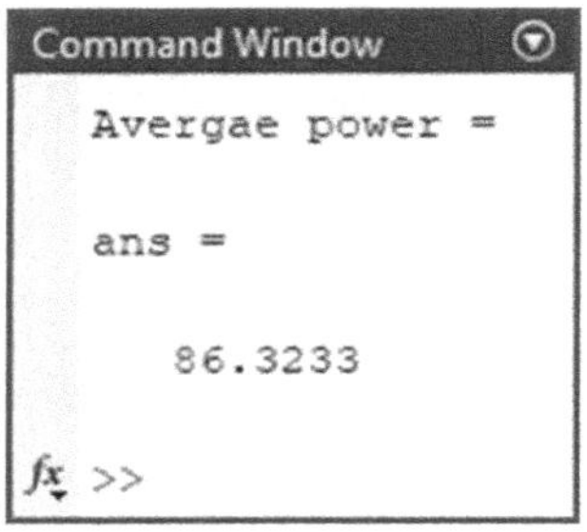

```
Command Window                    ⊙

    Avergae power =

    ans =

       86.3233

fx >>
```
Fig. 5.117.

5.13 Example 12: VAR meter block

You can measure the reactive power of the circuit using the VAR meter block.

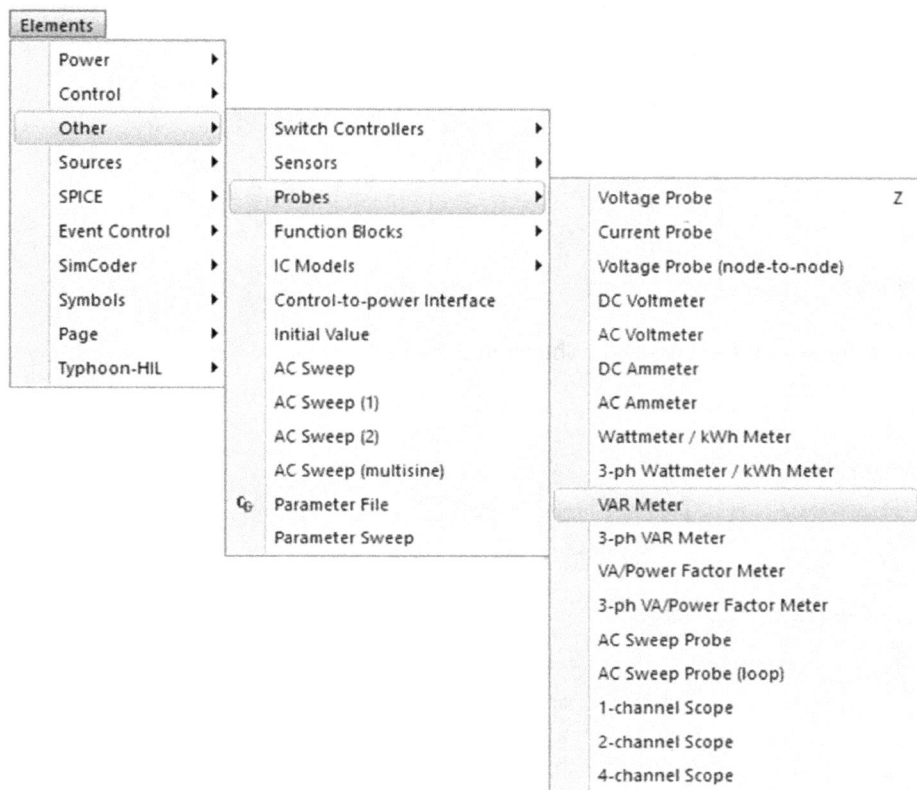

Fig. 5.118.

Consider the circuit shown in Fig. 5.119.

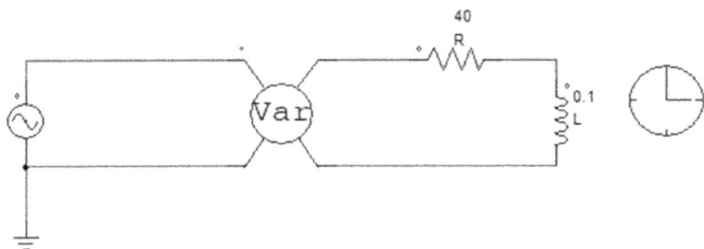

Fig. 5.119.

The settings of the elements are shown in Figs. 5.120–5.124.

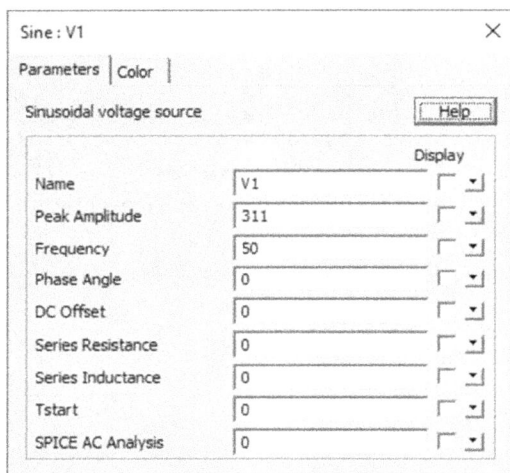

Fig. 5.120.

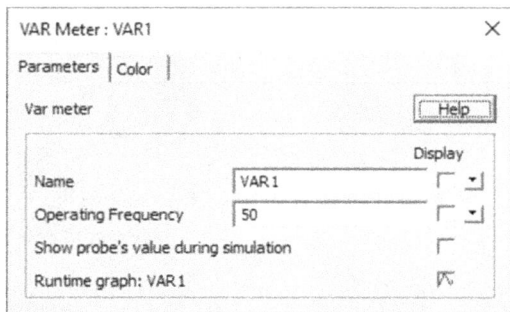

Fig. 5.121.

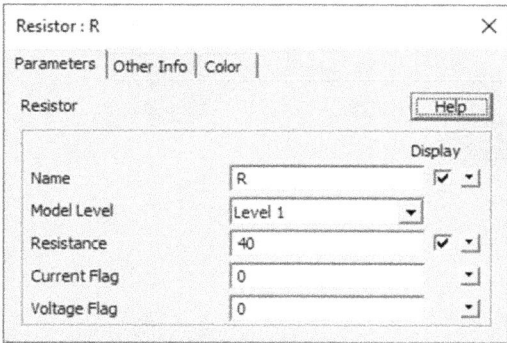

Fig. 5.122.

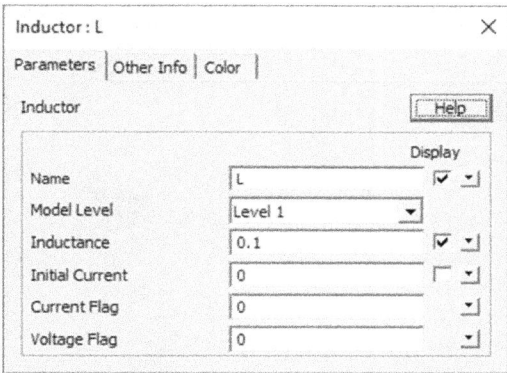

Fig. 5.123.

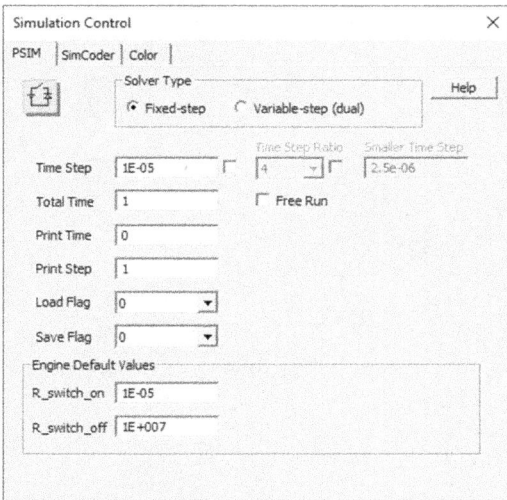

Fig. 5.124.

Run the simulation. The result shown in Fig. 5.125 will appear. Figure 5.126 is the zoomed view of Fig. 5.125. The reactive power is 587.27 VAR.

VAR1

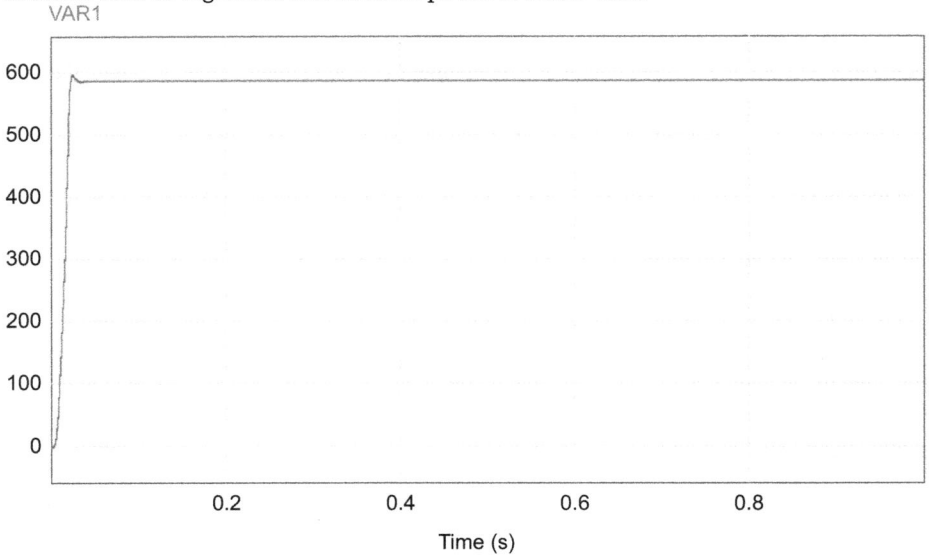

Fig. 5.125.

VAR1

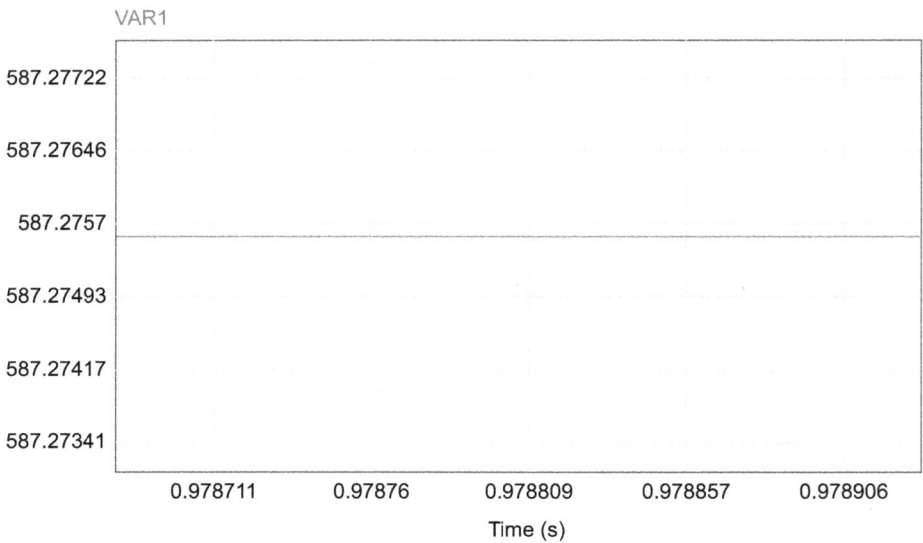

Fig. 5.126.

Let's use MATLAB to check the results. The MATLAB code shown below calculates the active and reactive powers of the circuit shown in Fig. 5.119.

```
clc

V=311*exp(i*0);
Vrms=V/sqrt(2);
f=50;

w=2*pi*f;
L=0.1;
R=40;
Z=40+i*L*w;
I=V/Z;
Irms=I/sqrt(2);

disp('Apparent power:')
S=Vrms*conj(Irms);
disp(S)
disp('Active power (W):')
disp(real(S))
disp('Reactive power (VAR):')
disp(imag(S))
```

After running the MATLAB code, the result shown in Fig. 5.127 will be obtained. This result is exactly the same as the one shown in Fig. 5.126.

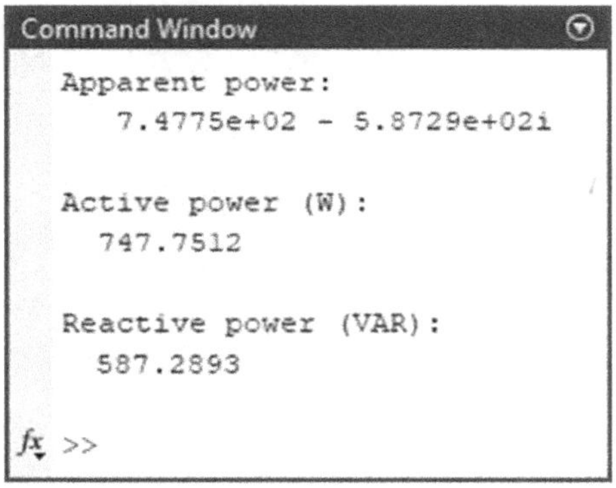

```
Command Window

    Apparent power:
        7.4775e+02 - 5.8729e+02i

    Active power (W):
        747.7512

    Reactive power (VAR):
        587.2893

fx >>
```

Fig. 5.127.

According to the basic circuit theory, the reactive power of inductive loads is positive and the reactive power of capacitive loads is negative. For instance, the reactive power of the circuit shown in Fig. 5.128 is negative. Analysis result is shown in Fig. 5.129. Figure 5.130 is the zoomed view of Fig. 5.129.

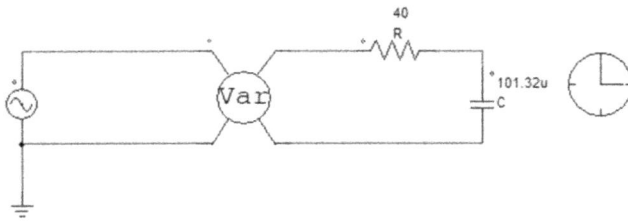

Fig. 5.128.

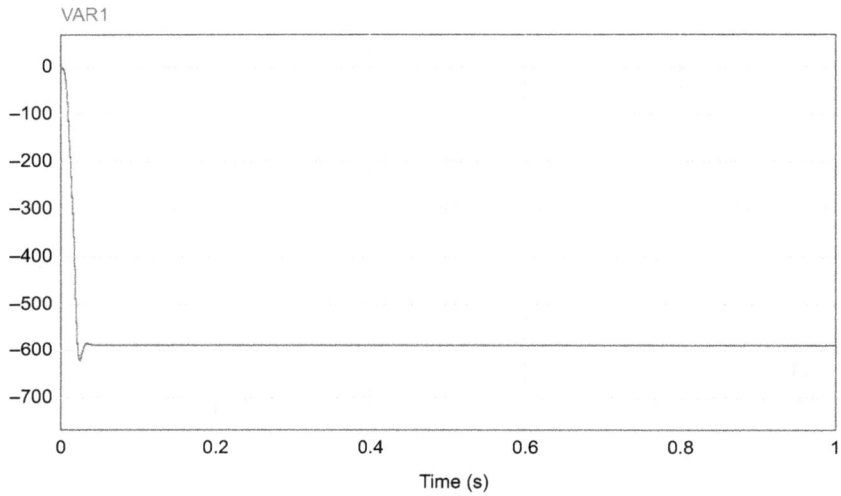

Fig. 5.129.

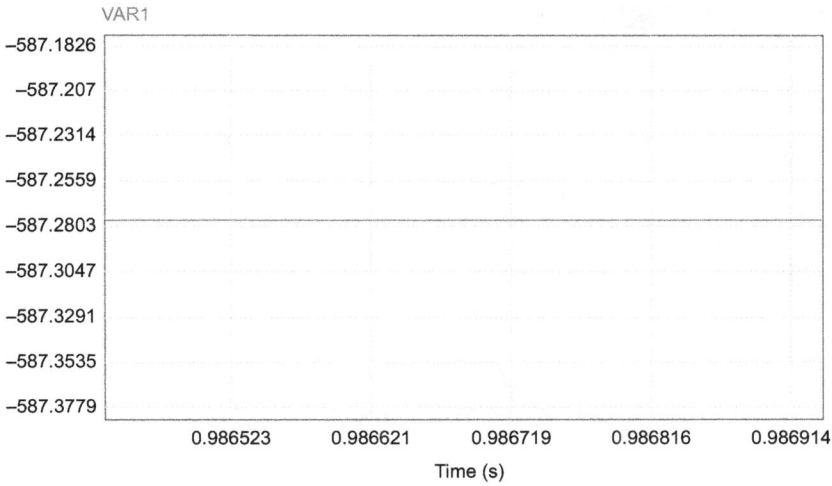

Fig. 5.130.

The following MATLAB code calculates the reactive power for circuit shown in Fig. 5.128.

```
clc
V=311*exp(i*0);

Vrms=V/sqrt(2);
f=50;
w=2*pi*f;
C=101.32e-6;
R=40;

Z=40+1/(i*w*C);
I=V/Z;
Irms=I/sqrt(2);
disp('Apparent power:')
S=Vrms*conj(Irms);

disp(S)
disp('Active power (W):')
disp(real(S))
disp('Reactive power (VAR):')
disp(imag(S))
```

After running the MATLAB code, the result shown in Fig. 5.131 will appear. The result is the same as in Fig. 5.130.

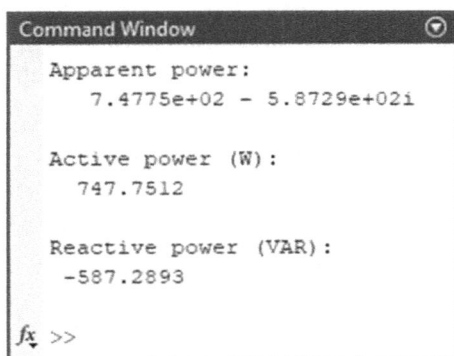

Fig. 5.131.

5.14 Example 13: reactive power calculation in presence of harmonics

If the voltages or currents contain harmonics, the meter will give the reactive power at the fundamental frequency. Consider the circuit shown in Fig. 5.132. The settings of blocks are shown in Figs. 5.133–5.138.

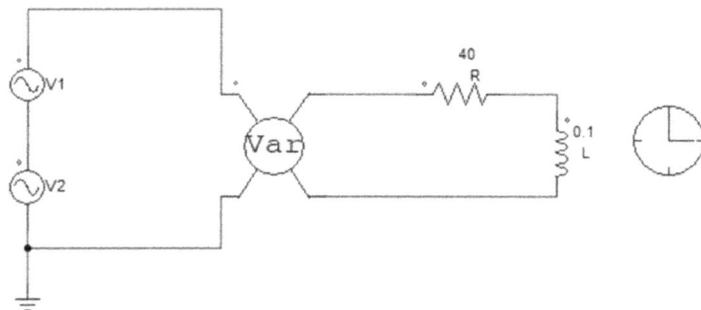

Fig. 5.132.

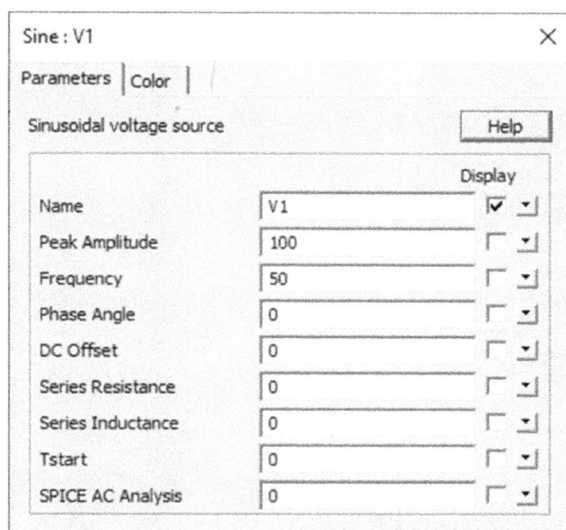

Fig. 5.133.

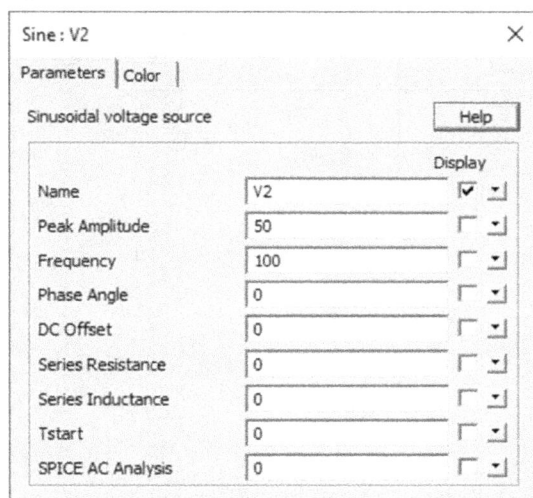

Fig. 5.134.

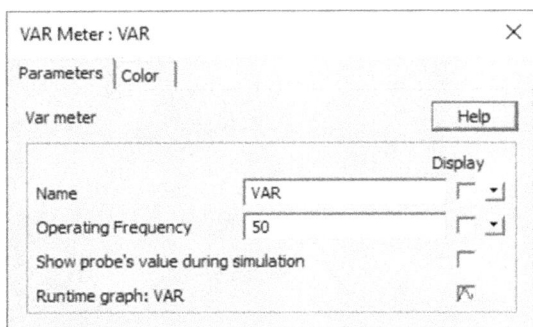

Fig. 5.135.

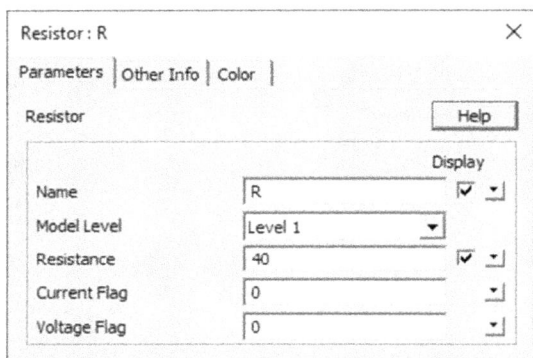

Fig. 5.136.

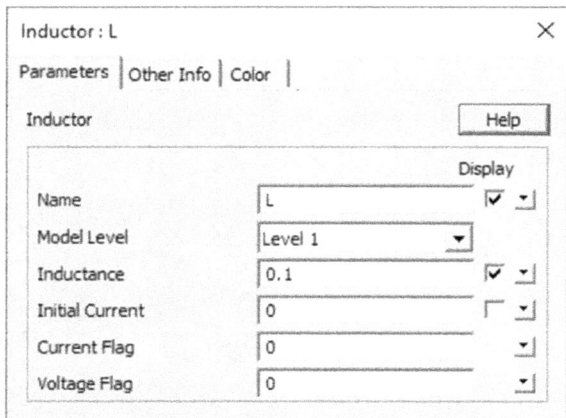

Fig. 5.137.

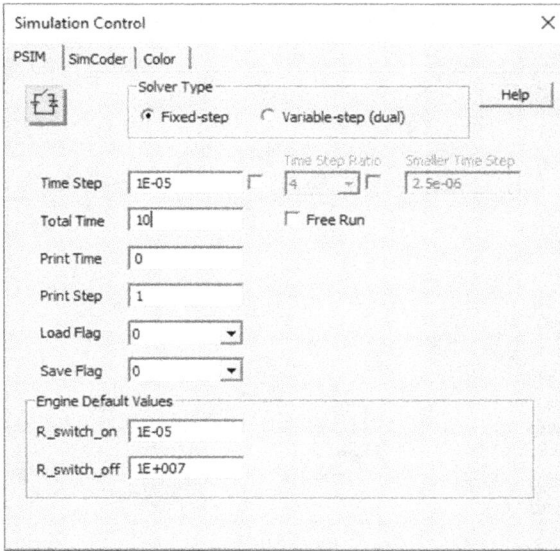

Fig. 5.138.

After running the simulation, the result shown in Fig. 5.139 will be obtained. Figure 5.140 is the zoomed view of Fig. 5.139. According to Fig. 5.140, the reactive power is 60.72 VAR.

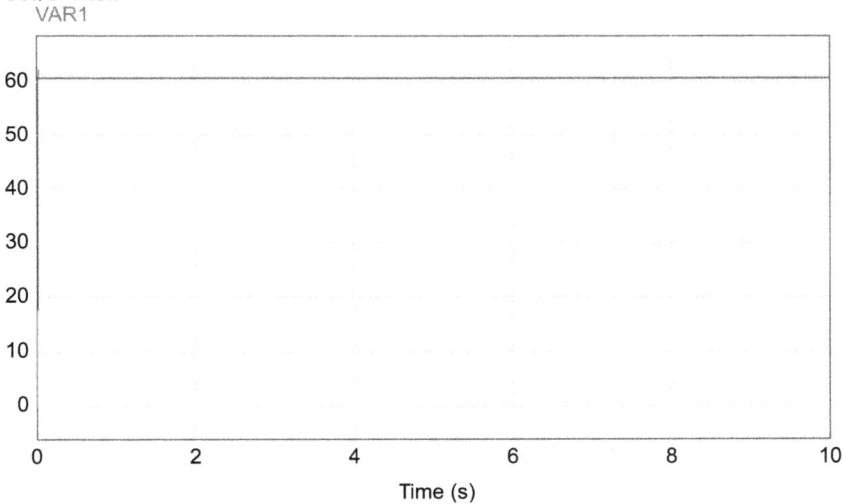

Fig. 5.139.

VAR

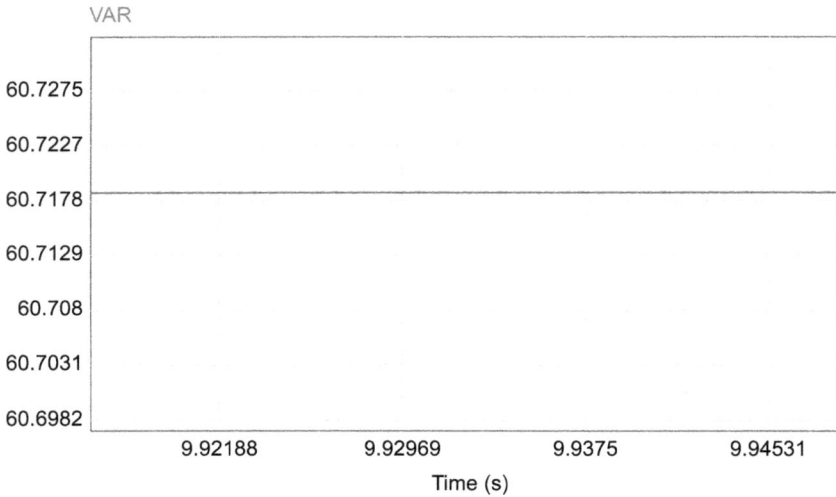

Fig. 5.140.

Following MATLAB code calculates the reactive power of sources.

```
clc

V1=100*exp(0*i);
V1rms=V1/sqrt(2);
V2=50*exp(0*i);
V2rms=V2/sqrt(2);
f1=50;

T=1/f1;
w1=2*pi*f1;
f2=100;
w2=2*pi*f2;
L=0.1;
R=40;

Z1=40+i*2*pi*f1*L;
Z2=40+i*2*pi*f2*L;

I1=V1/Z1;
I1rms=I1/sqrt(2);
I2=V2/Z2;
I2rms=I2/sqrt(2);

S1=V1rms*conj(I1rms);
S2=V2rms*conj(I2rms);

S=S1+S2;
disp ('Complex power for 50 Hz source (S1)')
disp (S1)
```

```
disp ('Complex power for 100 Hz source (S2)')
disp (S2)
disp ('Total Complex Power (S1+S2):')
disp (S)
```

After running the MATLAB code, the result shown in Fig. 5.141 will appear. When we compare this result with the PSIM result (Fig. 5.140), we see that PSIM calculates the reactive power at fundamental frequency only.

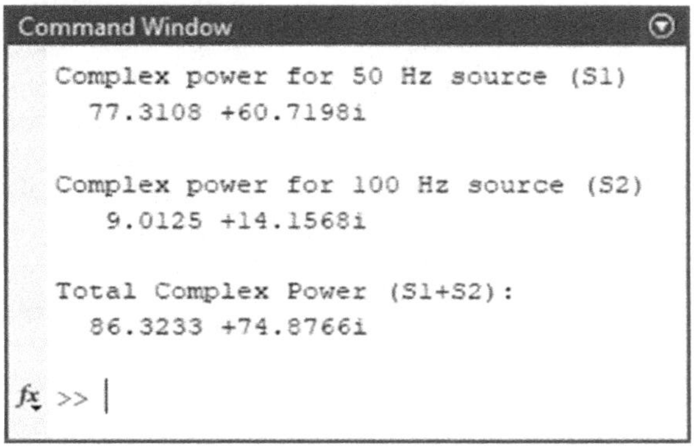

```
Command Window                              ⊙

  Complex power for 50 Hz source (S1)
     77.3108 +60.7198i

  Complex power for 100 Hz source (S2)
     9.0125 +14.1568i

  Total Complex Power (S1+S2):
     86.3233 +74.8766i

fx >> |
```

Fig. 5.141.

5.15 Example 14: VA/power factor meter block

The apparent power (VA)/power factor meter block (Fig. 5.142) measures the VA, the power factor, and the displacement power factor of a circuit.

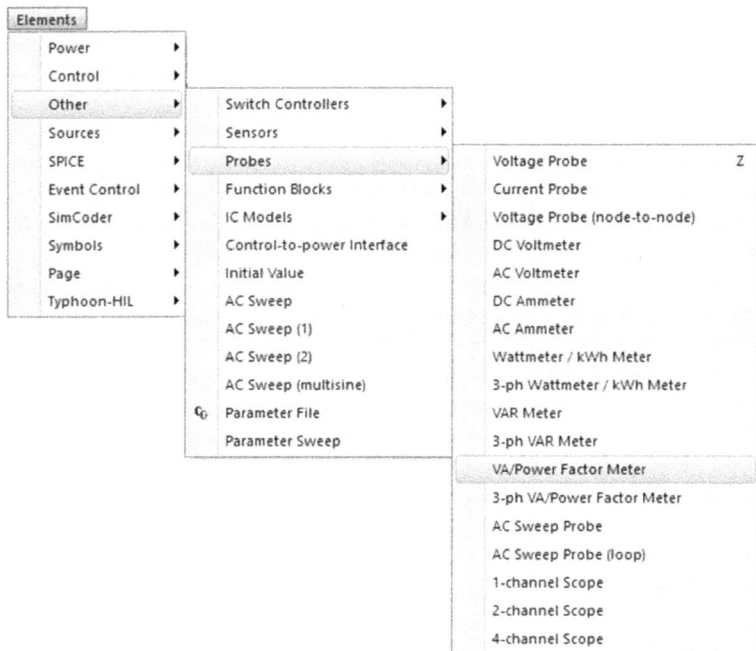

Fig. 5.142.

Consider the circuit shown in Fig. 5.143.

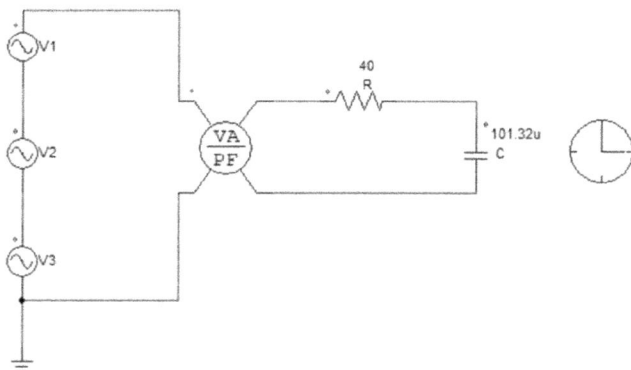

Fig. 5.143.

The settings of elements are shown in Figs. 5.144–5.150.

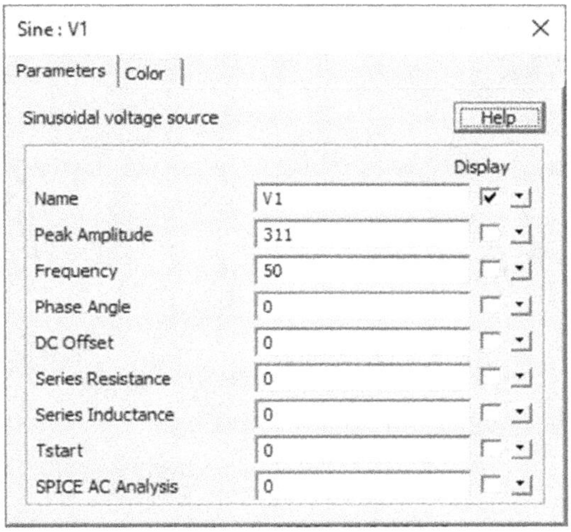

Sine : V1 ✕

Parameters | Color

Sinusoidal voltage source   [Help]

		Display
Name	V1	☑ ▾
Peak Amplitude	311	☐ ▾
Frequency	50	☐ ▾
Phase Angle	0	☐ ▾
DC Offset	0	☐ ▾
Series Resistance	0	☐ ▾
Series Inductance	0	☐ ▾
Tstart	0	☐ ▾
SPICE AC Analysis	0	☐ ▾

Fig. 5.144.

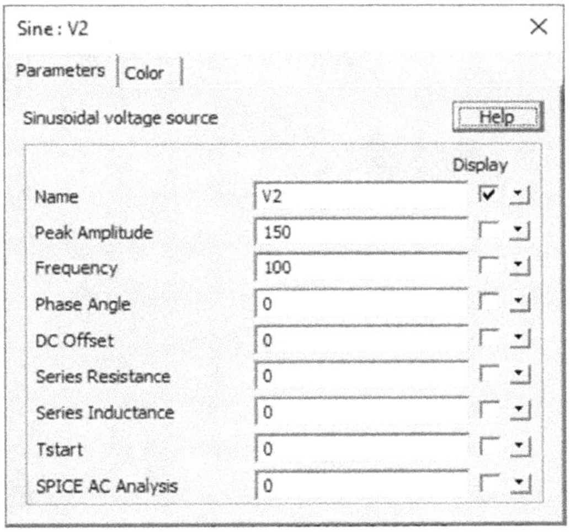

Sine : V2 ✕

Parameters | Color

Sinusoidal voltage source   [Help]

		Display
Name	V2	☑ ▾
Peak Amplitude	150	☐ ▾
Frequency	100	☐ ▾
Phase Angle	0	☐ ▾
DC Offset	0	☐ ▾
Series Resistance	0	☐ ▾
Series Inductance	0	☐ ▾
Tstart	0	☐ ▾
SPICE AC Analysis	0	☐ ▾

Fig. 5.145.

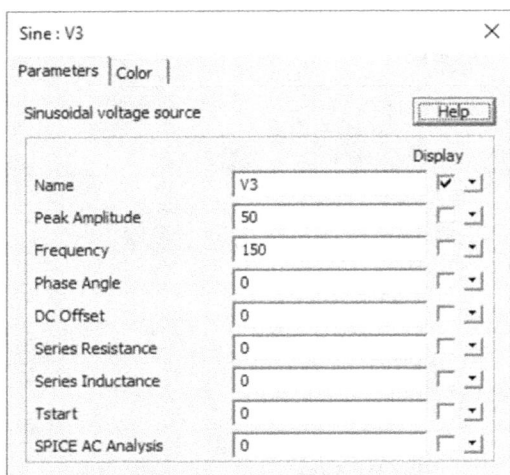

Fig. 5.146.

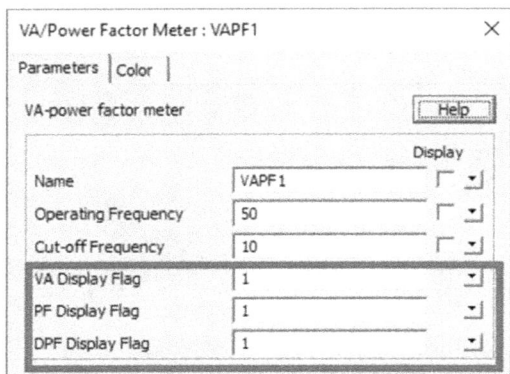

Fig. 5.147.

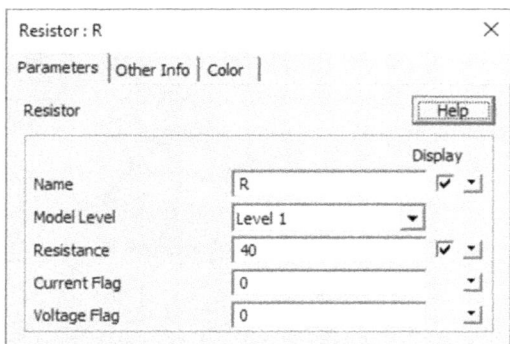

Fig. 5.148.

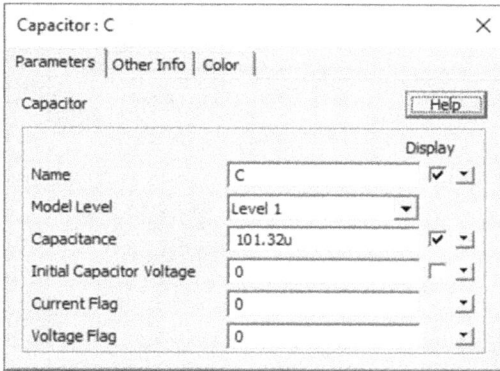

Fig. 5.149.

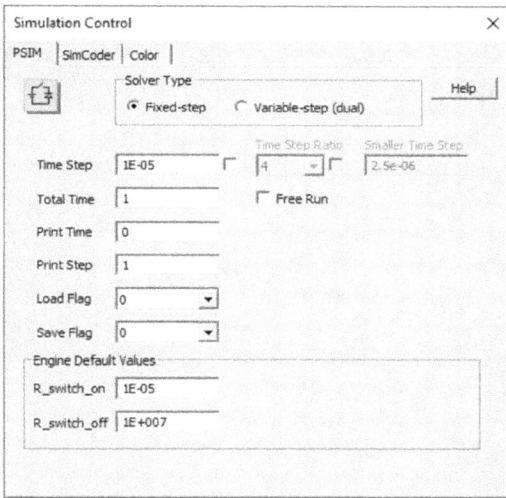

Fig. 5.150.

Run the simulation. Select the measured VA to be displayed. The VA and its zoomed view are shown in Figs. 5.151 and 5.152, respectively.

VAPF1_VA

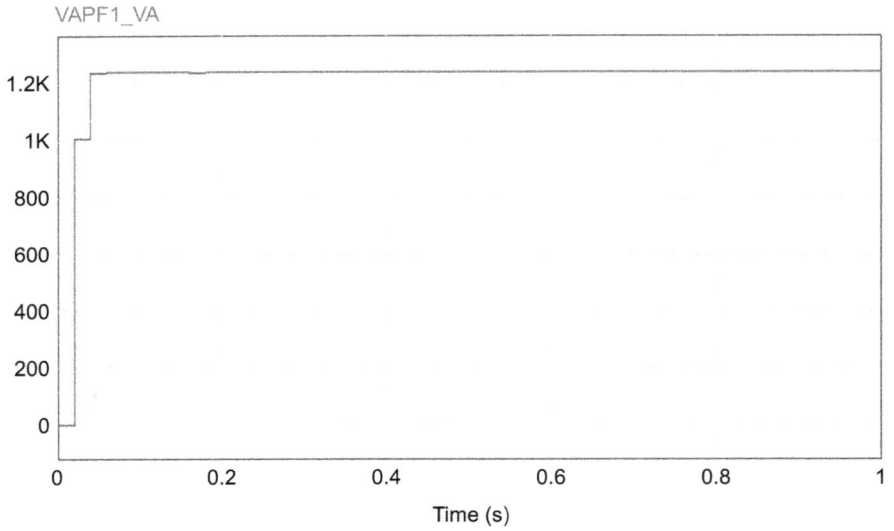

Fig. 5.151.

VAPF1_VA

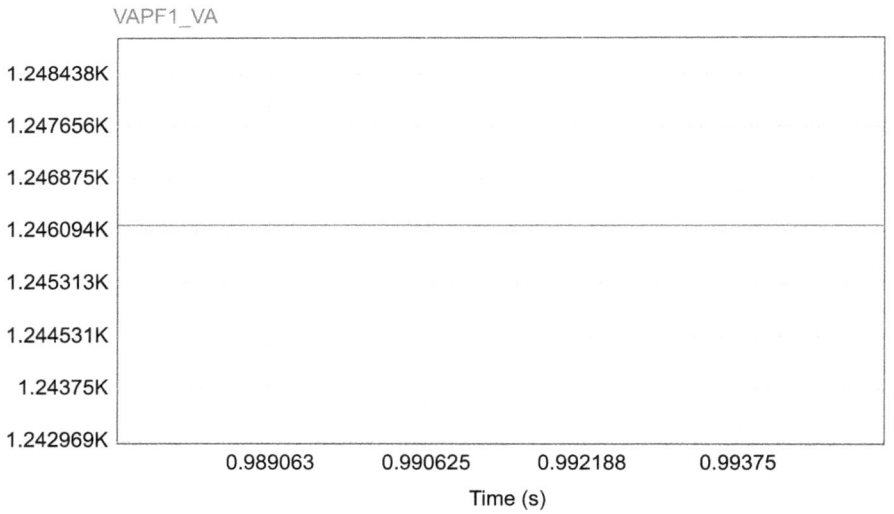

Fig. 5.152.

The measured power factor and its zoomed view are shown in Figs. 5.153 and 5.154, respectively.

VAPF1_PF

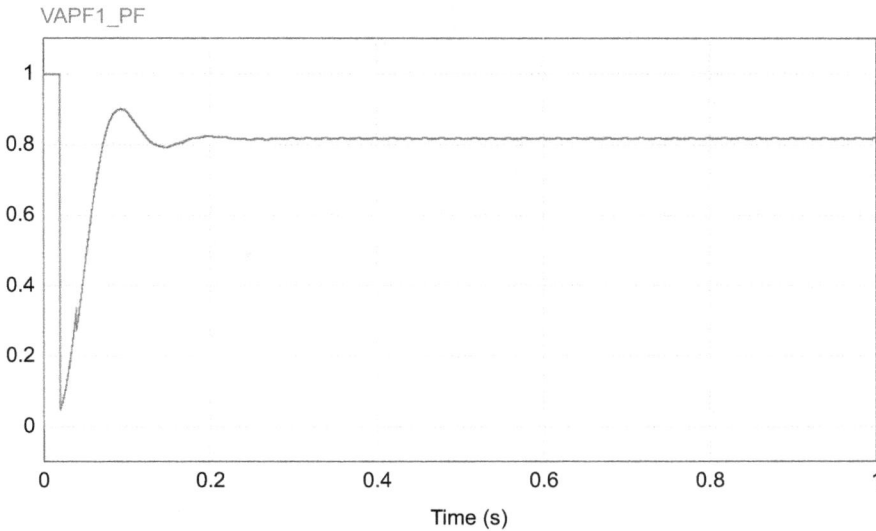

Fig. 5.153.

VAPF1_PF

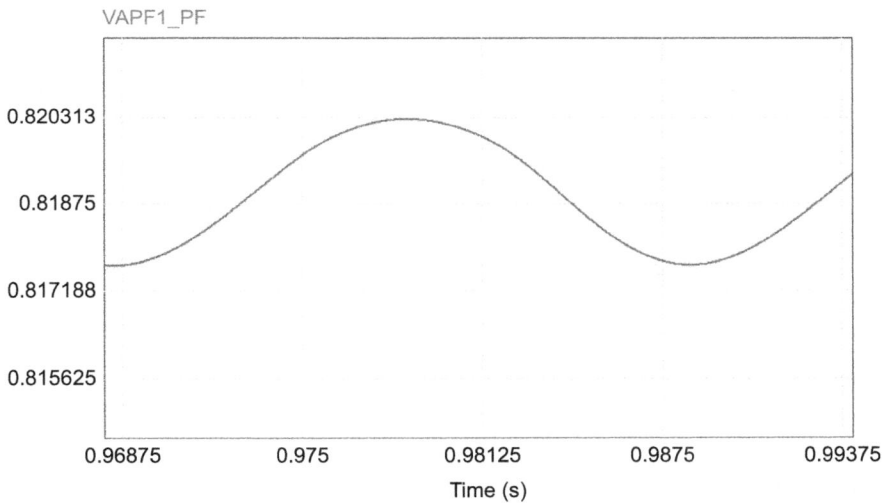

Fig. 5.154.

Remember that power factor is defined as $\frac{P}{S}$, where P is the average power and S is VA. PSIM uses a low-pass filter in order to extract the average power from instantaneous power waveform. When the value entered into the Cut-off Frequency box (Fig. 5.147) decreases, the accuracy of extracted average power increases and the ripple in the power factor plot decreases. However, the simulation must be run for a longer time period in order to see the steady-state section of the waveform. For instance, if we decrease the cut-off frequency to 1 Hz (Fig. 5.155) and increase the Total Time box to 5 (Fig. 5.156), then the result shown in Fig. 5.157 will be obtained. The ripple of steady-state waveform decreased considerably. A good value for Cut-off Frequency is around 20 Hz for a 60 Hz system.

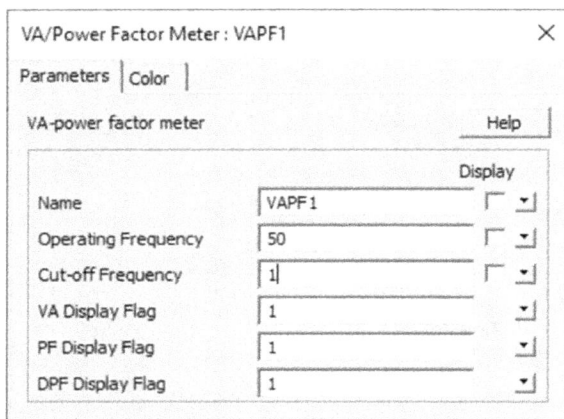

Fig. 5.155.

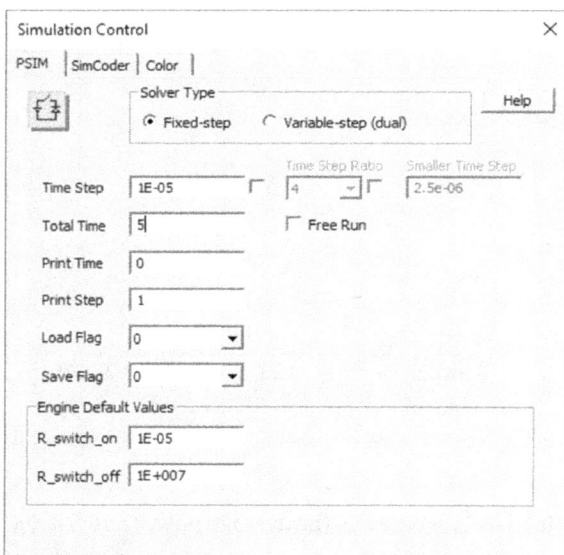

Fig. 5.156.

VAPF1_PF

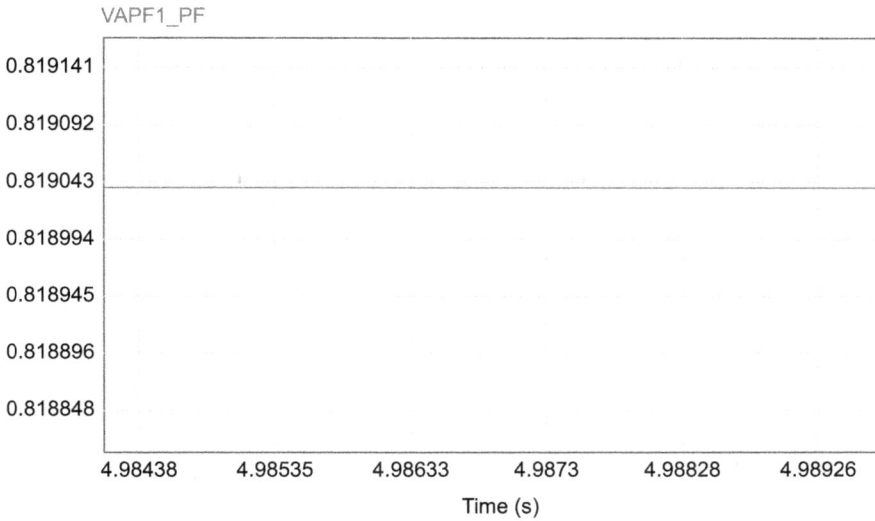

Fig. 5.157.

The displacement factor of this circuit is around 0.78 as shown in Fig. 5.158.

VAPF1_DPF

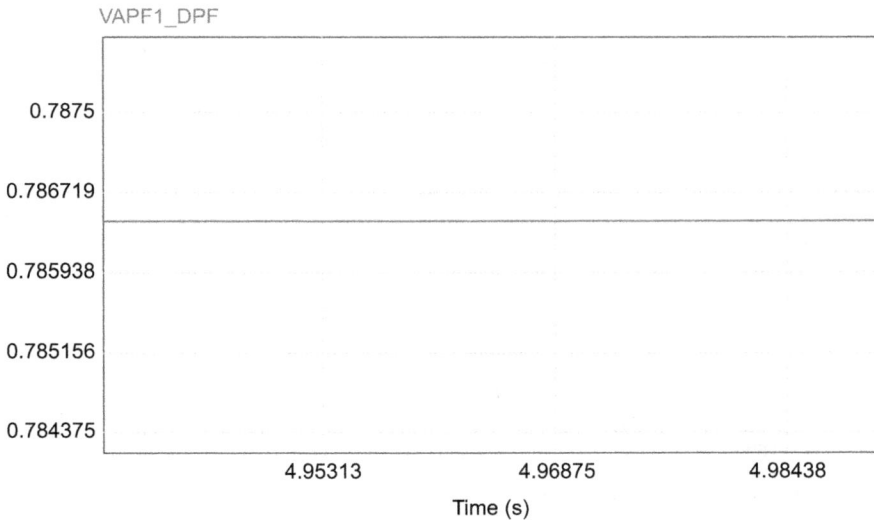

Fig. 5.158.

Let's use check the simulation results with MATLAB. The following MATLAB code analyzes the circuit shown in Fig. 5.143.

```
%This program calulates the circuits real power,
%power factor, apparent power and displacement factor
```

```
%clean the screen and variables in workspace
clc
clear all

%Circuit parameters
R=40;
C=101.32e-6;
f=50;
w=2*pi*f;
j=sqrt(-1);

%Input sources parameters
theta1=0;
theta2=0;
theta3=0;
V1=311*exp(j*theta1);
V2=150*exp(j*theta2);;
V3=50*exp(j*theta3);;

%Impedances at f=50 Hz, 2f=100Hz and 3f=150Hz
%in fact we use (Super Position) to find the current
Z1=R-j/2/pi/f/C;
Z2=R-j/2/pi/(2*f)/C;
Z3=R-j/2/pi/(3*f)/C;

%circuit's current components
i1=V1/Z1;
I1=abs(i1);
phi1=angle(i1);

i2=V2/Z2;
I2=abs(i2);
phi2=angle(i2);

i3=V3/Z3;
I3=abs(i3);
phi3=angle(i3);

%Apparent power calculation (VA)
IRMS=sqrt(.5*(I1^2+I2^2+I3^2))
VRMS=sqrt(.5*(V1^2+V2^2+V3^2))
ApparentPower=VRMS*IRMS

%computing the real power (Watts)
i= @(t) I1*sin(w*t+phi1)+I2*sin(2*w*t+phi2)+I3*sin(3*w*t+phi3);
v= @(t) V1*sin(w*t+theta1)+V2*sin(2*w*t+theta2)+V3*sin(3*w*t+theta3);
power=@(t) i(t).*v(t);
RealPower=f*integral(power,0,1/f)

%Power factor calculation
PowerFactor=RealPower/ApparentPower
%Displacement factor calculation
DisplacementFactor=cos(theta1-phi1)
```

After running the code, the result shown in Fig. 5.159 will be obtained. The result is the same as PSIM analysis.

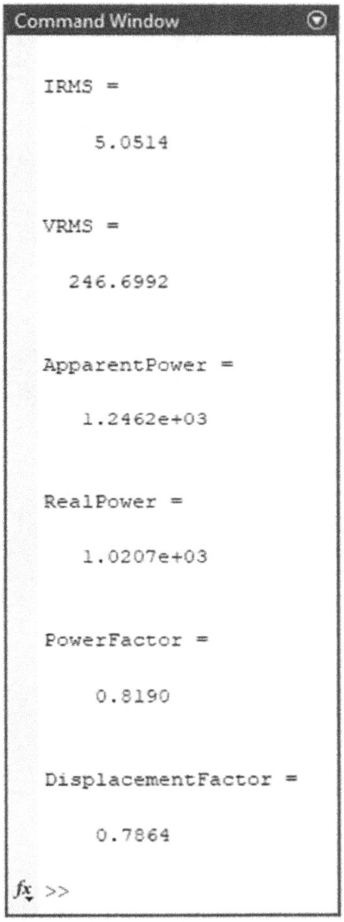

```
Command Window

  IRMS =

      5.0514


  VRMS =

    246.6992


  ApparentPower =

     1.2462e+03


  RealPower =

     1.0207e+03


  PowerFactor =

      0.8190


  DisplacementFactor =

      0.7864

fx >>
```

Fig. 5.159.

5.16 Example 15: half-wave rectifier

In this example, we will simulate a half-wave rectifier. The schematic is shown in Fig. 5.160. The voltage drop of the diode is assumed to be 0.7 V.

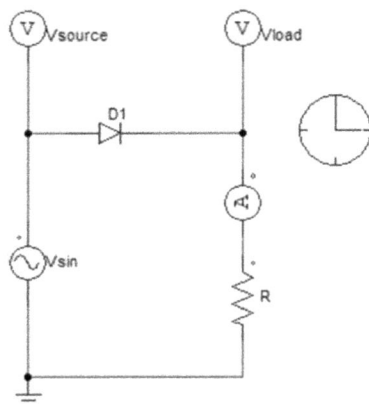

Fig. 5.160.

The settings of the blocks are shown in Figs. 5.161–5.167.

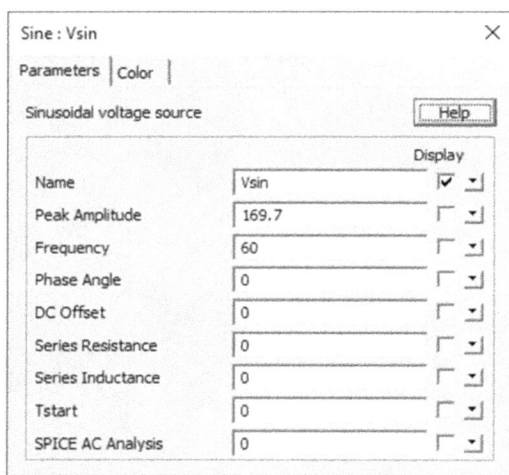

Sine : Vsin		✕
Parameters │ Color │		
Sinusoidal voltage source		[Help]
		Display
Name	Vsin	✓ ▾
Peak Amplitude	169.7	⌐ ▾
Frequency	60	⌐ ▾
Phase Angle	0	⌐ ▾
DC Offset	0	⌐ ▾
Series Resistance	0	⌐ ▾
Series Inductance	0	⌐ ▾
Tstart	0	⌐ ▾
SPICE AC Analysis	0	⌐ ▾

Fig. 5.161.

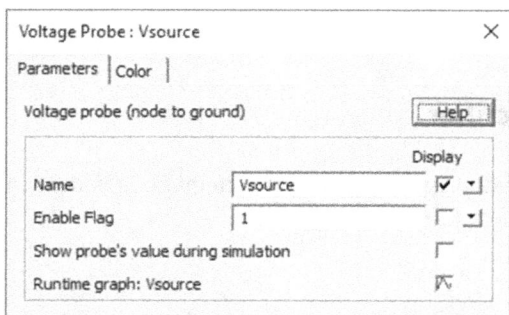

Voltage Probe : Vsource		✕
Parameters │ Color │		
Voltage probe (node to ground)		[Help]
		Display
Name	Vsource	✓ ▾
Enable Flag	1	⌐ ▾
Show probe's value during simulation		⌐
Runtime graph: Vsource		〃

Fig. 5.162.

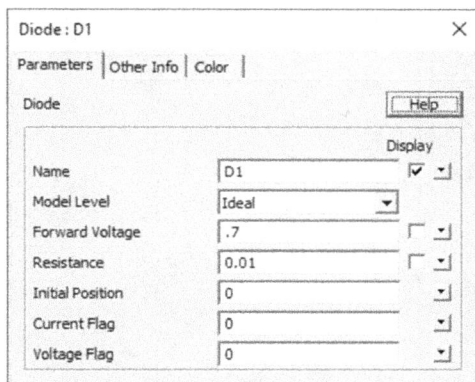

Fig. 5.163.

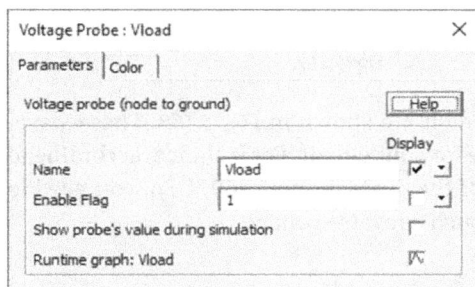

Fig. 5.164.

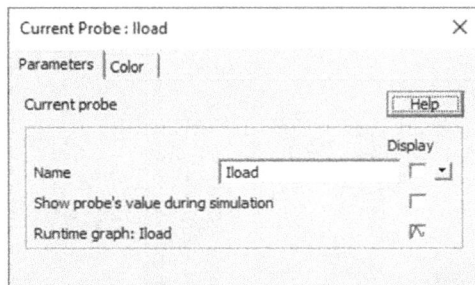

Fig. 5.165.

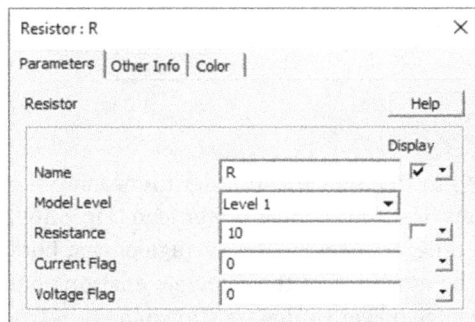

Fig. 5.166.

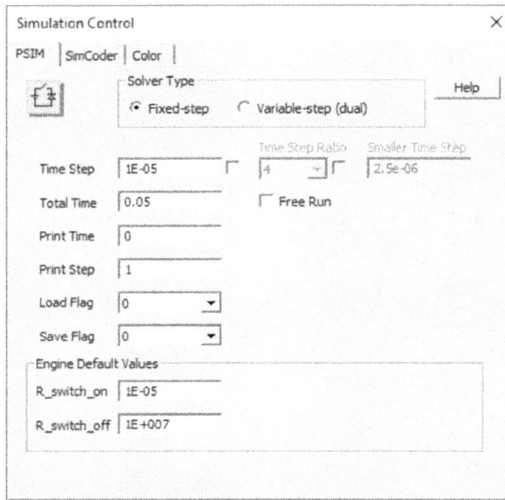

Fig. 5.167.

Run the simulation. The waveforms of circuit are shown in Fig. 5.168. These waveforms help you to select the suitable diode for your circuit. For instance, according to Fig. 5.168, the maximum reverse voltage of the diode is about 169 V. So, you need to select a diode which is able to withstand such a reverse voltage.

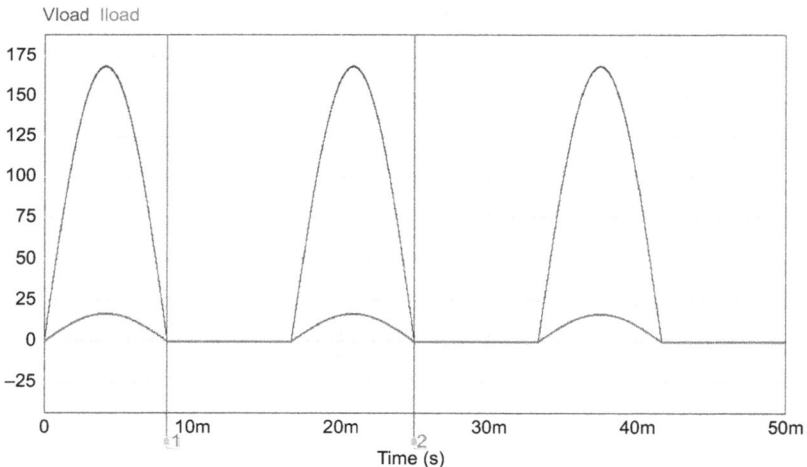

Fig. 5.168.

Use the average and RMS icons (Fig. 5.169) in the Simview in order to measure the average value and RMS of shown waveforms. Use the average power icon (Fig. 5.169) in order to measure the average power. In order to measure the average power, both the current and voltage waveforms must be available in the Simview environment simultaneously. The average, RMS, and average power values are shown in Fig. 5.170.

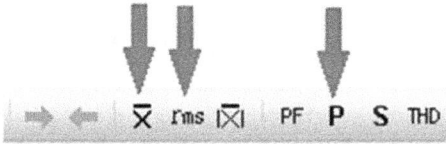

Fig. 5.169.

Measure							
⋮		X1	X2	Δ	Average	RMS	P
	Time	8.25635e-03	2.49423e-02	1.66859e-02			
	Iload	4.22005e-01	2.99044e-01	-1.22961e-01	5.35141e+00	8.42363e+00	7.09148e+02
	Vload	4.22005e+00	2.99044e+00	-1.22961e+00	5.35141e+01	8.42363e+01	

Fig. 5.170.

In order to measure the power factor, both the voltage source and current drawn from the source must be available in the Simview environment simultaneously. Use the power factor icon (Fig. 5.171) to measure the power factor of circuit.

Fig. 5.171.

After clicking the power factor icon, two cursors appear on the screen. Use these cursors to select one full period, that is, put one of the cursors in the beginning of a period and the other one in the end of that period (Fig. 5.172). The PSIM calculates the power factor and shows the value in the Measure window (Fig. 5.173).

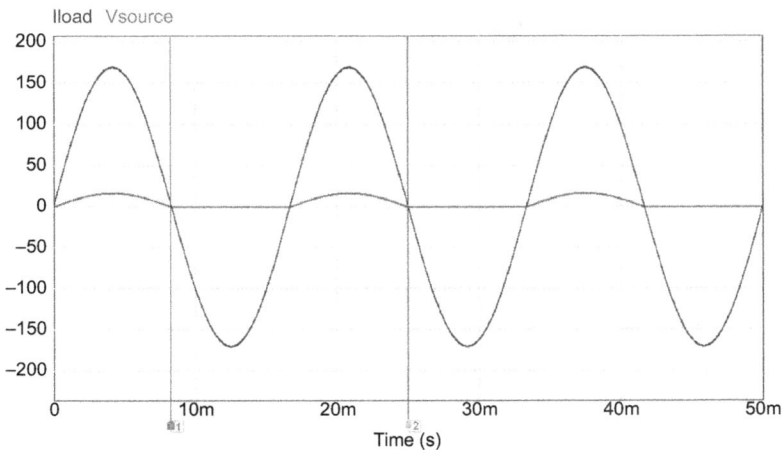

Fig. 5.172.

Measure				x
	X1	X2	Δ	PF
Time	8.25635e-03	2.49423e-02	1.66859e-02	
Iload	4.22005e-01	2.99044e-01	-1.22961e-01	7.06681e-01
Vsource	4.92427e+00	3.69343e+00	-1.23084e+00	

Fig. 5.173.

5.17 Example 16: measurement of source power and diode losses for half-wave rectifier circuit

We want to measure the power drawn from the source and diode losses for the circuit shown in Fig. 5.160. In order to see the graph of power drawn from the source, we need to draw the product of source voltage into the current drawn from the source (Fig. 5.174). The waveform of power drawn from the source is shown in Fig. 5.175. You can use the average icon (Fig. 5.169) to measure the average power drawn from the source.

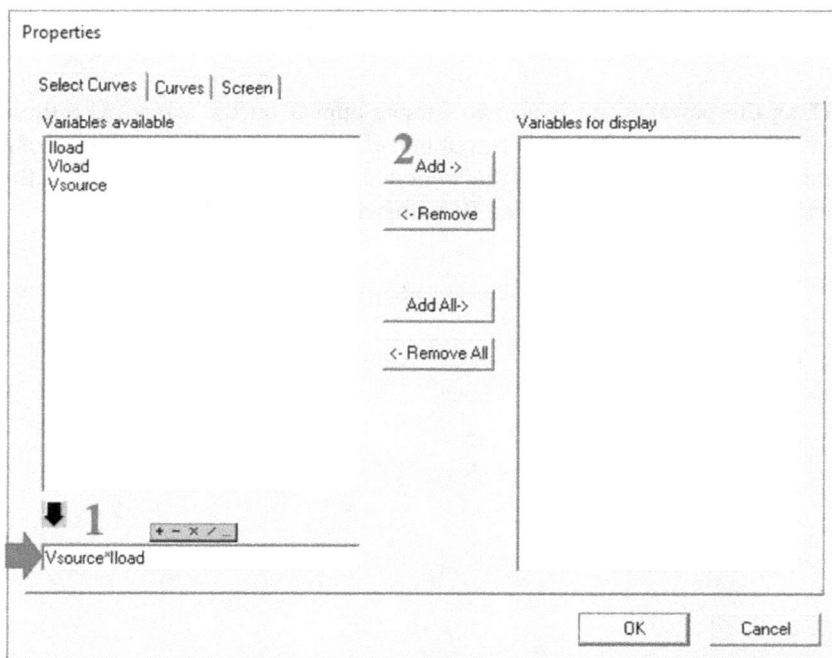

Fig. 5.174.

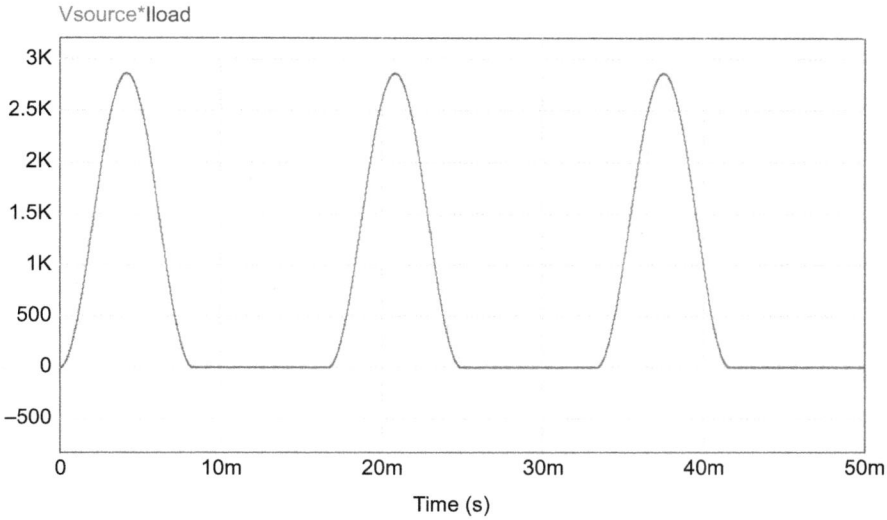

Fig. 5.175.

We need the voltage and current of diode in order to measure the diode losses. Double click on the diode and set the Current Flag and Voltage Flag to 1 (Fig. 5.176). Then, draw the product of voltage and current of diode (Fig. 5.177).

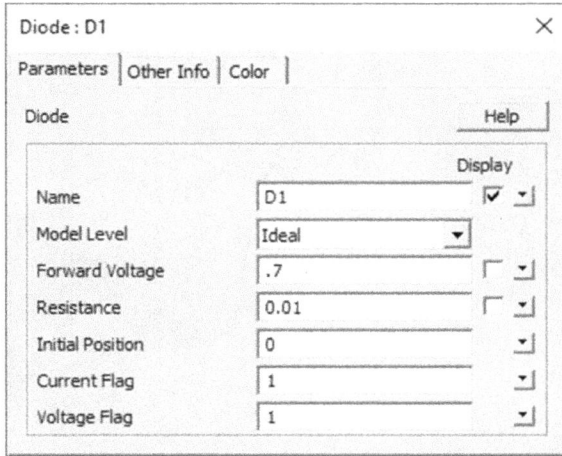

Fig. 5.176.

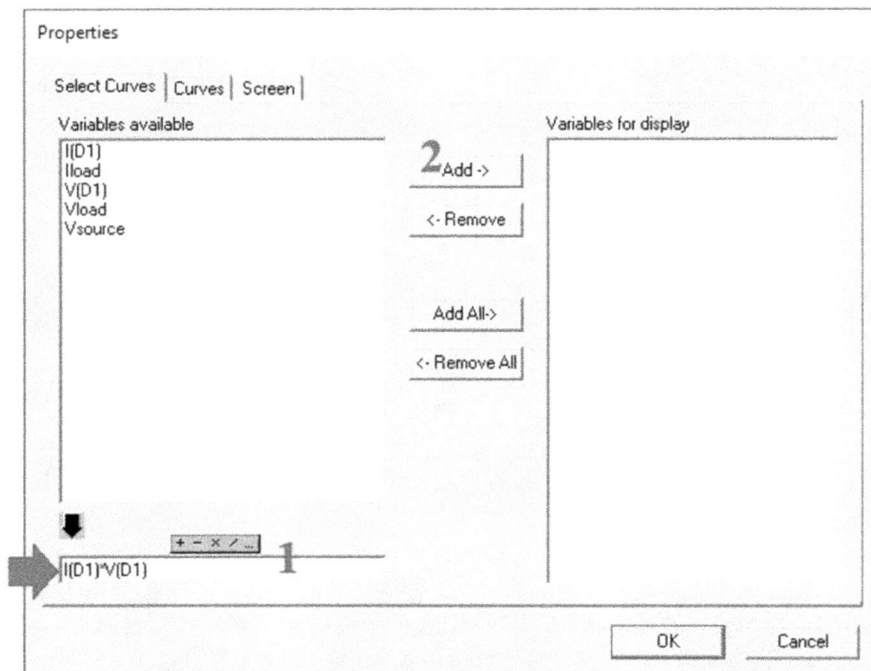

Fig. 5.177.

The waveform for power dissipated in the diode is shown in Fig. 5.178.

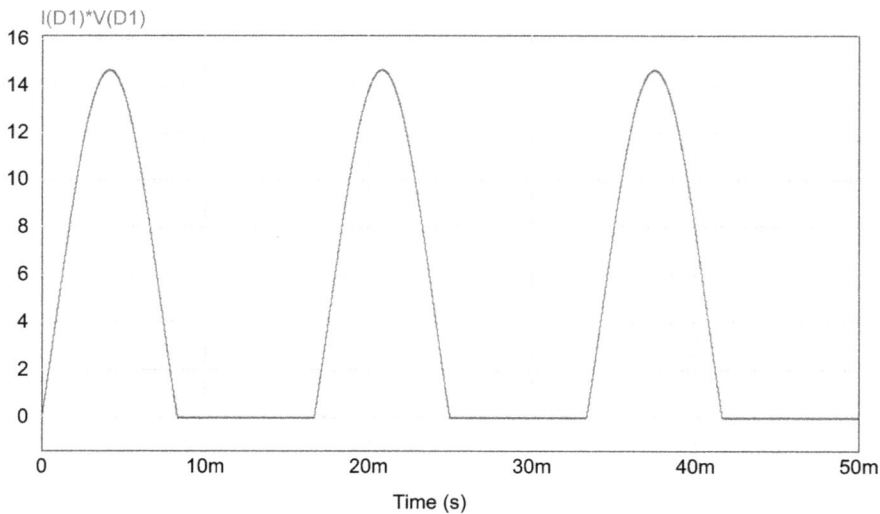

Fig. 5.178.

You can use the average icon in order to measure the dissipated average power in the diode. According to Fig. 5.179, the average power dissipated in the diode is about 0.45 W.

Measure		X1	X2	Δ	Average
	Time	8.29519e-03	2.50000e-02	1.67048e-02	
	I(D1)*V(D1)	1.21981e-01	0.00000e+00	-1.21981e-01	4.44988e+00

Fig. 5.179.

Let's use the hand analysis in order to check the results. The load voltage waveform is shown in Fig. 5.180.

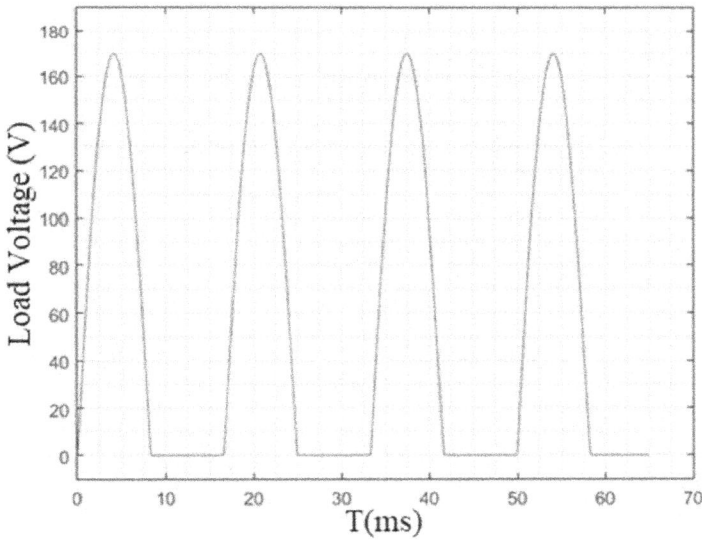

Fig. 5.180.

$$V_{load,AVERAGE} = \frac{V_m}{\pi} = \frac{169.7}{\pi} = 54.02\,V$$

$$I_{load,AVERAGE} = \frac{V_m}{\pi R} = \frac{169.7}{10\pi} = 5.4\,A$$

$$V_{load,\,RMS}=\frac{V_m}{2}=\frac{169.7}{2}=84.9\text{ V}$$

$$I_{load,\,RMS}=\frac{V_m}{2R}=\frac{169.7}{2\times10}=8.5\text{ A}$$

$$P_{load,\,AVERAGE}=RI_{RMS}^{\,2}=10\times8.5^2=722.5\text{ W}$$

$$\text{P.F.}=\frac{P}{S}=\frac{722.5}{120\times8.5}=0.708$$

Result of hand analysis is quite close to the PSIM values. Note that in hand analysis we ignore the voltage drop of diode for simplicity. So, there is a small difference between the hand analysis and PSIM results.

5.18 Example 17: single-phase full-wave uncontrolled rectifier

In this example, we will simulate a single-phase full-wave rectifier. Assume the circuit shown in Fig. 5.181.

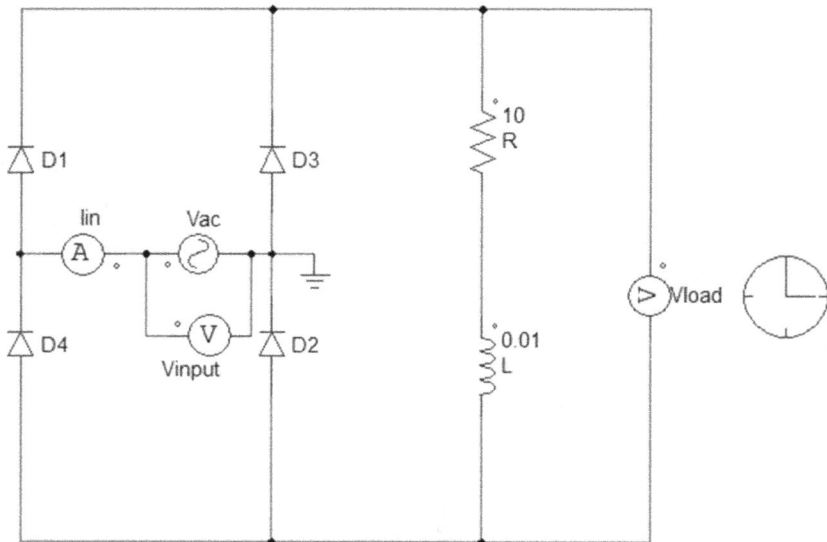

Fig. 5.181.

The settings of the elements are shown in Figs. 5.182–5.188.

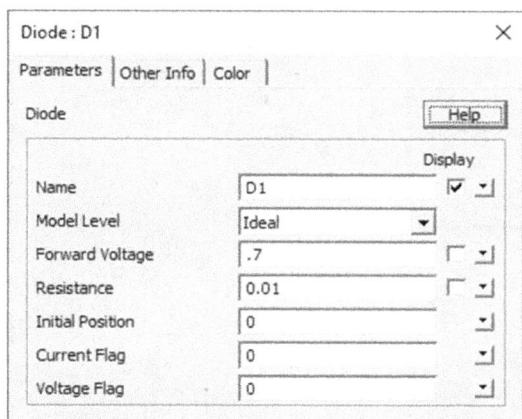

Fig. 5.182.

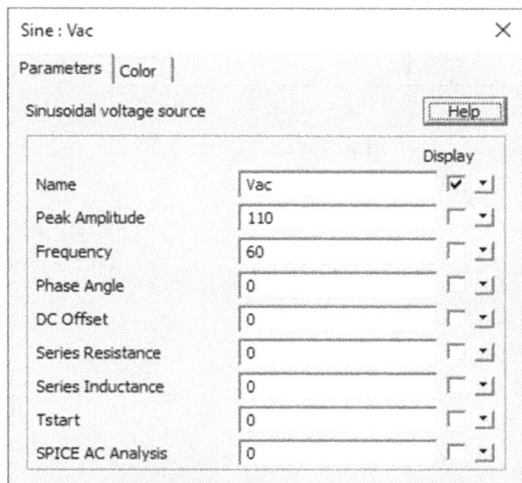

Fig. 5.183.

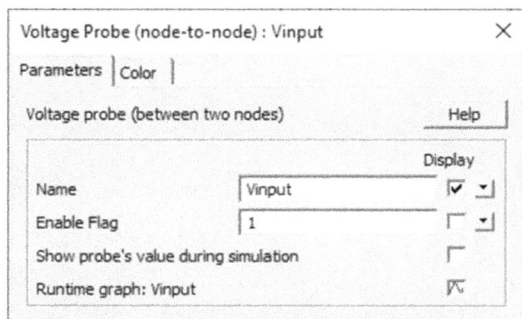

Fig. 5.184.

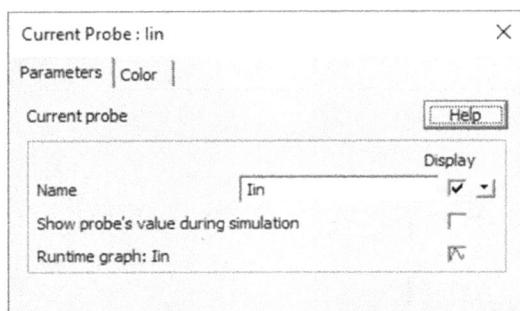

Fig. 5.185.

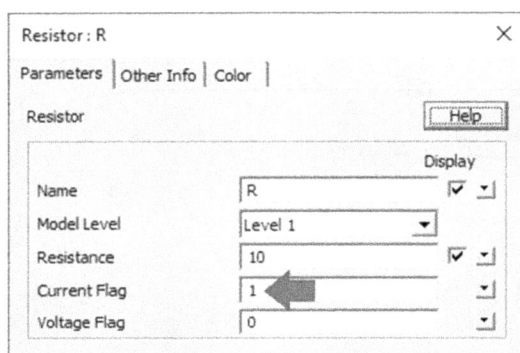

Fig. 5.186.

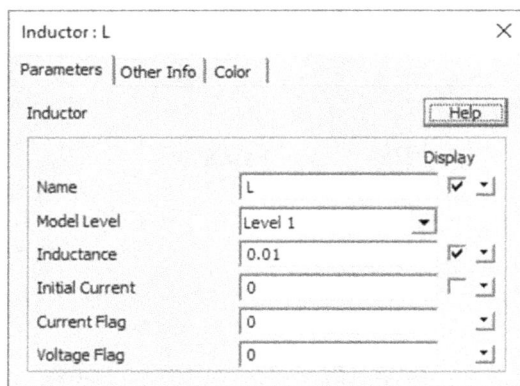

Fig. 5.187.

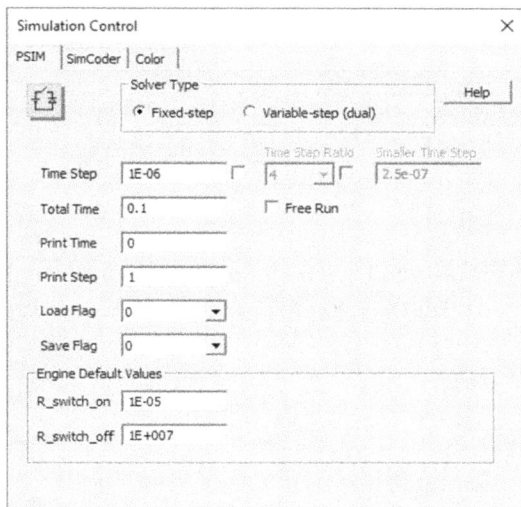

Fig. 5.188.

Run the simulation. The waveform for load voltage is shown in Fig. 5.189. Use two cursors in order to measure the frequency of this waveform. As shown in Fig. 5.190, frequency of load voltage is 120 Hz, which is two times bigger than the frequency of input source.

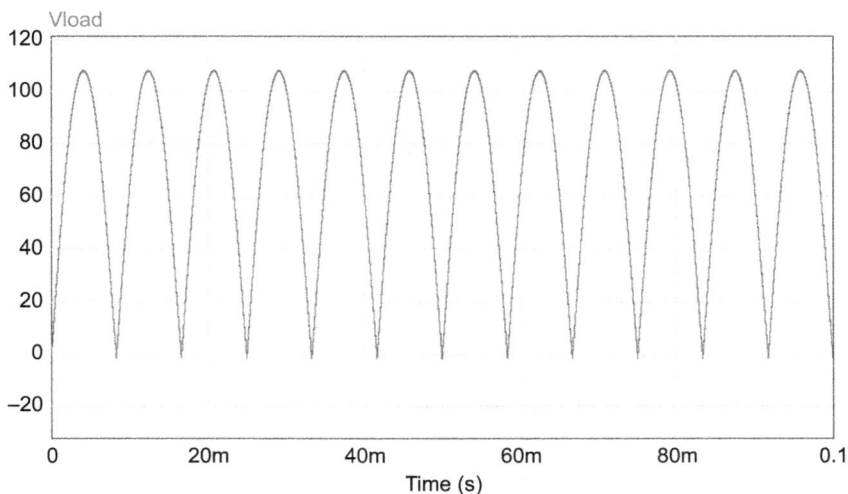

Fig. 5.189.

Measure		X1	X2	Δ	1/Δ
⋮	Time	2.50135e-02	3.33340e-02	8.32049e-03 🔒	1.20185e+02
	Vload	-9.10168e-01	-1.43478e+00	-5.24613e-01	

Fig. 5.190.

In order to measure the power factor, we need to have both the source voltage and current on the screen simultaneously (Fig. 5.191). According to Fig. 5.192, the power factor of this circuit is 0.96.

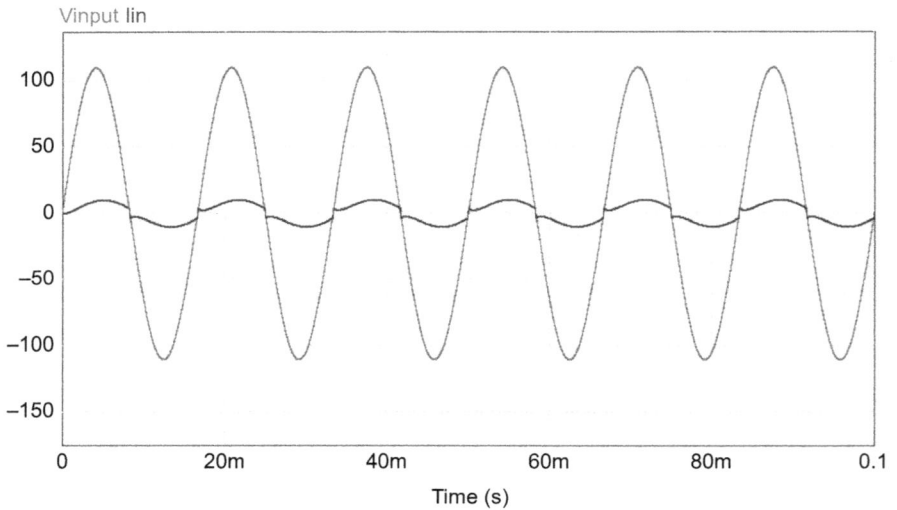

Fig. 5.191.

Measure		X1	X2	Δ	PF
⋮	Time	2.07657e-02	5.40603e-02	3.32947e-02 🔒	
	Vinput	1.09964e+02	1.09912e+02	-5.25894e-02	9.63309e-01
	Iin	9.50071e+00	9.44810e+00	-5.26102e-02	

Fig. 5.192.

If you increase the inductor value to 0.1 H, then the power factor will be decreased to 0.90 (Fig. 5.193). The input current for this case is shown in Fig. 5.194.

Measure		X1	X2	Δ	PF
	Time	2.07657e-02	3.75870e-02	1.68213e-02 🔒	
	Vinput	1.09964e+02	1.09941e+02	-2.33724e-02	9.02437e-01
	Iin	6.05565e+00	6.80396e+00	7.48317e-01	

Fig. 5.193.

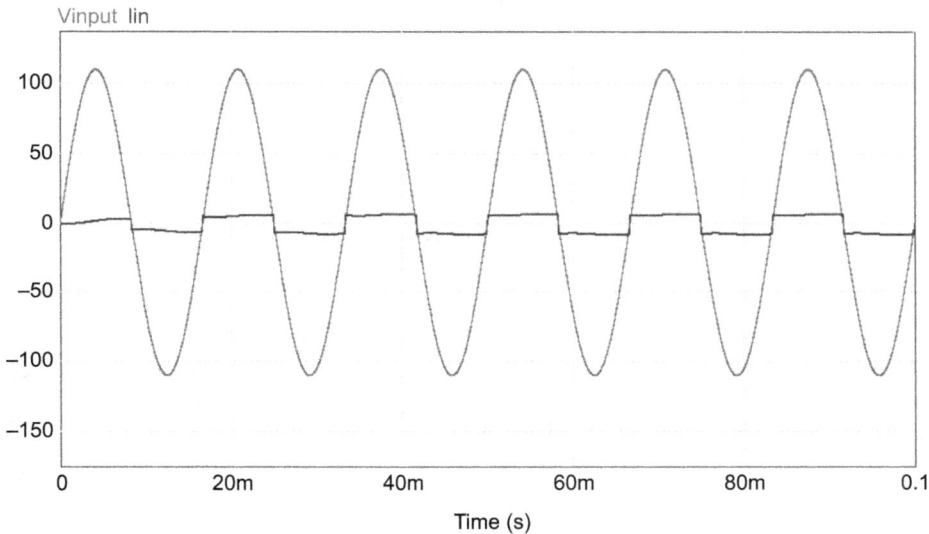

Fig. 5.194.

The load current and input current are shown in Fig. 5.195. The average values of these two currents are shown in Fig. 5.196. The average value of input current is about zero as expected. When diodes voltage drop is zero, the average load current is equal to $I = \frac{2V_m}{\pi}$. V_m and R show the peak of input voltage and the output load, respectively. For circuit of Fig. 5.181, this value equals to $I = \frac{\frac{2V_m}{\pi}}{R} = \frac{\frac{2 \times 110}{\pi}}{10} = 7.003\,A$. PSIM value (6.84885 A) is a little bit less than this value because PSIM considers the voltage drop of diodes. The formula $I = \frac{\frac{2V_m}{\pi}}{R}$ assumes that the voltage drop of diodes is zero.

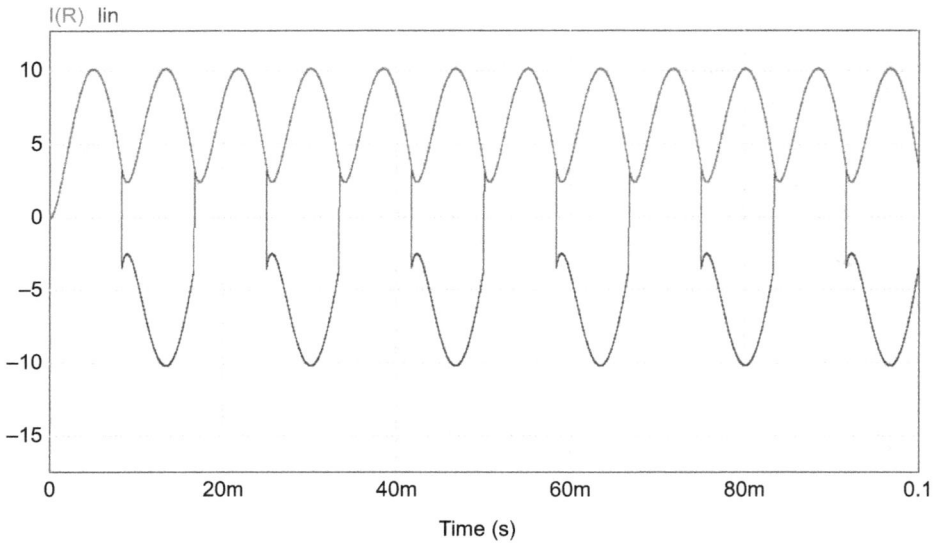

Fig. 5.195.

Measure	X1	X2	Δ	Average
Time	1.66667e-02	3.33333e-02	1.66667e-02	
I(R)	3.48018e+00	3.48018e+00	6.61489e-07	6.84885e+00
lin	5.91794e-08	1.11602e-07	5.24221e-08	1.38912e-07

Fig. 5.196.

5.19 Example 18: diode bridge block

PSIM has a ready-to-use single-phase diode bridge (Fig. 5.197). The schematic of previous example can be drawn with the aid of this block as shown in Fig. 5.198.

Fig. 5.197.

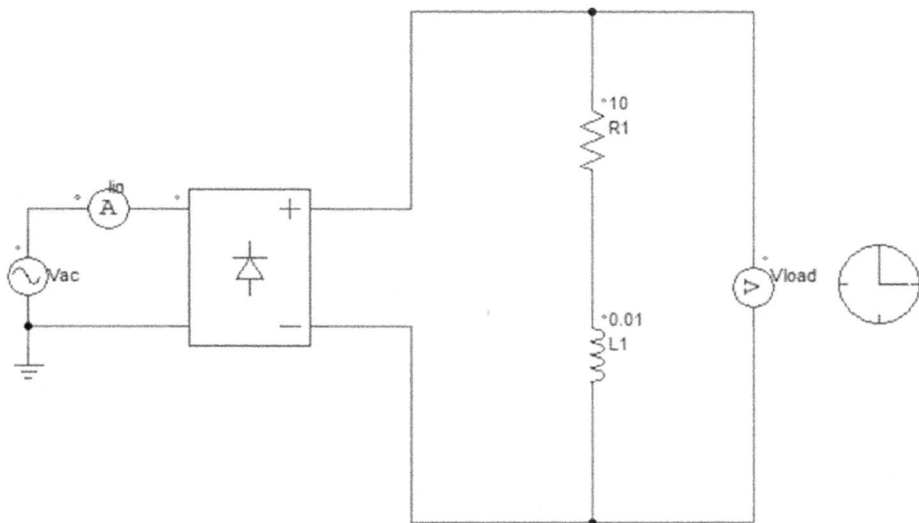

Fig. 5.198.

The equivalent circuit of this block is shown in Fig. 5.199.

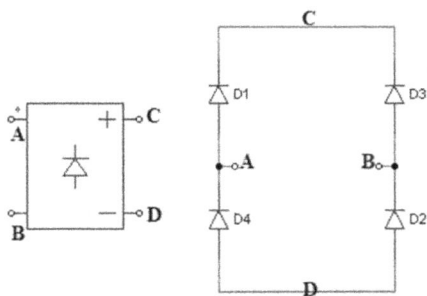

Fig. 5.199.

When you double click on the single-phase bridge block, its dialog window appears. You can set the voltage drop and series resistance of diode in this window. The current/ voltage flags permit you to monitor the current and voltage of each diode.

For instance, assume that we want to monitor the voltage and current of D3 in Fig. 5.199. In order to do this, enter the 0,0,1,0 to the Current Flag and Voltage Flag boxes. If you want to monitor diode D1 and D3, then enter 1,0,1,0 to the Current Flag and Voltage Flag boxes.

Sometimes, a circuit has non-zero initial inductor current. The initial inductor current needs a path. So, PSIM allow users to set the initial status of diodes. For instance, in order to set the initial status shown in Fig. 5.200, you can enter 0,0,1,1 to the Init. Position 1 . . . 4 box (Fig. 5.201).

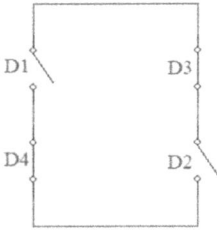

Fig. 5.200.

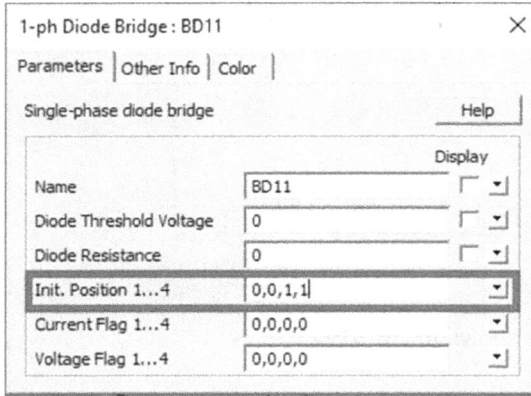

Fig. 5.201.

PSIM has a ready-to-use three-phase diode bridge as well.

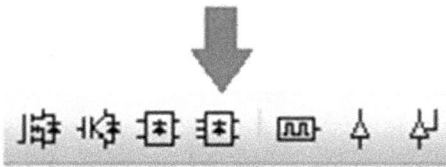

Fig. 5.202.

The equivalent circuit of this block is shown in Fig. 5.203.

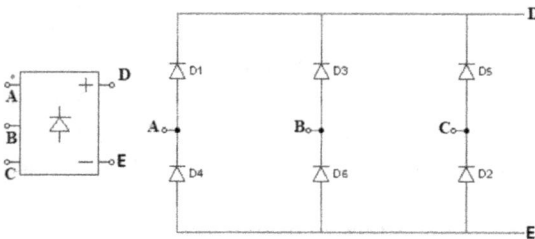

Fig. 5.203.

Beside the single-/three-phase diode bridges, PSIM has many ready-to-use switch modules. The modules are in the Switch Modules section Elements menu.

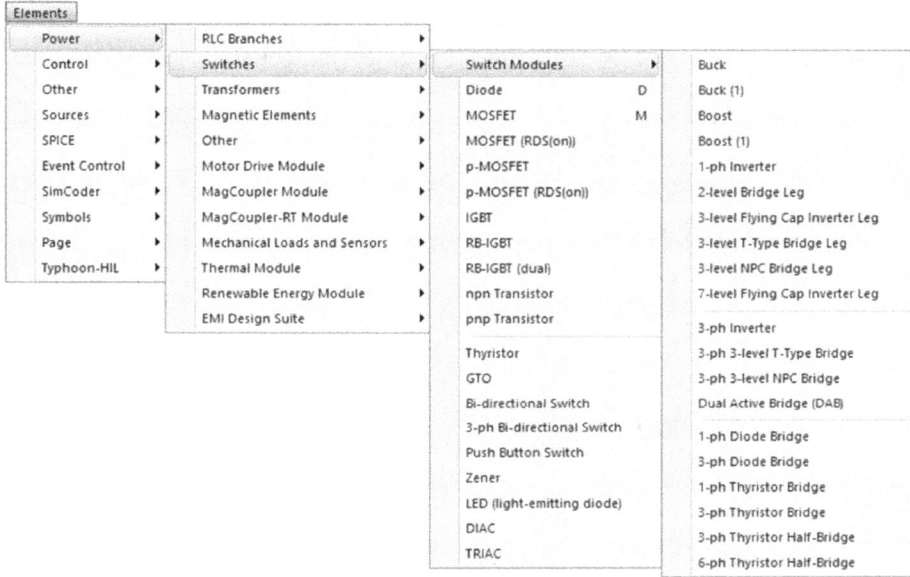

Elements				
Power ▸	RLC Branches ▸			
Control ▸	Switches ▸	Switch Modules ▸	Buck	
Other ▸	Transformers ▸	Diode D	Buck (1)	
Sources ▸	Magnetic Elements ▸	MOSFET M	Boost	
SPICE ▸	Other ▸	MOSFET (RDS(on))	Boost (1)	
Event Control ▸	Motor Drive Module ▸	p-MOSFET	1-ph Inverter	
SimCoder ▸	MagCoupler Module ▸	p-MOSFET (RDS(on))	2-level Bridge Leg	
Symbols ▸	MagCoupler-RT Module ▸	IGBT	3-level Flying Cap Inverter Leg	
Page ▸	Mechanical Loads and Sensors ▸	RB-IGBT	3-level T-Type Bridge Leg	
Typhoon-HIL ▸	Thermal Module ▸	RB-IGBT (dual)	3-level NPC Bridge Leg	
	Renewable Energy Module ▸	npn Transistor	7-level Flying Cap Inverter Leg	
	EMI Design Suite ▸	pnp Transistor	3-ph Inverter	
		Thyristor	3-ph 3-level T-Type Bridge	
		GTO	3-ph 3-level NPC Bridge	
		Bi-directional Switch	Dual Active Bridge (DAB)	
		3-ph Bi-directional Switch	1-ph Diode Bridge	
		Push Button Switch	3-ph Diode Bridge	
		Zener	1-ph Thyristor Bridge	
		LED (light-emitting diode)	3-ph Thyristor Bridge	
		DIAC	3-ph Thyristor Half-Bridge	
		TRIAC	6-ph Thyristor Half-Bridge	

Fig. 5.204.

5.20 Example 19: three-phase uncontrolled rectifier

In this example, we will study a three-phase uncontrolled (diode) rectifier. As shown in Fig. 5.205, PSIM has a ready-to-use three-phase four-wire source. This block will be used in this example.

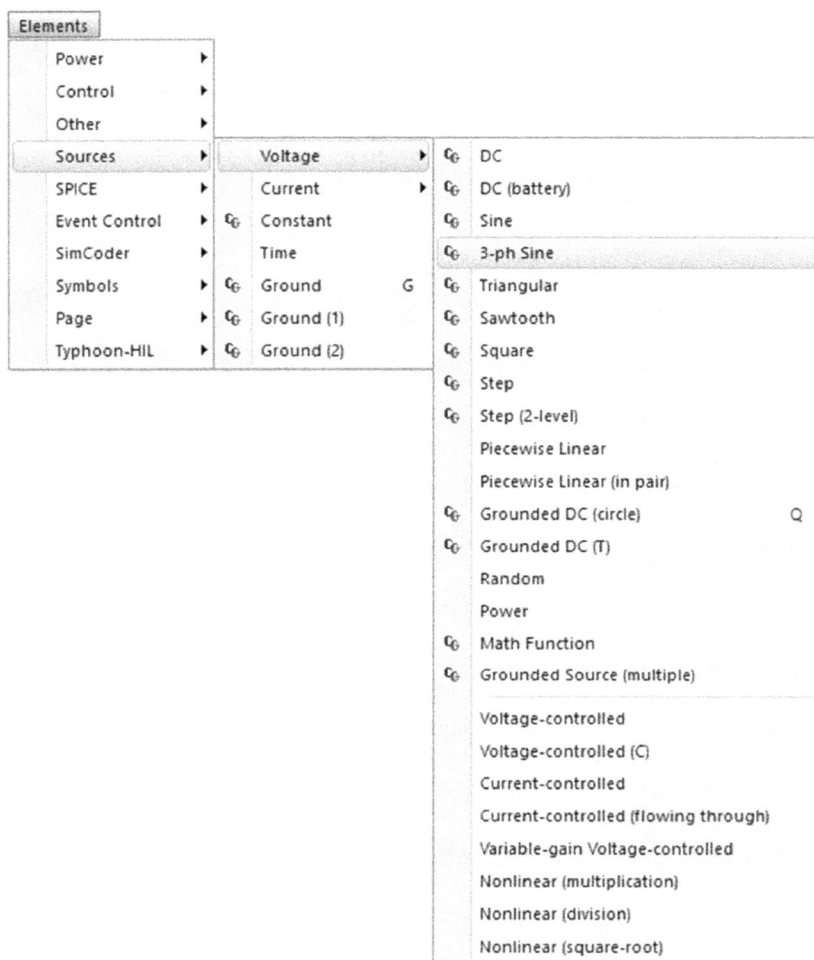

Elements

Power	▸		
Control	▸		
Other	▸		
Sources	▸	Voltage	▸	⚉ DC
SPICE	▸	Current	▸	⚉ DC (battery)
Event Control	▸	⚉ Constant		⚉ Sine
SimCoder	▸	Time		⚉ 3-ph Sine
Symbols	▸	⚉ Ground	G	⚉ Triangular
Page	▸	⚉ Ground (1)		⚉ Sawtooth
Typhoon-HIL	▸	⚉ Ground (2)		⚉ Square

⚉ Step

⚉ Step (2-level)

Piecewise Linear

Piecewise Linear (in pair)

⚉ Grounded DC (circle)   Q

⚉ Grounded DC (T)

Random

Power

⚉ Math Function

⚉ Grounded Source (multiple)

Voltage-controlled

Voltage-controlled (C)

Current-controlled

Current-controlled (flowing through)

Variable-gain Voltage-controlled

Nonlinear (multiplication)

Nonlinear (division)

Nonlinear (square-root)

Fig. 5.205.

Assume the schematic shown in Fig. 5.206.

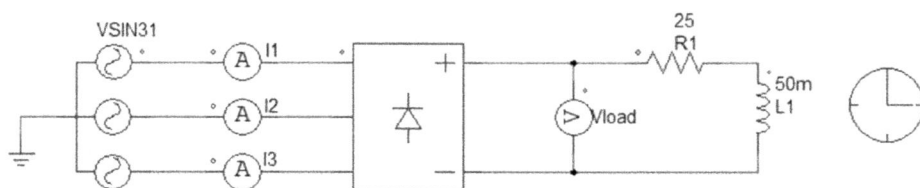

Fig. 5.206.

The settings of blocks are shown in Figs. 5.207–5.215.

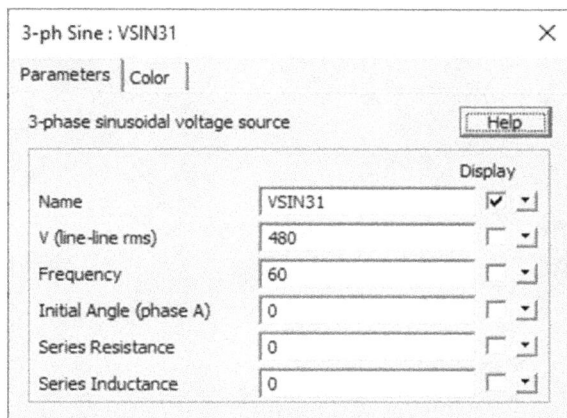

Fig. 5.207.

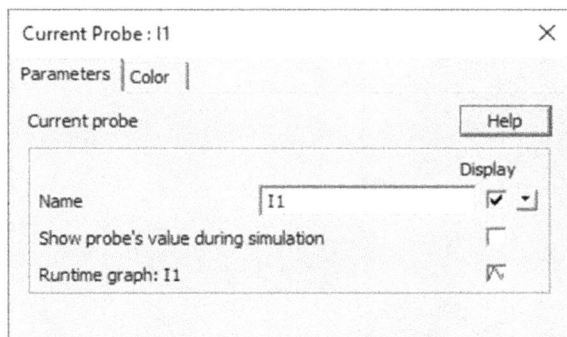

Fig. 5.208.

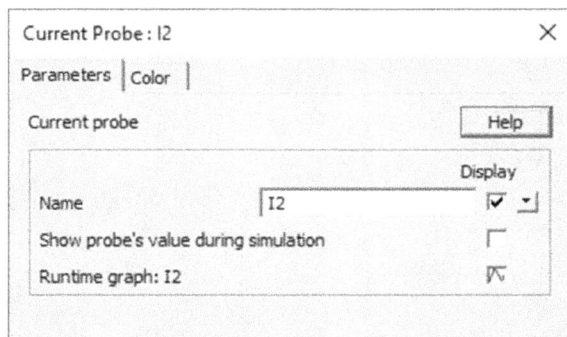

Fig. 5.209.

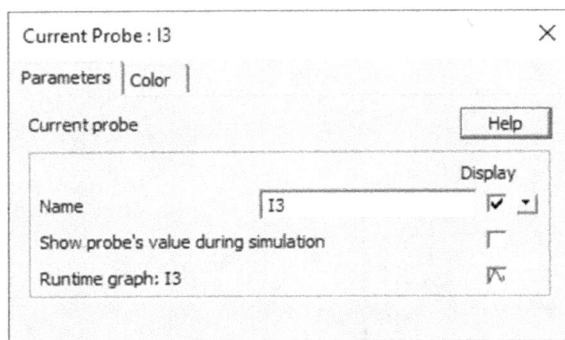

Fig. 5.210.

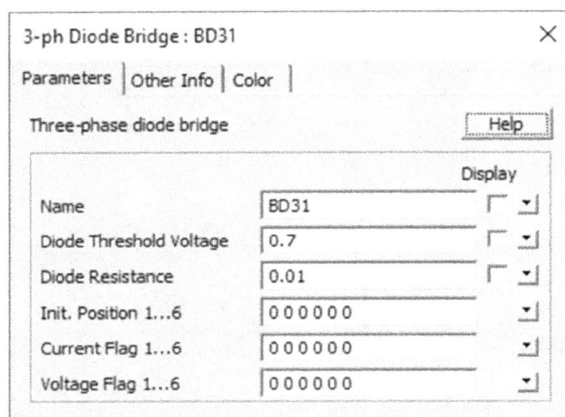

Fig. 5.211.

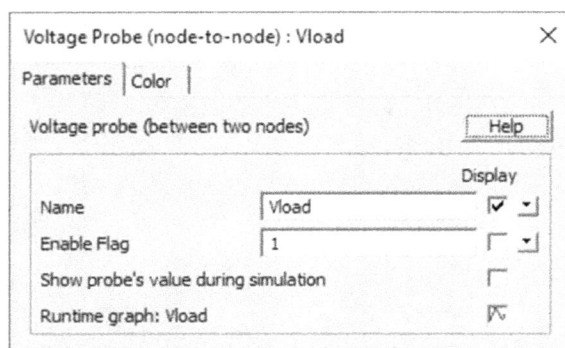

Fig. 5.212.

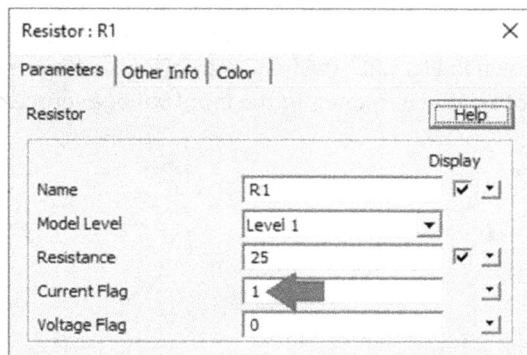

Fig. 5.213.

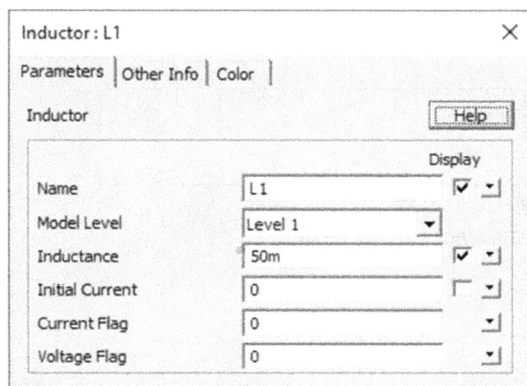

Fig. 5.214.

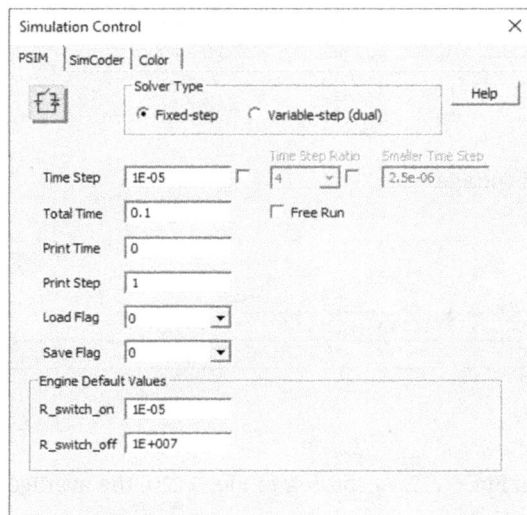

Fig. 5.215.

Run the simulation. Figure 5.216 shows the load voltage. Measure the frequency of this waveform with two cursors. As shown in Fig. 5.217, the frequency of load voltage is about 360 Hz. So, it is six times bigger than the frequency of the input voltage source.

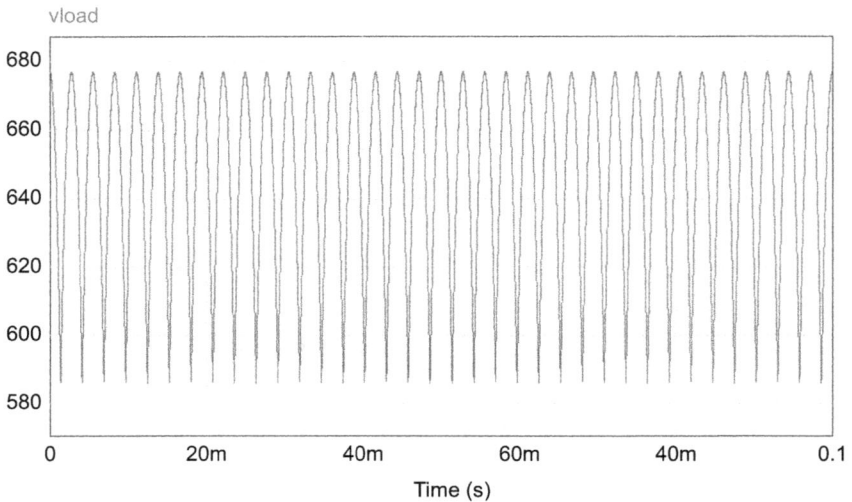

Fig. 5.216.

Fig. 5.217.

Measure the average and RMS of load voltage.

Fig. 5.218.

Waveform of load current is shown in Fig. 5.219. According to Fig. 5.220, the average value of load current is 25.85 A.

I(R1)

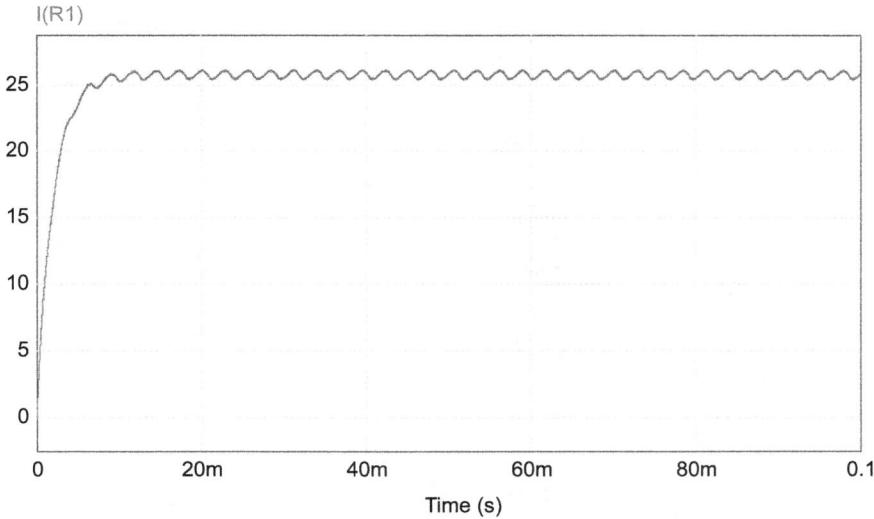

Fig. 5.219.

Measure					⊠
⋮		X1	X2	Δ	Average
	Time	2.01259e-02	2.56764e-02	5.55042e-03 ⟳	
	I(R1)	2.61578e+01	2.61588e+01	9.69035e-04	2.58524e+01

Fig. 5.220.

Let's check the results given by PSIM using hand analysis:

$$V_{Load} = \frac{3 \times V_{m,L-L}}{\pi} = \frac{3 \times \sqrt{2} \times 480}{\pi} = 648\,V$$

$$I_{load,DC} = \frac{V_{Load}}{R} = \frac{648}{25} = 25.9\,A$$

The obtained results are very close to values given by PSIM values. Note that in hand analysis we ignore the voltage drop of diodes, so the real values are a little bit smaller than what we obtained.

You can calculate the average value of diodes current and RMS value of current drawn from the source using the following equations.

$$I_{Diode,\,average} = \frac{I_{load,DC}}{3} = \frac{25.9}{3} = 8.63\,A$$

$$I_{Source,RMS} = \sqrt{\frac{2}{3}} I_{load,\,RMS} = \sqrt{\frac{2}{3}} \times 25.9 = 21.2$$

5.21 Example 20: single-phase half-wave controlled rectifier

In this example, we will simulate a single-phase half-wave controlled (thyristor) rectifier. Assume the circuit shown in Fig. 5.221.

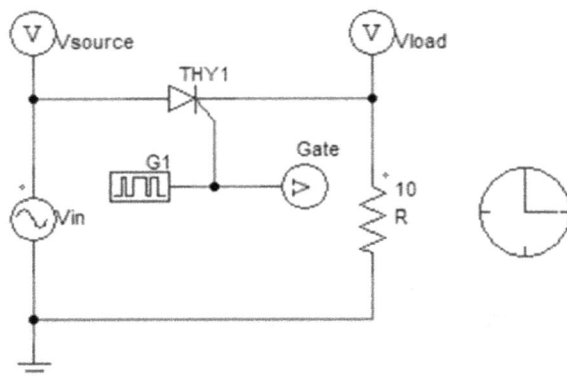

Fig. 5.221.

The settings of elements are shown in Figs. 5.222–5.229. The thyristor is controlled with a gating block. The triggering angle of thyristor is 30°. The thyristor requires a small pulse to be triggered. Because of this, the width of pulse is only 2° (=111 μs). You can use any other small values for width of trigger pulse.

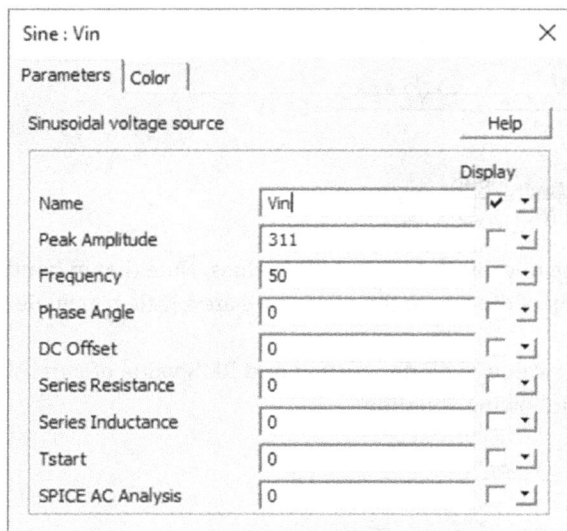

Fig. 5.222.

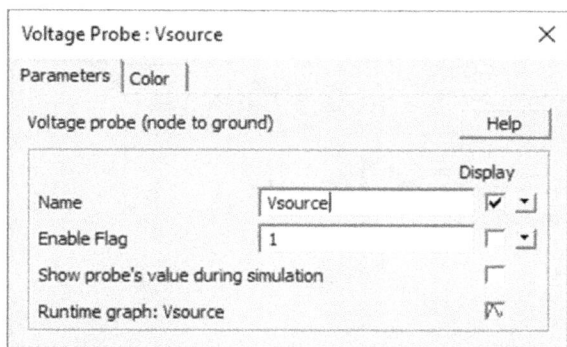

Fig. 5.223.

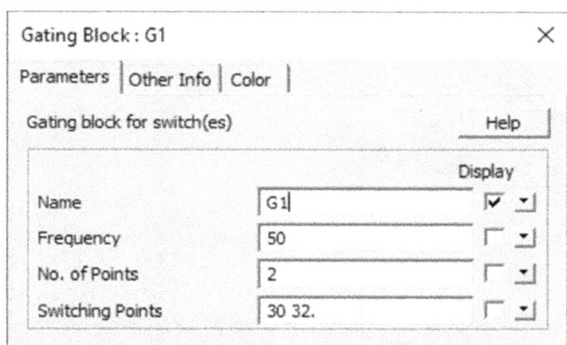

Fig. 5.224.

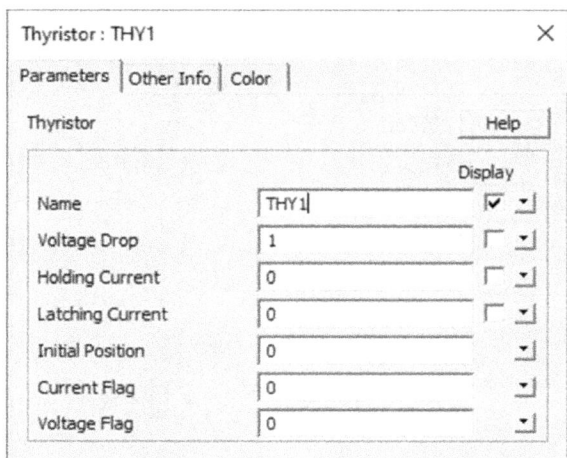

Fig. 5.225.

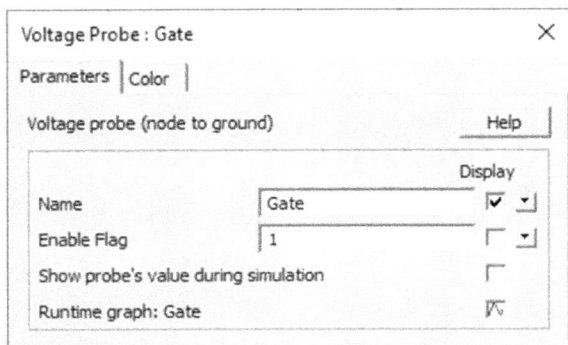

Fig. 5.226.

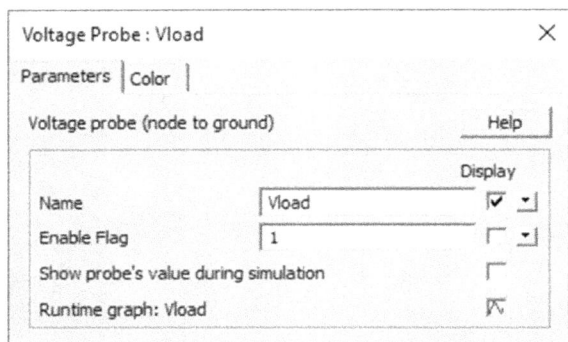

Fig. 5.227.

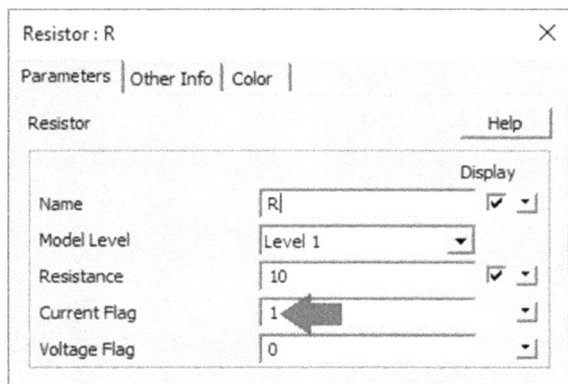

Fig. 5.228.

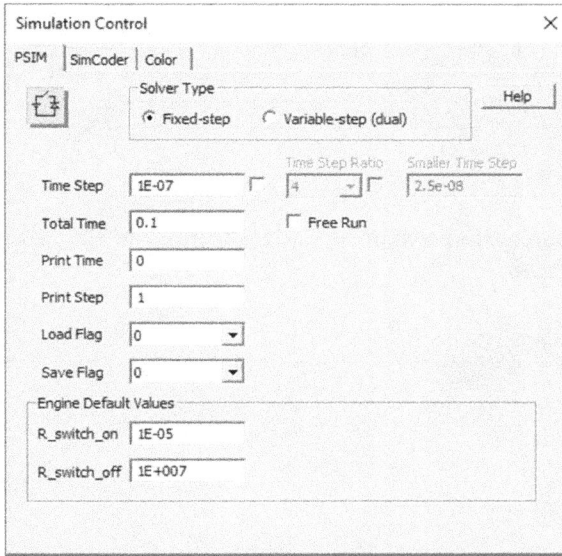

Fig. 5.229.

Run the simulation. Figure 5.230 shows the load voltage. The average and RMS values of this waveform are shown in Fig. 5.231.

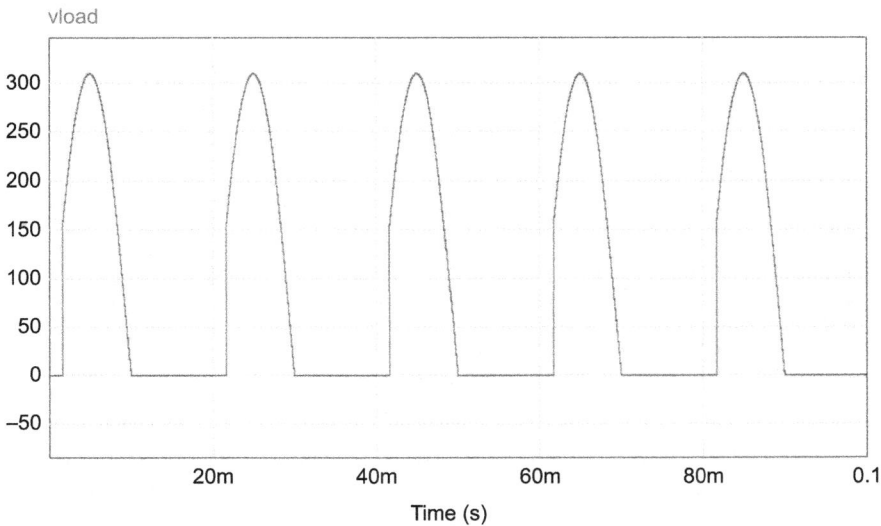

Fig. 5.230.

Measure

:		X1	X2	Δ	Average	RMS
	Time	2.15935e-02	6.16628e-02	4.00693e-02		
	Vload	1.49271e-04	1.55174e-04	5.90276e-06	9.17873e+01	1.52507e+02

Fig. 5.231.

Voltage and current of input AC source are shown in Fig. 5.232. According to Fig. 5.233, the power factor of this circuit is 0.696

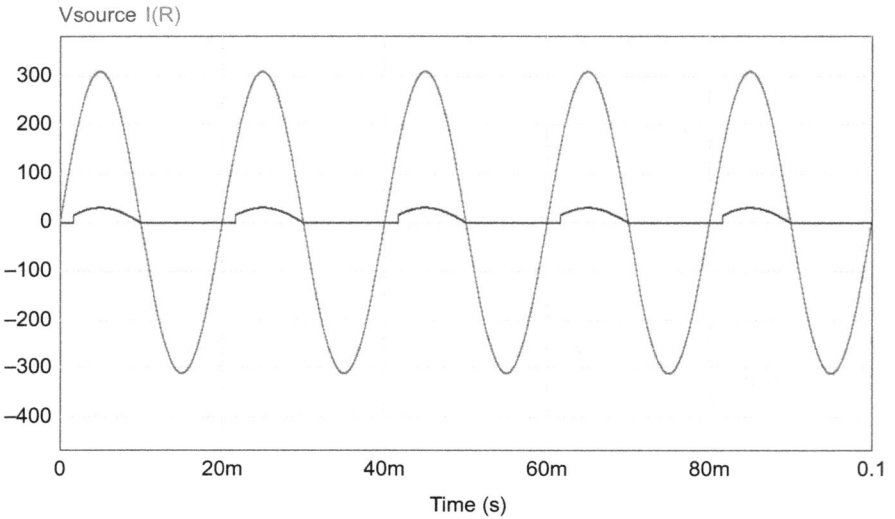

Vsource I(R)

Fig. 5.232.

Measure

:		X1	X2	Δ	PF
	Time	2.00001e-02	8.00000e-02	5.99999e-02	
	Vsource	7.81632e-03	1.95422e-03	-5.86210e-03	6.96837e-01
	I(R)	7.61158e-10	1.74949e-10	-5.86210e-10	

Fig. 5.233.

Let's check the PSIM results with hand analysis. In hand analysis, we ignore the voltage drop of thyristor, so our values are a little bit bigger than the PSIM values:

$$V_{o,AVERAGE} = \frac{1}{2\pi}\int_{\frac{\pi}{6}}^{\pi} 311\sin(\theta)\,d\theta = 92.36 \text{ V}$$

$$V_{o,RMS} = \sqrt{\frac{1}{2\pi}\int_{\frac{\pi}{6}}^{\pi}(311\sin(\theta))^2 d\theta} = 153.24\,\text{V}$$

$$P_{LOAD} = \frac{V_{RMS}^{2}}{R} = 2.348\,\text{kW}$$

$$I_{o,RMS} = \frac{V_{O,RMS}}{R} = \frac{153.24}{10} = 15.32\,\text{A}$$

$$S = V_{in,RMS} \times I_{in,RMS} = \frac{311}{\sqrt{2}} \times 15.32 = 3.37\,\text{kVA}$$

$$V_{in,RMS} = \frac{P_{LOAD}}{S} = 0.69$$

The above integrals could be calculated easily with MATLAB.

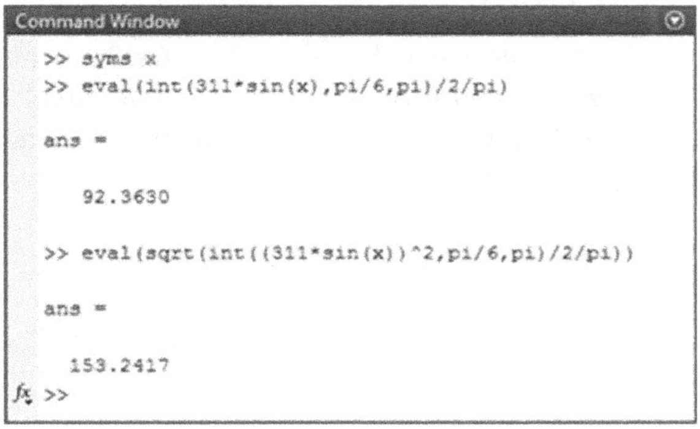

```
Command Window
>> syms x
>> eval(int(311*sin(x),pi/6,pi)/2/pi)

ans =

    92.3630

>> eval(sqrt(int((311*sin(x))^2,pi/6,pi)/2/pi))

ans =

   153.2417
fx >>
```

Fig. 5.234.

5.22 Example 21: single-phase full-wave controlled rectifier

In this example, we will study a single-phase full-wave controlled rectifier. We use alpha controller block (Fig. 5.235) in order to trigger the thyristors.

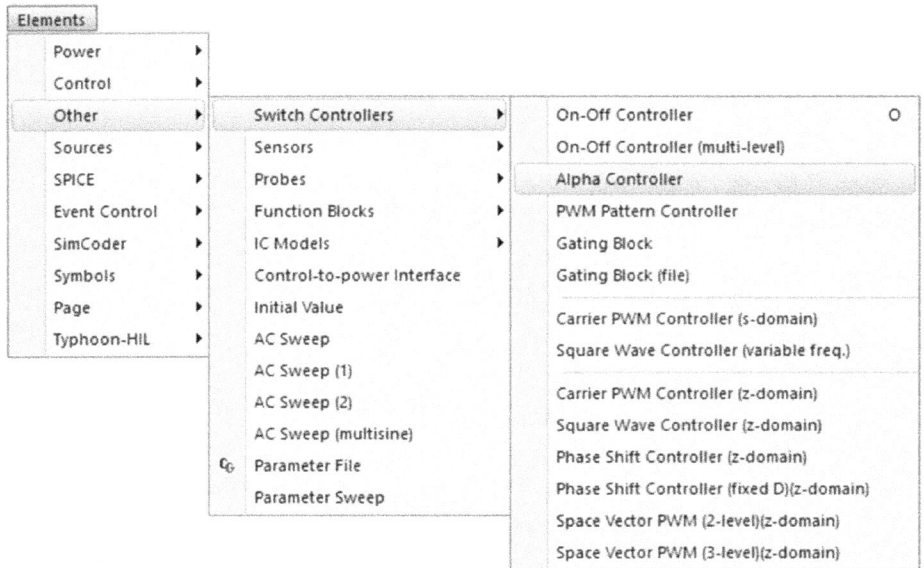

Fig. 5.235.

The alpha controller block is shown in Fig. 5.236. It has four terminals. Terminals 1, 2, and 3 are input terminals and terminal 4 is connected to the gate of thyristor. Synchronization signal enters to terminal 1, triggering angle (in degrees) enters to terminal 2 and enable/disable signal enters terminal 3. The alpha controller is enabled (or disabled) if the enable/disable signal is high (or low).

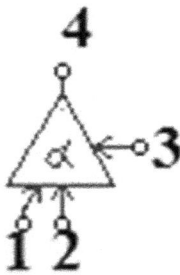

Fig. 5.236.

Assume the circuit shown in Fig. 5.237.

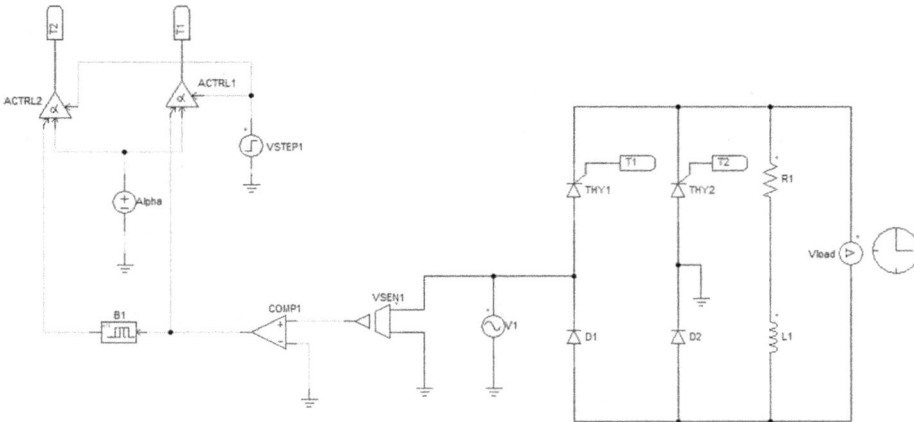

Fig. 5.237.

In the above schematic, we used Label block (Fig. 5.238) to connect the output of alpha controllers to the thyristors.

Fig. 5.238.

You can draw the schematic shown in Fig. 5.238 similar to the one shown in Fig. 5.239. Both of them are the same from PSIM viewpoint. However, Fig. 5.238 is easier to understand from user viewpoint.

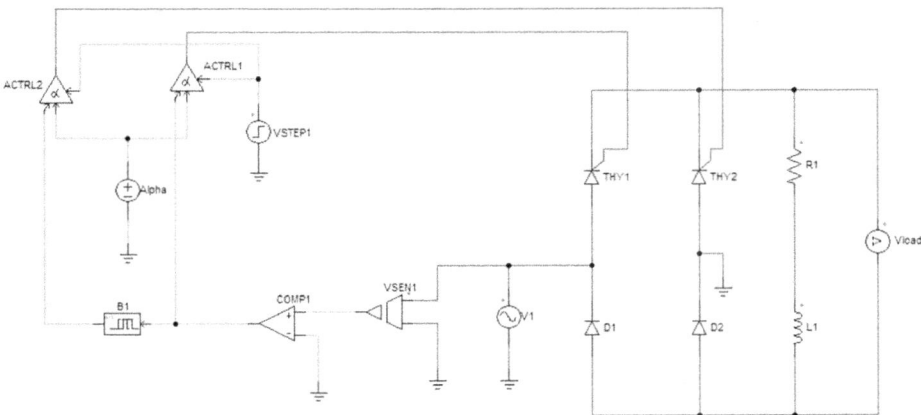

Fig. 5.239.

We used a time delay block in order give a delayed synchronization signal to alpha control 2 (block with ACTRL2 label). PSIM has two time delay blocks (Figs. 5.240 and 5.241). The Time Delay (analog) block delays the input signal by a specified amount of time interval. It can be used to delay analog and logic signals. However, for logic signals, the block Time Delay (logic) provides faster simulation speed. Both of them can be used in this simulation. The trigger angle of thyristors is 20°. The triggering angle of the thyristors is determined by the value of Alpha DC source.

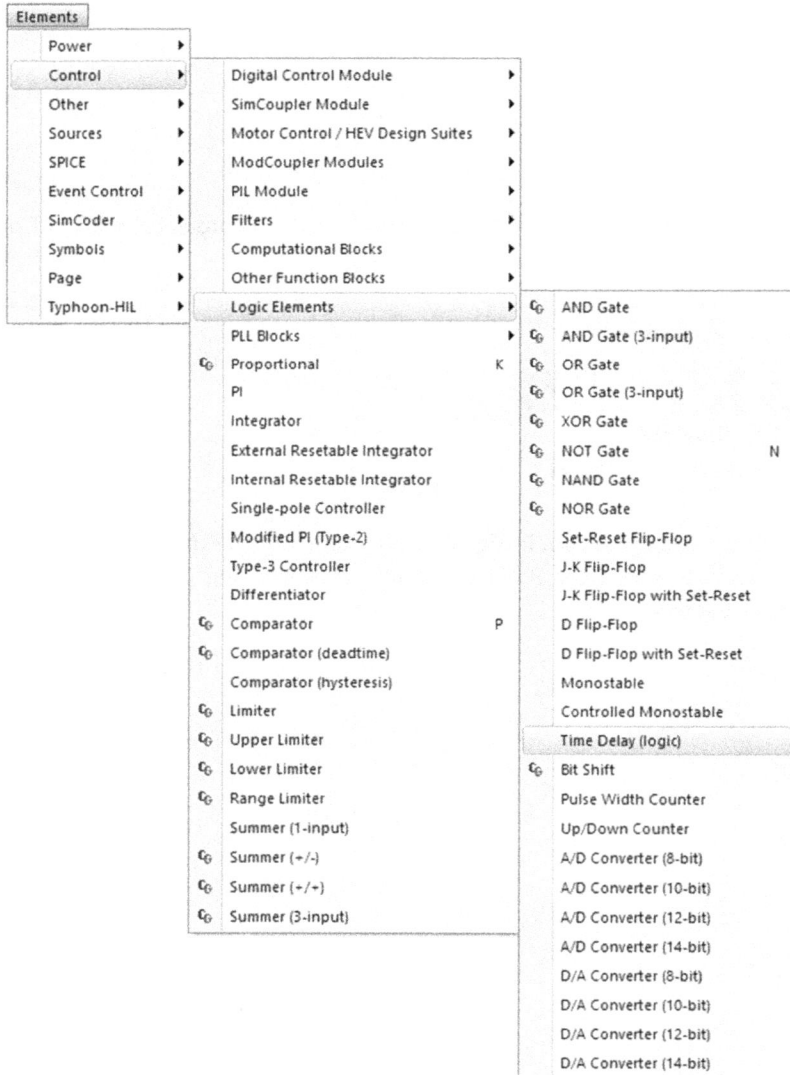

Fig. 5.240.

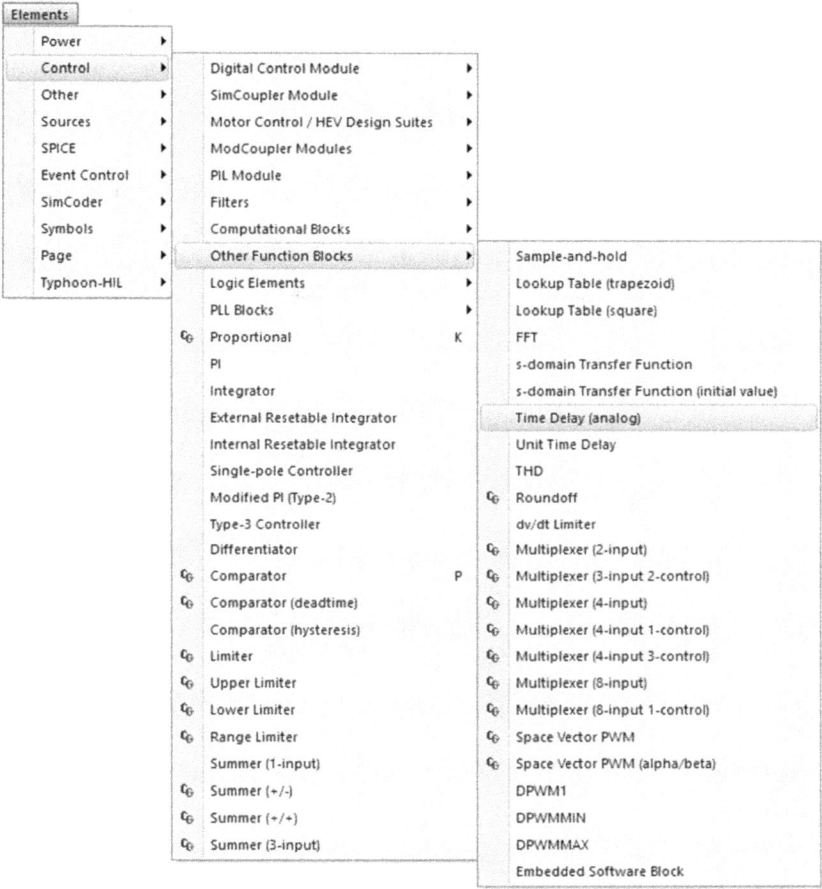

Elements		
Power ▶		
Control ▶	Digital Control Module ▶	
Other ▶	SimCoupler Module ▶	
Sources ▶	Motor Control / HEV Design Suites ▶	
SPICE ▶	ModCoupler Modules ▶	
Event Control ▶	PIL Module ▶	
SimCoder ▶	Filters ▶	
Symbols ▶	Computational Blocks ▶	
Page ▶	Other Function Blocks ▶	Sample-and-hold
Typhoon-HIL ▶	Logic Elements ▶	Lookup Table (trapezoid)
	PLL Blocks ▶	Lookup Table (square)
	Proportional K	FFT
	PI	s-domain Transfer Function
	Integrator	s-domain Transfer Function (initial value)
	External Resetable Integrator	Time Delay (analog)
	Internal Resetable Integrator	Unit Time Delay
	Single-pole Controller	THD
	Modified PI (Type-2)	Roundoff
	Type-3 Controller	dv/dt Limiter
	Differentiator	Multiplexer (2-input)
	Comparator P	Multiplexer (3-input 2-control)
	Comparator (deadtime)	Multiplexer (4-input)
	Comparator (hysteresis)	Multiplexer (4-input 1-control)
	Limiter	Multiplexer (4-input 3-control)
	Upper Limiter	Multiplexer (8-input)
	Lower Limiter	Multiplexer (8-input 1-control)
	Range Limiter	Space Vector PWM
	Summer (1-input)	Space Vector PWM (alpha/beta)
	Summer (+/-)	DPWM1
	Summer (+/+)	DPWMMIN
	Summer (3-input)	DPWMMAX
		Embedded Software Block

Fig. 5.241.

Settings of used elements are shown in Figs. 5.242–5.256. Settings of thyristor THY 2 and diode D2 is similar to Figs. 5.252 and 5.253, respectively.

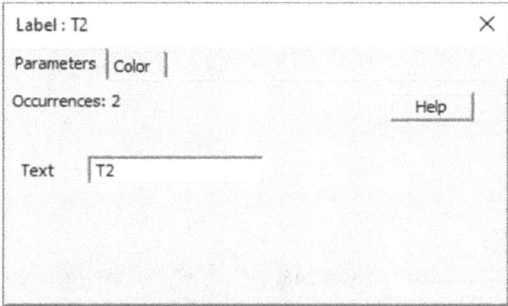

Label : T2   ×

Parameters | Color |

Occurrences: 2   Help |

Text | T2

Fig. 5.242.

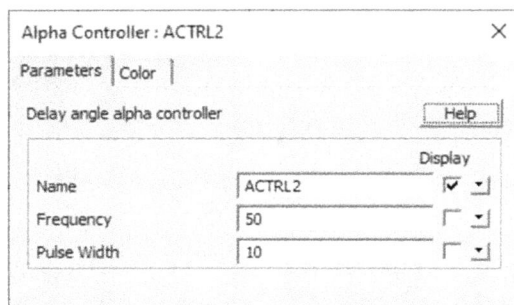

Fig. 5.243.

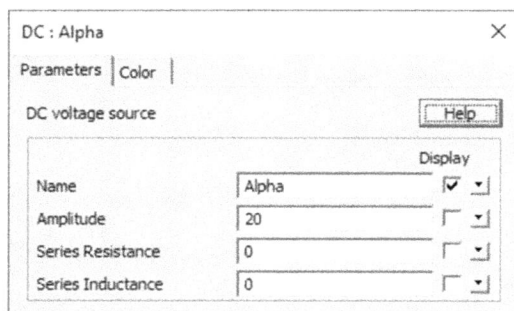

Fig. 5.244.

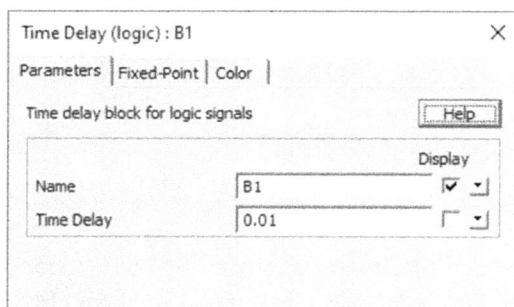

Fig. 5.245.

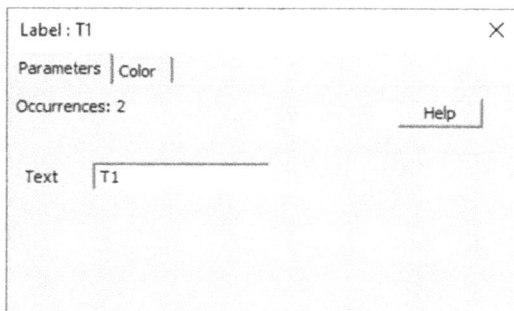

Fig. 5.246.

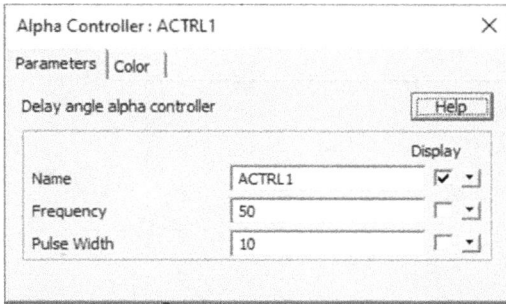

Fig. 5.247.

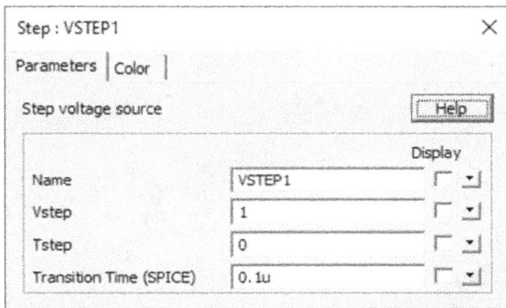

Fig. 5.248.

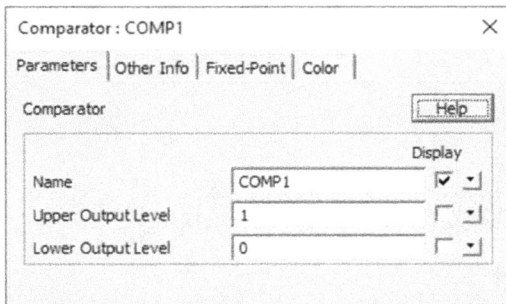

Fig. 5.249.

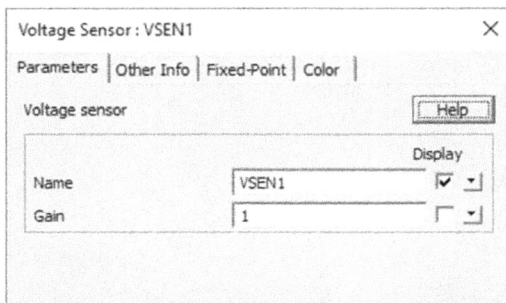

Fig. 5.250.

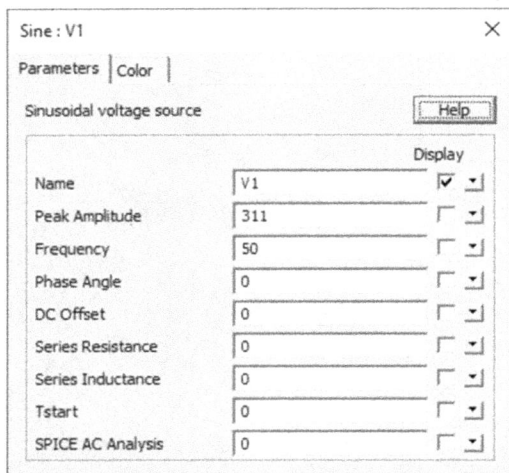

Fig. 5.251.

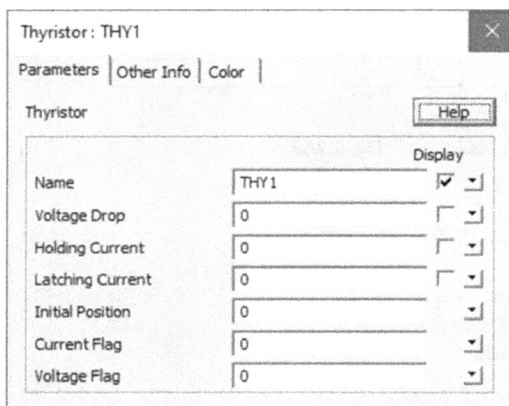

Fig. 5.252.

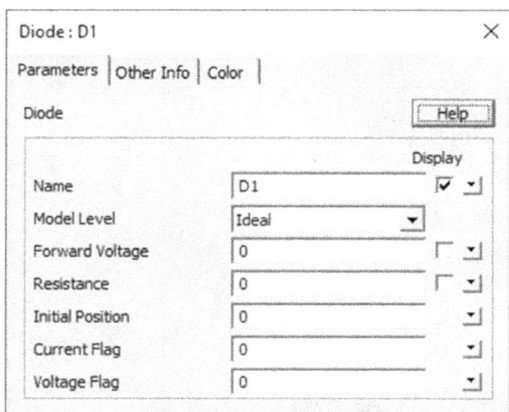

Fig. 5.253.

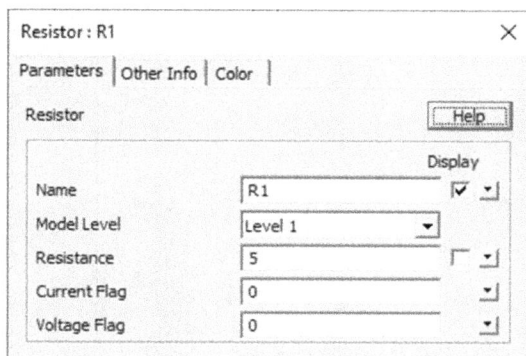

Fig. 5.254.

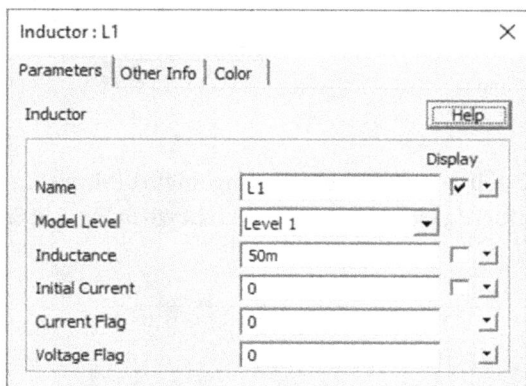

Fig. 5.255.

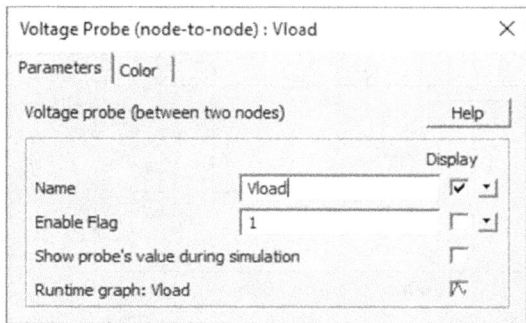

Fig. 5.256.

Run the simulation. The load voltage is shown in Fig. 5.257.

Vload

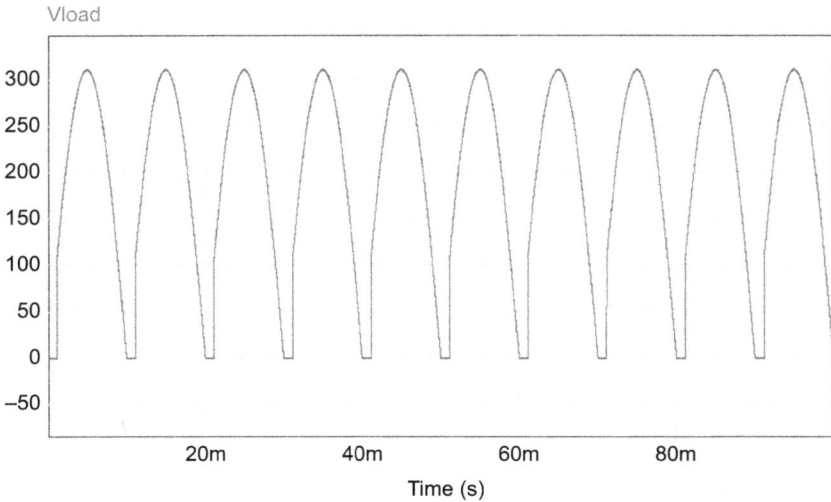

Fig. 5.257.

If you increase the value of Alpha DC source to 90, the triggering angle of thyristors will change to 90°. The output voltage for triggering angle of 90° is shown in Fig. 5.258.

Vload

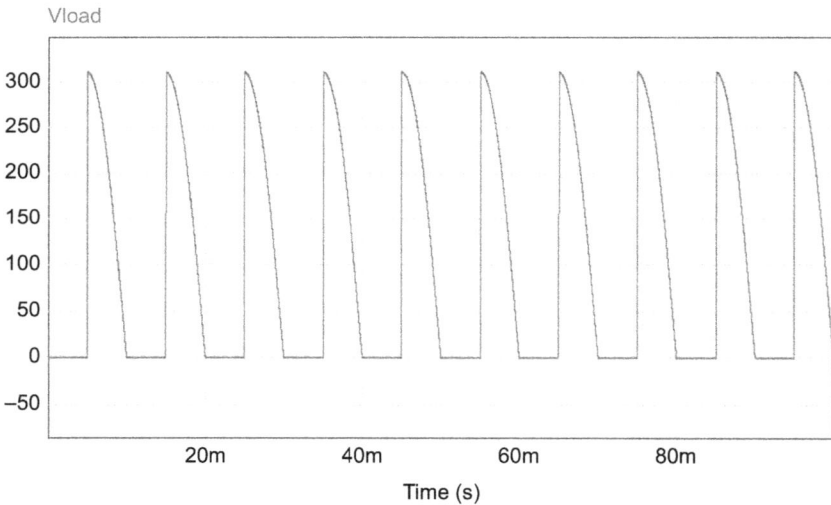

Fig. 5.258.

You can trigger the thyristors using the Gating block as well (Fig. 5.259). For instance, for triggering angle of α, Switching Points box of thyristor 1 must be α $\alpha + \epsilon$ and Switching Points box of thyristor 2 must be $180 + \alpha$ 180 $\alpha + \epsilon$ where ϵ is a small number. Figures 5.260 and 5.261 show the settings of gating block for trigger angle of 20°.

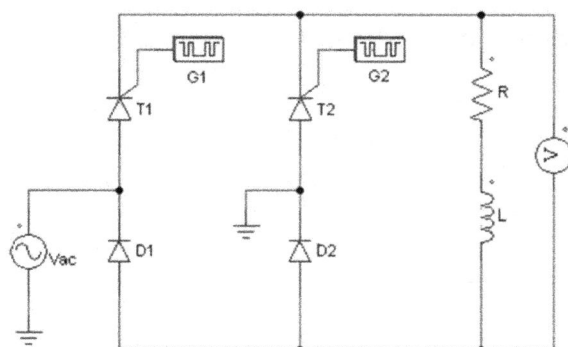

Fig. 5.259.

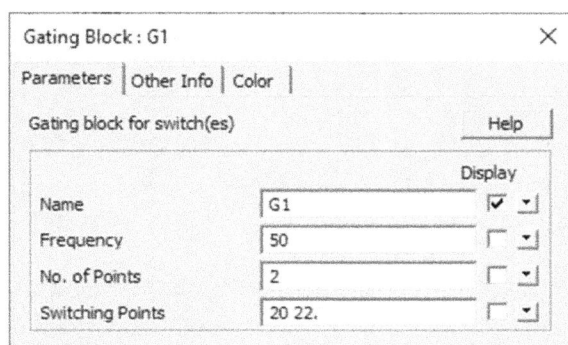

Fig. 5.260.

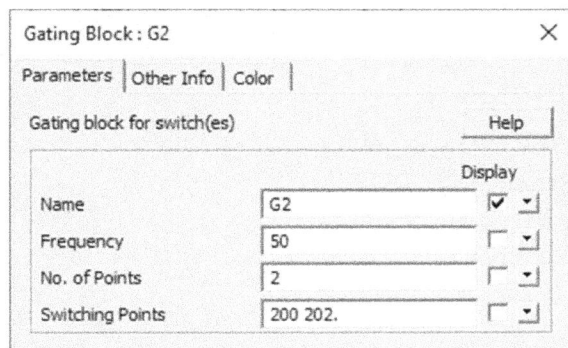

Fig. 5.261.

5.23 Example 22: measurement of power factor for different triggering angles

In this example, we want to study the effect of triggering angle on the power factor of the circuit. Add a VA/Power Factor Meter block to the previous circuit (Fig. 5.262).

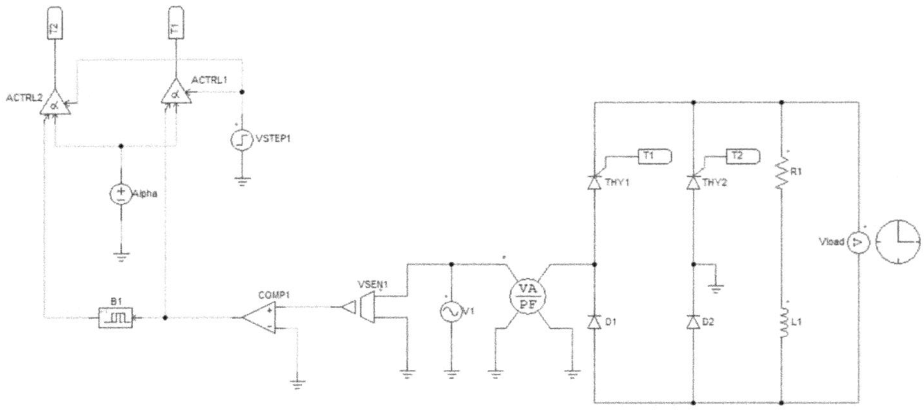

Fig. 5.262.

Settings of VA/Power Factor Meter and Simulation Control blocks are shown in Figs. 5.263 and 5.264, respectively.

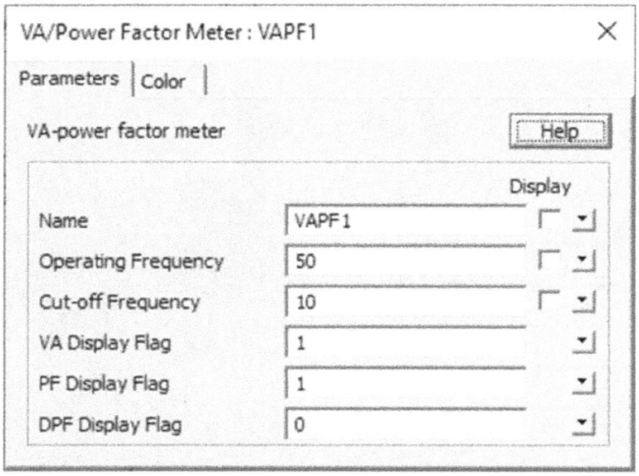

Fig. 5.263.

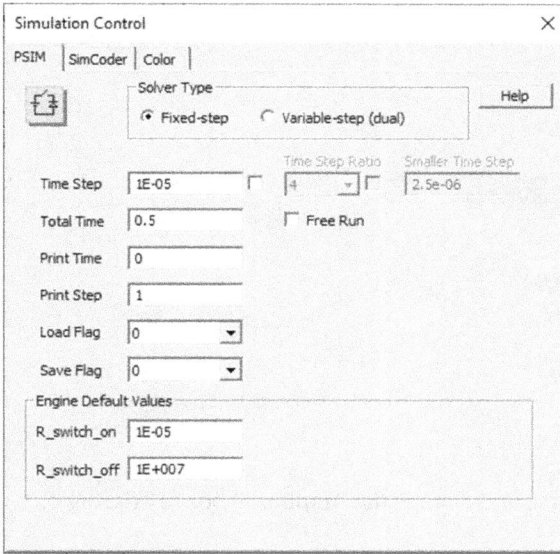

Fig. 5.264.

Add a parameter sweep block (Fig. 5.265) to the schematic (Fig. 5.266).

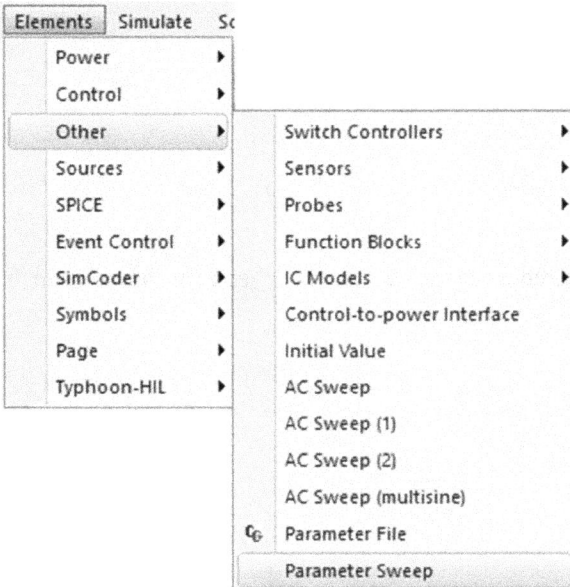

Fig. 5.265.

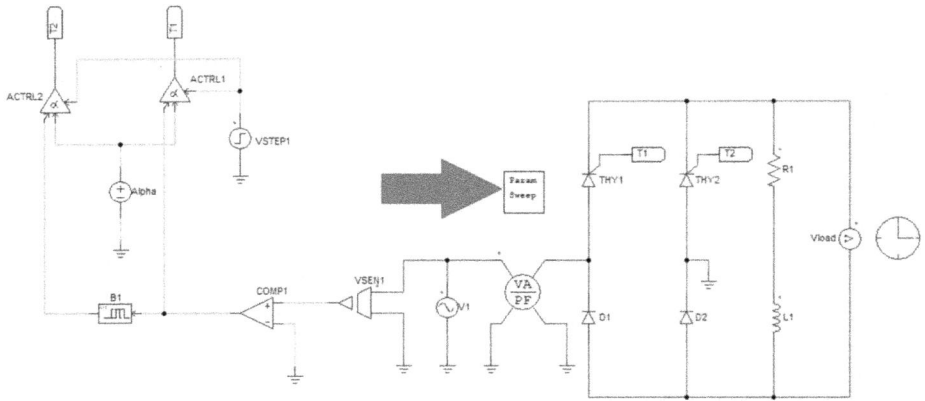

Fig. 5.266.

Double click on the Alpha DC source and change the Amplitude box to TrigAngle.

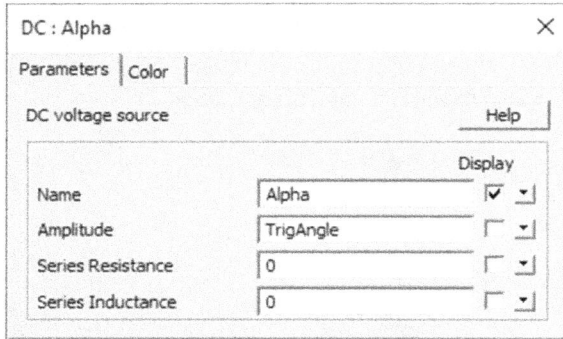

Fig. 5.267.

Double click on the added Parameter Sweep block and check the box shown in Fig. 5.268.

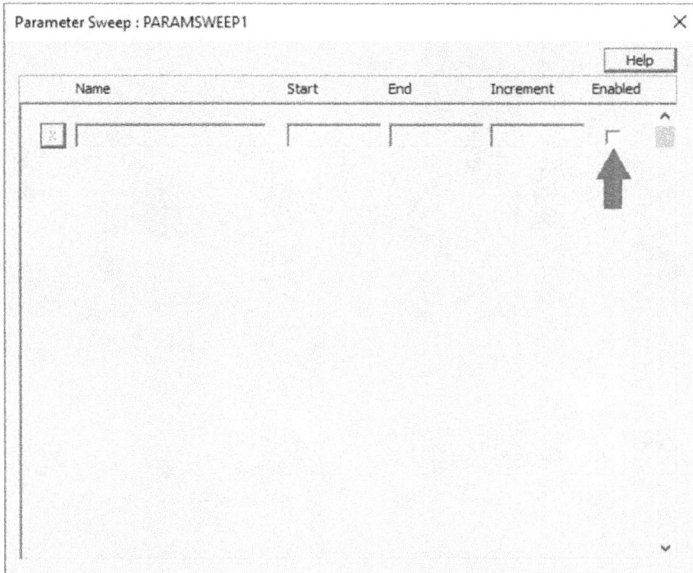

Fig. 5.268.

Add the TrigAngle to the Name box. Then enter 20 to the Start box, 90 to the End box and 10 to the Increment box. With these settings, PSIM changes the TrigAngle from 20 to 90 with 10 steps.

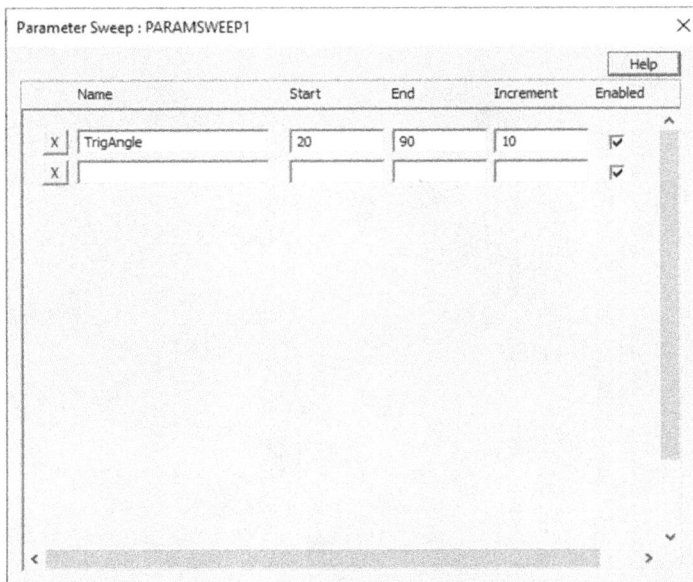

Fig. 5.269.

Now run the simulation. The properties window shown in Fig. 5.270 will appear.

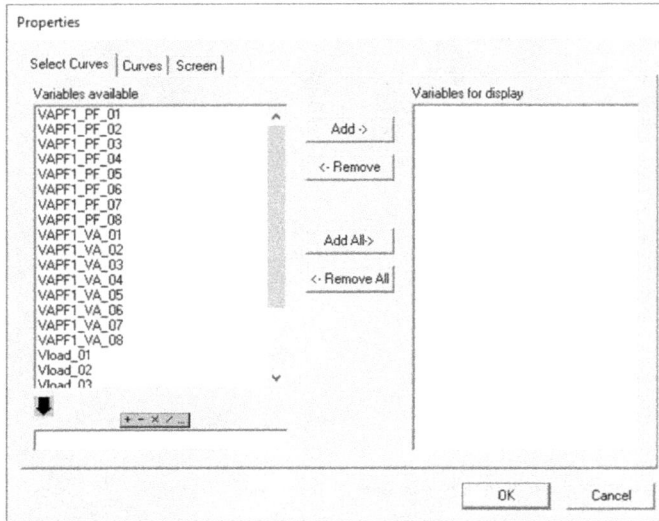

Fig. 5.270.

Add the VAPF1_PF_01, VAPF1_PF_02, VAPF1_PF_03, . . ., VAPF1_PF_08 to the Variables for display list (Fig. 5.271). VAPF1_PF_01 is the power factor for TrigAngle = 20, VAPF1_PF_02 is the power factor for TrigAngle = 30, . . .

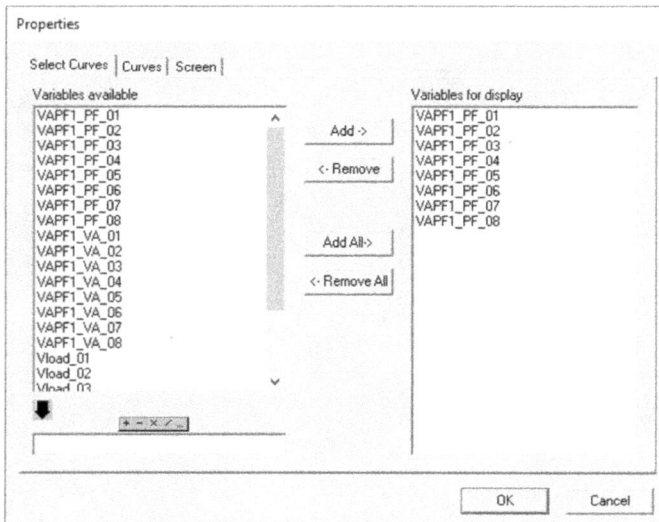

Fig. 5.271.

The simulation result is shown in Fig. 5.272. The zoomed view of Fig. 5.272 is shown in Fig. 5.273. As triggering angle of thyristors increases, the power factor decreases.

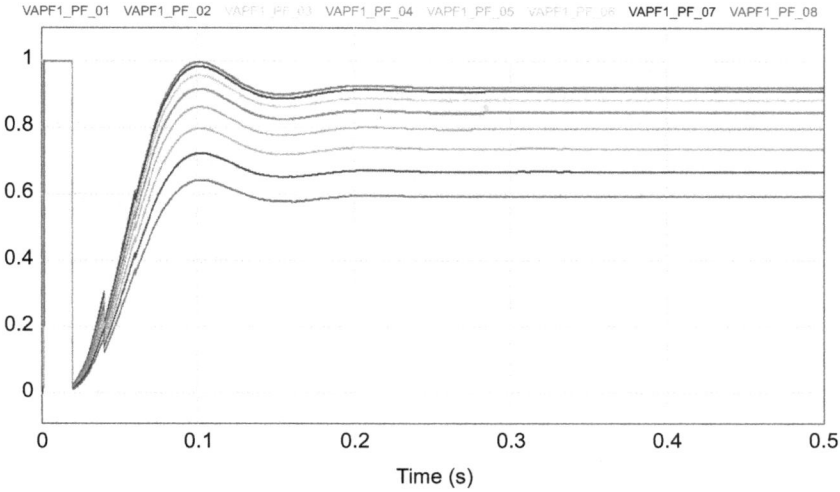

Fig. 5.272.

Fig. 5.273.

5.24 Example 23: three-phase controlled rectifier

In this example, we will study the three-phase controlled rectifier. Assume the circuit shown in Fig. 5.274.

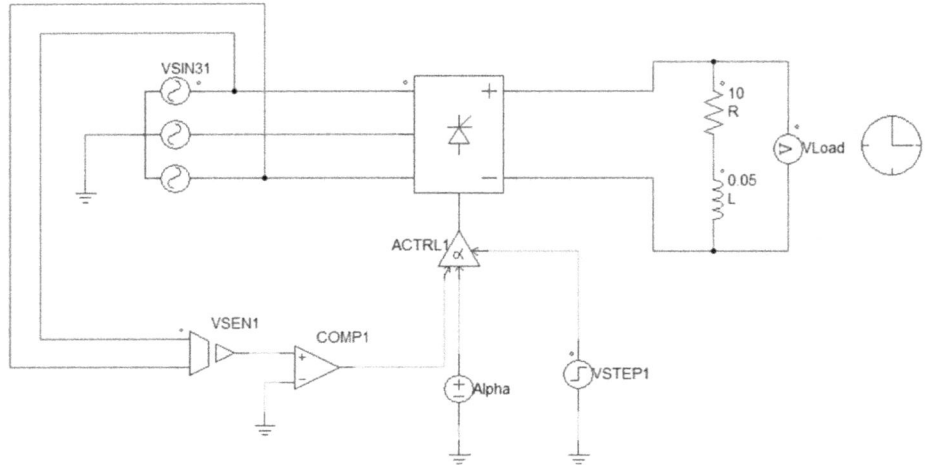

Fig. 5.274.

Settings of the blocks are shown in Figs. 5.275–5.285.

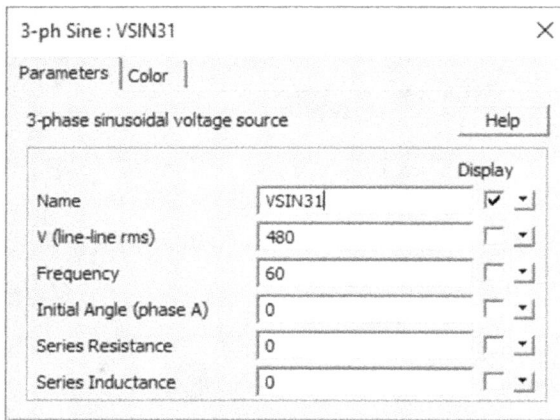

Fig. 5.275.

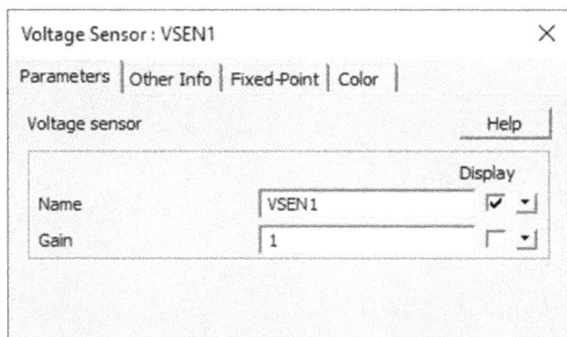

Fig. 5.276.

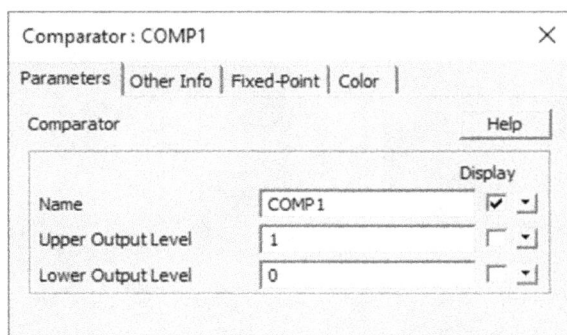

Fig. 5.277.

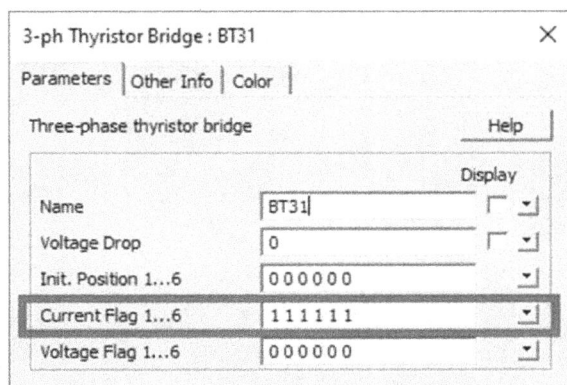

Fig. 5.278.

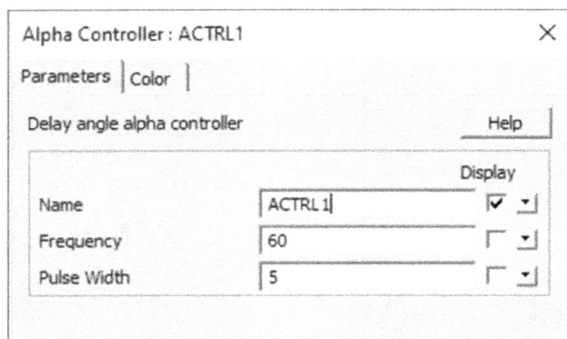

Fig. 5.279.

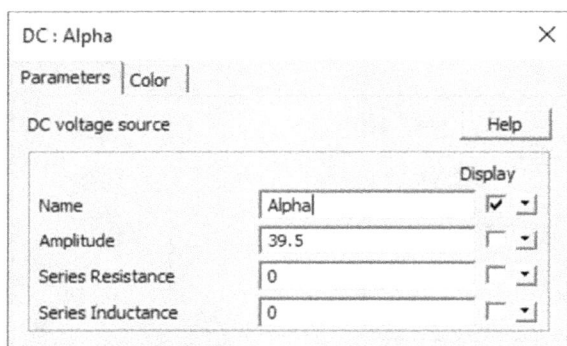

Fig. 5.280.

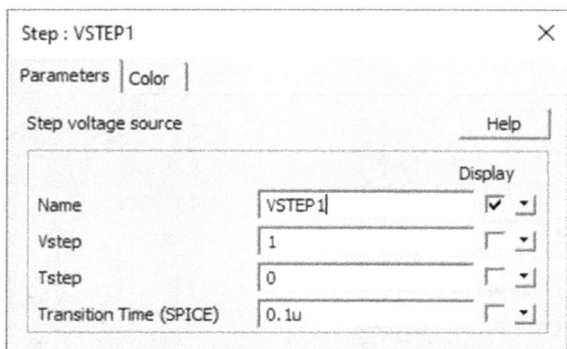

Fig. 5.281.

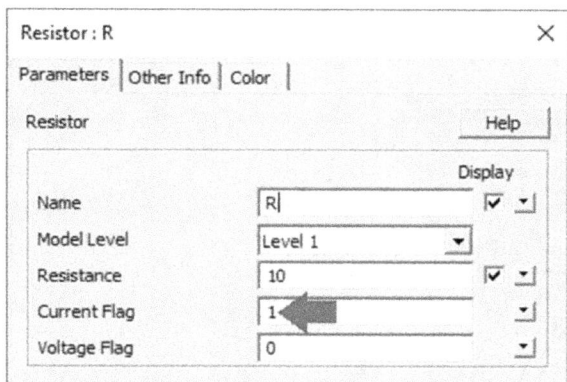

Fig. 5.282.

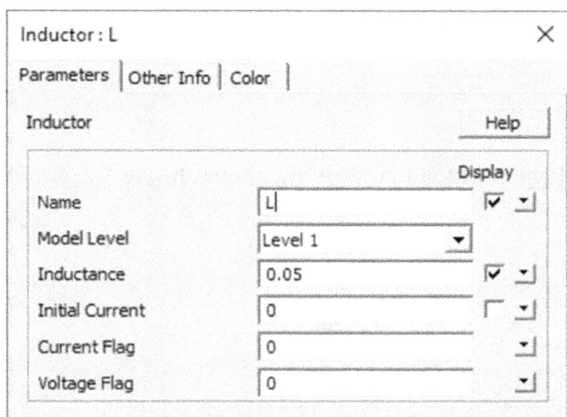

Fig. 5.283.

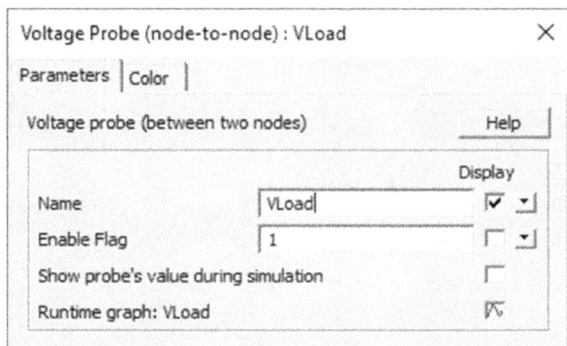

Fig. 5.284.

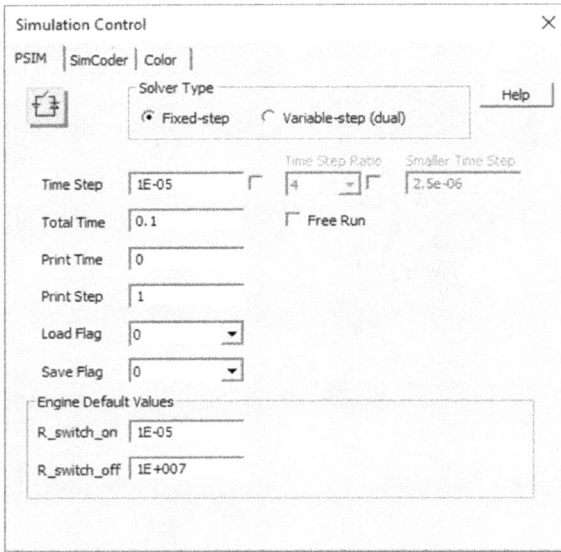

Fig. 5.285.

Run the simulation. The load voltage and load current are shown in Fig 5.286 and 5.287, respectively.

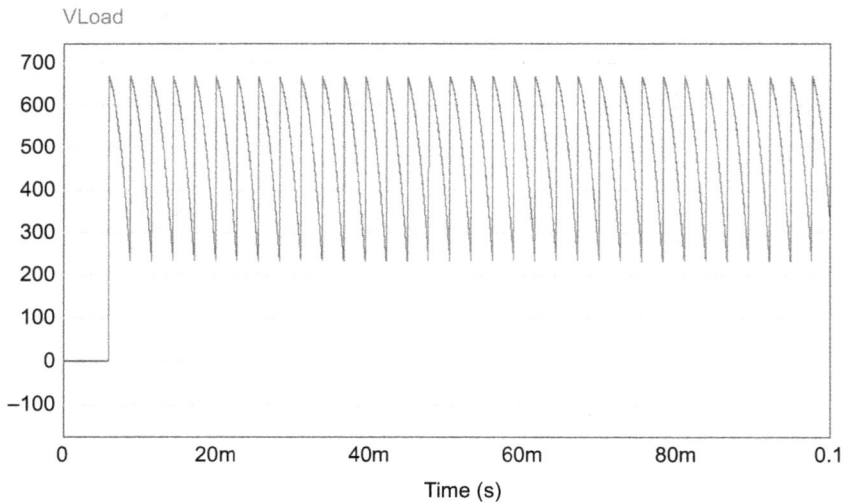

Fig. 5.286.

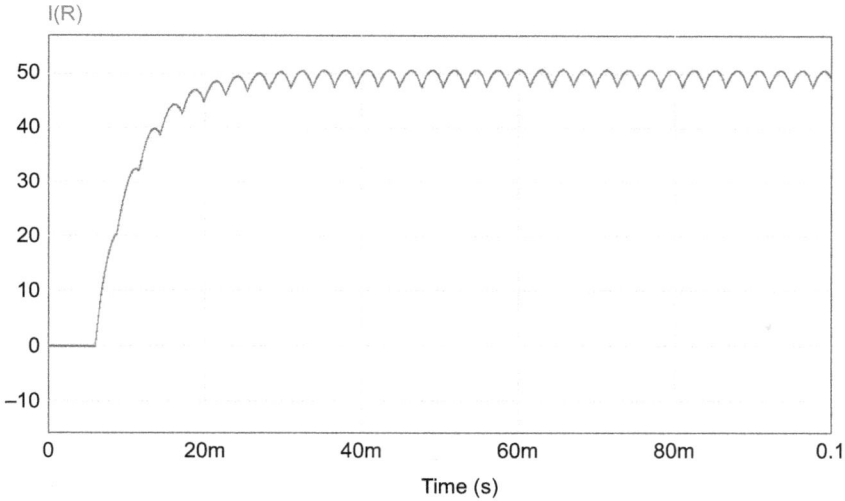

Fig. 5.287.

Frequency of ripple in the load current is 360 Hz. The average value of load current is 49.73.

Measure		X1	X2	Δ	1/Δ	Average
	Time	4.76838e-02	5.04552e-02	2.77141e-03	3.60828e+02	
	I(R)	4.76830e+01	4.76933e+01	1.03149e-02		4.97341e+01

Fig. 5.288.

The current of thyristor 1 (Fig. 5.289) is shown in Fig. 5.290.

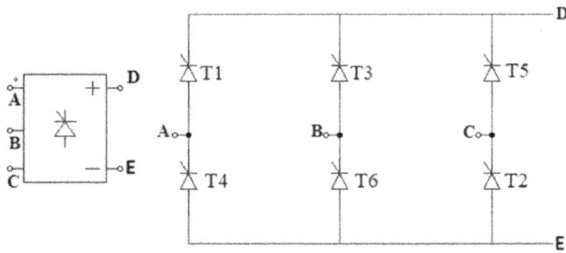

Fig. 5.289.

I(BT31_1)

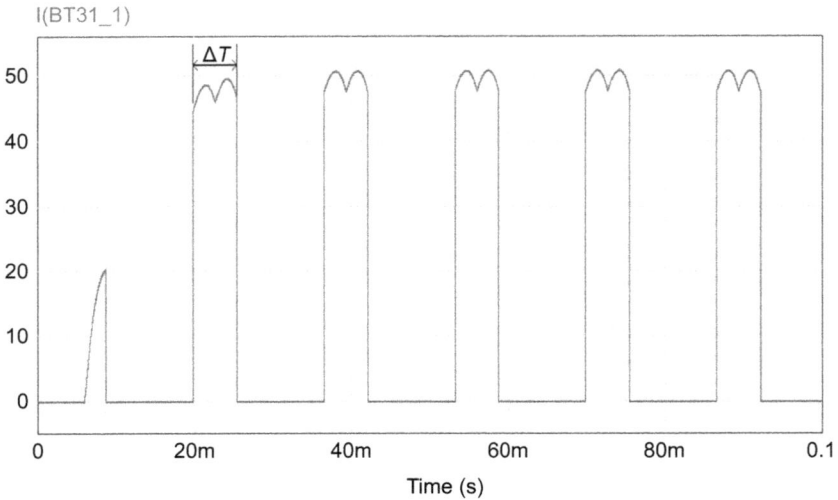

Fig. 5.290.

According to Fig. 5.291, the ΔT shown in Fig. 5.290 is equal to 5.556 ms. This is expected since each thyristor conducts for one third of period, so $\frac{1}{3} \times \frac{1}{60} = 5.6$ ms .

Measure				
		X1	X2	Δ
	Time	3.65640e-02	4.21204e-02	5.55640e-03
	I(BT31_1)	1.92823e+01	4.55340e+01	2.62518e+01

Fig. 5.291.

The average value of thyristor 1 current is 16.665 A. Each thyristor conducts the load current for one third of period, so the average value of each thyristor current must be one third of average load current.

Measure					
		X1	X2	Δ	1/Δ	Average
	Time	3.65640e-02	5.32675e-02	1.67035e-02	5.98677e+01	
	I(BT31_1)	1.92823e+01	4.78528e+01	2.85706e+01		1.66656e+01

Fig. 5.292.

Let's check the obtained results with hand analysis:

$$V_{o,DC} = \frac{1}{\frac{\pi}{3}} \int_{\frac{\pi}{3}+á}^{\frac{2\pi}{3}+á} V_{m,L-L} \times \sin(\omega t)d(\omega t) = \frac{3V_{m,L-L}}{\pi}\cos(\alpha)$$

$$V_{o,DC} = \frac{3\times480\sqrt{2}}{\pi}\cos(39.5°) = 500\,\text{V}$$

$$I_{o,Average} = \frac{V_{o,DC}}{R} = 50\,\text{A}$$

$$I_{Thyristor,\,Average} = \frac{1}{3}\times\frac{V_{o,DC}}{R} = 16.66\,\text{A}$$

PSIM has a three-phase rectifier example which uses closed loop control in order to keep the output voltage constant. In order to see that example, click the File>Open Examples. Then open the ac–dc folder.

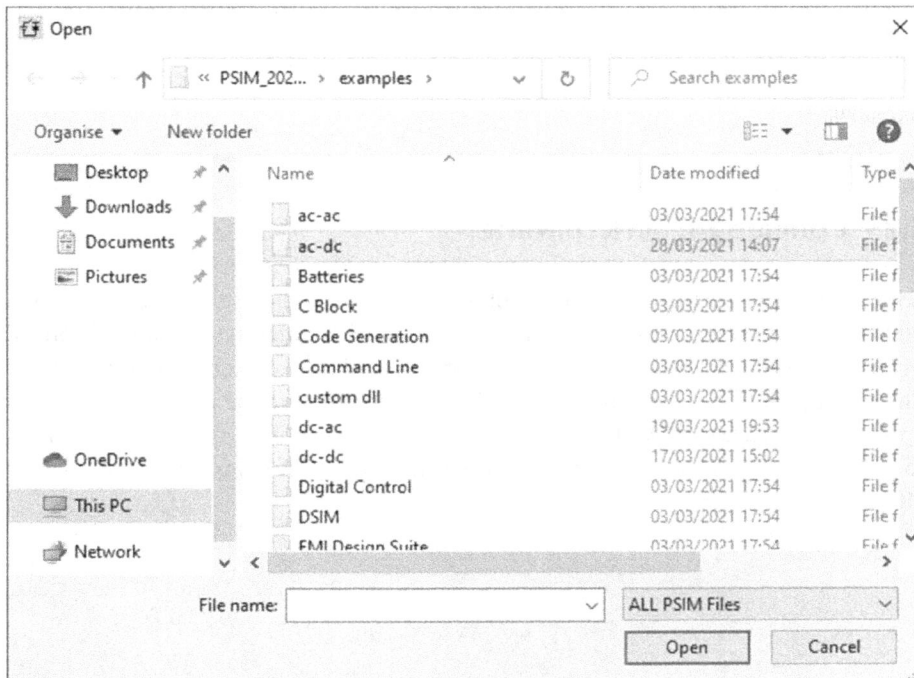

Fig. 5.293.

Select the thy-3 f in order to see the closed loop three-phase rectifier circuit.

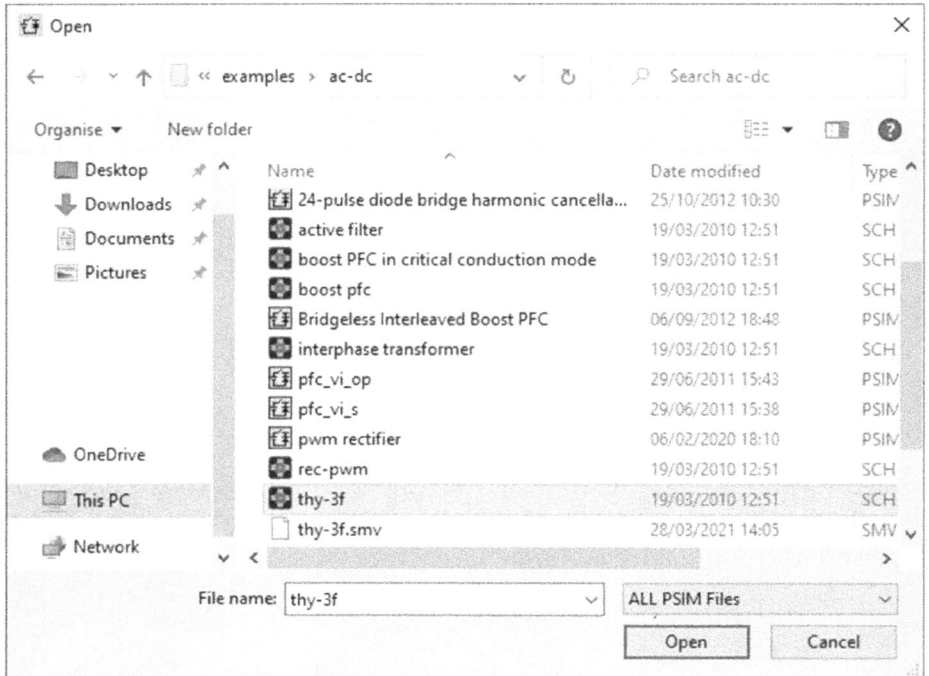

Fig. 5.294.

5.25 Example 24: buck converter

A buck converter decreases the input voltage. In this example, we study an open loop buck converter. Assume the circuit shown in Fig. 5.295. In this circuit, required control pulses are produced by comparing a reference signal with a high frequency carrier. The duty cycle, that is, ratio of time which the switch is on to the period, of the pulse applied to the gate of MOSFET is 0.25. Switching frequency is 100 kHz. Do not forget to put the on–off switch controller between the output of comparator and gate of MOSFET.

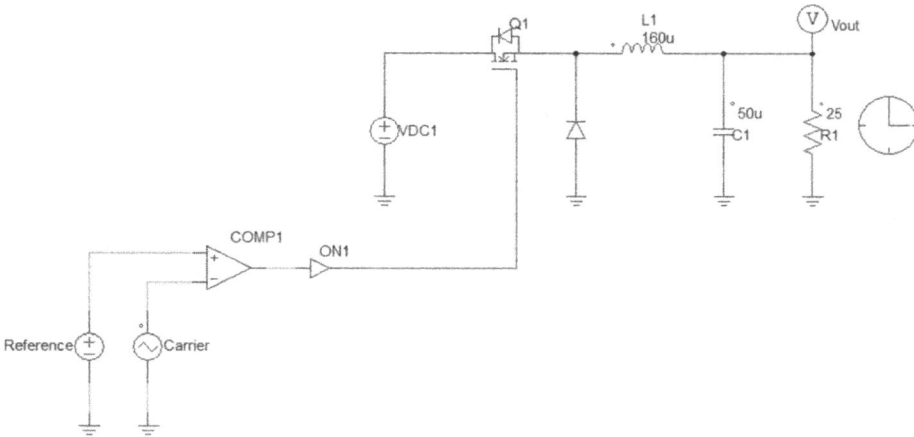

Fig. 5.295.

Settings of blocks are shown in Figs. 5.296–5.307.

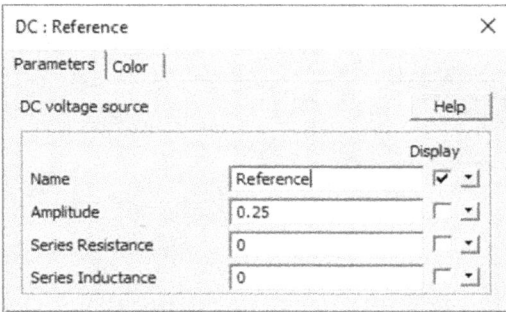

Fig. 5.296.

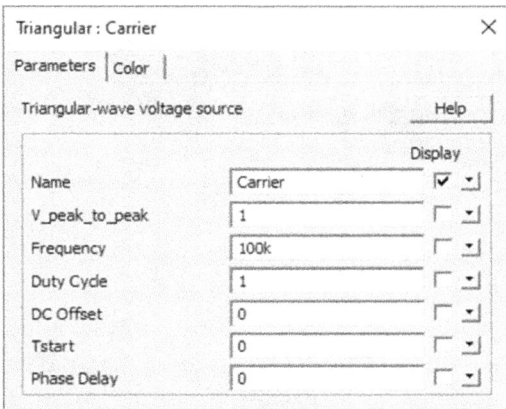

Fig. 5.297.

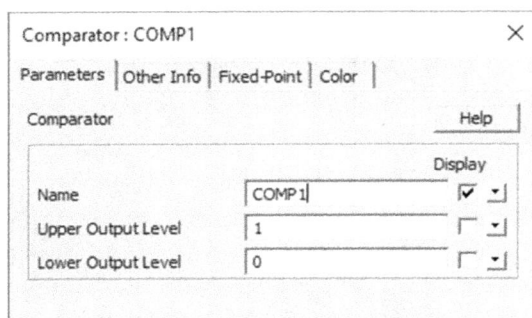

Fig. 5.298.

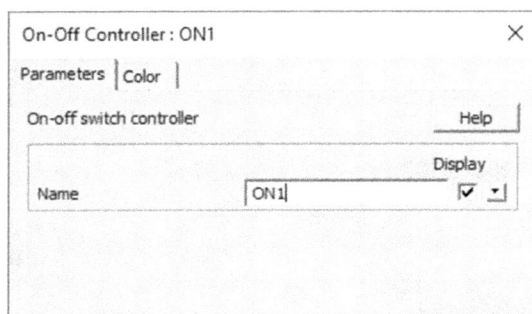

Fig. 5.299.

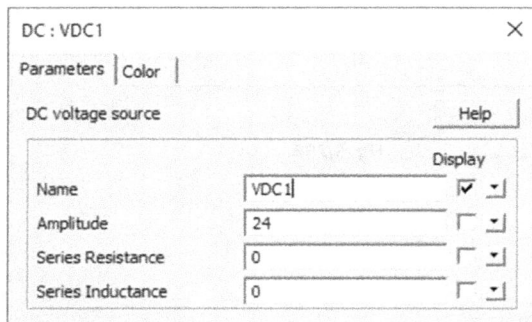

Fig. 5.300.

MOSFET : Q1   ✕

Parameters | Other Info | Color

MOSFET switch   Help

		Display
Name	Q1	☑ ▾
Model Level	Ideal ▾	
On Resistance RDS(on)	100m	☐ ▾
Diode Forward Voltage	3	☐ ▾
Diode Resistance	1m	☐ ▾
Initial Position	0	▾
Current Flag	0	▾
Voltage Flag	0	▾

Fig. 5.301.

Diode : D1   ✕

Parameters | Other Info | Color

Diode   Help

		Display
Name	D1	☐ ▾
Model Level	Ideal ▾	
Forward Voltage	0.7	☐ ▾
Resistance	0.01	☐ ▾
Initial Position	0	▾
Current Flag	0	▾
Voltage Flag	0	▾

Fig. 5.302.

Inductor : L1   ✕

Parameters | Other Info | Color

Inductor   Help

		Display
Name	L1	☑ ▾
Model Level	Level 1 ▾	
Inductance	160u	☑ ▾
Initial Current	0	☐ ▾
Current Flag	0	▾
Voltage Flag	0	▾

Fig. 5.303.

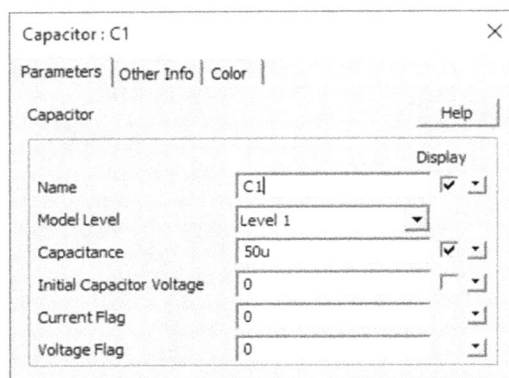

Fig. 5.304.

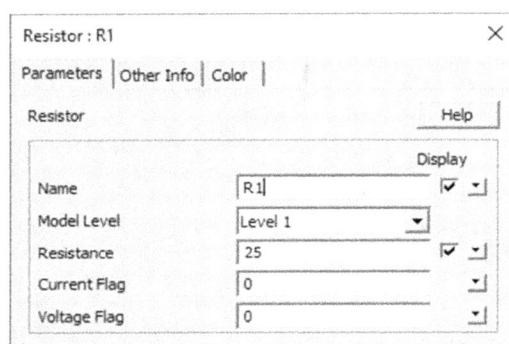

Fig. 5.305.

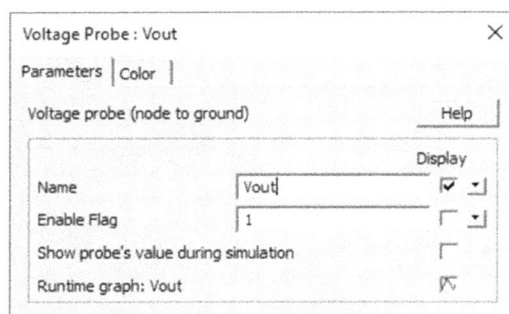

Fig. 5.306.

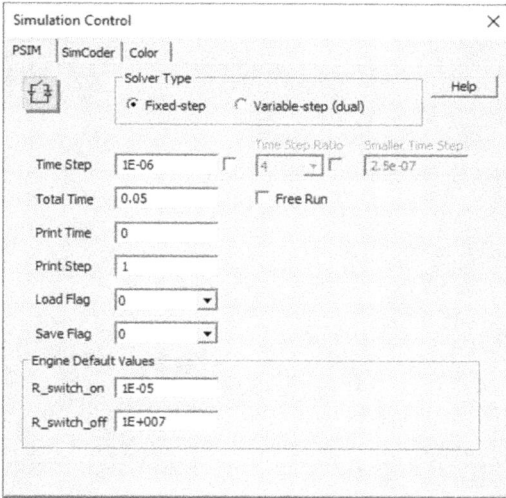

Fig. 5.307.

Run the simulation. The output voltage is shown in Fig. 5.308. Figure 5.309 shows the zoomed view of steady-state part of waveform shown in Fig. 5.308. The output voltage is about 5.46 V.

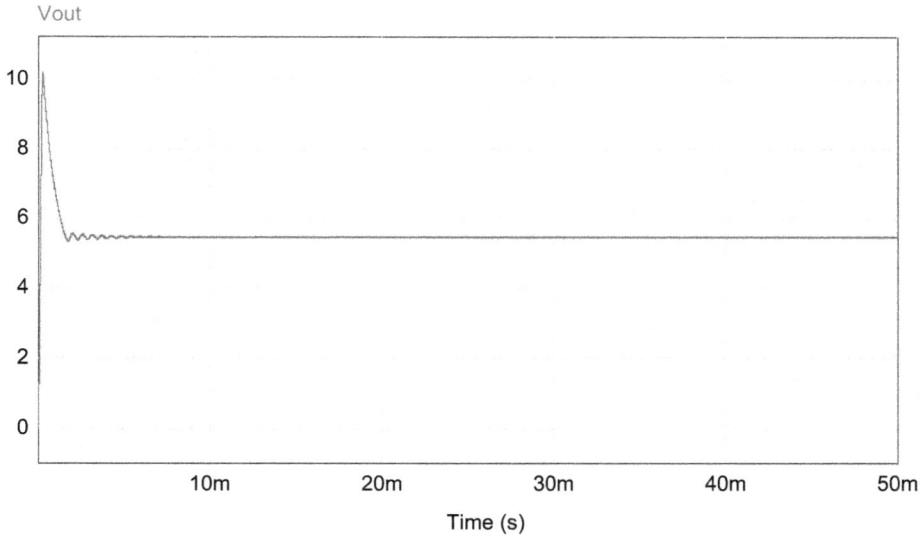

Fig. 5.308.

Vout

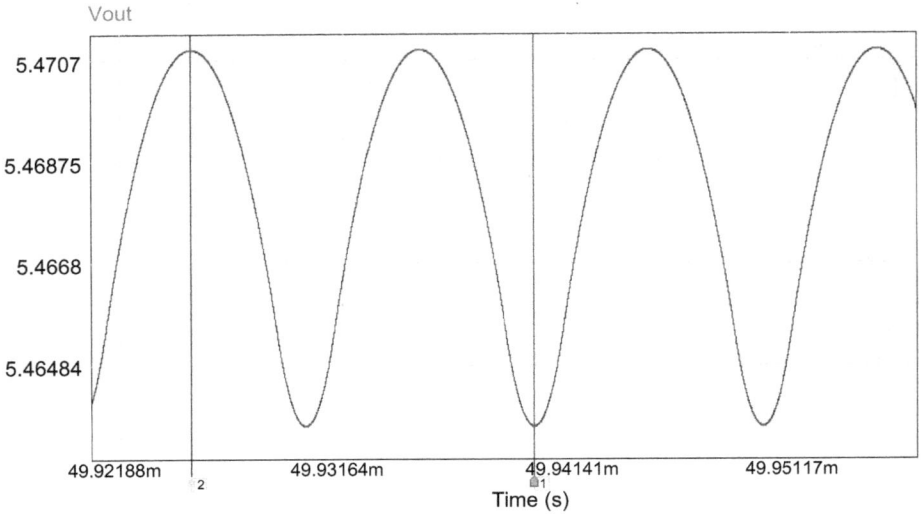

Fig. 5.309.

Use the cursors in order to measure the output voltage ripple. According to Fig. 5.310, the output voltage ripple is 7.23 mV. Average value of output voltage is about 5.46 V.

Measure			[x]
	X1	X2	Δ
Time	4.99400e-02	4.99250e-02	-1.49987e-05
Vout	5.46367e+00	5.47090e+00	7.23276e-03

Fig. 5.310.

Duty cycle of Triangular-wave voltage source block in Fig. 5.295 is 1. With these settings, the carrier will be a saw tooth signal. If you want to use a triangular carrier, change the Duty Cycle box of Triangular-wave voltage source block to 0.5.

Carrier PWM Refrence

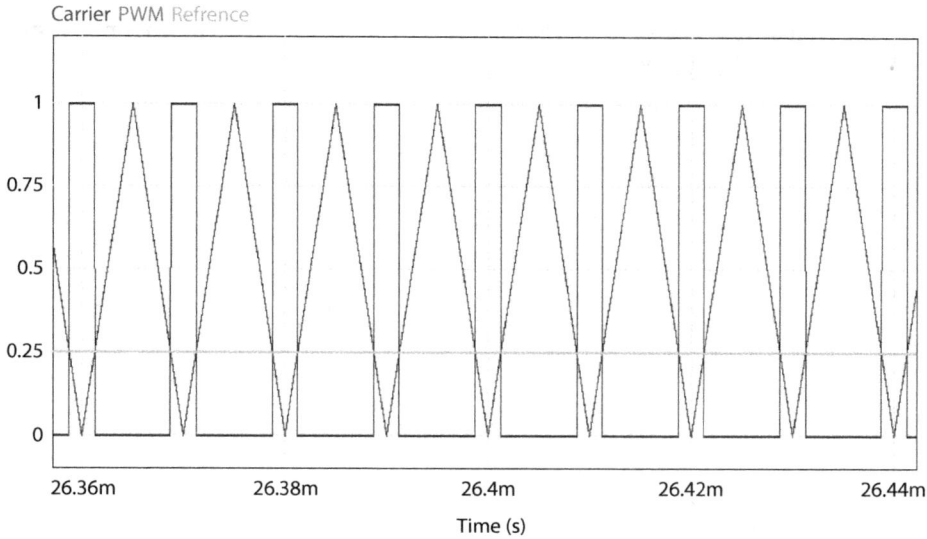

Fig. 5.311.

You can use the Gating block in order to simulate the buck converter as well (Fig. 5.312). Settings of Gating block must be similar to Fig. 5.313 in order to apply a pulse with duty cycle of 0.25 to the gate of MOSFET.

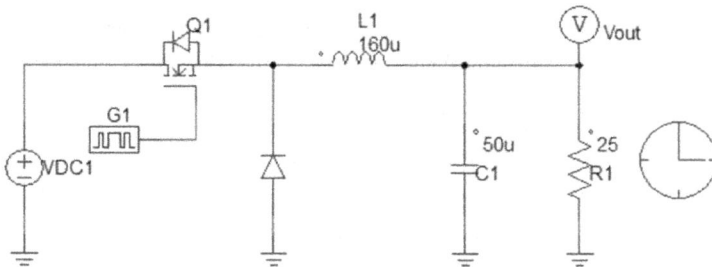

Fig. 5.312.

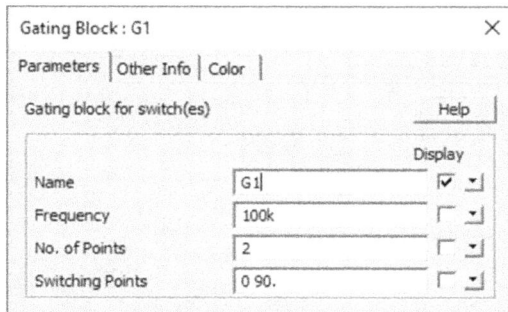

Fig. 5.313.

Run the simulation. The simulation result is shown in Fig. 5.314. The obtained result is the same as in Fig. 5.308.

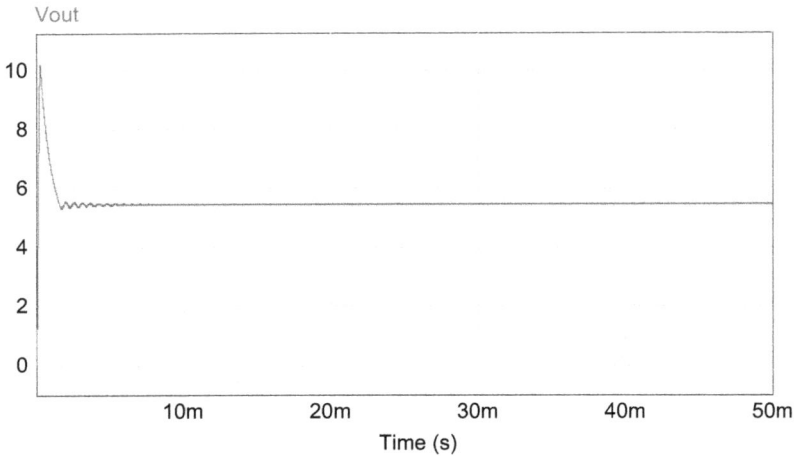

Fig. 5.314.

5.26 Example 25: effect of time step on edges of signals

In this example, we see the effect of Time Step box of Simulation Control block on the edges of control signals. Add three probes to the modulator section of previous example.

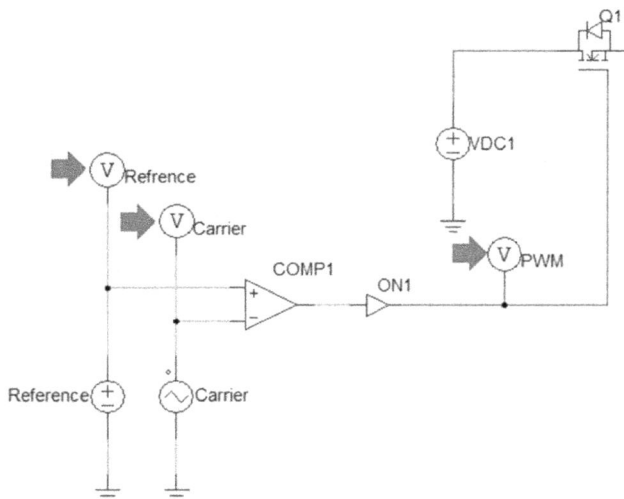

Fig. 5.315.

Run the simulation. The Carrier, PWM, and Reference waveforms are shown in Fig. 5.316. Note that the Carrier signal does not reach to zero and the edges of PWM signal are not sharp. But this is only a visual effect and does not mean that results are not accurate.

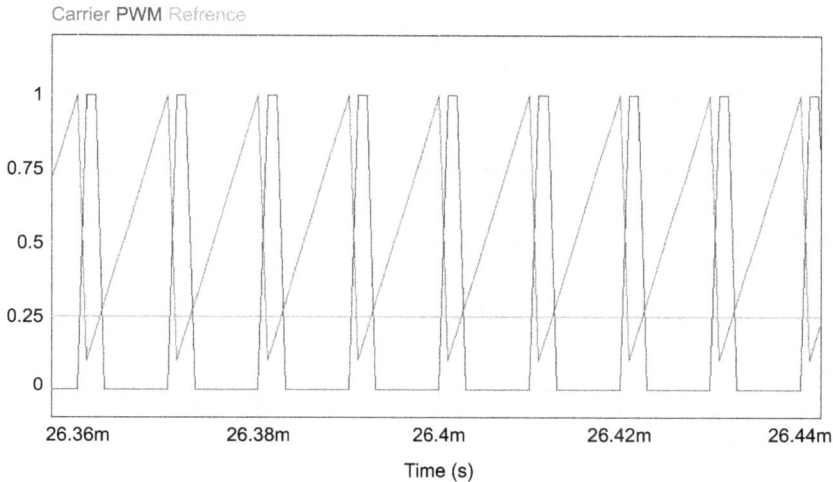

Fig. 5.316.

Now double click on the Simulation Control block. Decrease the Time Step box to 1E-8 which means 10^{-8}.

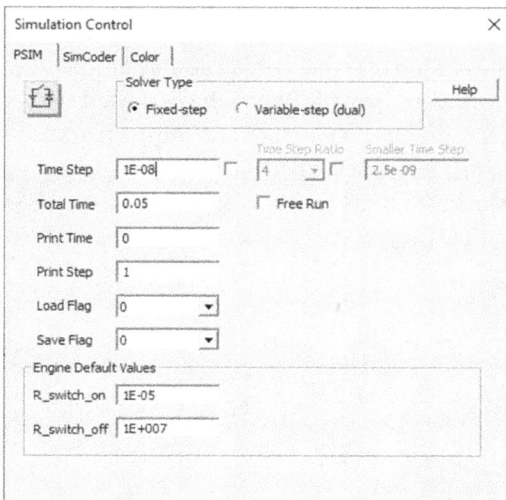

Fig. 5.317.

Run the simulation. The result is shown in Fig. 5.318. Note that the Carrier signal starts from zero and the edges of PWM signal are sharp. So, when you decrease the Time Step, accuracy increases. However, the simulation requires more time in order to be done.

Carrier PWM Refrence

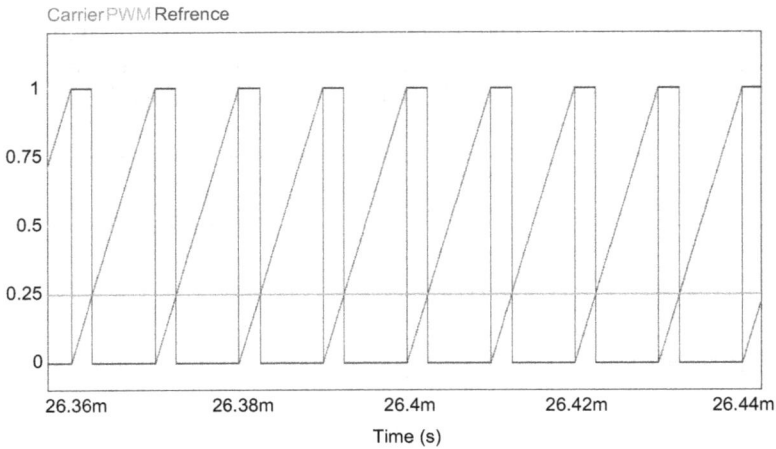

Fig. 5.318.

When you decrease the Time Step, accuracy increases. However, the simulation requires more time in order to be done.

5.27 Example 26: making subcircuit

When you want to simulate a complex circuit, putting all the elements on the schematic make your schematic crowded and difficult to understand. It is recommended to put related blocks in subcircuits in order to make it more understandable for the user. For instance, in the previous buck converter example, we can put the modulator in a subcircuit in order to hide the details of modulator and simplify the schematic. When the user wants to see the details of modulator, he can simply click the subcircuit block and the details will be shown. In this example, we will learn how to add a subcircuit to our schematic.

First of all, remove the modulator blocks from the previous example.

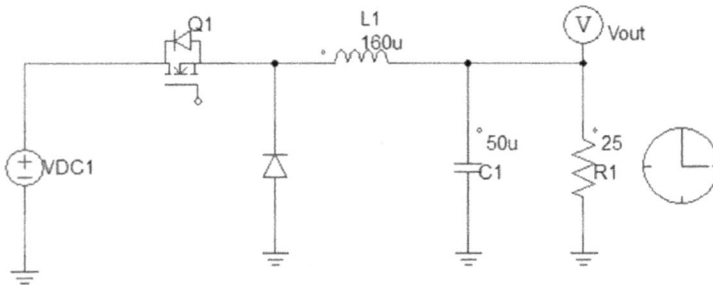

Fig. 5.319.

Click on New Subcircuit (Fig. 5.320) and add a subcircuit to the schematic (Fig. 5.321).

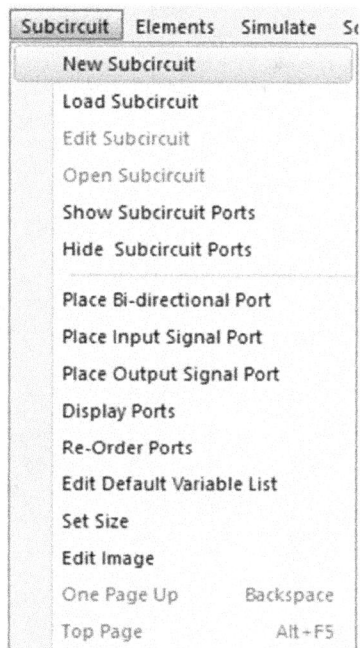

Subcircuit	Elements	Simulate	S(

New Subcircuit

Load Subcircuit

Edit Subcircuit

Open Subcircuit

Show Subcircuit Ports

Hide Subcircuit Ports

Place Bi-directional Port

Place Input Signal Port

Place Output Signal Port

Display Ports

Re-Order Ports

Edit Default Variable List

Set Size

Edit Image

One Page Up   Backspace

Top Page   Alt+F5

Fig. 5.320.

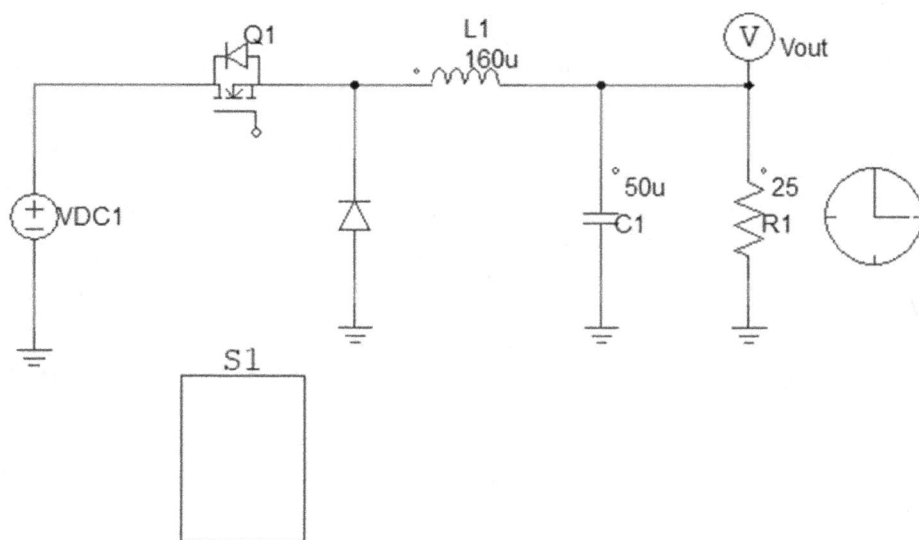

Fig. 5.321.

Right click on the subcircuit block and click the Attributes from appeared menu.

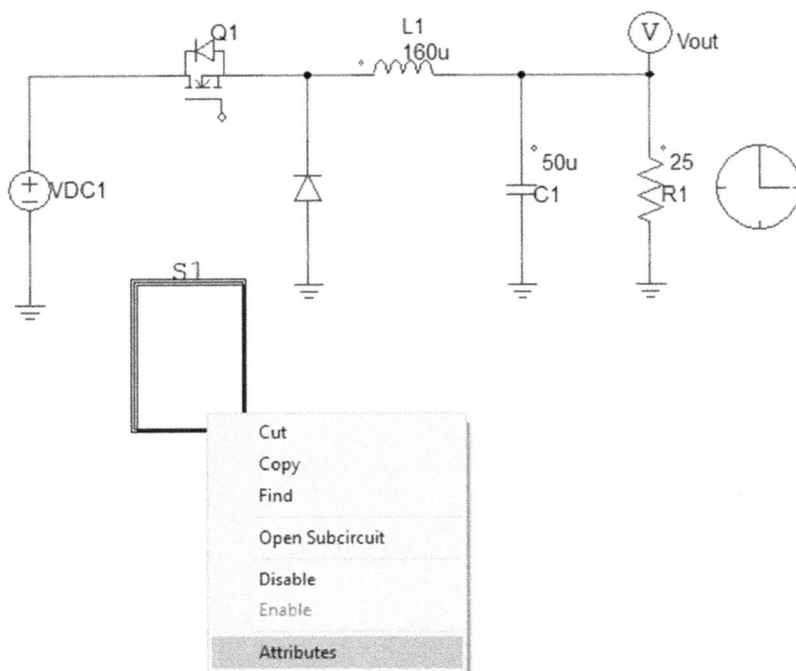

Fig. 5.322.

Enter your desired name in the Name box. Select a name which describes the subcircuit.

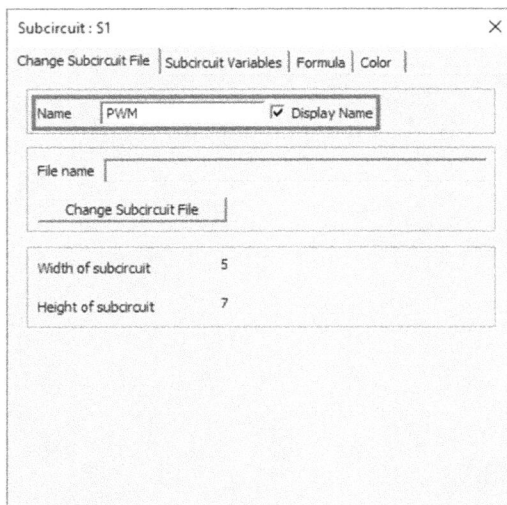

Fig. 5.323.

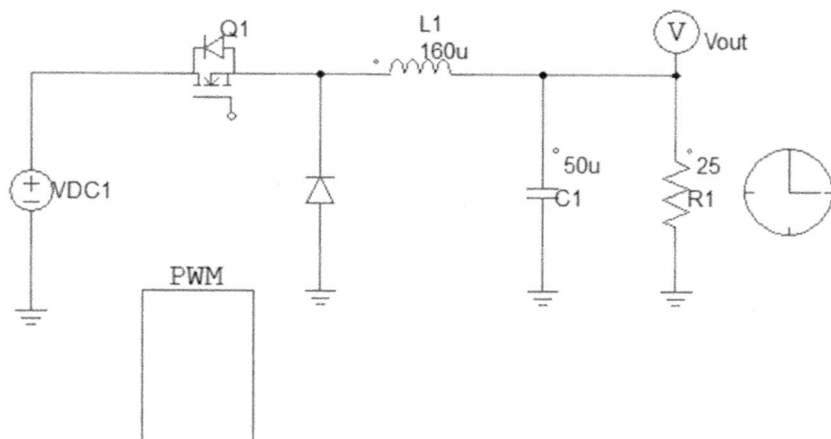

Fig. 5.324.

Double click on the subcircuit block in order to open it. Draw the schematic shown in Fig. 5.325.

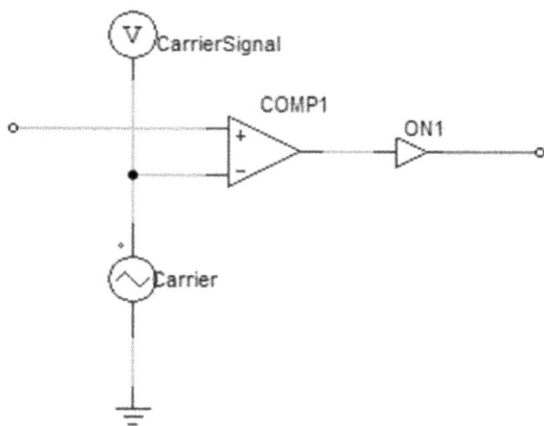

Fig. 5.325.

Settings of elements of Fig. 5.325 are shown in Figs. 5.326–5.329.

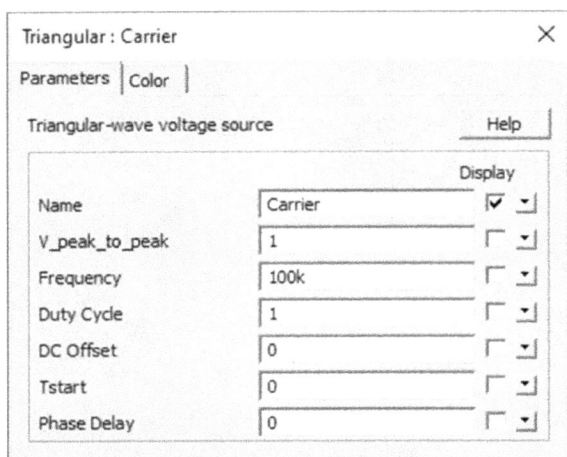

Fig. 5.326.

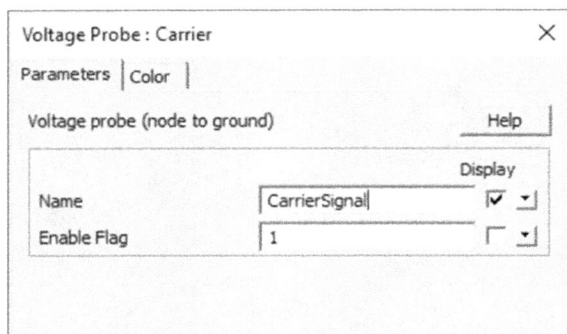

Fig. 5.327.

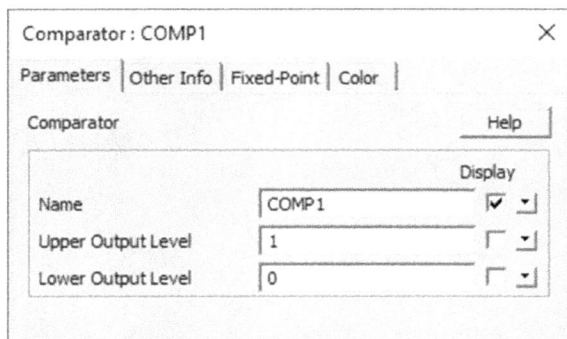

Fig. 5.328.

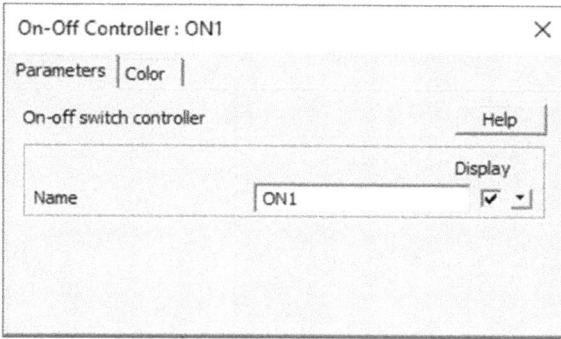

On-Off Controller : ON1 ✕

Parameters | Color |

On-off switch controller   Help

                       Display

Name   ON1   ☑ ▾

Fig. 5.329.

Now add input and output ports to the subcircuit (Fig. 5.330). The input port (Fig. 5.331) is an entrance for a signal which comes from outside into the subcircuit block. The output port (Fig. 5.332) is used to send a signal from the subcircuit block to outside.

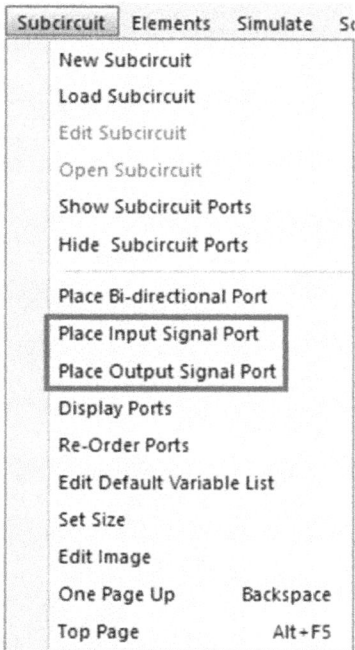

Subcircuit | Elements   Simulate   S‹

     New Subcircuit

     Load Subcircuit

     Edit Subcircuit

     Open Subcircuit

     Show Subcircuit Ports

     Hide Subcircuit Ports

     Place Bi-directional Port

     Place Input Signal Port

     Place Output Signal Port

     Display Ports

     Re-Order Ports

     Edit Default Variable List

     Set Size

     Edit Image

     One Page Up   Backspace

     Top Page   Alt+F5

Fig. 5.330.

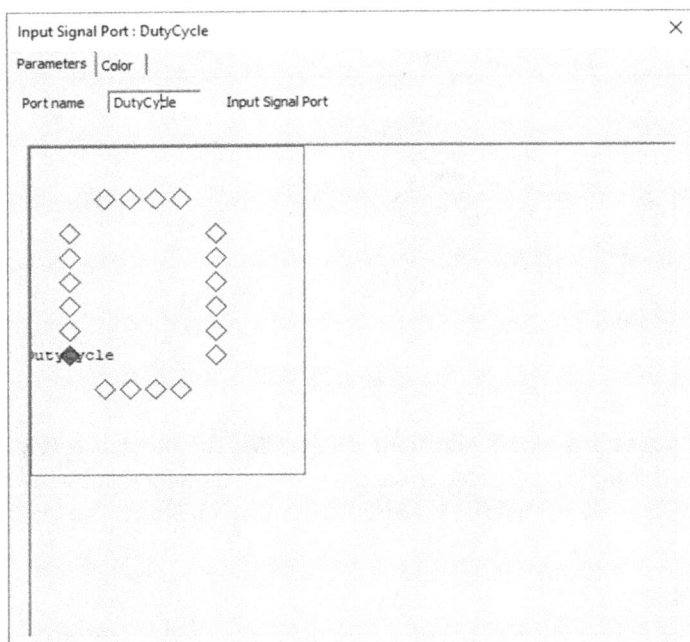

Fig. 5.331.

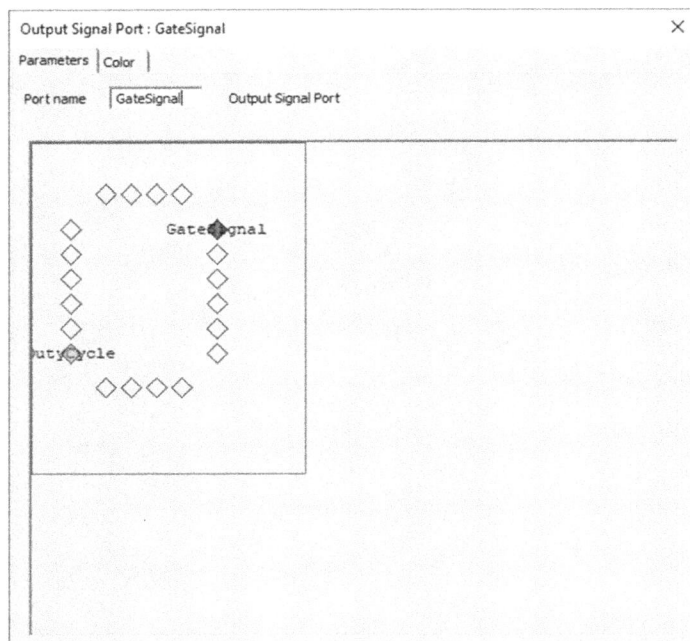

Fig. 5.332.

After adding the input and output ports to the subcircuit, save it and return to the schematic of power stage of converter. The subcircuit block must have the name of ports on it.

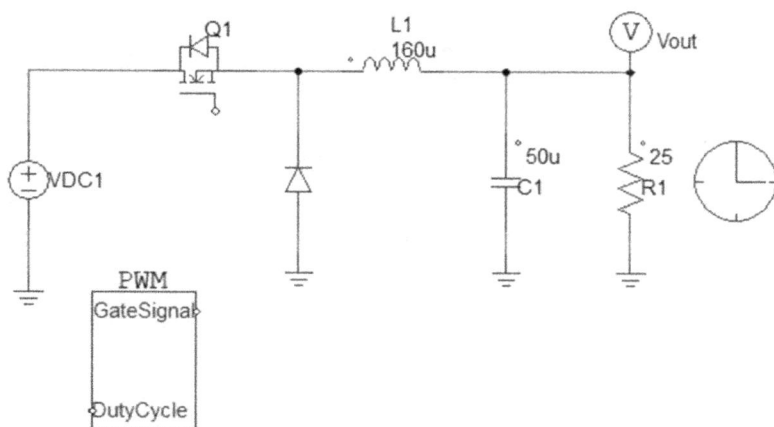

Fig. 5.333.

Connect the subcircuit to the rest of circuit.

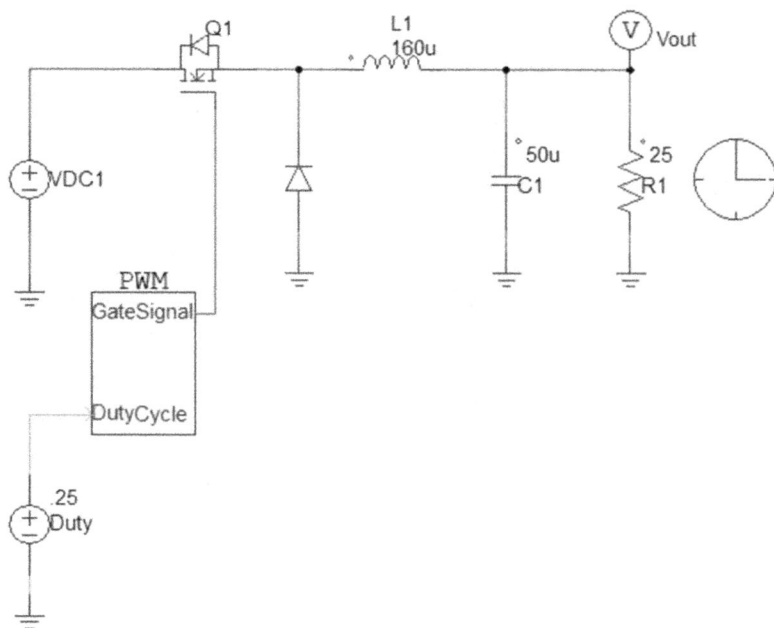

Fig. 5.334.

Run the simulation. Note that signal inside of subcircuit has the A.B format, where A is the name of subcircuit and B is the name of probe which monitors the signal.

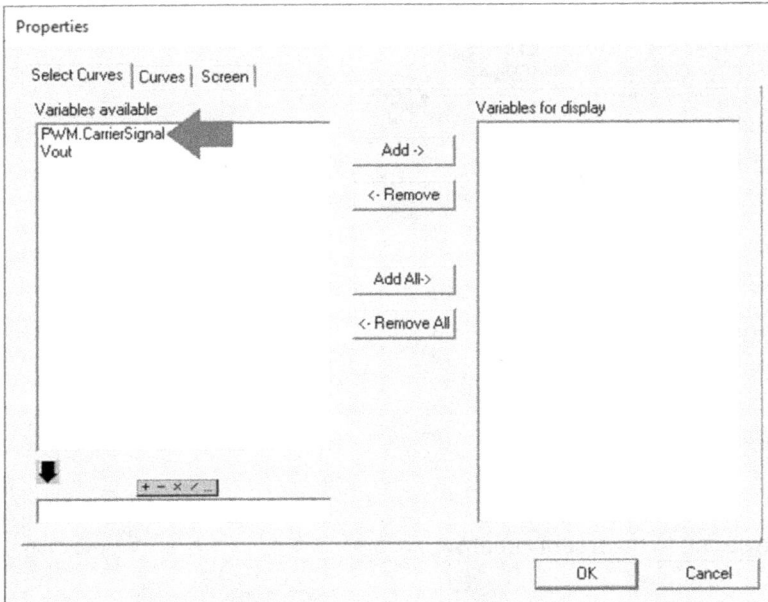

Fig. 5.335.

The load voltage is shown in Fig. 5.336. This result is the same as previous results (Figs. 5.308 and 5.314).

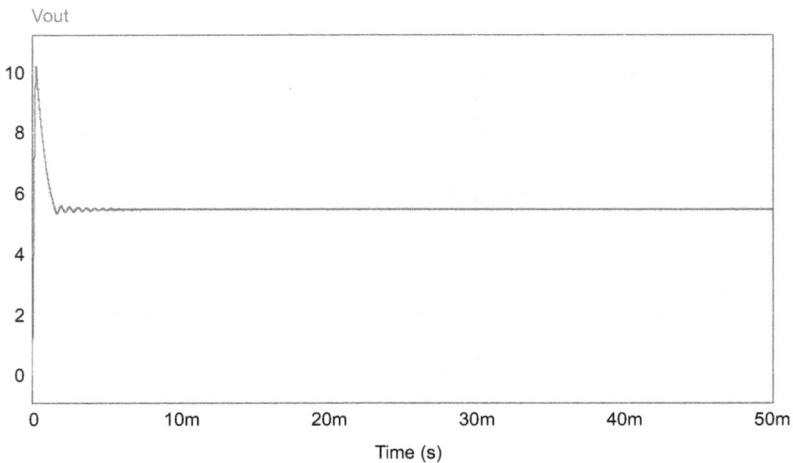

Fig. 5.336.

5.28 Example 27: PWM generation with carrier PWM controller block

In this example, we will introduce the Carrier PWM Controller (s-domain) block (Fig. 5.337). You can use the Carrier PWM Controller (s-domain) block in order to produce the required pulses for the switches in the converter.

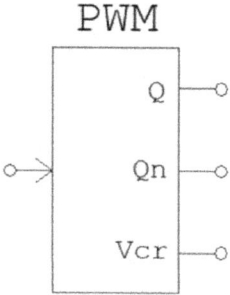

Fig. 5.337.

The Carrier PWM Controller (s-domain) block can be found in the Switch Controllers section.

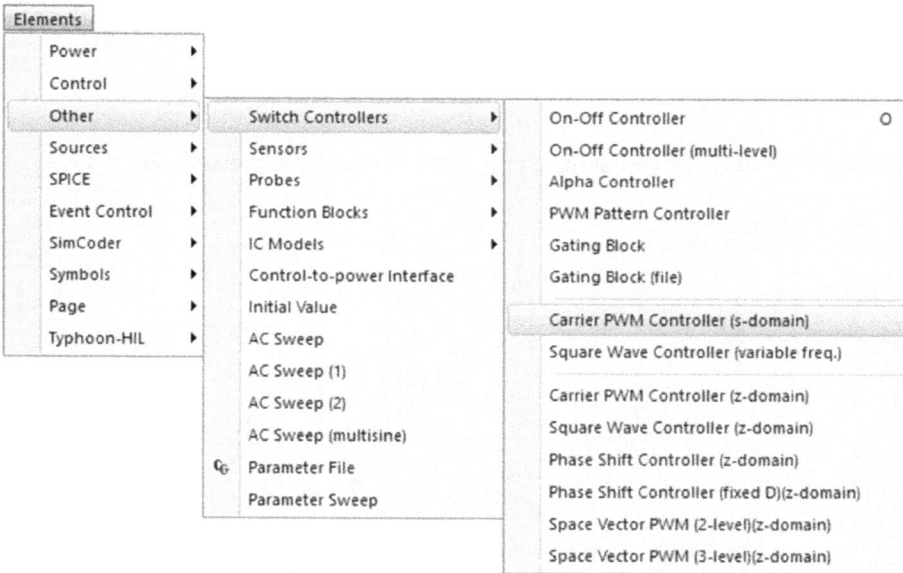

Fig. 5.338.

Figure 5.339 shows a buck converter. The required pulses for MOSFET Q1 are produced with a Carrier PWM Controller (s-domain) block. The Carrier PWM Controller (s-domain) block has one input and three outputs. The only input is reference signal. The reference signal is a low-frequency signal which is compared with the high-frequency carrier. The Qn output is the complement of Q output. You can even define a dead time between them. The third output is the carrier signal which is used in the pulse width modulation process.

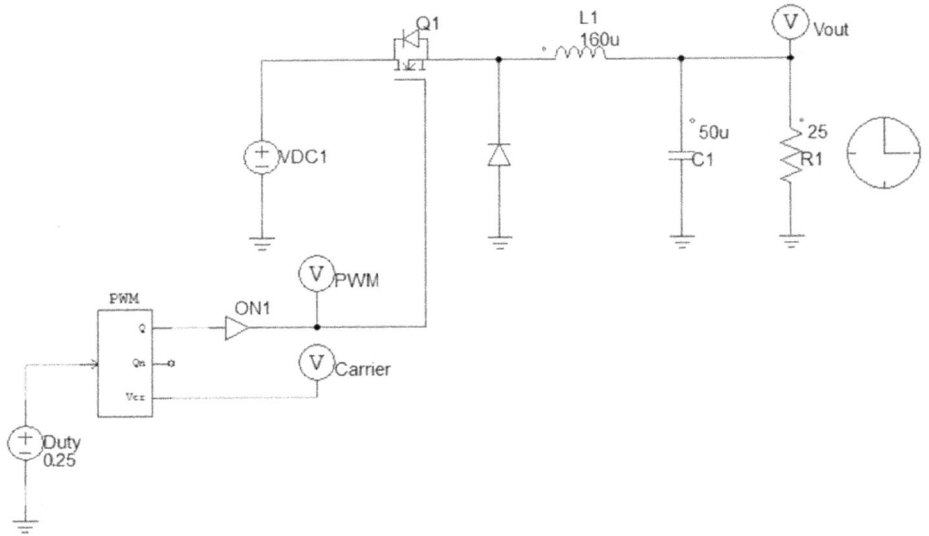

Fig. 5.339.

Settings of the Carrier PWM Controller (s-domain) block are shown in Fig. 5.340.

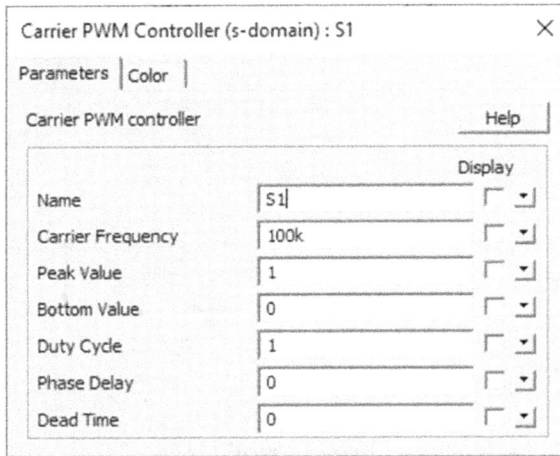

Fig. 5.340.

The Carrier PWM Controller (s-domain) block has six settings:

- **Carrier frequency:** The frequency of the triangular carrier wave, in Hz.
- **Peak value:** Peak value of the triangular wave. In Fig. 5.341, the peak value is b.
- **Bottom value:** Bottom value of the triangular wave. In Fig. 5.341, the bottom value is a.
- **Duty cycle:** Duty cycle of the rising slope interval must be in the range between 0 and 1.
- **Phase delay:** Phase delay of the triangular waveform, in degree. When the value is 0, the wave starts at the bottom of the rising slope, as shown in Fig. 5.341.
- **Dead time:** The dead time between the output Q and Qn, in seconds.

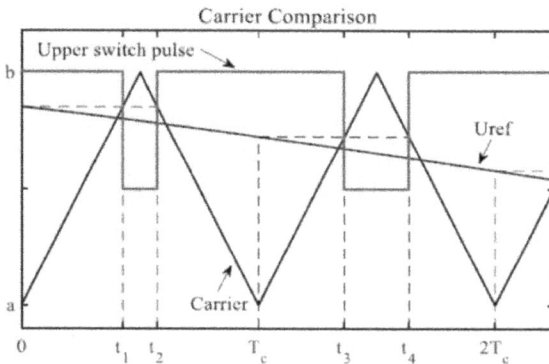

Fig. 5.341.

Different types of modulations can be produced with the aid of Carrier PWM Controller (s-domain) block. PSIM has ready examples for different types of modulations. To see these examples, click the File>Open Examples. Then open the "other" folder. For unipolar modulation, Carrier_PWM_Controller_UMod.psimsch; for bipolar sine pulse width modulation (SPWM), Carrier_PWM_Controller_BSPWM.psimsch; for three-phase SPWM, Carrier_PWM_Controller_3-Ph_SPWM.psimsch; and for harmonic injection pulse width modulation (HIPWM), see Carrier_PWM_Controller_3-Ph_HIPWM. psimsch.

5.29 Example 28: efficiency of SEPIC converter

In this example we want to measure the efficiency of a single-ended primary-inductor converter (SEPIC) converter. Assume the SEPIC converter shown in Fig. 5.342. The switching frequency is 100 kHz and the duty cycle of MOSFET is 0.4.

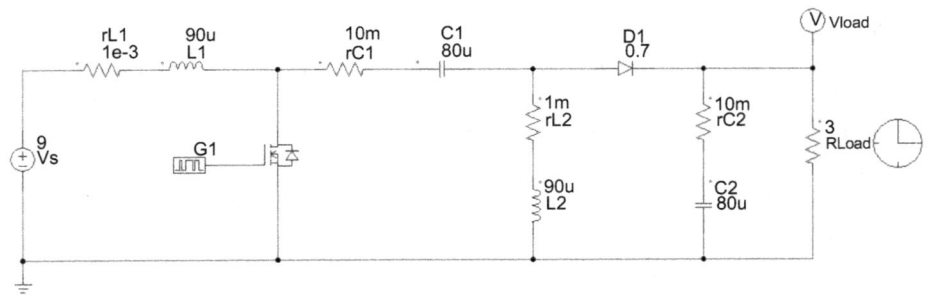

Fig. 5.342.

Settings of the blocks are shown in Figs. 5.343–5.346.

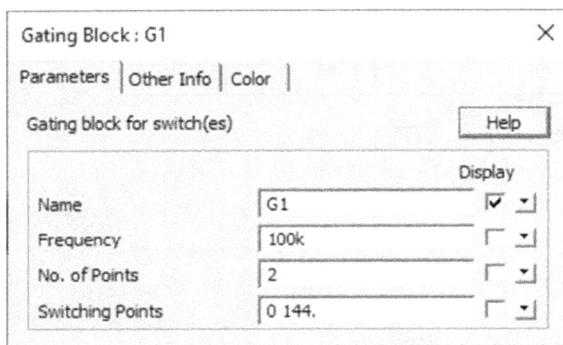

Fig. 5.343.

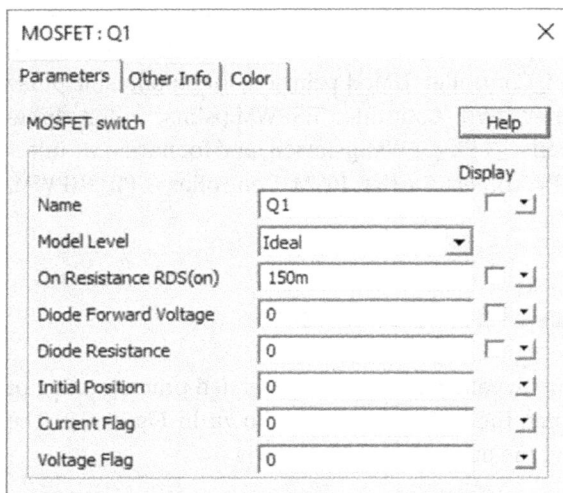

Fig. 5.344.

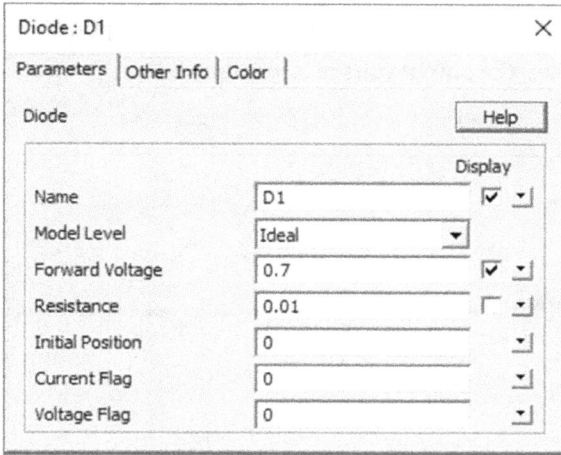

Fig. 5.345.

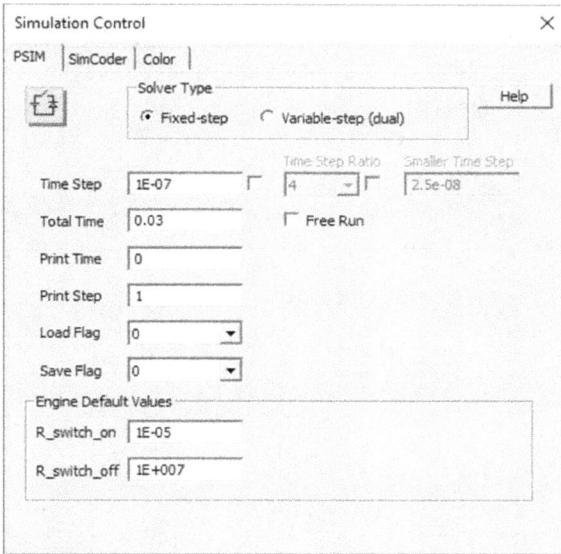

Fig. 5.346.

Run the simulation. The load voltage is shown in Fig. 5.347. Figure 5.348 is the zoomed view of Fig. 5.347. According to Fig. 5.348, the average value of output voltage is about 4.96 V.

The formula for output voltage of a SEPIC converter which is operated in continuous conduction mode (CCM) and has no losses is $V_o = \frac{D}{1-D} V_s$. So, a lossless CCM

SEPIC converter with duty cycle of 0.4 must produce $V_o = \frac{0.4}{1-0.4} \times 9 = 6$ V in its output. Since the switches of Fig. 5.342 have losses, the output voltage is less than 6 V.

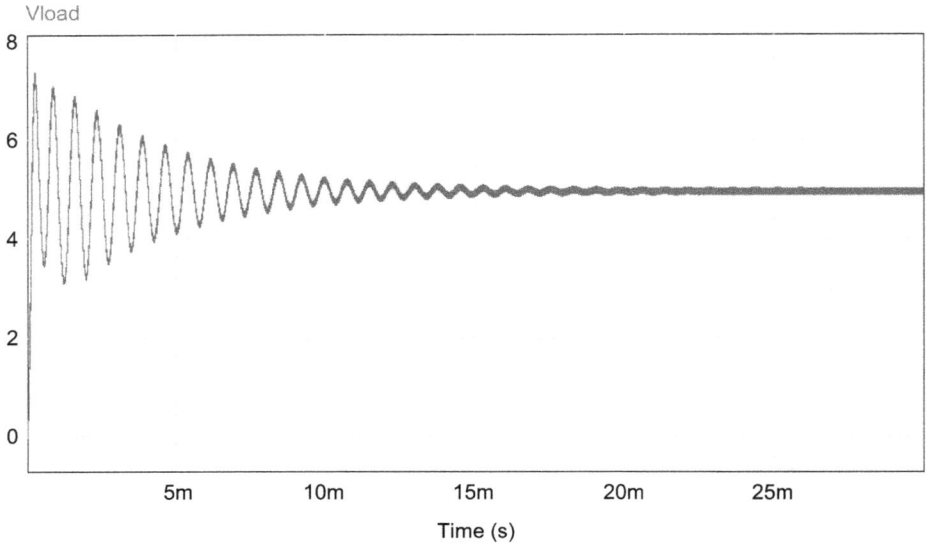

Fig. 5.347.

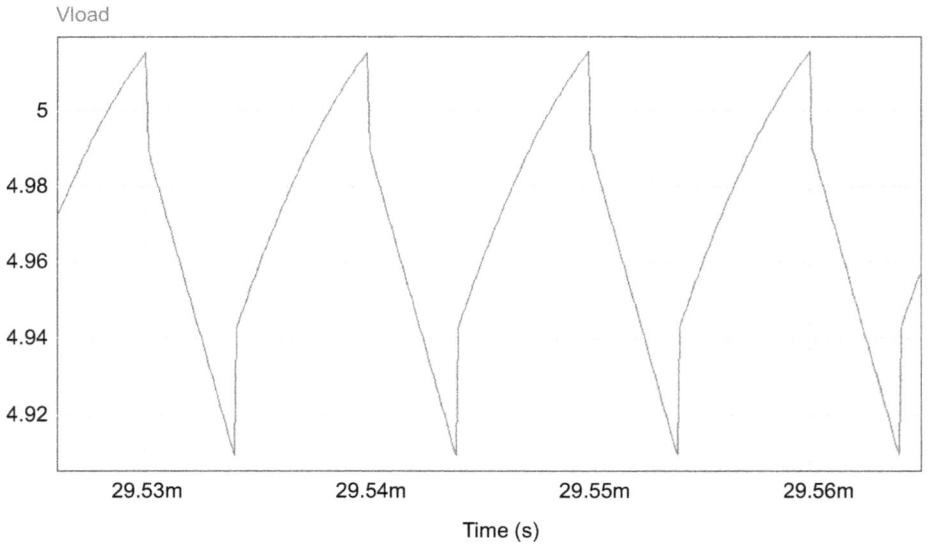

Fig. 5.348.

Now we want to measure the efficiency of the converter. In order to measure the efficiency, we need to measure the input power and output power. The ratio of output power to input power gives the efficiency.

In the schematic shown in Fig. 5.349, the input and output power are measured with the aid of Wattmeter blocks. A division block is used to calculate the ratio of output power to input power. The result of division is multiplied with 100 in order to convert the result into percent.

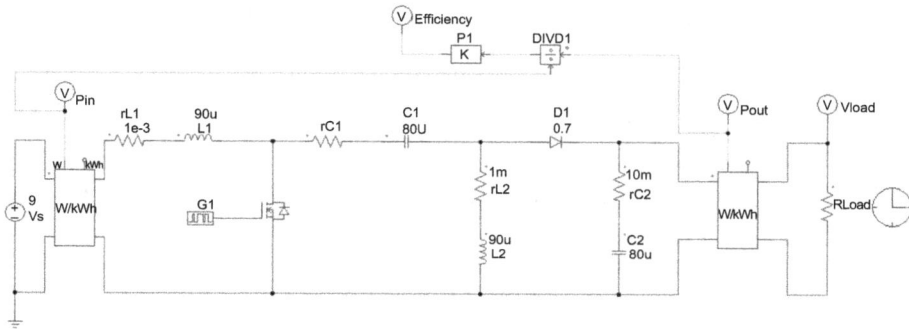

Fig. 5.349.

The settings of both of the Wattmeter blocks are similar to Fig. 5.350. The settings of Simulation Control block are shown in Fig. 5.351.

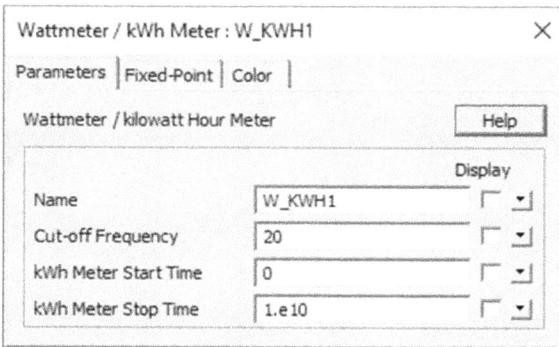

Fig. 5.350.

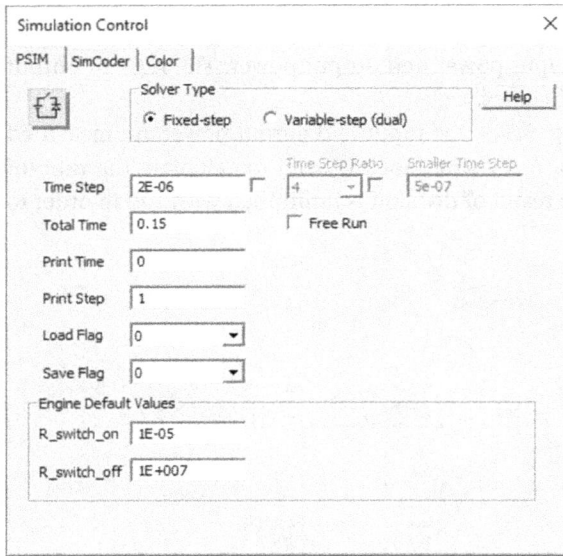

Fig. 5.351.

Run the simulation. The result shown in Fig. 5.352 will be obtained. The zoomed view of this figure is shown in Fig. 5.353.

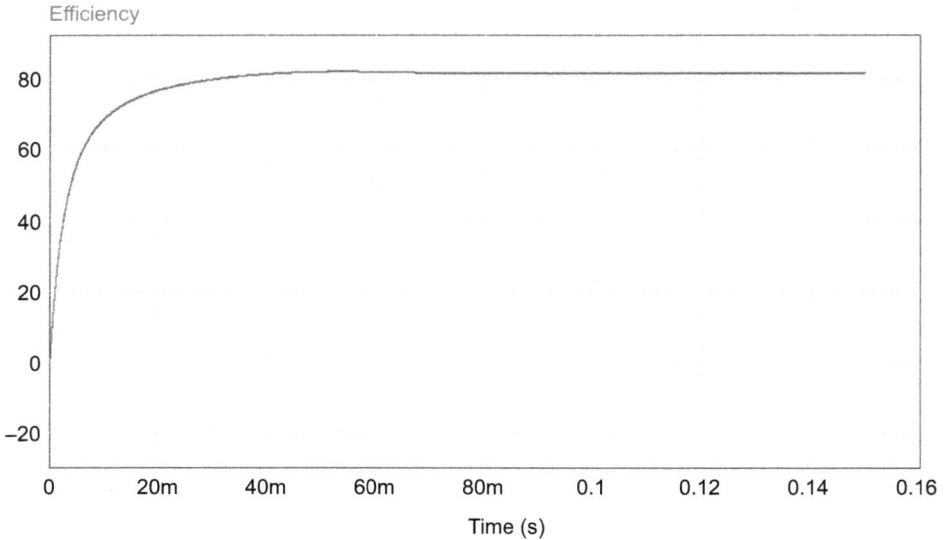

Fig. 5.352.

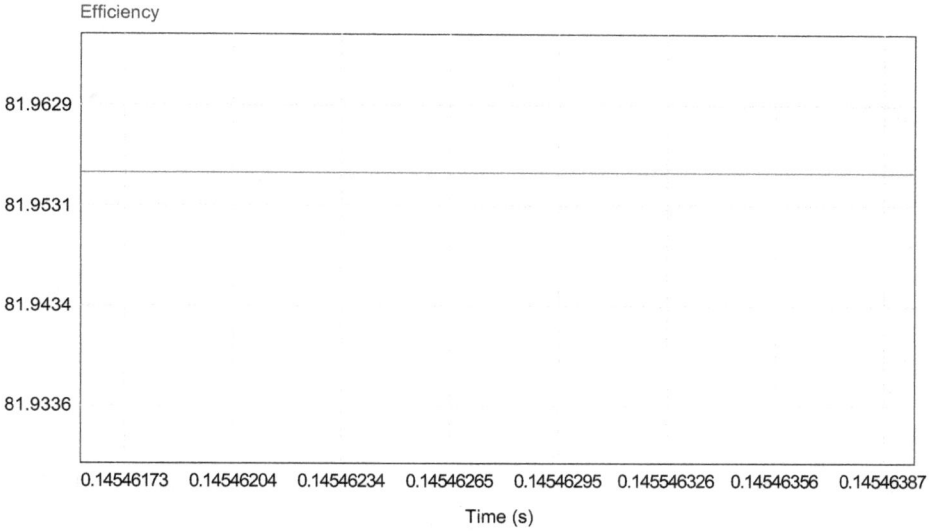

Fig. 5.353.

The input and output power are shown in Figs. 5.354 and 5.355, respectively.

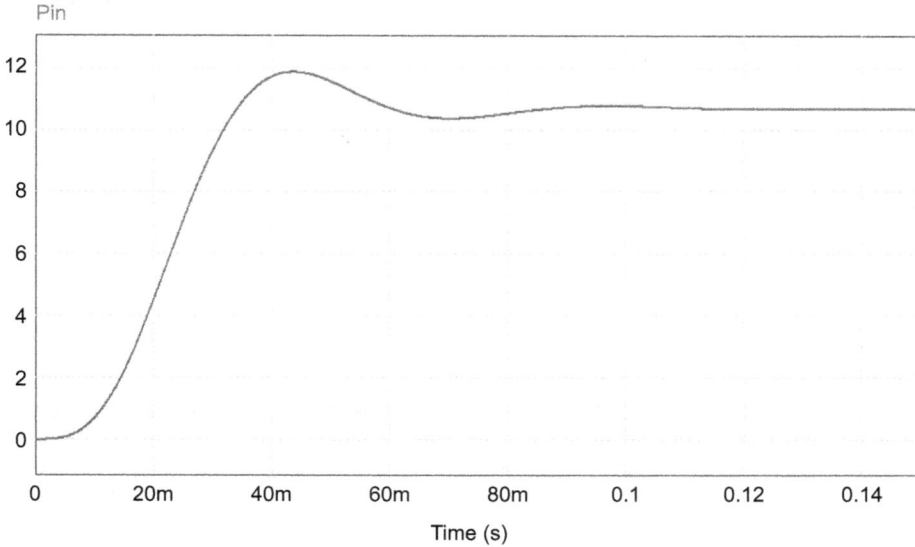

Fig. 5.354.

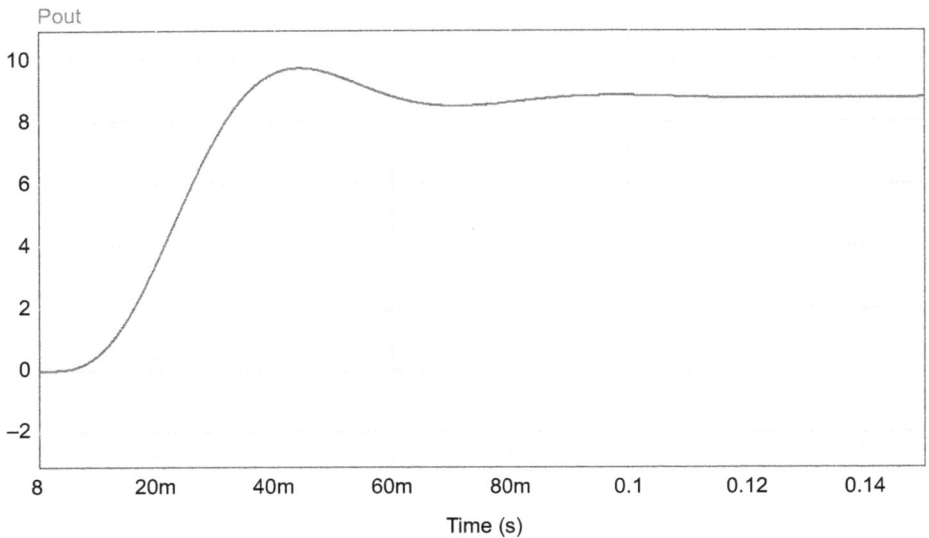

Fig. 5.355.

You can use the schematic shown in Fig. 5.356 in order to measure the efficiency as well. In this schematic, we used voltage and current sensors in order to measure the input and output power. The required division for obtaining the efficiency will be done by the Simview.

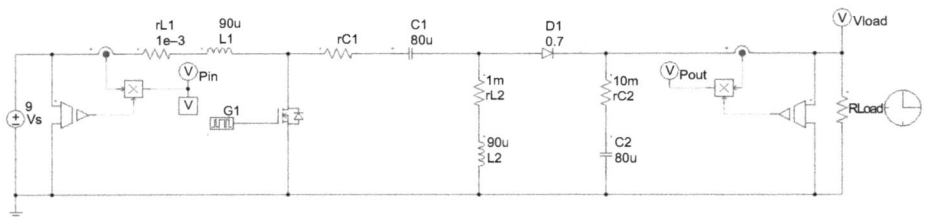

Fig. 5.356.

Run the simulation. When the Properties window appears, enter the expression shown in Fig. 5.357. Then press the Add and OK buttons.

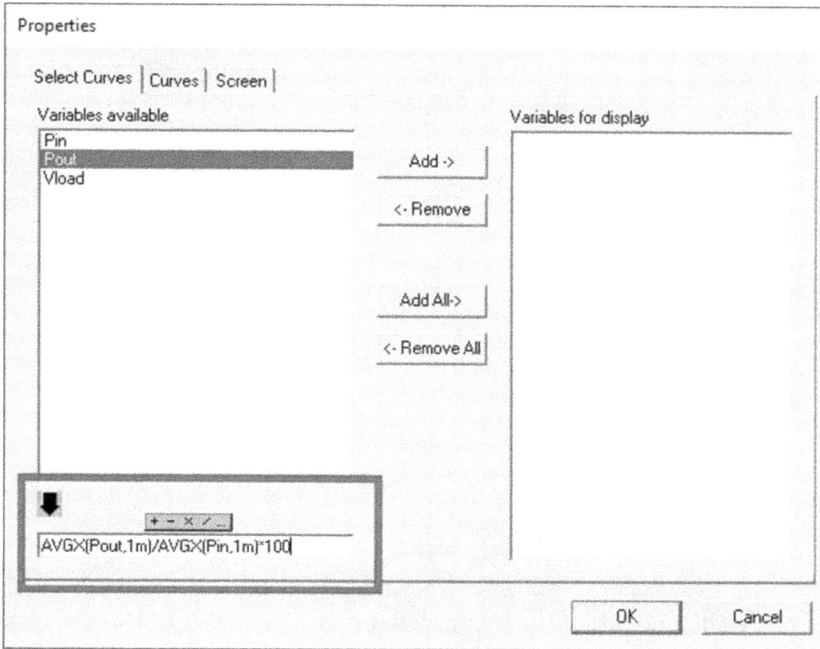

Fig. 5.357.

The graph shown in Fig. 5.358 will appear. The zoomed view is shown in Fig. 5.359. The efficiency is about 82%. This simulation does not consider the switching losses. It considers only the conduction losses. If you want to consider the switching losses, you need to use the Thermal Module (see Example 45).

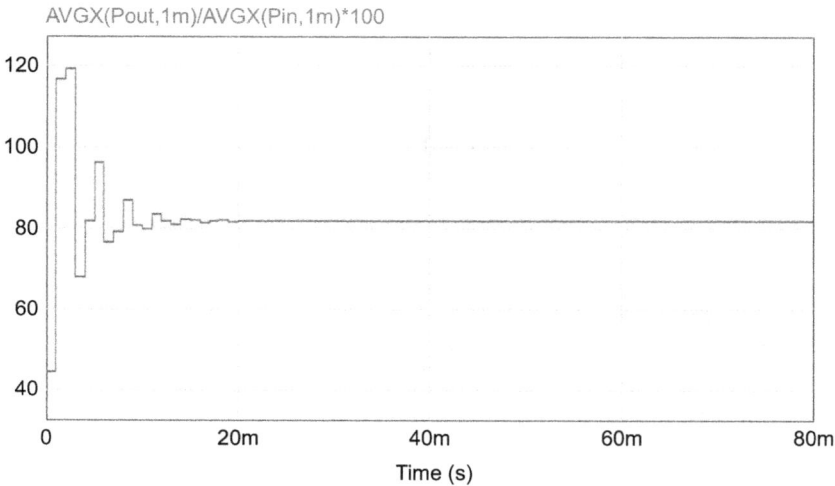

Fig. 5.358.

AVGX(Pout,1m)/AVGX(Pin,1m)*100

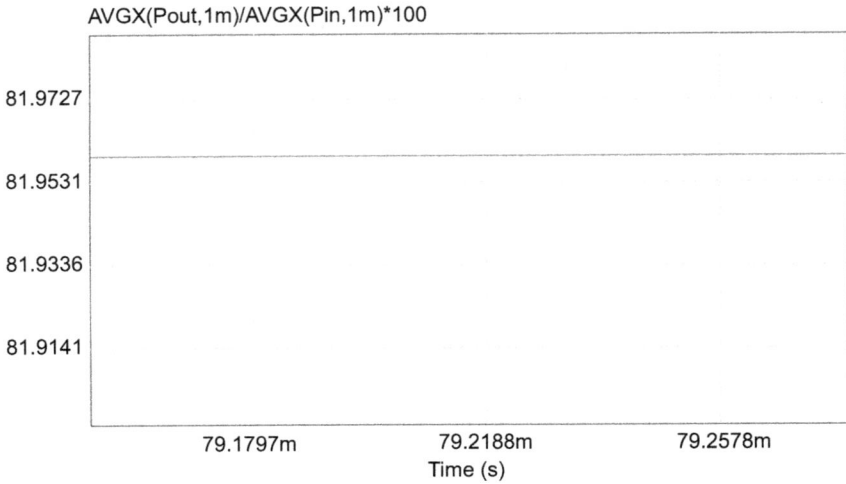

Fig. 5.359.

Note that if you enter the simple expression shown in Fig. 5.360, you will get the error message shown in Fig. 5.361. So, use the expression shown in Fig. 5.357 to do the division.

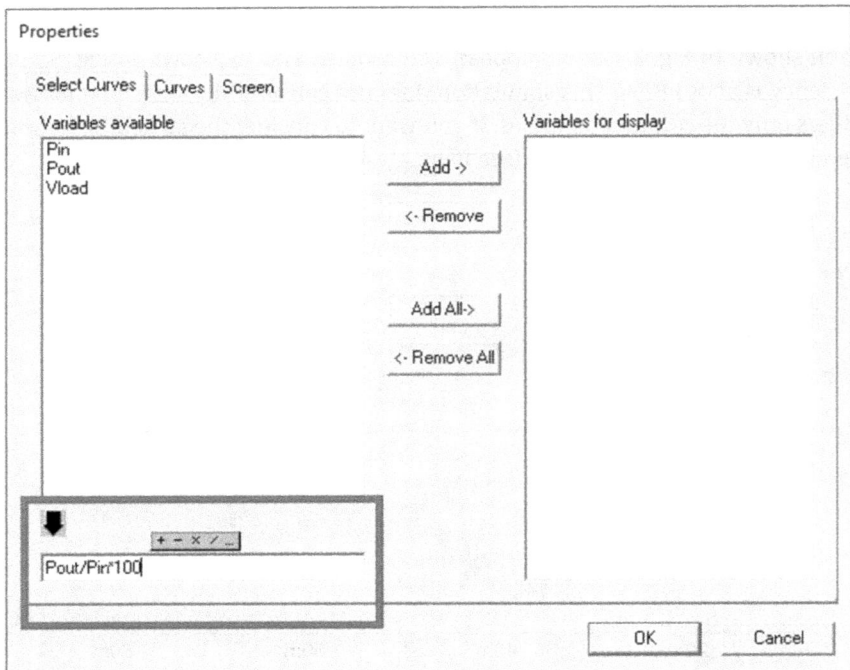

Fig. 5.360.

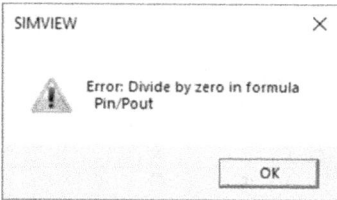

Fig. 5.361.

5.30 Example 29: limiter blocks

PSIM has different types of limiters (Fig. 5.362) to limit the minimum or maximum of signals. The upper limiter, limits the maximum of signal and the lower limiter, limits the minimum of signal.

For example, assume a signal is entered into an Upper Limiter block with upper limit of 0.8. In this case, when the input signal is bigger than 0.8, the output will be clipped and the output is equal to 0.8. When the value of input signal is less than 0.8, then the Upper Limiter blocks permits it to pass without any change.

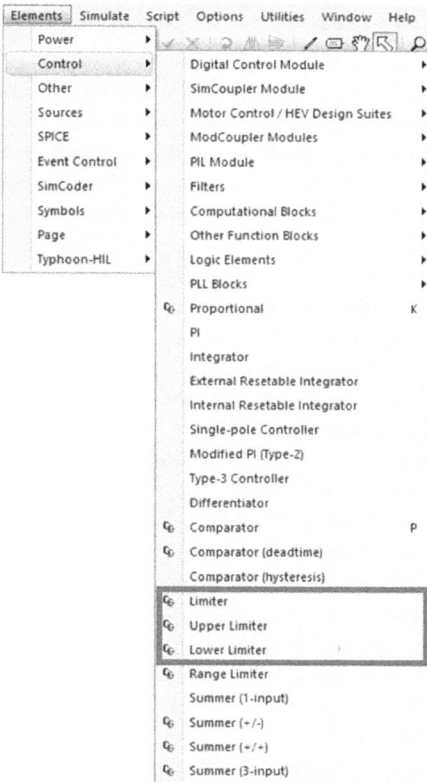

Fig. 5.362.

Assume the simple schematic shown in Fig. 5.363.

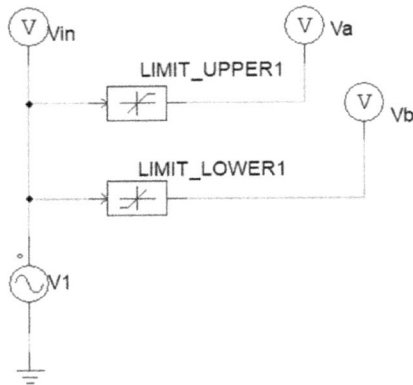

Fig. 5.363.

The settings of blocks are shown in Figs. 5.364–5.367.

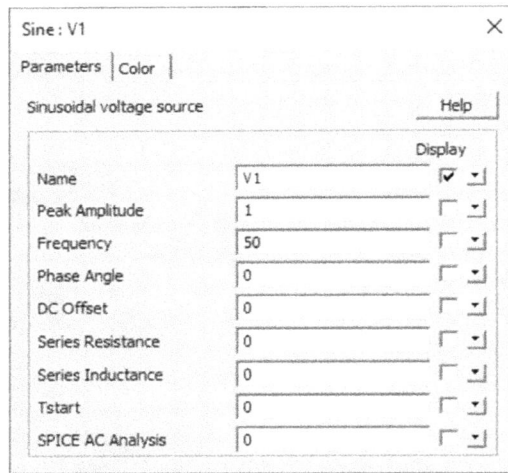

Fig. 5.364.

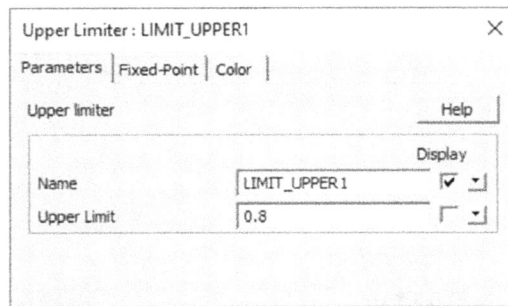

Fig. 5.365.

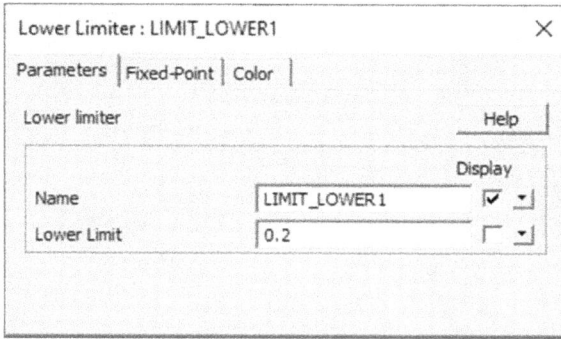

Fig. 5.366.

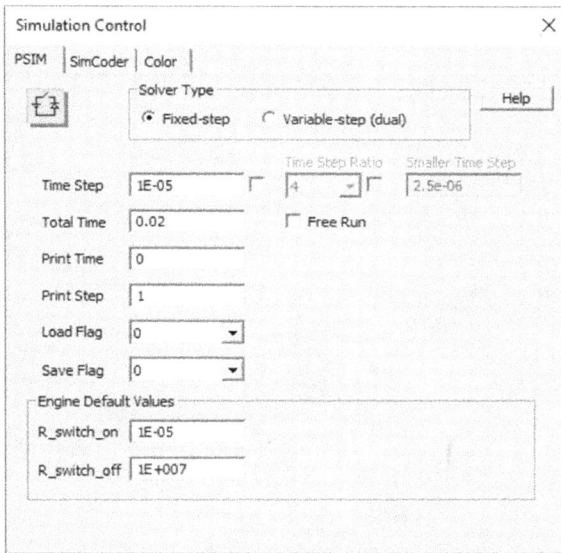

Fig. 5.367.

Run the simulation. The output a and b waveforms are shown in Figs. 5.368 and 5.369, respectively. The upper limiter does not permit output a to go beyond 0.8 and the lower limiter does not permit output b to go lower than 0.8.

Va

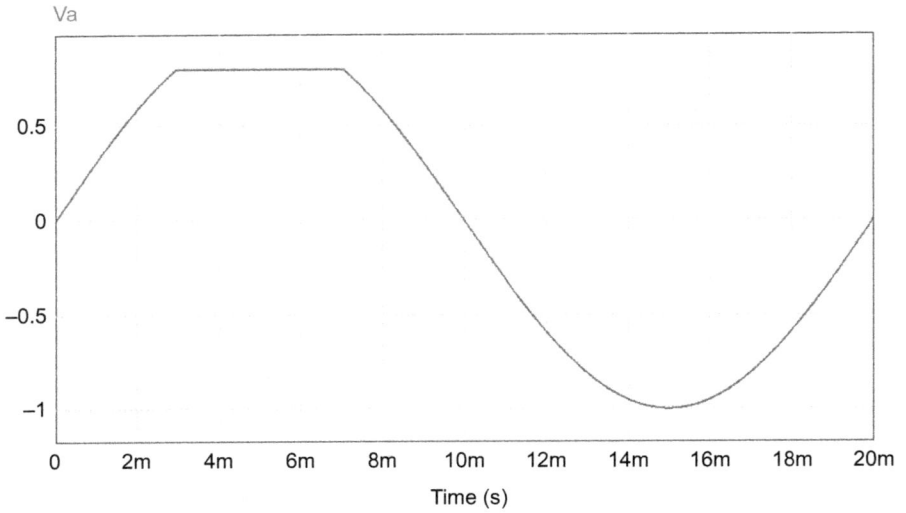

Fig. 5.368.

Vb

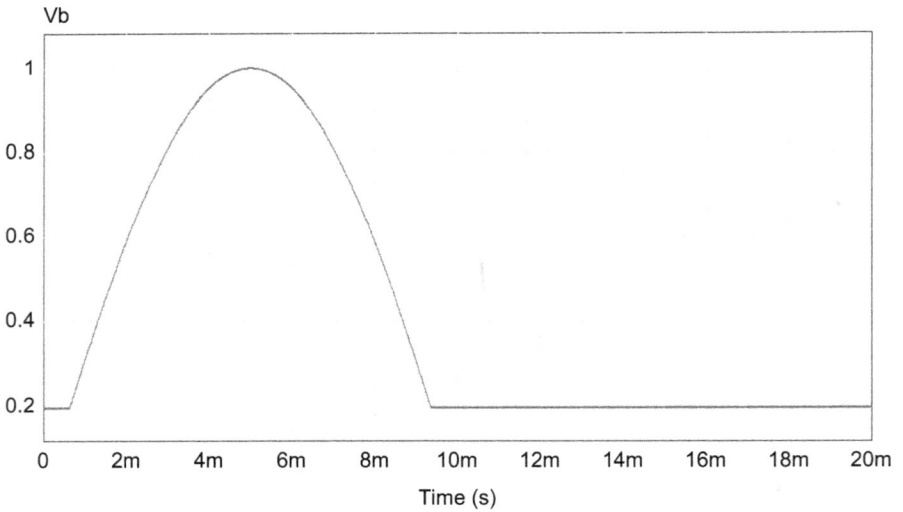

Fig. 5.369.

The Limiter block, limits both the maximum and minimum of input signal. So, it does the job of both of Upper Limiter and Lower Limiter blocks at the same time. Assume the simple schematic shown in Fig. 5.370.

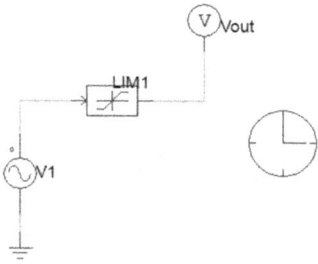

Fig. 5.370.

The settings of input voltage source V1 and Simulation Control blocks are shown in Figs. 5.364 and 5.367, respectively. The settings of Limiter block are shown in Fig. 5.371.

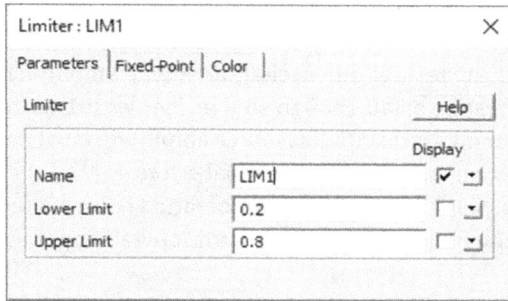

Fig. 5.371.

Run the simulation. The result shown in Fig. 5.372 will be obtained. As shown, the signal is limited between 0.2 and 0.8.

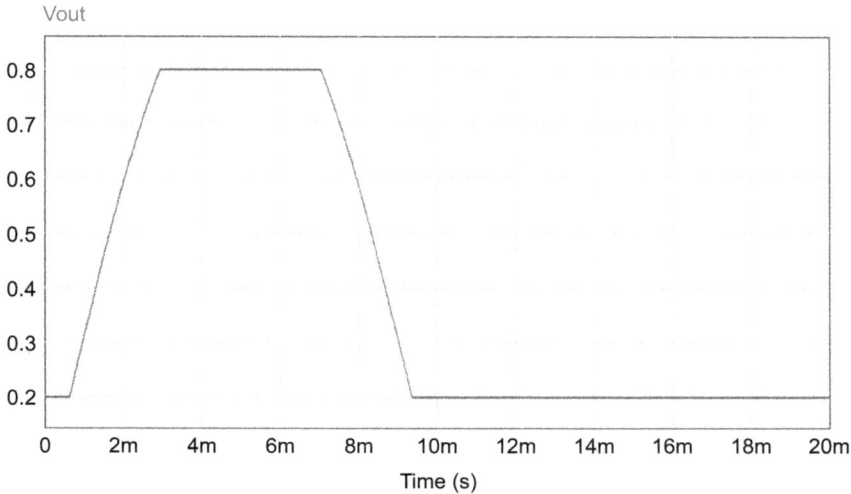

Fig. 5.372.

The limiter block is generally used at the input of pulse modulators in order to ensure that the signal that enters the modulator is in the allowed interval.

5.31 Example 30: frequency response of control to output voltage for buck converter

In this example we will study the dynamics of a buck converter. We want to extract the frequency response of control-to-output $\left(\frac{\tilde{v}_o (s)}{\tilde{d} (s)}\right)$ transfer function. $\tilde{v}_o (s)$ shows the Laplace transform of small signal part of output voltage and $\tilde{d} (s)$ shows the small signal part of duty cycle ($d = D + \tilde{d}$ and $v_o = V_o + \tilde{v}_o$).

The principle of the ac analysis is that a small ac excitation signal is injected into the system as the perturbation, and the signal at the same frequency is extracted at the output. To obtain accurate ac analysis results, the excitation source amplitude must be set properly. The amplitude must be small enough so that the perturbation stays in the linear region. On the other hand, the excitation source amplitude must be large enough so that the output signal is not affected by numerical errors.

Assume the schematic shown in Fig. 5.373. The steady-state operating point of this converter is $D = 0.6$. We want to obtain the frequency response of this operating point.

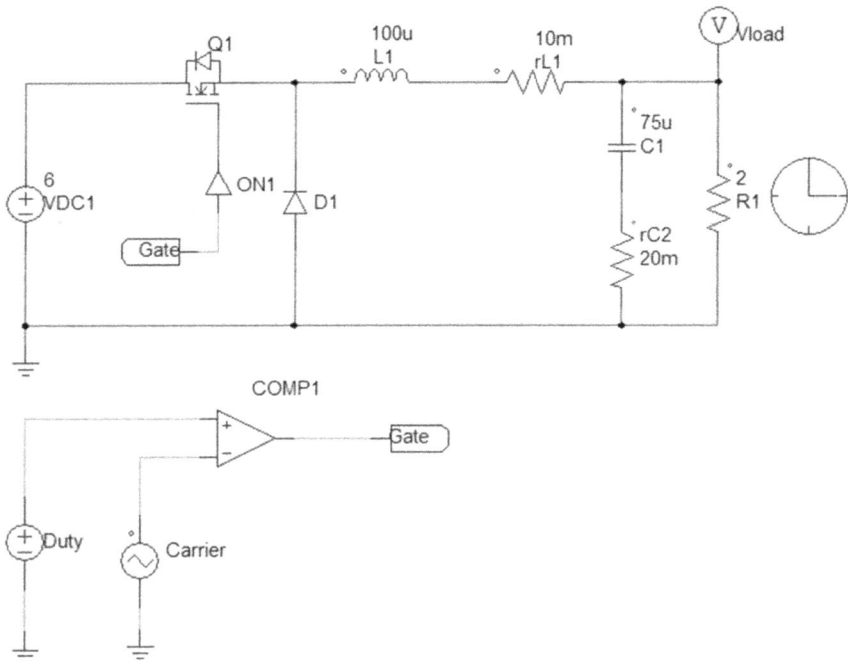

Fig. 5.373.

The settings of the used elements are shown in Figs. 5.374–5.378.

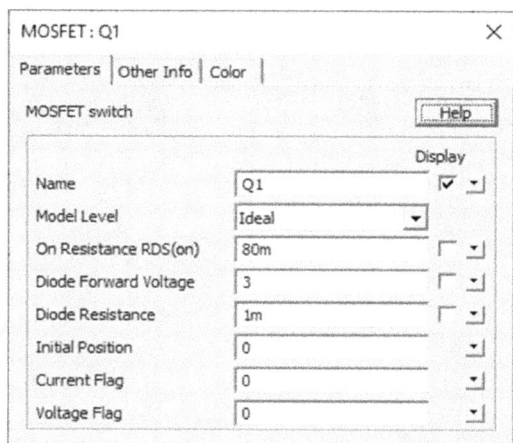

Fig. 5.374.

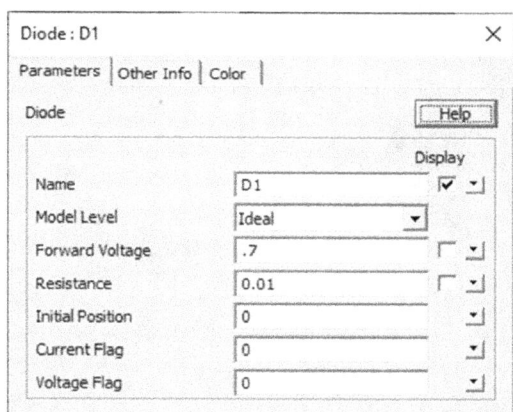

Fig. 5.375.

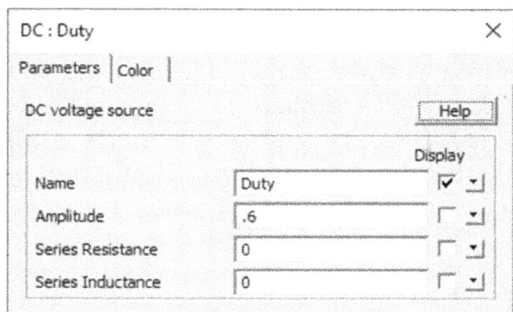

Fig. 5.376.

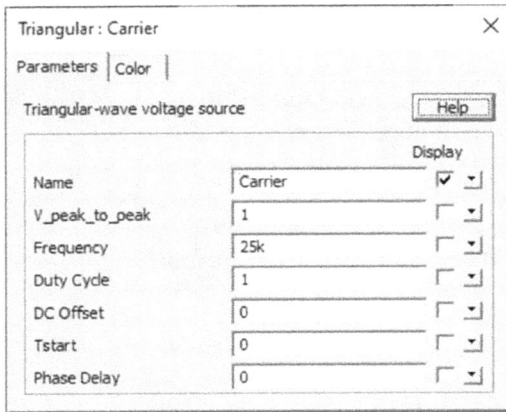

Fig. 5.377.

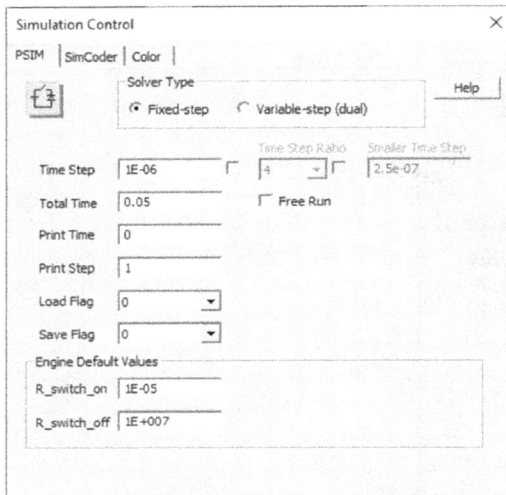

Fig. 5.378.

We need to add an AC Sweep (Fig. 5.379) and AC Sweep Probe (Fig. 5.380) blocks the schematic. These two blocks are added to the schematic as shown in Fig. 5.381.

We want to obtain the frequency response between the control (duty cycle) and output voltage. Because of these, the AC Sweep Probe is connected to the output node to measure its voltage. In the schematic of Fig. 5.381, a sinusoidal source (Fig. 5.382) connected in series to Duty source block. This adds the required small signal component (perturbation) to the duty cycle.

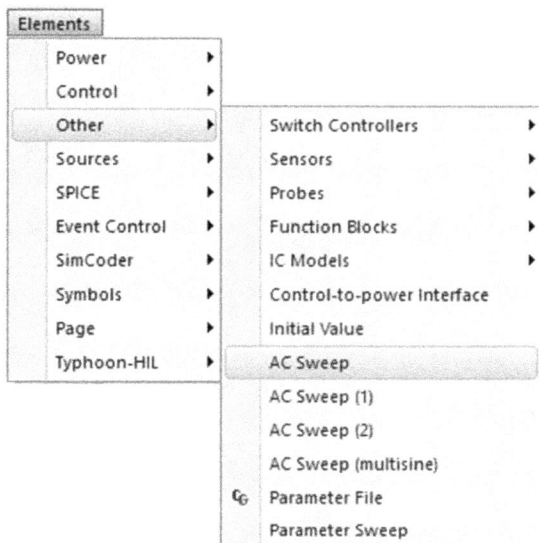

Fig. 5.379.

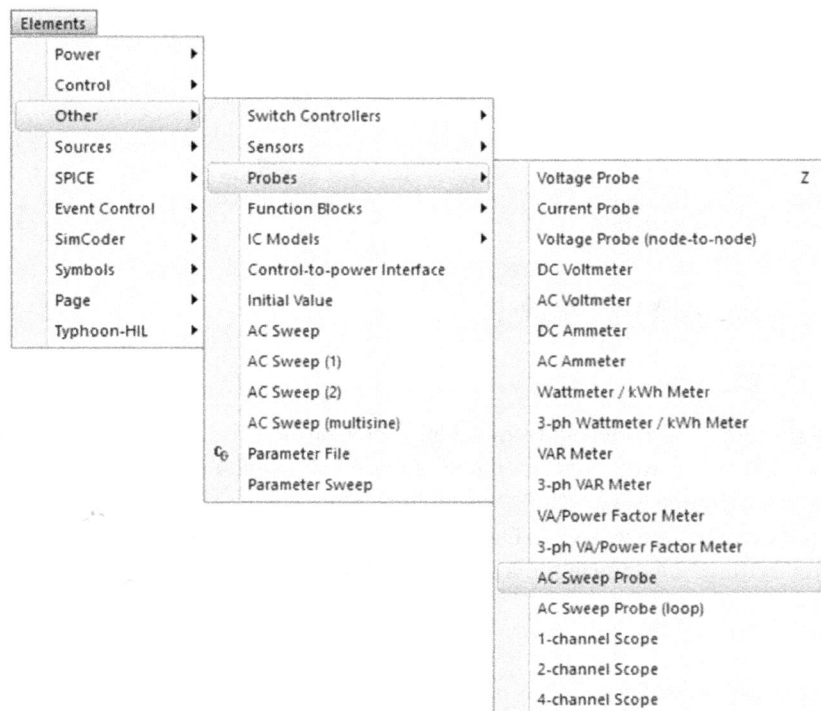

Fig. 5.380.

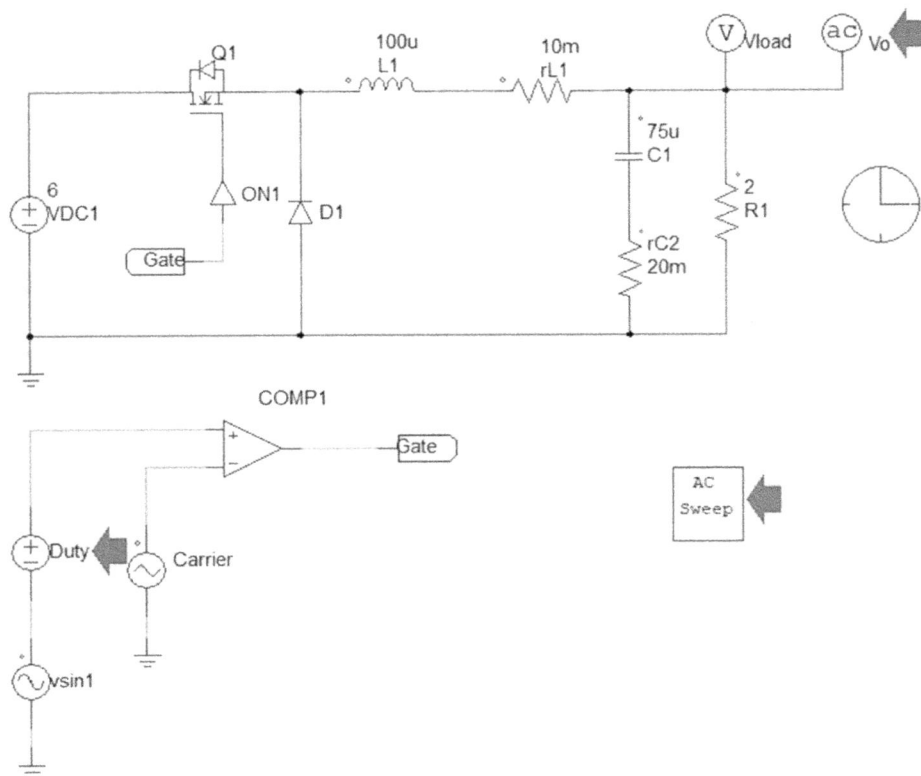

Fig. 5.381.

Fig. 5.382.

Settings of the added AC Sweep and sinusoidal source blocks are shown in Figs. 5.383 and 5.384, respectively. Note that the name of added sinusoidal source block must be entered to the Source Name of AC Sweep block. The settings of sinusoidal source (Peak Amplitude, Frequency, etc., in Fig. 5.384) have no effect on AC Sweep analysis. The required settings are done in the AC Sweep dialog (Fig. 5.383).

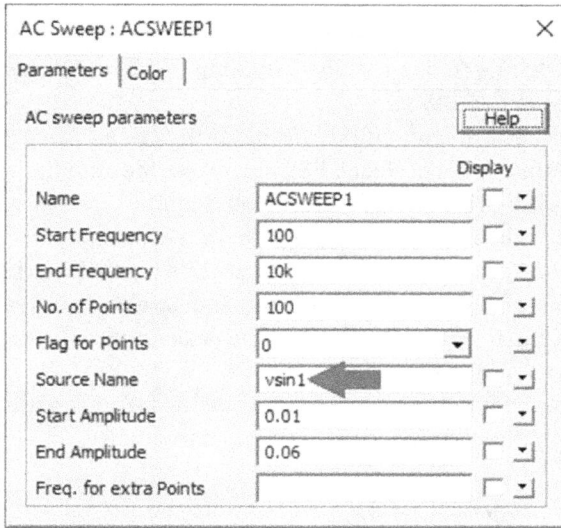

Fig. 5.383.

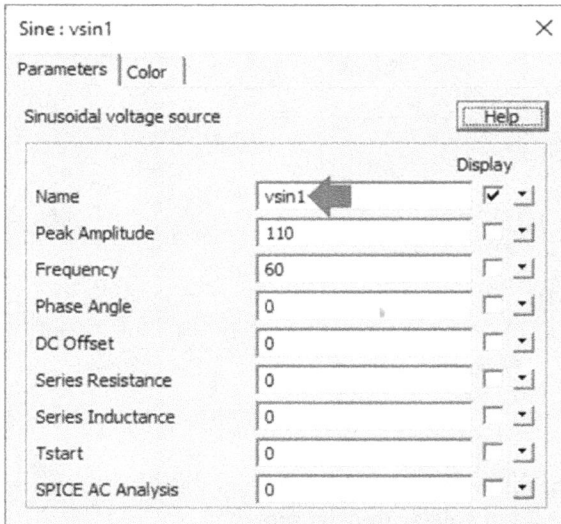

Fig. 5.384.

According to Fig. 5.383, the frequency response will be drawn from 100 Hz up to 10 kHz. No. of Points determines number of different frequencies between the minimum (100 Hz) and maximum (in this example 10 kHz) which will be used to form the output graph. Increasing the number entered into the No. of Points box increases the accuracy of analysis however the required time for the simulation will be increased as well. A number between 50 and 200 is well enough for most of circuits. Generally, the sweep is done up to half of the switching frequency. The switching frequency of this converter is 25 kHz, so we can sweep up to 12.5 kHz.

Start Amplitude determines the starting amplitude for perturbation. It must be small in comparison to the steady-state value of signal. PSIM increases the amplitude of perturbation as frequency increases. This gradual increase of amplitude provides a smooth output graph especially in the high frequency region. In this example, the 100 Hz perturbation has amplitude of 0.01 and the 10 kHz perturbation has amplitude of 0.06. Note that the values entered to the Start Amplitude and End amplitude boxes will be used during the simulation and the value entered into the Peak Amplitude box (Fig. 5.384) has no effect on the simulation result.

Run the simulation. The result shown in Fig. 5.385 will be obtained. You can use the cursors in order to read the amplitude and phase at the desired frequency.

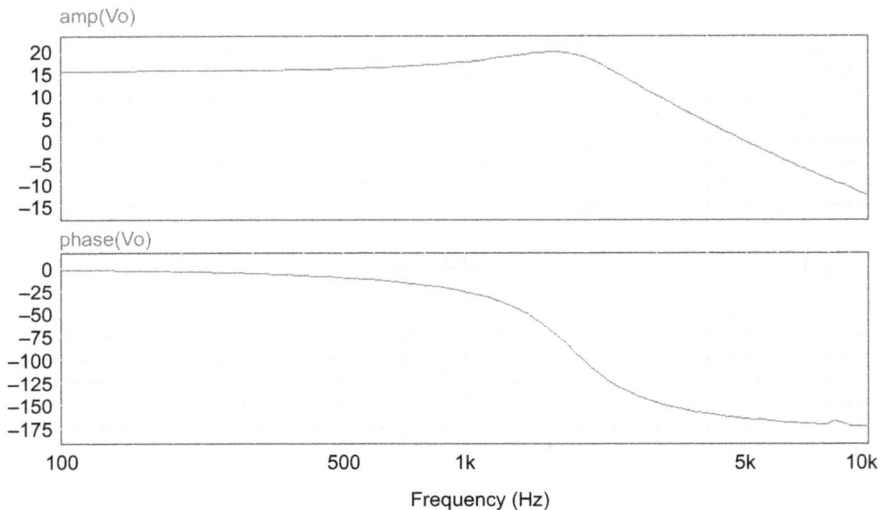

Fig. 5.385.

Let's check the obtained result. The following MATLAB code uses state-space averaging (SSA) to extract the control-to-output voltage transfer function.

```
%This program calculates the small signal transfer
%functions for Buck converter
R=2;

VIN=6;
rin=1e-9;

L=100e-6;
rL=10e-3;
```

```
C=75e-6;
rC=20e-3;

rD=.01;
VD=.7;

rds=.1;
D=.6;

R1=rin+rds+rL+R*rC/(R+rC);
R2=rD+rL+R*rC/(R+rC);

IL=(R+rC)*(D*VIN-(1-D)*VD)/((R+rC)*R2+R^2+D*(R+rC)*(R1-R2));

A=[(R2*(D-1)-R1*D)/L -R/(R+rC)/L;R/(R+rC)/C -1/(R+rC)/C];
B=[(VIN+VD+(R2-R1)*IL)/L D/L;0 0];
CC=[R*rC/(rC+R) R/(R+rC)]; %C shows the capacitance so CC is used for matrix
H=tf(ss(A,B,CC,0));
vO_d=H(1)% transfer function between output voltage and duty ratio
vO_vin=H(2) %transfer function between output voltage and input source

bode(vO_d), grid on
```

Run the MATLAB code. The result shown in Fig. 5.386 is obtained. You can use cursors in order to compare the two curves. The MATLAB graph verify PSIM result. Note that in MATLAB graph, horizontal axis has unit of $\frac{Rad}{s}$, however horizontal axis of PSIM graph has unit of Hertz (remember that $f = \frac{\omega}{2\pi}$.

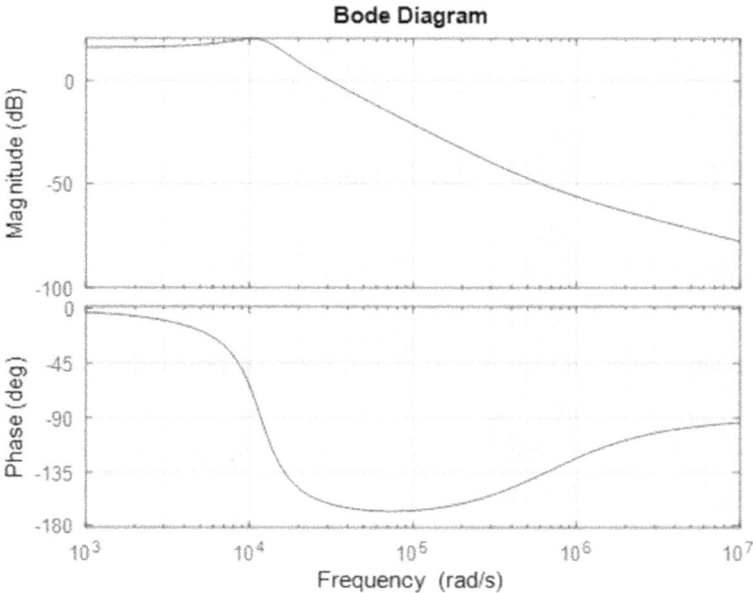

Fig. 5.386.

5.32 Example 31: frequency response of control to inductor current for buck converter

In this example, we measure the frequency response of control to inductor current for the buck converter of previous example. In order to do this, we need to connect the AC sweep probe to a current sensor which measures the current of the inductor.

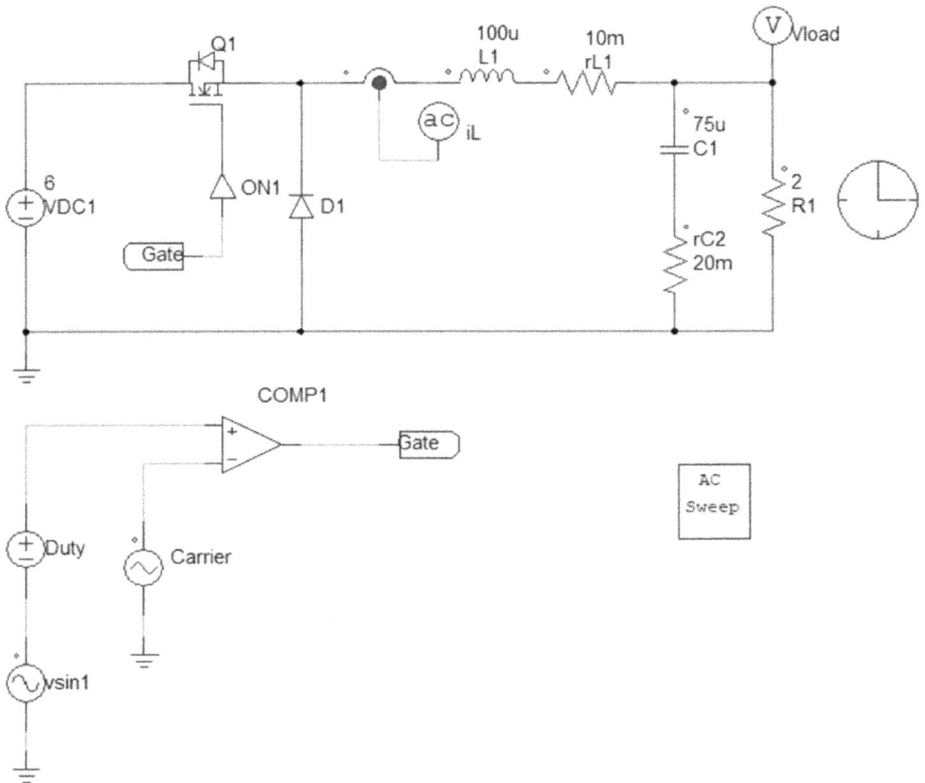

Fig. 5.387.

After running the simulation, the result shown in Fig. 5.388 will be obtained.

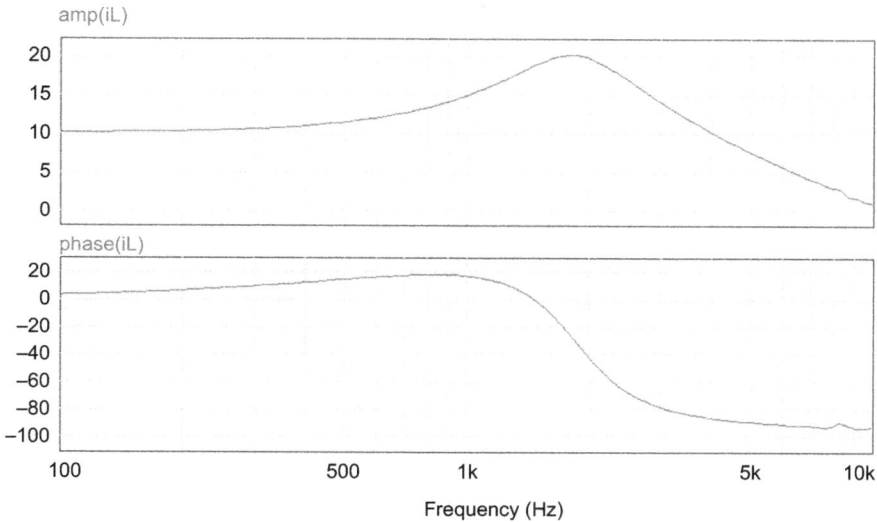

Fig. 5.388.

5.33 Example 32: frequency response of electric circuit

The AC sweep is not limited to switching circuits. You can use AC sweep in order to obtain the frequency response of other type of circuits. Assume the circuit shown in Fig. 5.389.

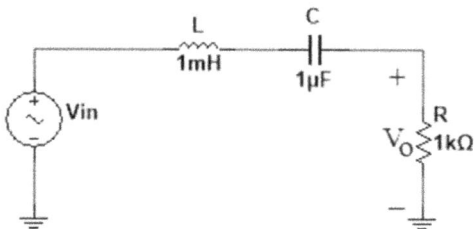

Fig. 5.389.

According to basic circuit theory:

$$\frac{V_o(s)}{V_{in}(s)} = \frac{R.C.s}{L.C.s^2 + R.C.s + 1} = \frac{10^{-3}s}{10^{-9}s^2 + 10^{-3}s + 1}$$

The frequency response of this transfer function can be drawn with the aid of MATLAB commands shown in Fig. 5.390. The result is shown in Fig. 5.391. Note that the unit of horizontal axis is not but $\frac{Rad}{s}$.

```
Command Window
>> H=tf([1e-3 0],[1e-9 1e-3 1]);
>> w=logspace(-1,4,300);%upper freq. about 1.6 KHz
>> bode(H,w), grid on
fx >>
```

Fig. 5.390.

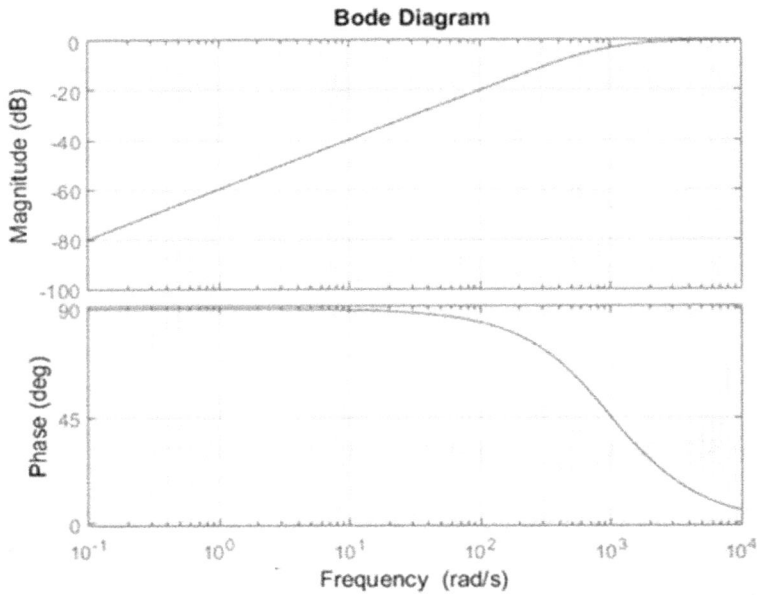

Fig. 5.391.

Let's analyze this circuit with PSIM. The required schematic is shown in Fig. 5.392.

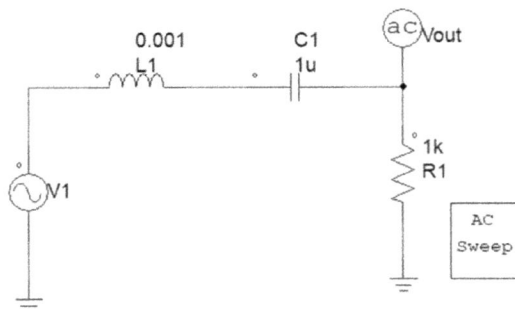

Fig. 5.392.

The settings of elements are shown in Figs. 5.393–5.398.

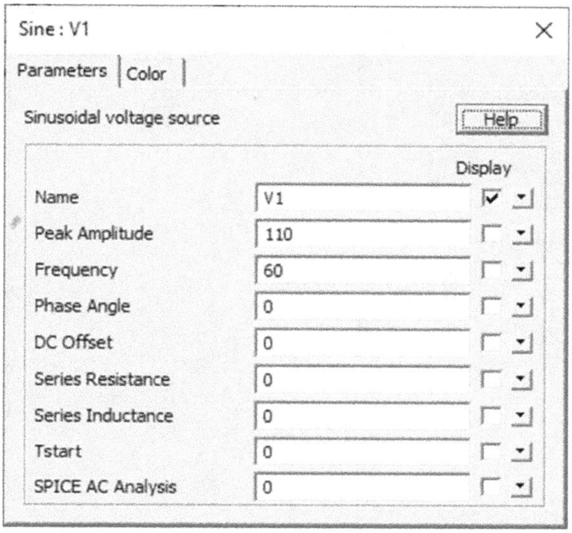

Sine : V1		X
Parameters	Color	

Sinusoidal voltage source · Help

		Display
Name	V1	☑ ▾
Peak Amplitude	110	☐ ▾
Frequency	60	☐ ▾
Phase Angle	0	☐ ▾
DC Offset	0	☐ ▾
Series Resistance	0	☐ ▾
Series Inductance	0	☐ ▾
Tstart	0	☐ ▾
SPICE AC Analysis	0	☐ ▾

Fig. 5.393.

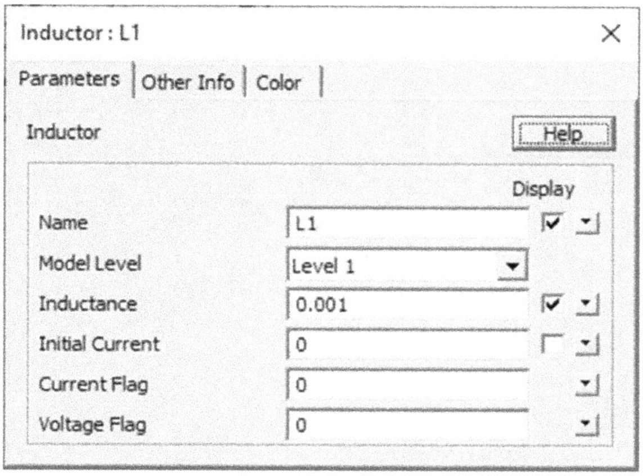

Inductor : L1			X
Parameters	Other Info	Color	

Inductor · Help

		Display
Name	L1	☑ ▾
Model Level	Level 1	▾
Inductance	0.001	☑ ▾
Initial Current	0	☐ ▾
Current Flag	0	▾
Voltage Flag	0	▾

Fig. 5.394.

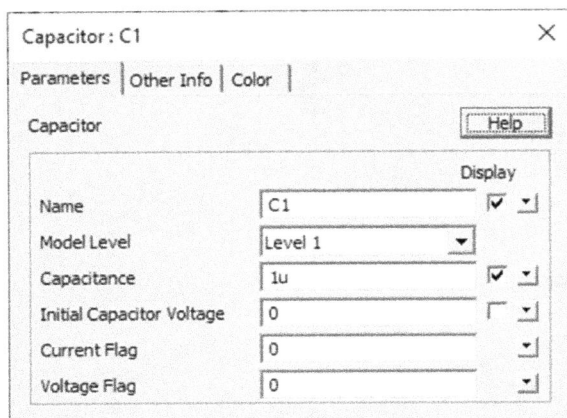

Fig. 5.395.

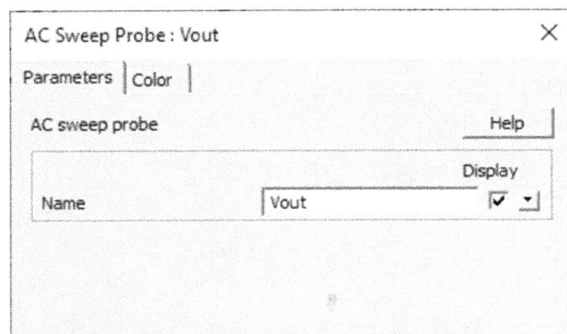

Fig. 5.396.

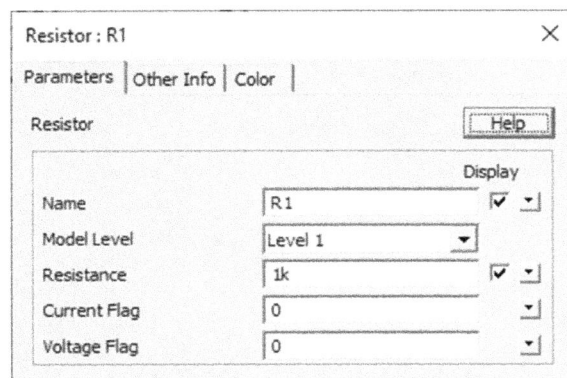

Fig. 5.397.

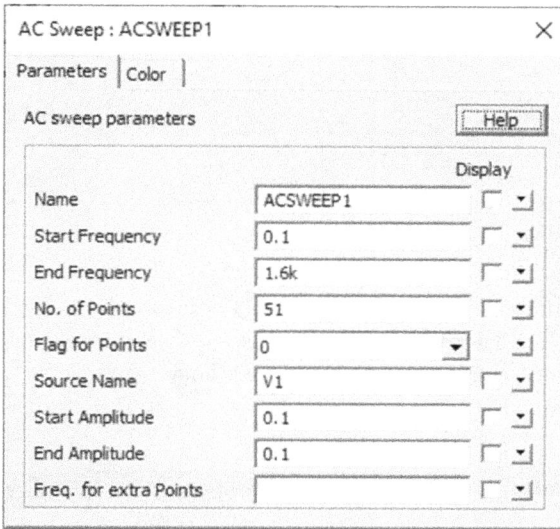

Fig. 5.398.

Run the simulation. The result shown in Fig. 5.399 will be obtained.

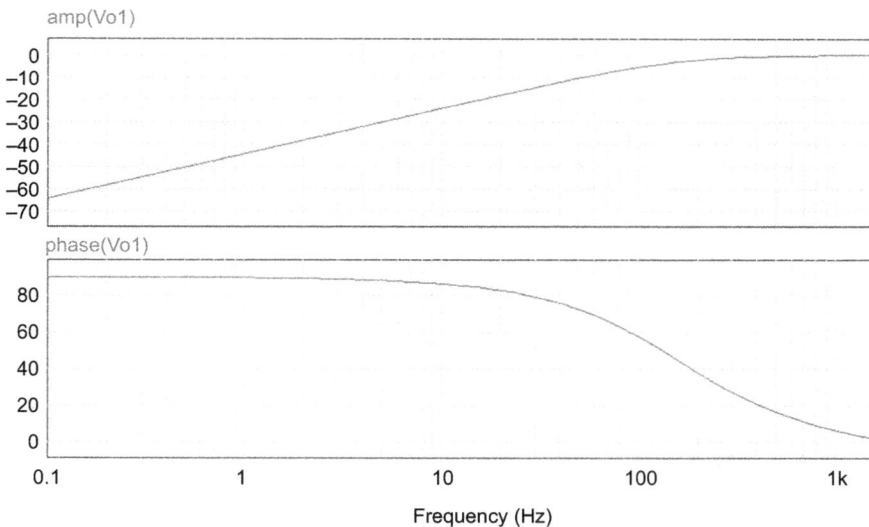

Fig. 5.399.

You can use cursors in order to compare the obtained result with the result obtained by MATLAB. For instance, according to Fig. 5.400, for 19.43 Hz the amplitude and phase angle are –18.33 dB and 82.98°, respectively. For 343.69 Hz, the amplitude is –8.378 dB and phase angle is 24.13°.

Measure				
⋮		X1	X2	Δ
Frequency		1.94398e+01	3.43697e+02	3.24257e+02
amp(Vo1)		-1.83292e+01	-8.37869e-01	1.74913e+01
phase(Vo1)		8.29857e+01	2.41292e+01	-5.88565e+01

Fig. 5.400.

The MATLAB commands shown in Figs. 5.401 and 5.402 calculate the amplitude and phase angle for 19.43 and 343.69 Hz. The results are very close to the PSIM results. Note that PSIM uses numerical algorithms to extract the frequency response so a small amount of difference is expected.

```
Command Window

>> H=tf([1e-3 0],[1e-9 1e-3 1]);
>> [mag,ph]=bode(H,2*pi*19.4398)

mag =

    0.1212


ph =

   83.0361

>> mag_db=20*log10(mag)

mag_db =

  -18.3268

fx >>
```

Fig. 5.402.

```
Command Window

>> H=tf([1e-3 0],[1e-9 1e-3 1]);
>> [mag,ph]=bode(H,2*pi*343.697)

mag =

    0.9082


ph =

   24.7454

>> mag_db=20*log10(mag)

mag_db =

   -0.8366

fx >>
```

Fig. 5.401.

5.34 Example 33: frequency response of filters

Let's study the frequency response of filter shown in Fig. 5.403.

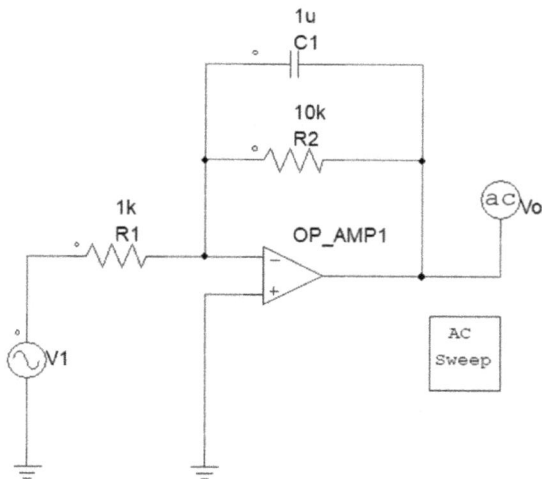

Fig. 5.403.

Settings of sinusoidal voltage source block, op-amp, and AC sweep blocks are shown in Figs. 5.404–5.406.

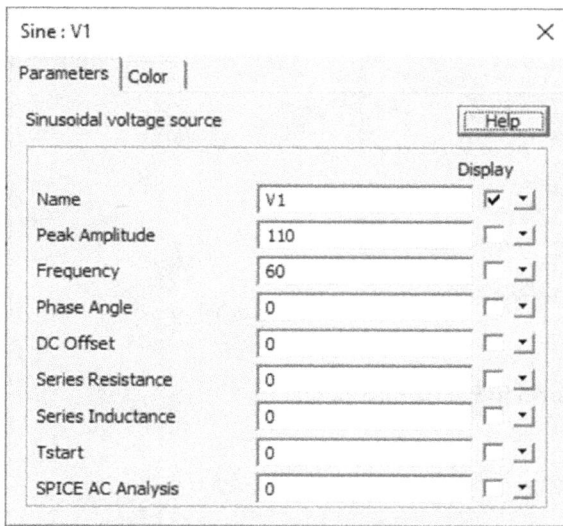

Fig. 5.404.

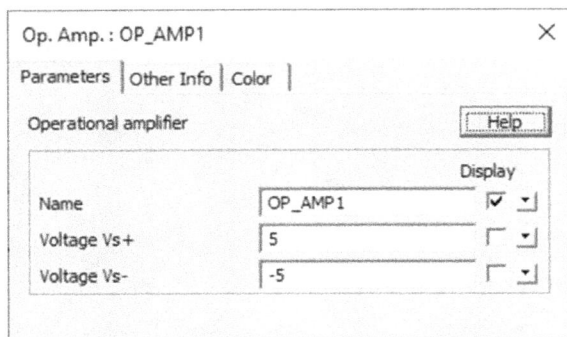

Fig. 5.405.

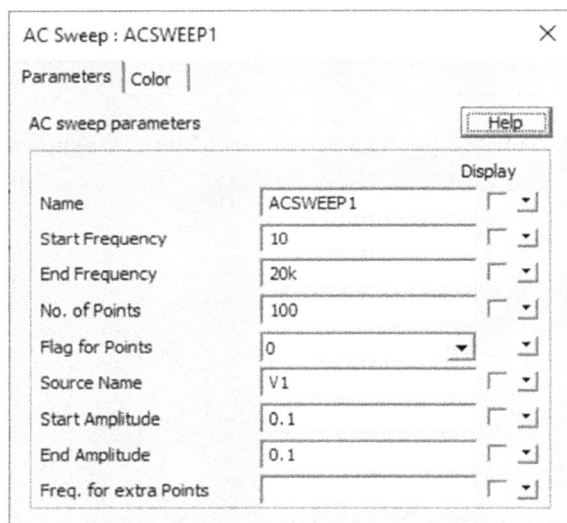

Fig. 5.406.

Run the simulation. The result shown in Fig. 5.407 is obtained.

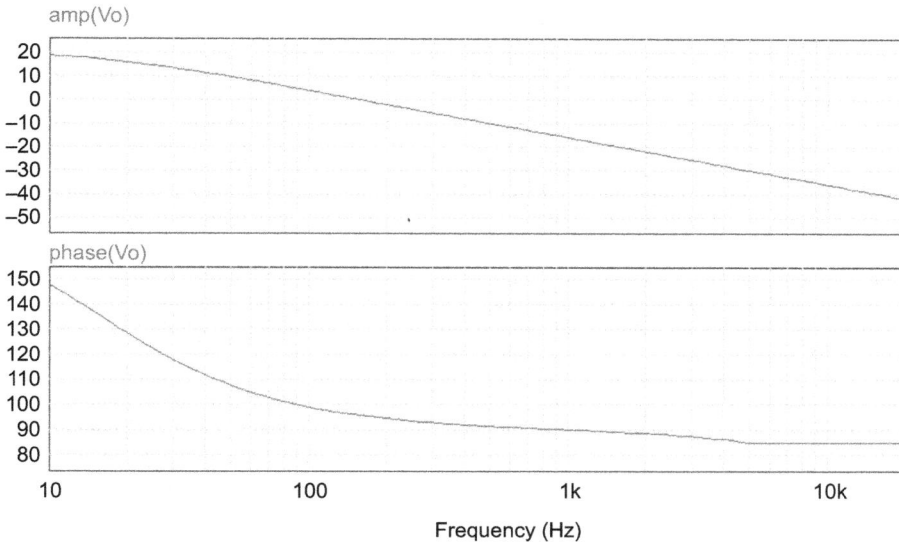

amp(Vo)

phase(Vo)

Frequency (Hz)

Fig. 5.407.

Let's analyze the circuit with the aid of MATLAB and compare the results with PSIM.

```
R1=1e3;
R2=1e4;
C1=1e-6;

s=tf('s');
Z2=R2*(1/C1/s)/(R2+(1/C1/s));
Z1=R1;
H=-Z2/Z1;
w=logspace(1.7982,5.0992,200); % w=[2*pi*10,2*pi*20e3]
bode(H,w), grid on
```

After running the above MATLAB code, the result shown in Fig. 5.408 is obtained. You can use cursors and compare different points together in order to ensure that both of the graphs are the same.

Note that the horizontal axis in PSIM has unit of Hz and the horizontal axis in MATLAB has units of $\frac{Rad}{s}$. So, when you put your cursor on the point 1,000 in MATLAB environment, it is $\frac{Rad}{s}$ or $\frac{1000}{2\pi} = 159.15\ Hz$. So, point 1,000 of MATLAB must be compared to point 159.15 of PSIM.

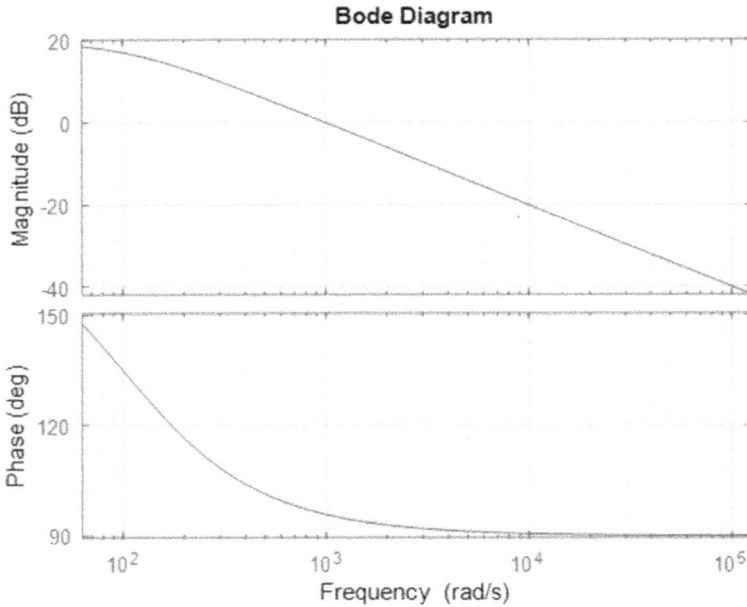

Fig. 5.408.

5.35 Example 34: input impedance of buck converter

We want to measure the input impedance of a buck converter. In order to do this, we need to add a perturbation to the input DC voltage. Then we need to measure both the voltage ($\tilde{v}_{in}$) and current ($\tilde{i}_{in}$) perturbations. The input impedance is nothing more than $\frac{\tilde{v}_{in}(s)}{\tilde{i}_{in}(s)}$.

The schematic shown in Fig. 5.409 measures the input impedance of buck converter of Example 31.

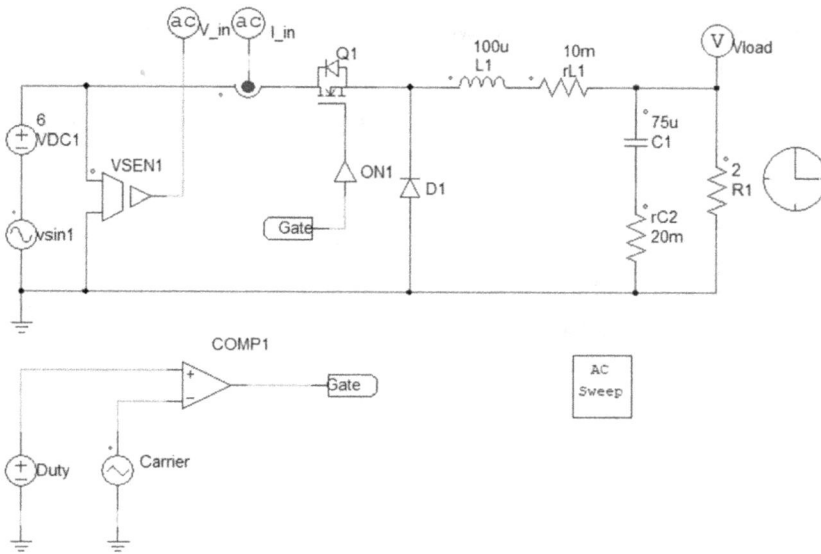

Fig. 5.409.

Settings of the AC sweep block is shown in Fig. 5.410. In this example, we want to add the perturbation to the DC source with value of 6 V. So, the amplitude of perturbation must be small in comparison to the steady state value of DC source (in this example, 6 V). So, the Start Amplitude and End Amplitude boxes are filled with 0.1 and 0.6, respectively. These values are small in comparison to the steady state value of input DC source.

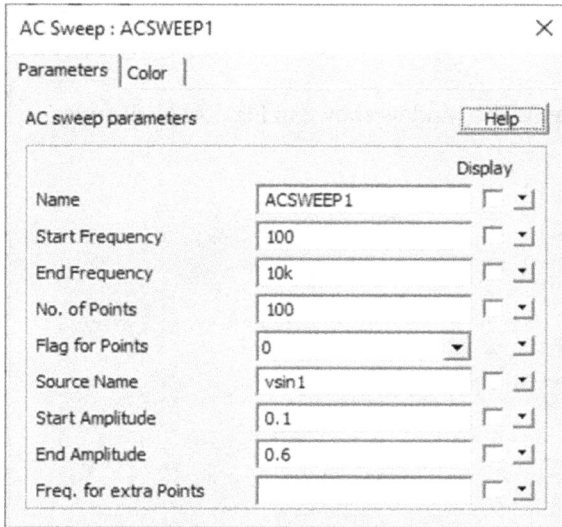

Fig. 5.410.

Run the simulation. When the Simview window opens, click the Auto-Scale Y-Axis (Fig. 5.411) or Re-draw (Fig. 5.412).

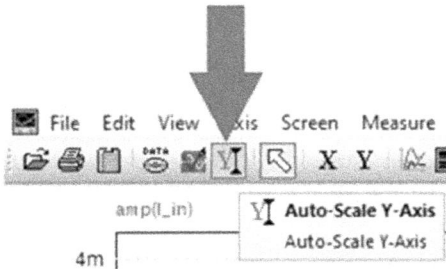

Fig. 5.411.

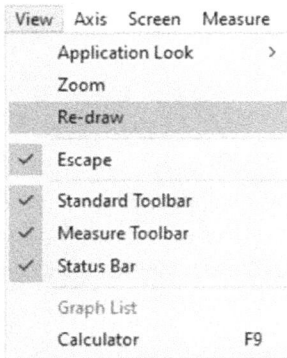

Fig. 5.412.

Double click on the amplitude graph. The window shown in Fig. 5.413 will appear.

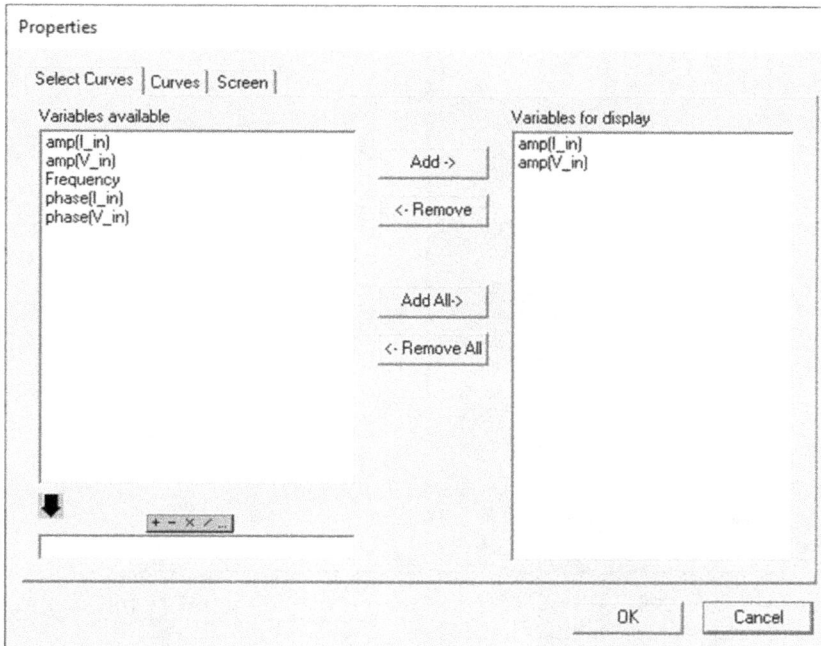

Fig. 5.413.

Remove the amp(I_in) and amp(V_in) from the right list (click on them and click the Remove button). Write the amp(V_in)–amp(I_in) in the bottom box (Fig. 5.414) and press Add in order to add it to the right list (415). After adding to the right hand side list, click the OK button. PSIM draws the graph shown in Fig. 5.416. Now, the amplitude graph is correct. It shows the amplitude of input impedance.

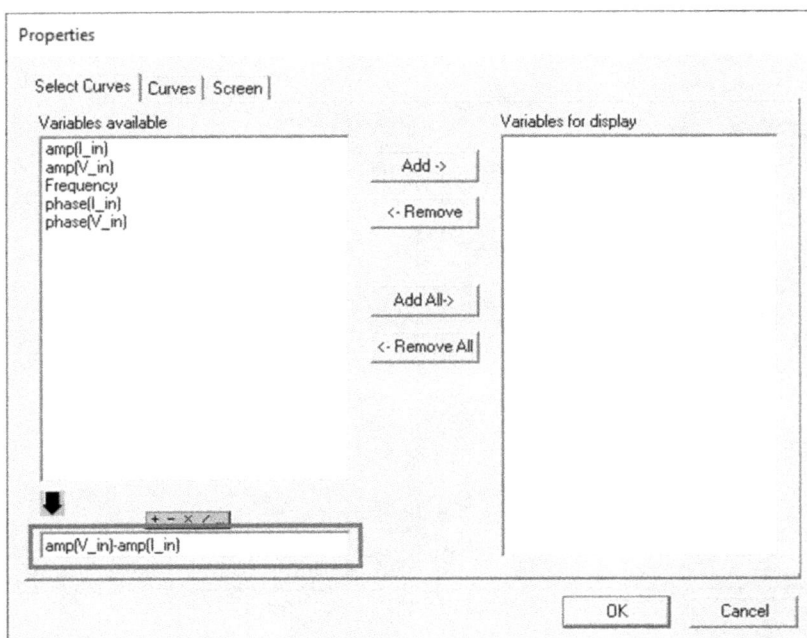

Fig. 5.414.

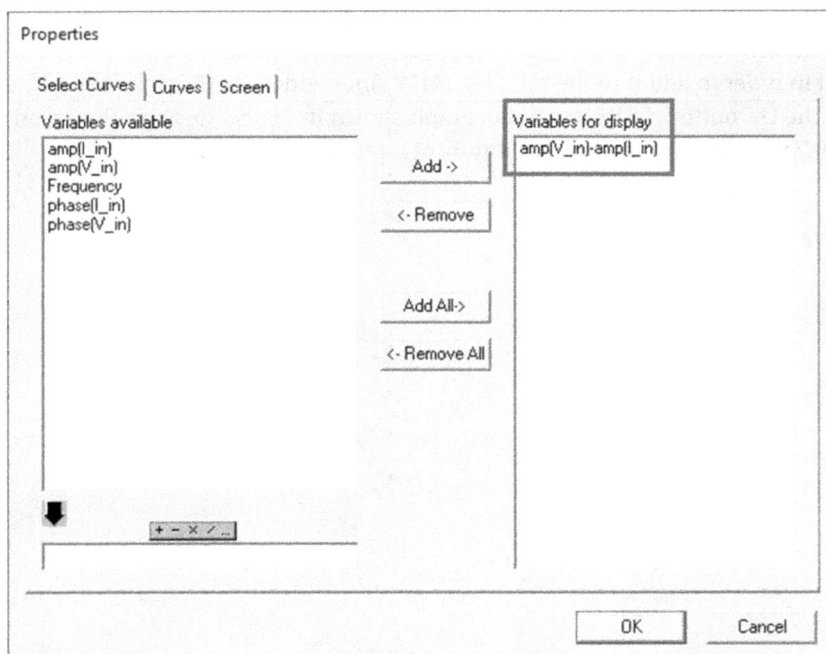

Fig. 5.415.

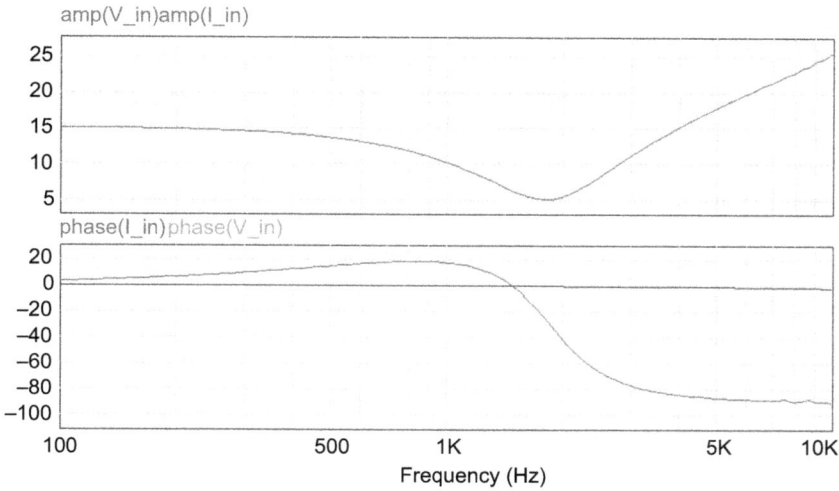

Fig. 5.416.

Now, double click on the phase graph. The window shown in Fig. 5.417 will appear.

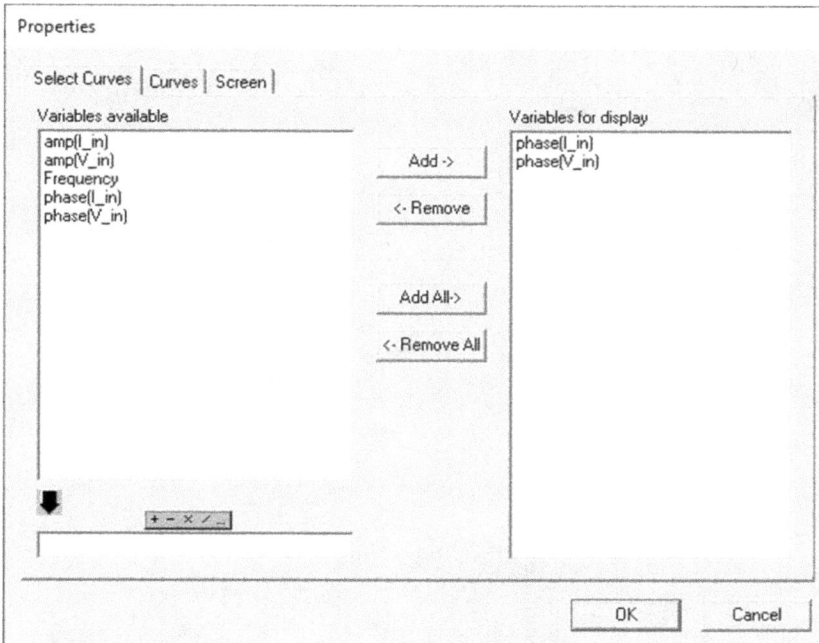

Fig. 5.417.

Remove the phase(I_in) and phase(V_in) from the right list (click on them and click the Remove button). Write the phase(V_in)–phase(I_in) in the bottom box (Fig. 5.418) and press Add in order to add it to the right list (419). After adding to the right hand side list, click the OK button. PSIM draws the graph shown in Fig. 5.420. Now, both the amplitude graph and phase graph are correct. Figure 5.420 shows the input impedance of the converter.

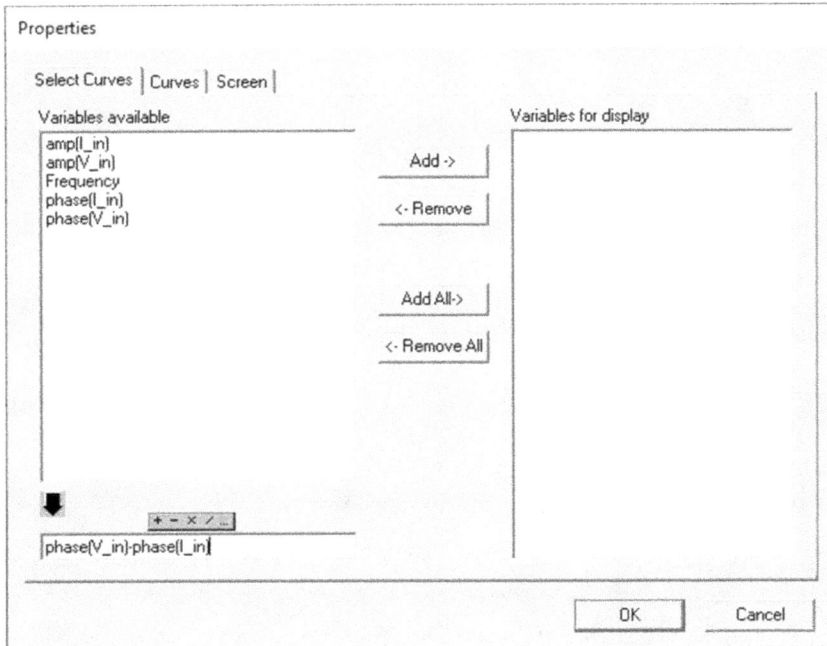

Fig. 5.418.

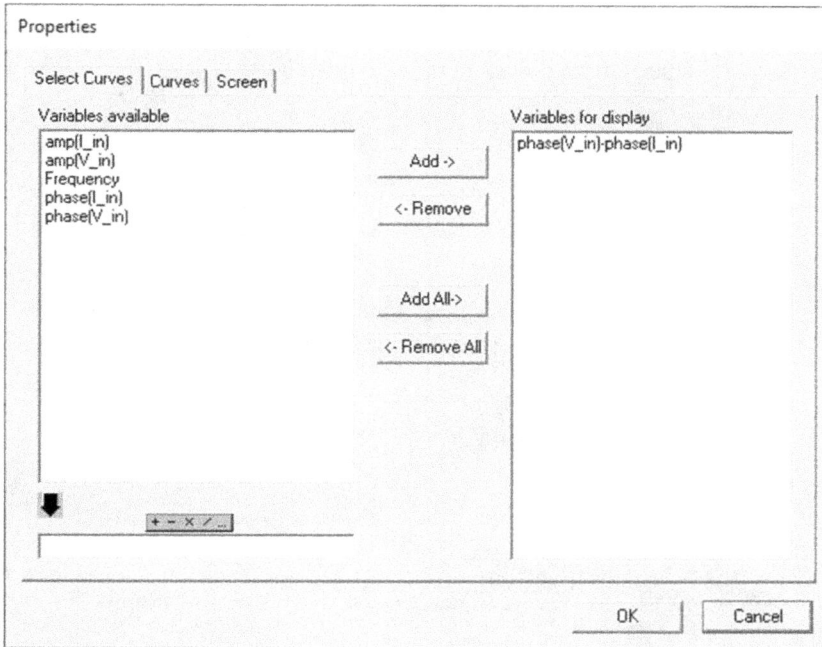

Fig. 5.419.

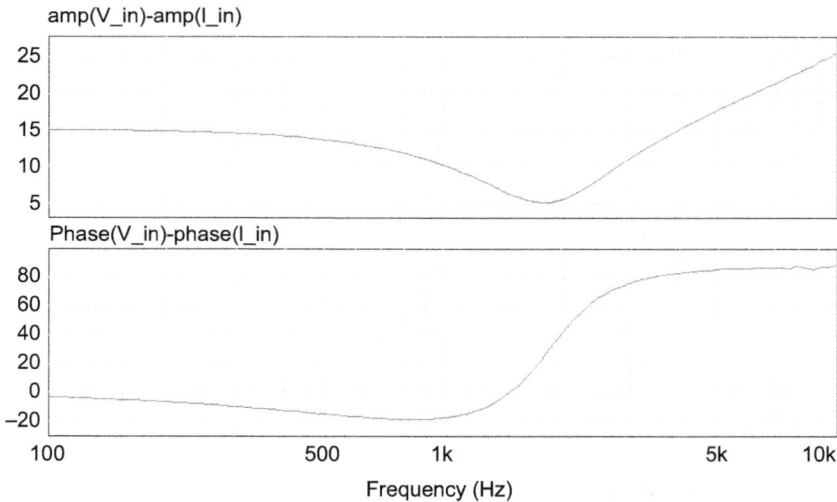

Fig. 5.420.

PSIM has an example about measurement of input impedance. In order to see the example, click the File>Search Examples and then enter "input impedance" to the search box and click the Find button. Select the UC3842 buck converter – input impedance (level-1).psimsch from the list in order to open it.

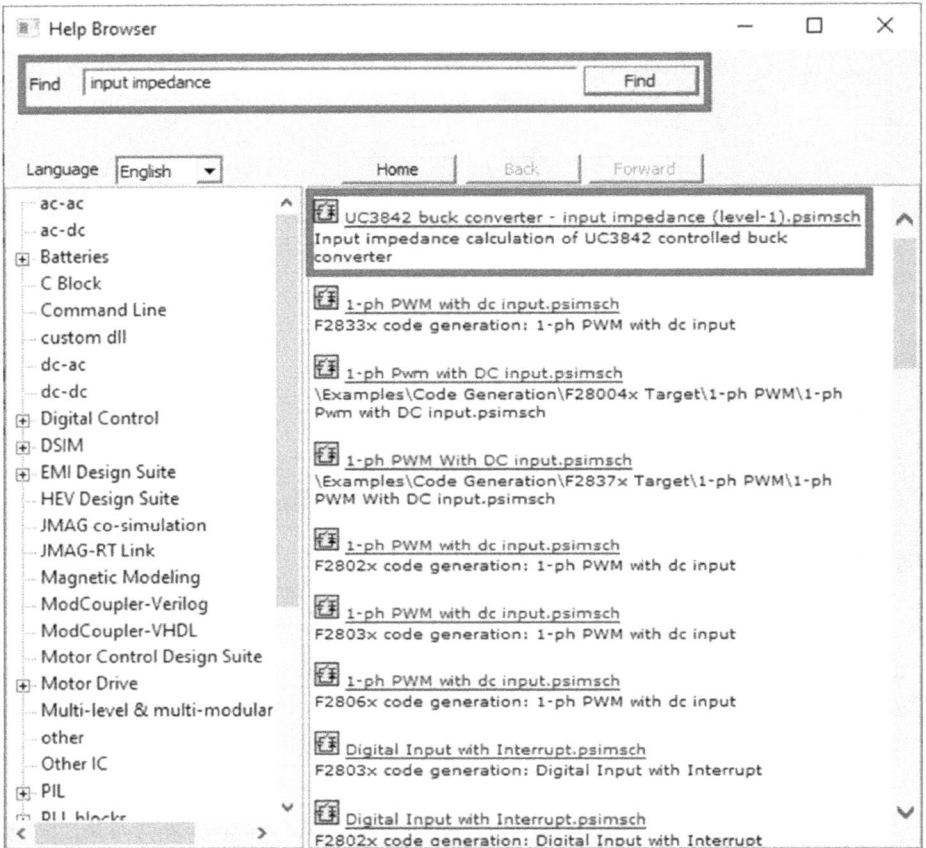

Fig. 5.421.

5.36 Example 35: output impedance of buck converter

In this example, we will extract the output impedance of the buck converter of Example 31. In order to do this, we need to enter the perturbation to the output of the system. Then we measure the voltage and current perturbations. The output impedance equals to $\frac{\tilde{v}_o(s)}{\tilde{i}_o(s)}$ where $\tilde{v}_o(s)$ and $\tilde{i}_o(s)$ show the output voltage and current perturbations, respectively.

Assume the schematic shown in Fig. 5.422. In this schematic, we used a voltage-controlled current source in order to inject a small current to the output. The injected current makes a small voltage perturbation. The AC Sweep blocks measure the voltage and current perturbations and permits us to calculate the $Z_o(s) \dfrac{\tilde{v}_o(s)}{\tilde{i}_o(s)}$

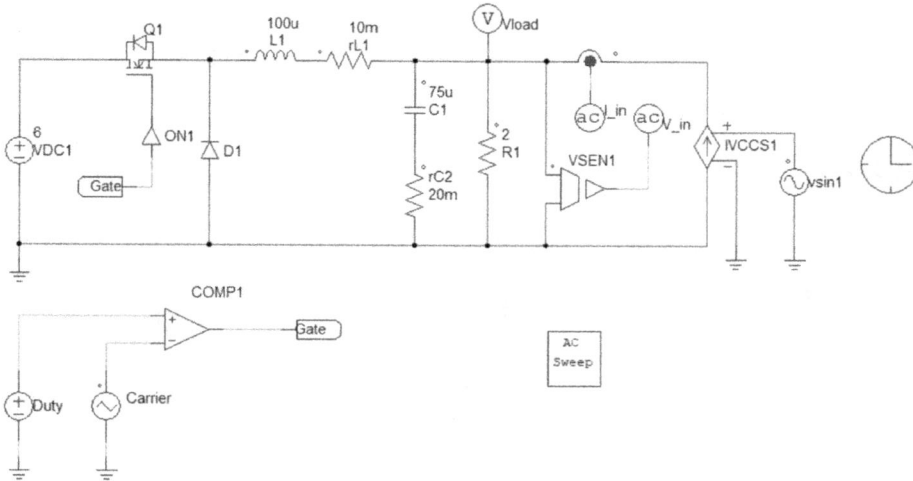

Fig. 5.422.

The settings of blocks are shown in Figs. 5.423–5.428.

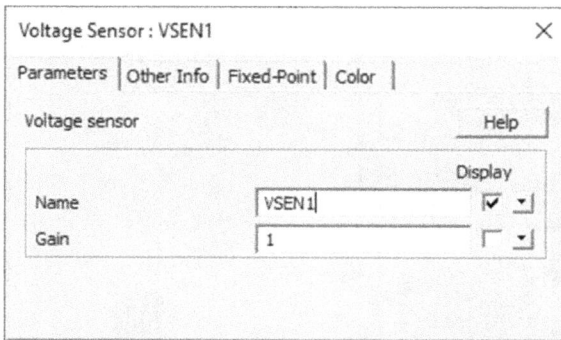

Fig. 5.423.

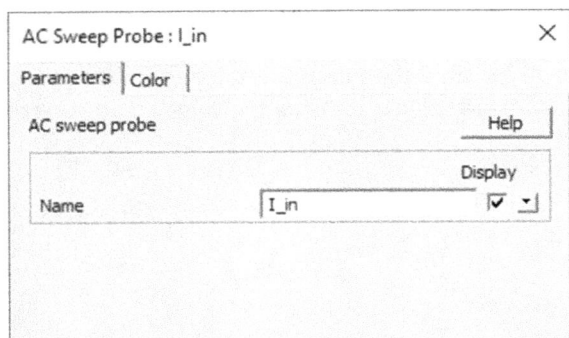

Fig. 5.424.

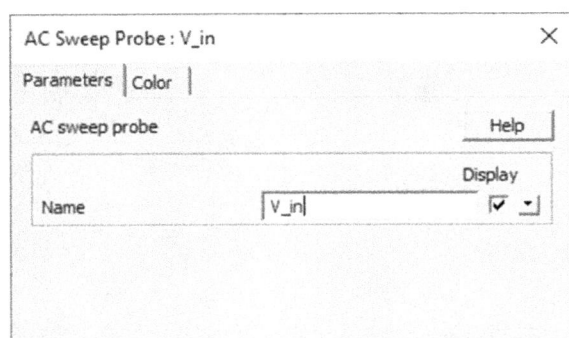

Fig. 5.425.

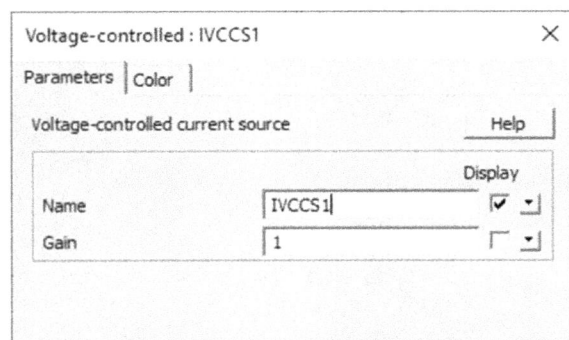

Fig. 5.426.

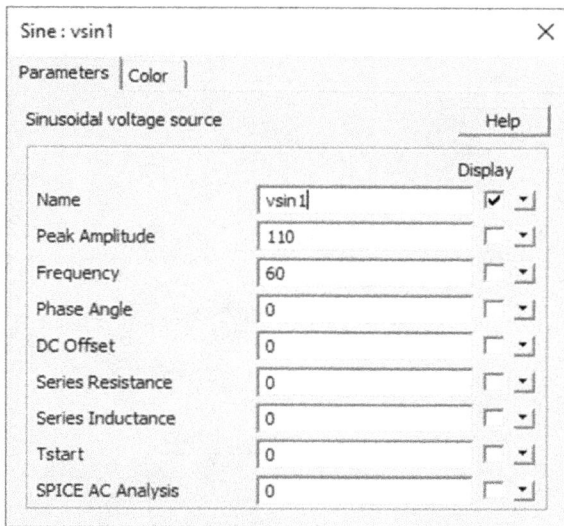

Fig. 5.427.

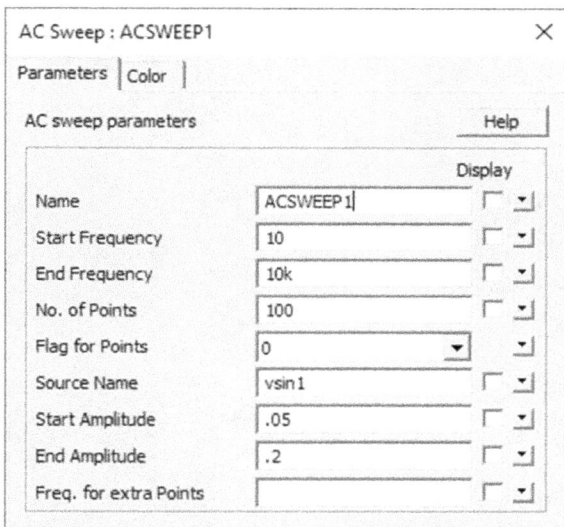

Fig. 5.428.

Run the simulation. The result shown in Fig. 5.429 appears.

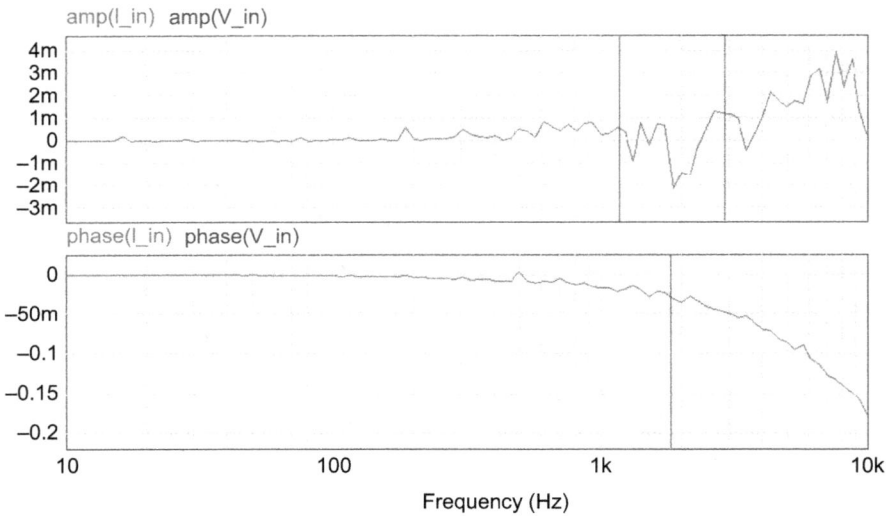

Fig. 5.429.

Click the Auto-Scale (Fig. 5.430) or Redraw (Fig. 5.431). The simulation result changes to what shown in Fig. 5.432.

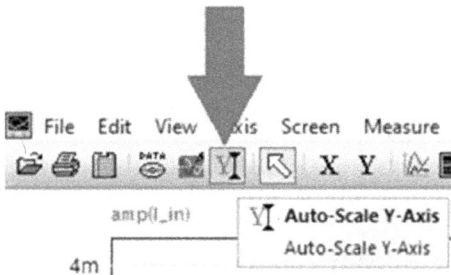

Fig. 5.430.

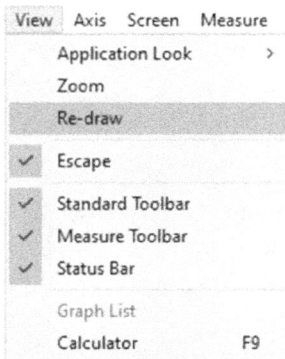

Fig. 5.431.

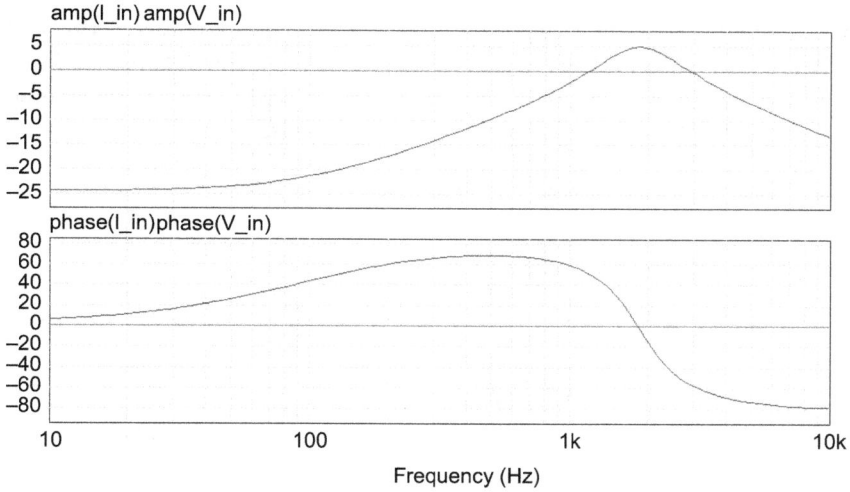

Fig. 5.432.

Double click on the magnitude plot. The properties window shown in Fig. 5.433 will appear. Click on the amp(I_in) and amp(V_in) from the right list. Then click the Remove button.

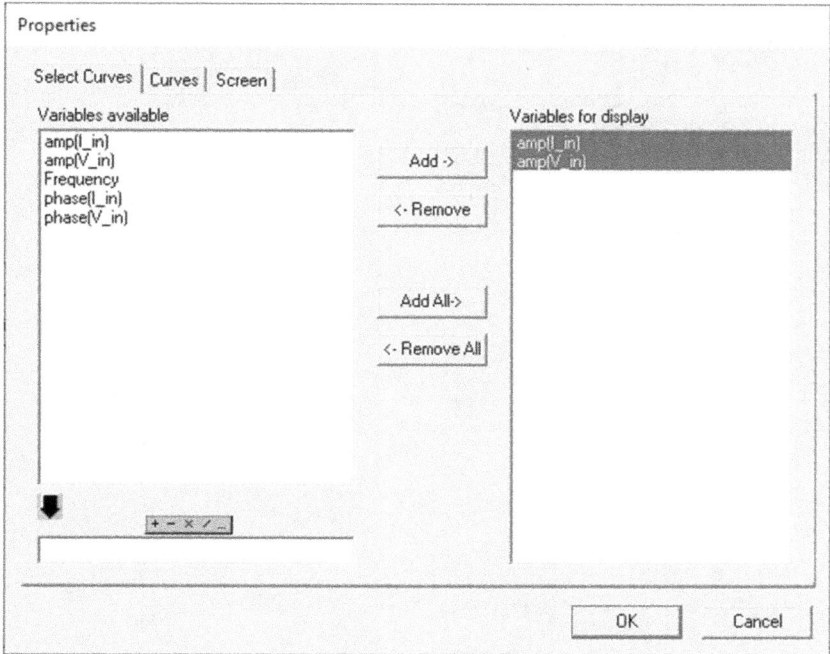

Fig. 5.433.

Now enter the amp(V_in)–amp(I_in) expression to the bottom box (Fig. 5.434) and click the Add button. After adding the amp(V_in)–amp(I_in) to the right list, click the OK button. Now the graph changes to what shown in Fig. 5.435. The amplitude graph shows the amplitude of output impedance. However, the phase graph is not correct yet.

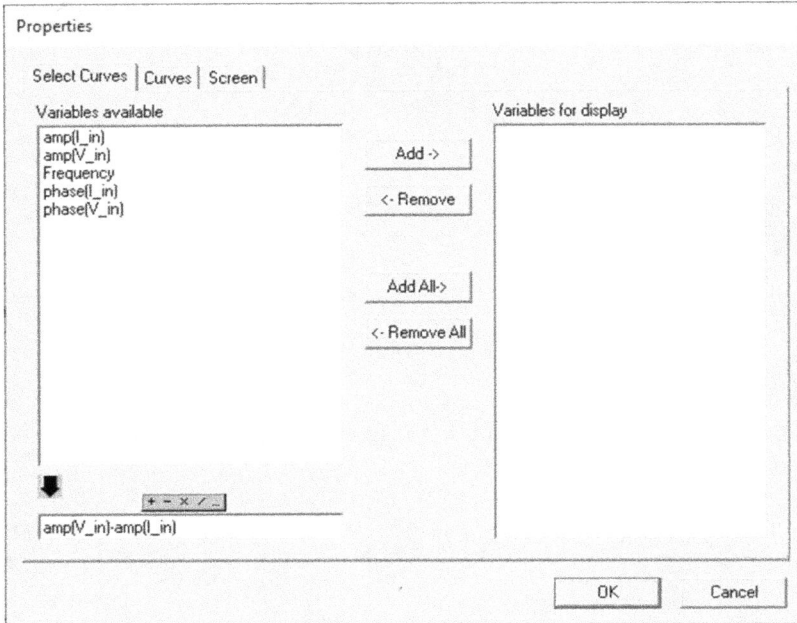

Fig. 5.434.

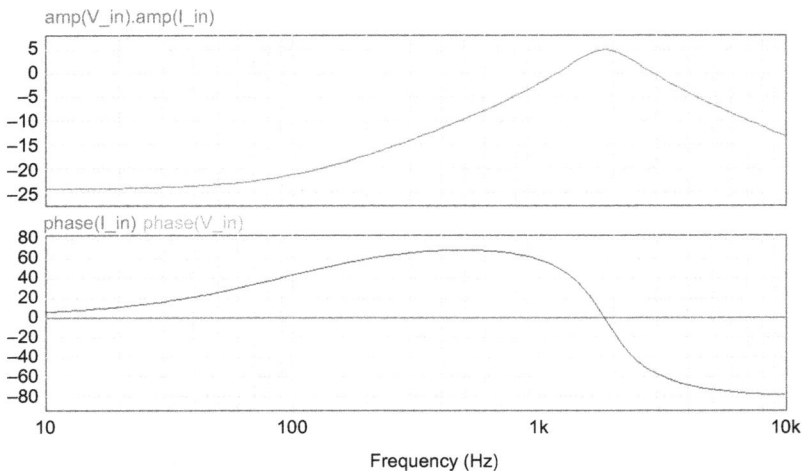

Fig. 5.435.

Now double click the phase plot. Remove the phase(V_in) and from phase(I_in) the right list. Enter the phase(V_in)–phase(I_in) expression to the bottom box (Fig. 5.436) and click the Add button in order to add it to the right list. Click OK to apply the changes. After a few seconds, plot shown in Fig. 5.437 will appear on the screen. This is the graph of output impedance of converter.

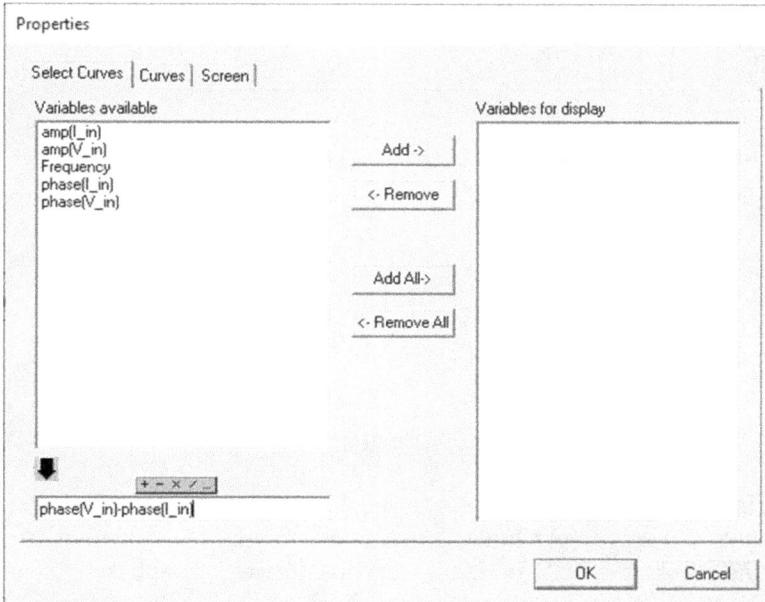

Fig. 5.436.

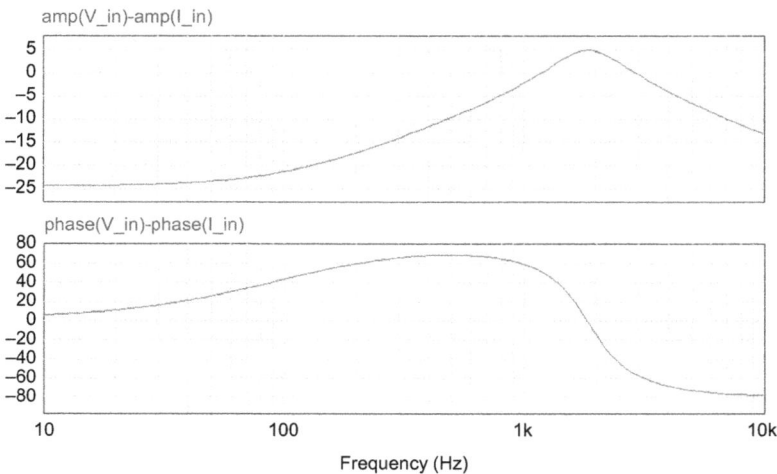

Fig. 5.437.

Using the AC sweep block to extract the output impedance is a little bit time consuming. You can obtain the same result using AC Sweep (multisine) block (Fig. 5.438) in a very shorter time.

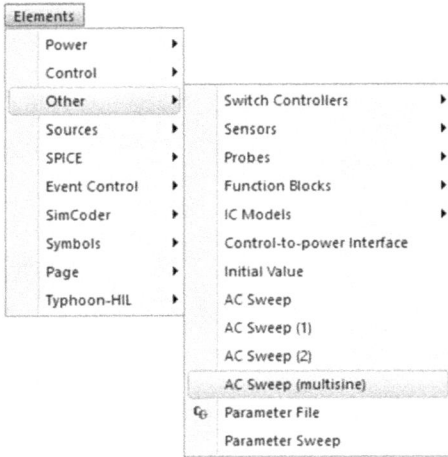

Fig. 5.438.

Simply replace the AC Sweep block in Fig. 5.422 with AC Sweep (multisine) block (Fig. 5.439) and do the settings as shown in Fig. 5.440. After running the simulation, you need to draw the graph of amp(V_in)–amp(I_in) and phase(V_in)–phase(I_in) in order to obtain the output impedance.

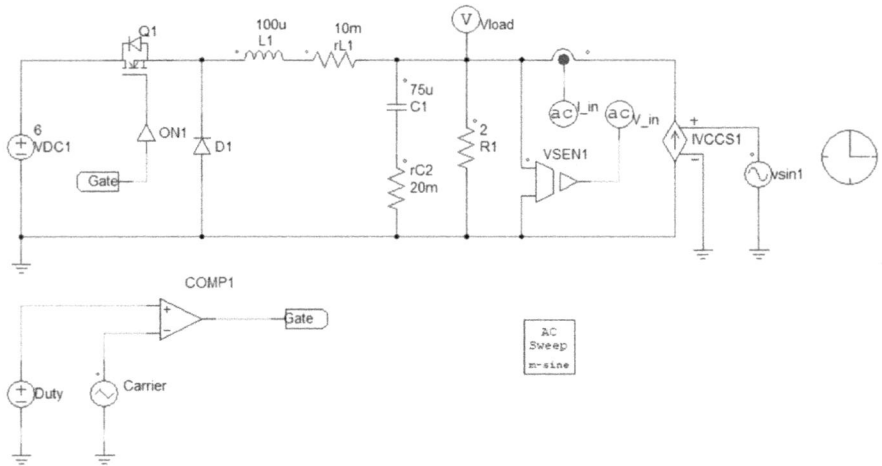

Fig. 5.439.

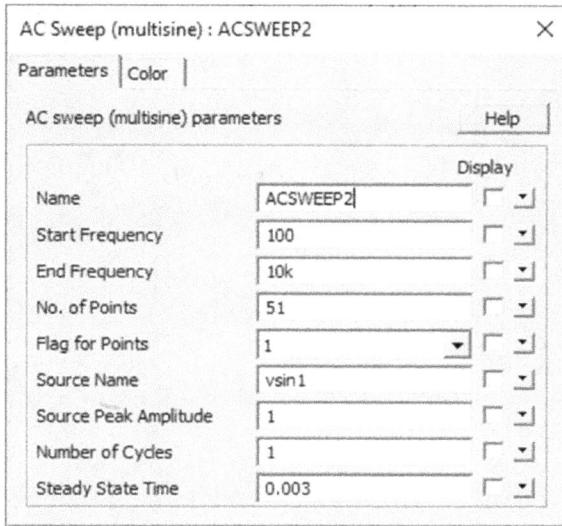

AC Sweep (multisine) : ACSWEEP2   ✕

Parameters | Color |

AC sweep (multisine) parameters   Help

		Display
Name	ACSWEEP2	▢ ▾
Start Frequency	100	▢ ▾
End Frequency	10k	▢ ▾
No. of Points	51	▢ ▾
Flag for Points	1 ▾	▢ ▾
Source Name	vsin 1	▢ ▾
Source Peak Amplitude	1	▢ ▾
Number of Cycles	1	▢ ▾
Steady State Time	0.003	▢ ▾

Fig. 5.440.

The Number of Cycles box of AC Sweep (multisine) is normally equals to 1. The Steady State Time box must be filled with estimated time when the transient is over and steady state starts. In order to obtain a good value for this box, disable the AC Sweep (multisine) block and run the simulation. Use probes to see when the converter reaches the steady state (e.g., you can monitor the output voltage).

5.37 Example 36: closed-loop control of buck converter

In this example, we will study the closed loop control of a buck converter. Assume the schematic shown in Fig. 5.441. In this schematic, two Summer blocks (Fig. 5.442) are used for doing the required addition and subtractions.

This schematic uses a simple proportional integral (PI) controller. The PI controller is made with discrete elements. However, you can use the ready-to-use PI controller block if you prefer. Note that with the aid of s-domain Transfer Function (Fig. 5.443) and z-domain Transfer Function (Fig. 5.444) blocks, you can apply desired (non-PI) controller to the system.

In this example, we assume that the proportional and integral gains are given ($k_p = 50$ and $k_i = 0.001$). In Chapter 8, we will study the problem of designing a controller for a converter.

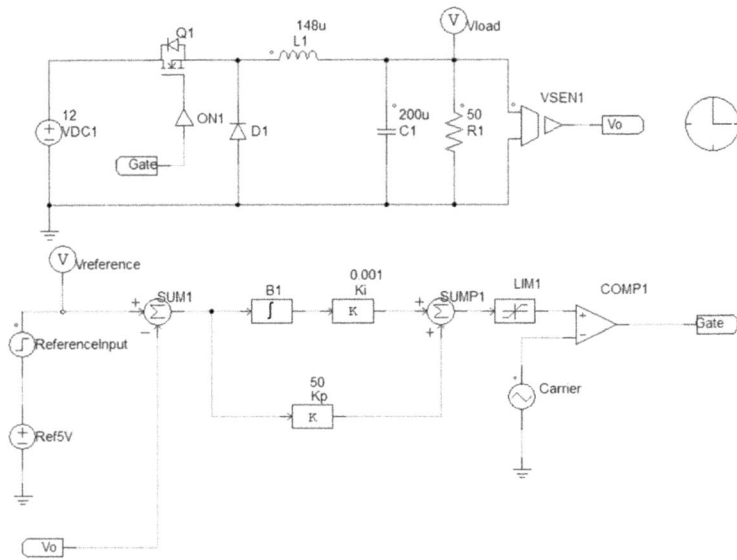

Fig. 5.441.

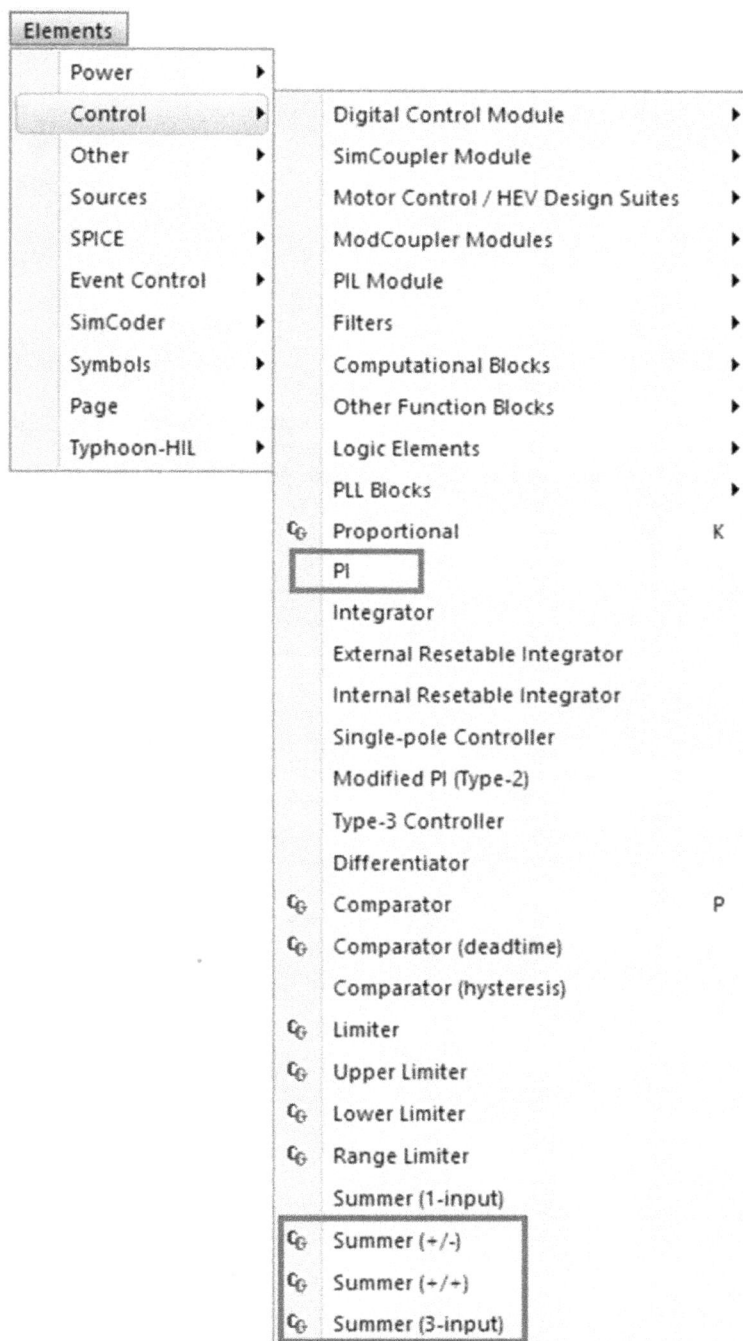

Elements				
Power ▶				
Control ▶	Digital Control Module	▶		
Other ▶	SimCoupler Module	▶		
Sources ▶	Motor Control / HEV Design Suites	▶		
SPICE ▶	ModCoupler Modules	▶		
Event Control ▶	PIL Module	▶		
SimCoder ▶	Filters	▶		
Symbols ▶	Computational Blocks	▶		
Page ▶	Other Function Blocks	▶		
Typhoon-HIL ▶	Logic Elements	▶		
	PLL Blocks	▶		

- Proportional   K
- PI
- Integrator
- External Resetable Integrator
- Internal Resetable Integrator
- Single-pole Controller
- Modified PI (Type-2)
- Type-3 Controller
- Differentiator
- Comparator   P
- Comparator (deadtime)
- Comparator (hysteresis)
- Limiter
- Upper Limiter
- Lower Limiter
- Range Limiter
- Summer (1-input)
- Summer (+/-)
- Summer (+/+)
- Summer (3-input)

Fig. 5.442.

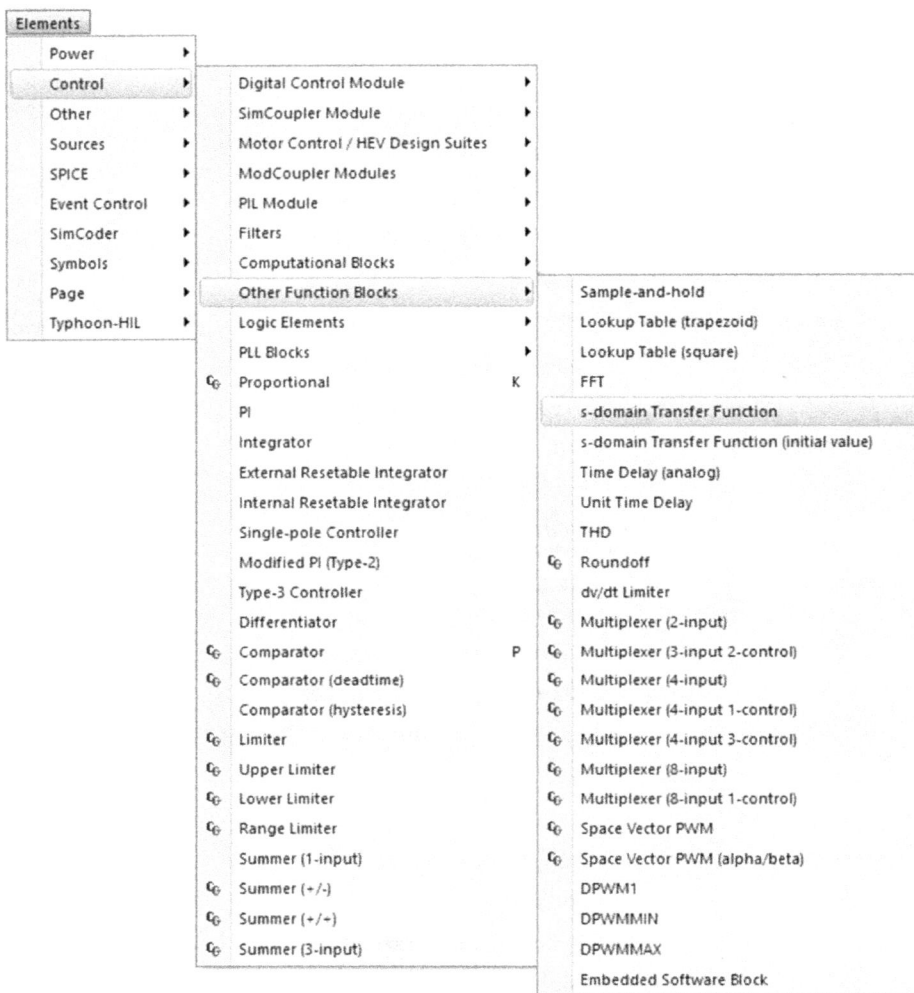

Elements			
Power ▶			
Control ▶	Digital Control Module ▶		
Other ▶	SimCoupler Module ▶		
Sources ▶	Motor Control / HEV Design Suites ▶		
SPICE ▶	ModCoupler Modules ▶		
Event Control ▶	PIL Module ▶		
SimCoder ▶	Filters ▶		
Symbols ▶	Computational Blocks ▶		
Page ▶	Other Function Blocks ▶	Sample-and-hold	
Typhoon-HIL ▶	Logic Elements ▶	Lookup Table (trapezoid)	
	PLL Blocks ▶	Lookup Table (square)	
	₲ Proportional K	FFT	
	PI	s-domain Transfer Function	
	Integrator	s-domain Transfer Function (initial value)	
	External Resetable Integrator	Time Delay (analog)	
	Internal Resetable Integrator	Unit Time Delay	
	Single-pole Controller	THD	
	Modified PI (Type-2)	₲ Roundoff	
	Type-3 Controller	dv/dt Limiter	
	Differentiator	₲ Multiplexer (2-input)	
	₲ Comparator P	₲ Multiplexer (3-input 2-control)	
	₲ Comparator (deadtime)	₲ Multiplexer (4-input)	
	Comparator (hysteresis)	₲ Multiplexer (4-input 1-control)	
	₲ Limiter	₲ Multiplexer (4-input 3-control)	
	₲ Upper Limiter	₲ Multiplexer (8-input)	
	₲ Lower Limiter	₲ Multiplexer (8-input 1-control)	
	₲ Range Limiter	₲ Space Vector PWM	
	Summer (1-input)	₲ Space Vector PWM (alpha/beta)	
	₲ Summer (+/-)	DPWM1	
	₲ Summer (+/+)	DPWMMIN	
	₲ Summer (3-input)	DPWMMAX	
		Embedded Software Block	

Fig. 5.443.

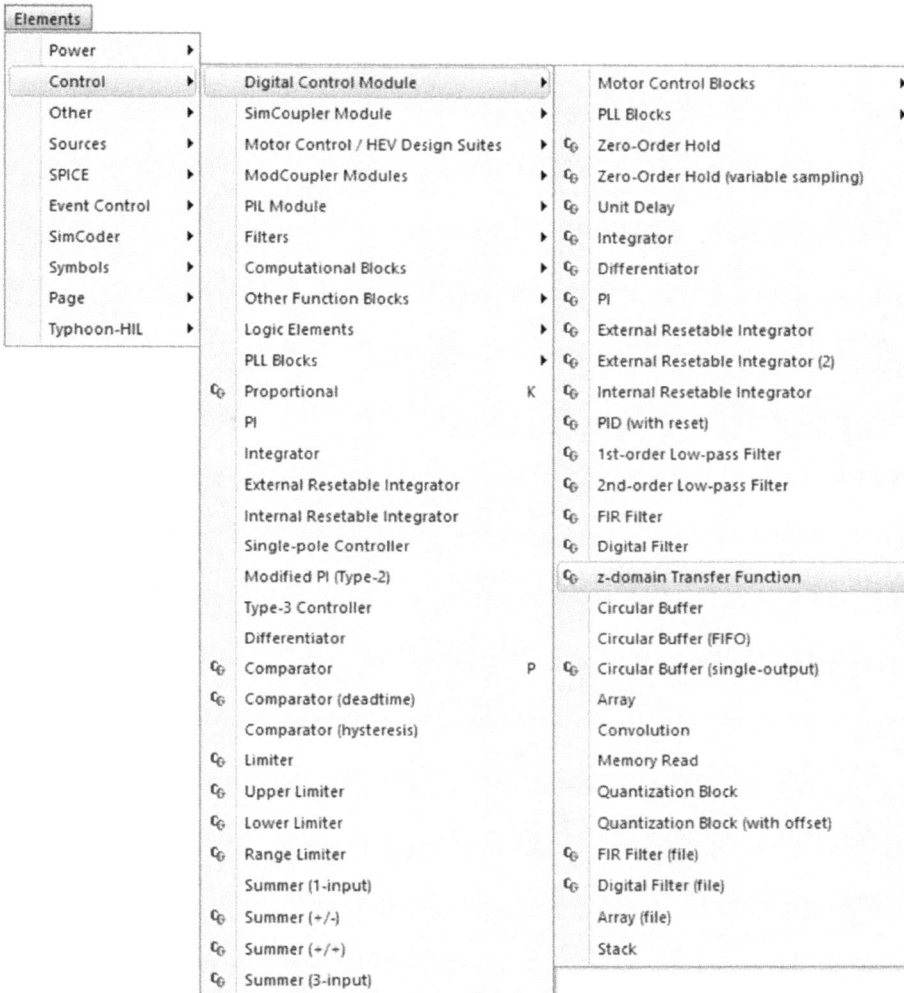

Elements			
Power ▶			
Control ▶	Digital Control Module ▶	Motor Control Blocks ▶	
Other ▶	SimCoupler Module ▶	PLL Blocks ▶	
Sources ▶	Motor Control / HEV Design Suites ▶	₲ Zero-Order Hold	
SPICE ▶	ModCoupler Modules ▶	₲ Zero-Order Hold (variable sampling)	
Event Control ▶	PIL Module ▶	₲ Unit Delay	
SimCoder ▶	Filters ▶	₲ Integrator	
Symbols ▶	Computational Blocks ▶	₲ Differentiator	
Page ▶	Other Function Blocks ▶	₲ PI	
Typhoon-HIL ▶	Logic Elements ▶	₲ External Resetable Integrator	
	PLL Blocks ▶	₲ External Resetable Integrator (2)	
	₲ Proportional K	₲ Internal Resetable Integrator	
	PI	₲ PID (with reset)	
	Integrator	₲ 1st-order Low-pass Filter	
	External Resetable Integrator	₲ 2nd-order Low-pass Filter	
	Internal Resetable Integrator	₲ FIR Filter	
	Single-pole Controller	₲ Digital Filter	
	Modified PI (Type-2)	₲ z-domain Transfer Function	
	Type-3 Controller	Circular Buffer	
	Differentiator	Circular Buffer (FIFO)	
	₲ Comparator P	₲ Circular Buffer (single-output)	
	₲ Comparator (deadtime)	Array	
	Comparator (hysteresis)	Convolution	
	₲ Limiter	Memory Read	
	₲ Upper Limiter	Quantization Block	
	₲ Lower Limiter	Quantization Block (with offset)	
	₲ Range Limiter	₲ FIR Filter (file)	
	Summer (1-input)	₲ Digital Filter (file)	
	₲ Summer (+/-)	Array (file)	
	₲ Summer (+/+)	Stack	
	₲ Summer (3-input)		

Fig. 5.444.

The settings of the blocks used in Fig. 5.441 are shown in Figs. 5.445–5.454. The controller is a simple PI controller and it is implemented with an integrator and a gain block. The limiter block ensures that the signal that enters the modulator has value between 0 and 1, that is, in the [0,1] interval. The ReferenceInput source makes a step change in the reference signal of control system. It is used to see the closed loop system behavior, that is, testing the system.

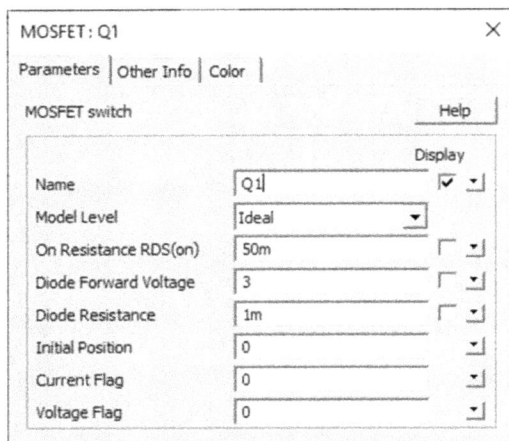

Fig. 5.445.

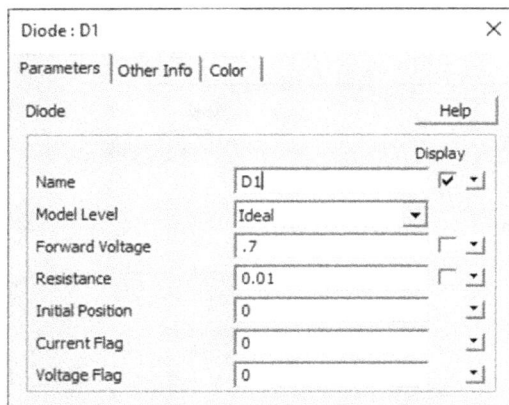

Fig. 5.446.

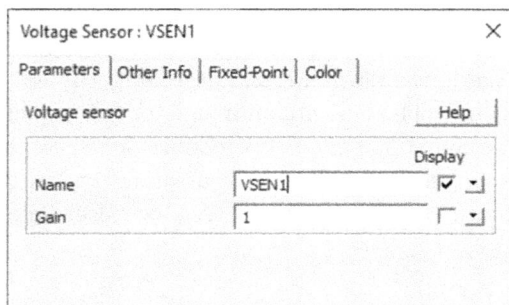

Fig. 5.447.

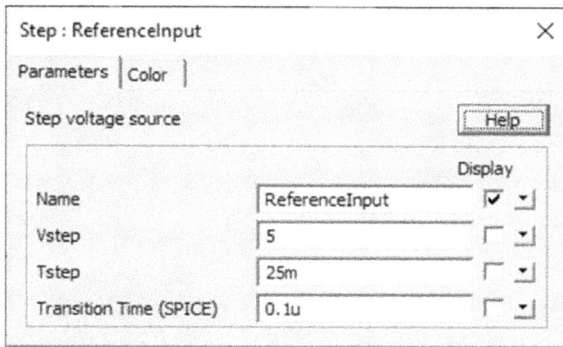

Fig. 5.448.

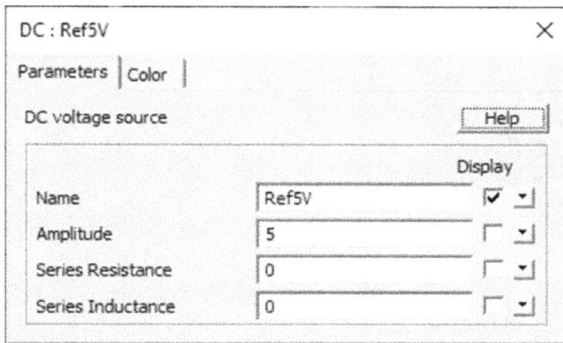

Fig. 5.449.

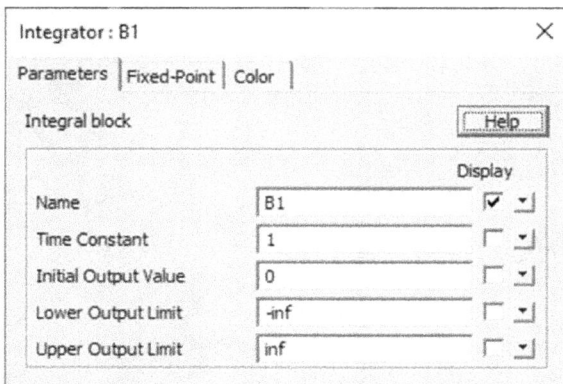

Fig. 5.450.

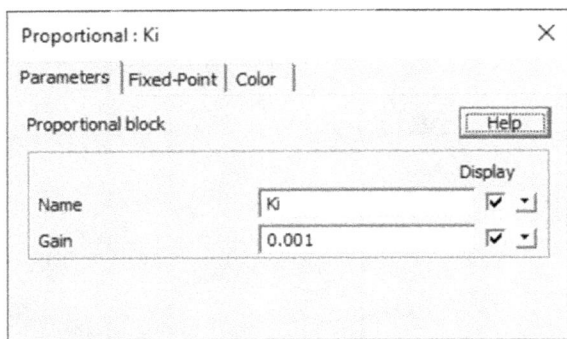

Fig. 5.451.

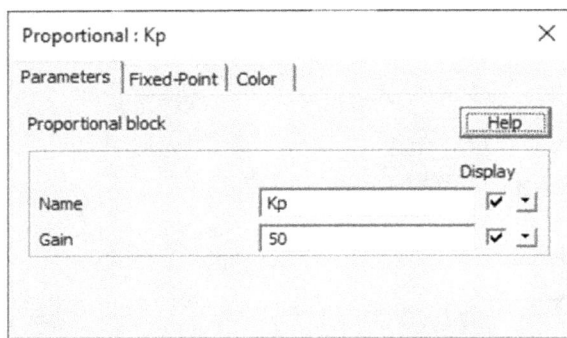

Fig. 5.452.

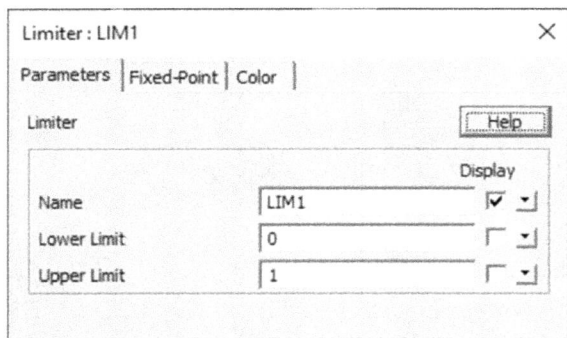

Fig. 5.453.

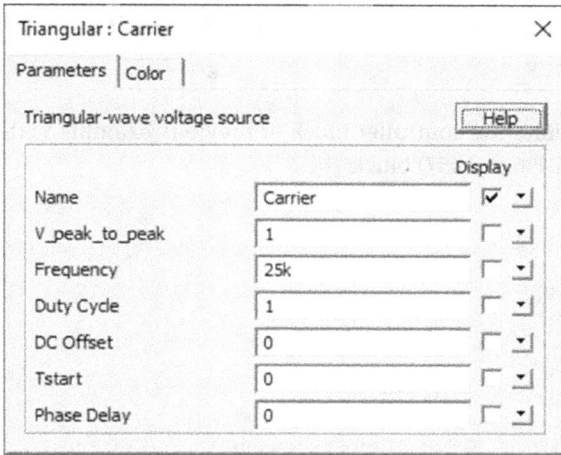

Triangular : Carrier			✕

Parameters | Color

Triangular-wave voltage source   [Help]

		Display	
Name	Carrier	☑ ▾	
V_peak_to_peak	1	☐ ▾	
Frequency	25k	☐ ▾	
Duty Cycle	1	☐ ▾	
DC Offset	0	☐ ▾	
Tstart	0	☐ ▾	
Phase Delay	0	☐ ▾	

Fig. 5.454.

Run the simulation. The result shown in Fig. 5.455 is obtained. The system followed the reference with zero error.

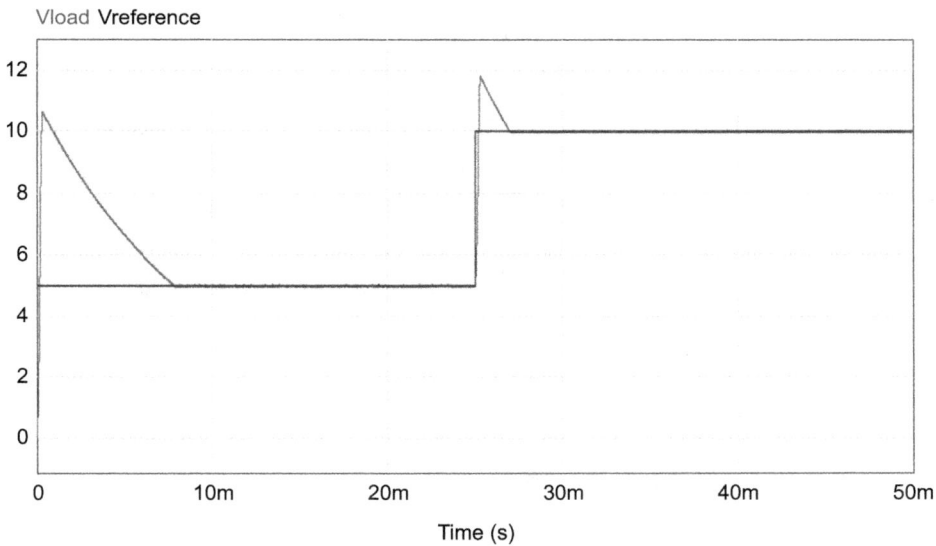

Fig. 5.455.

5.38 Example 37: closed-loop control of buck converter with simplified C code block

In this example, we want to simulate the controller block of previous example with the aid of simplified C code (Fig. 5.456 or 5.457) block.

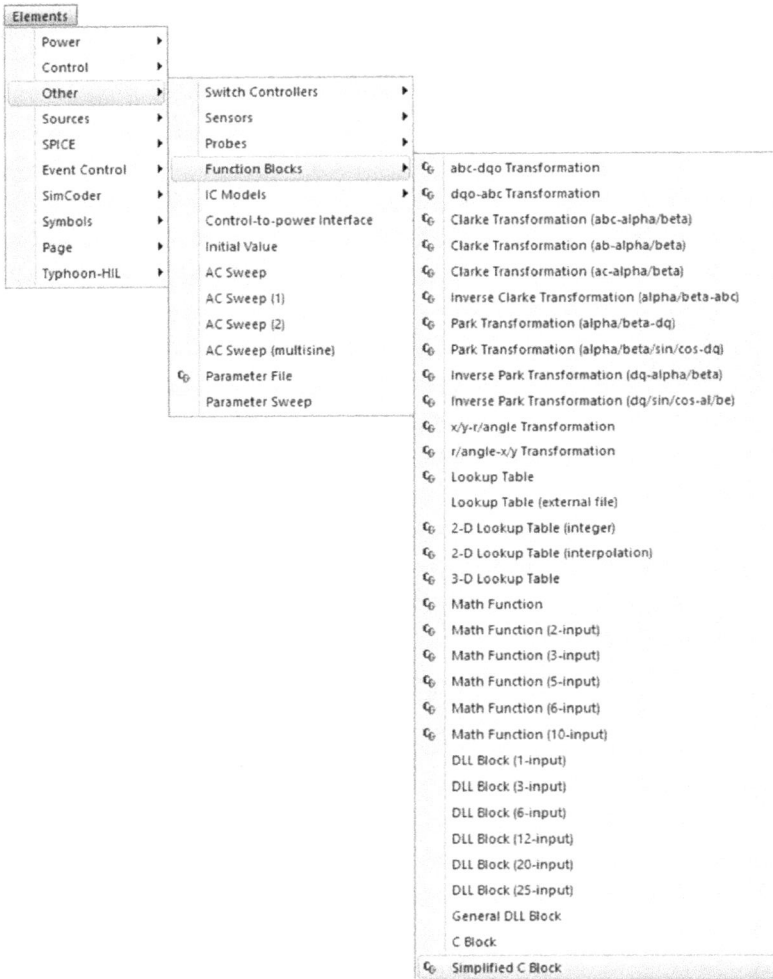

Fig. 5.456.

Fig. 5.457.

First of all, remove the controller blocks from the previous simulation.

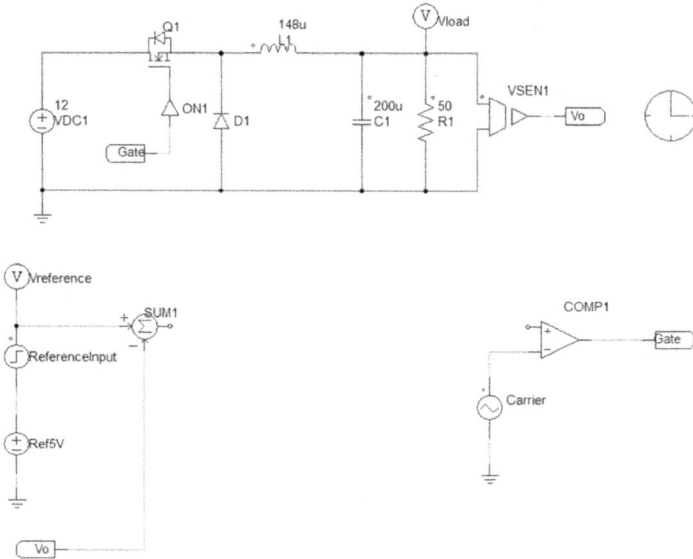

Fig. 5.458.

Now add a simplified C code block to the schematic. Connect the block to the output of comparator and modulator.

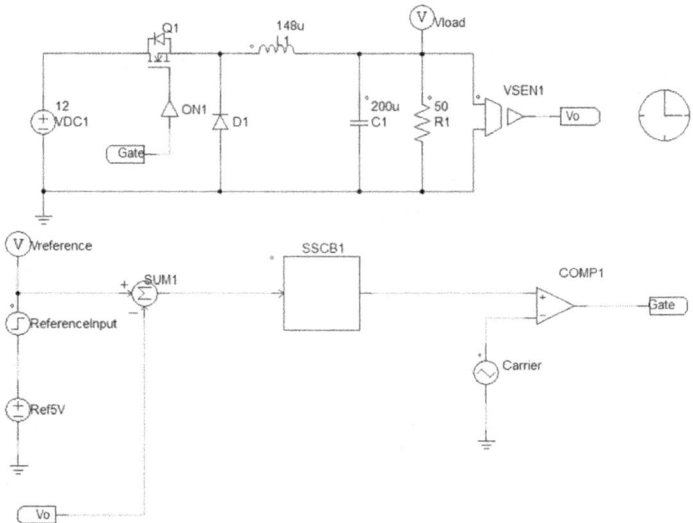

Fig. 5.459.

Double click the simplified C code block and type the code shown in Fig. 5.460.

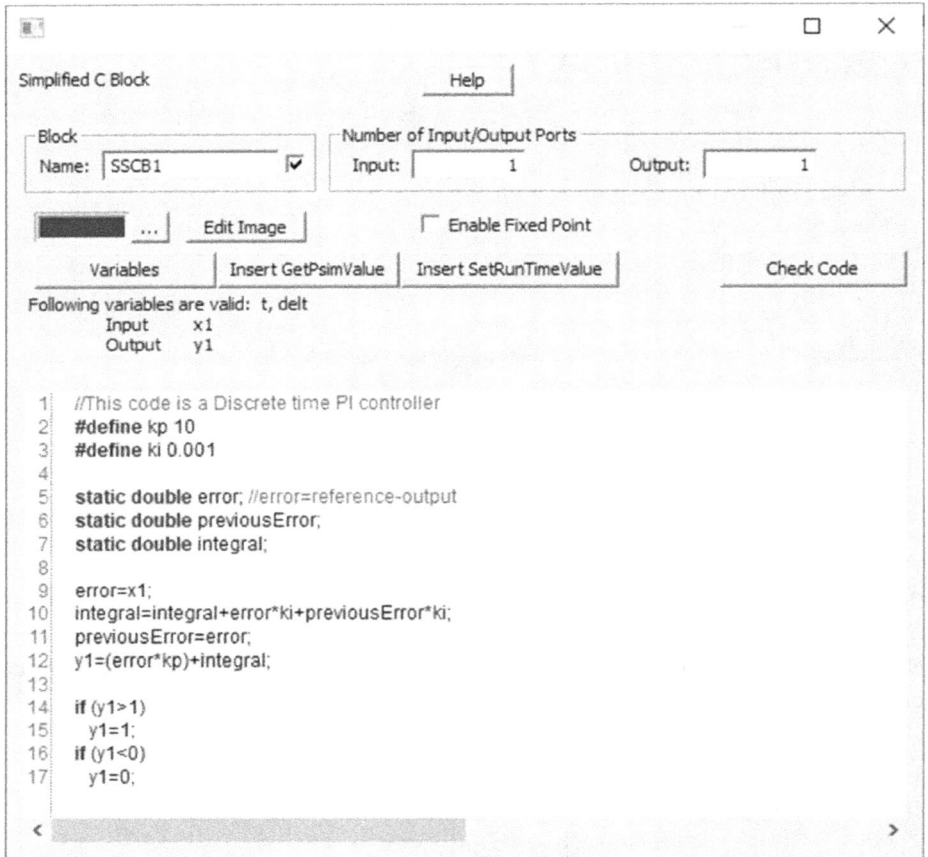

Fig. 5.460.

The Input and Output box (Fig. 5.460) determines the number of input and outputs. The default number of input/output is 1.

In addition to the variables that users define, the following variables can also be used in the code:
– t: time t passed from PSIM.
– delt: time step, passed from PSIM.
– x1, x2, . . ., xn: input x1, x2, . . ., xn passed from PSIM.
– y1, y2, . . ., yn: output y1, y2, . . ., yn passed back to PSIM.

The input/output node sequence of the C block is from the top to the bottom. For instance, node assignments of a block with 2 inputs and 3 outputs are shown in Fig. 5.461.

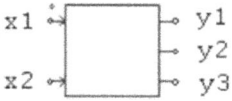

x1 ┤ ├ y1
    │ │ ├ y2
x2 ┤ ├ y3

Fig. 5.461.

All inputs and outputs are in double data type, unless the Enable Fixed Point box (double click the block in order to see this box) is checked and the input/output data types are explicitly defined otherwise.

When a C block is used in a z-domain circuit, PSIM will check if there are any Zero Order Hold (ZOH) blocks connected immediately at the inputs of the C block. If there are, it will take the sampling rate of the first ZOH block that it encounters starting from the first input, and use this sampling rate for the whole block. For example, if there are 2 inputs, and Input 1 is connected to a 10 kHz ZOH, and Input 2 is connected to a 20 kHz ZOH, it will take 10 kHz and ignore 20 kHz.

Run the simulation. The obtained result is shown in Fig. 5.462 which is the same as Fig. 5.455.

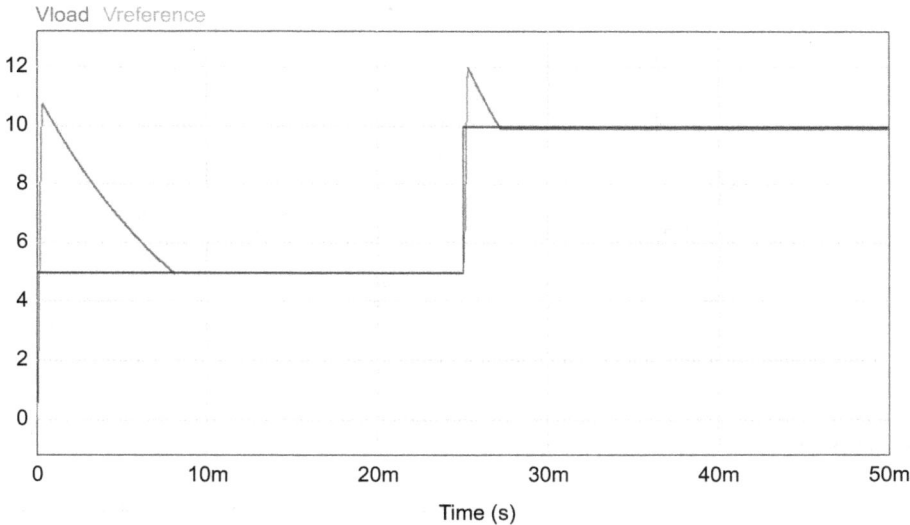

Fig. 5.462.

5.39 Example 38: multiplexer block

PSIM has different types of multiplexers (Fig. 5.463).

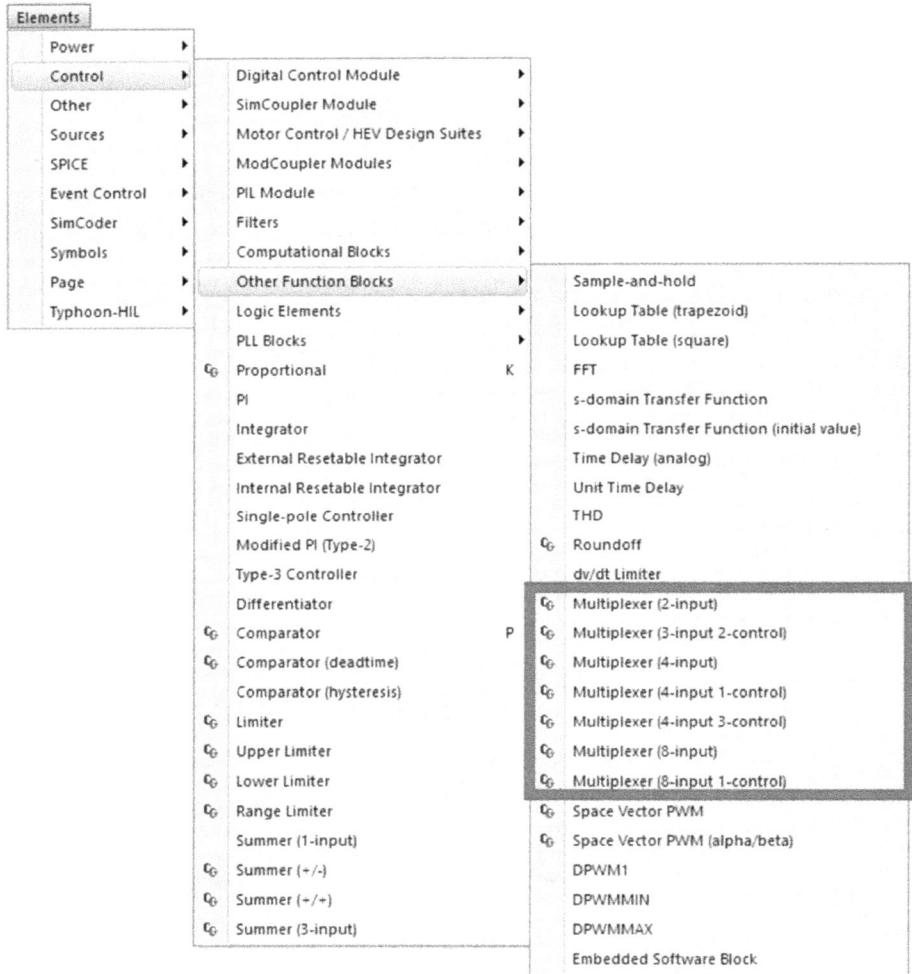

Fig. 5.463.

For instance, assume that you designed a closed loop control system for a buck converter and you want to compare the performance of designed system with the open loop case which duty cycle is constant despite of changes in the input voltage output load.

In this case, you can use a multiplexer block in order to switch between open loop and closed loop quickly (Fig. 5.464). In Fig. 5.464, the control line of multiplexer is controlled to a constant block (Fig. 5.465). When this constant block is zero (user can double click on this block and enter the desired value before running the simulation), the upper line (the signal that comes from op-amp) is selected. When it is one, the lower line (the constant duty cycle) is selected (Fig. 5.466).

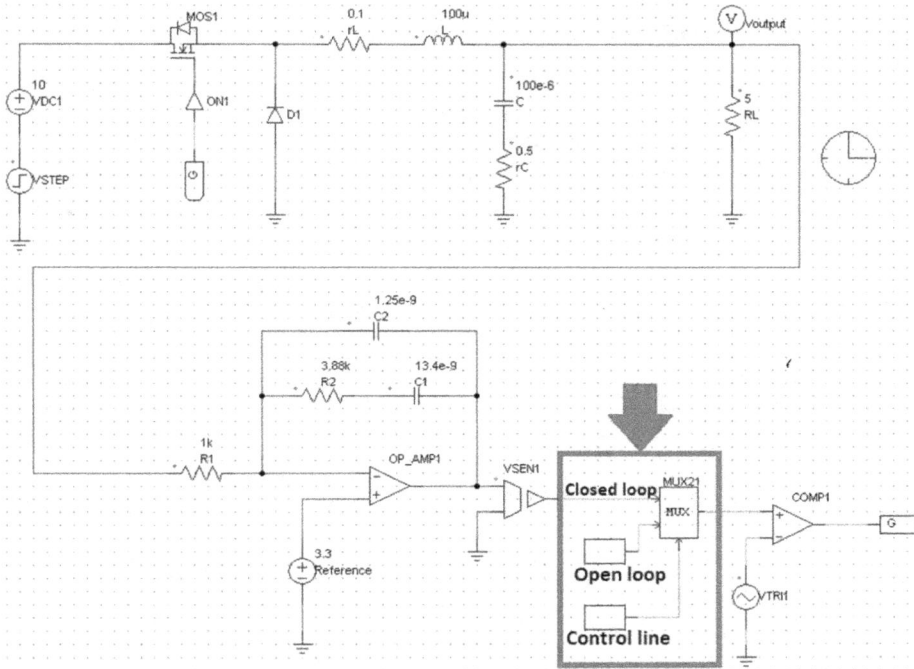

Fig. 5.464.

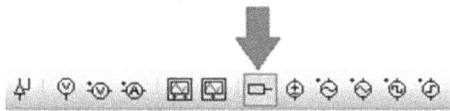

Fig. 5.465.

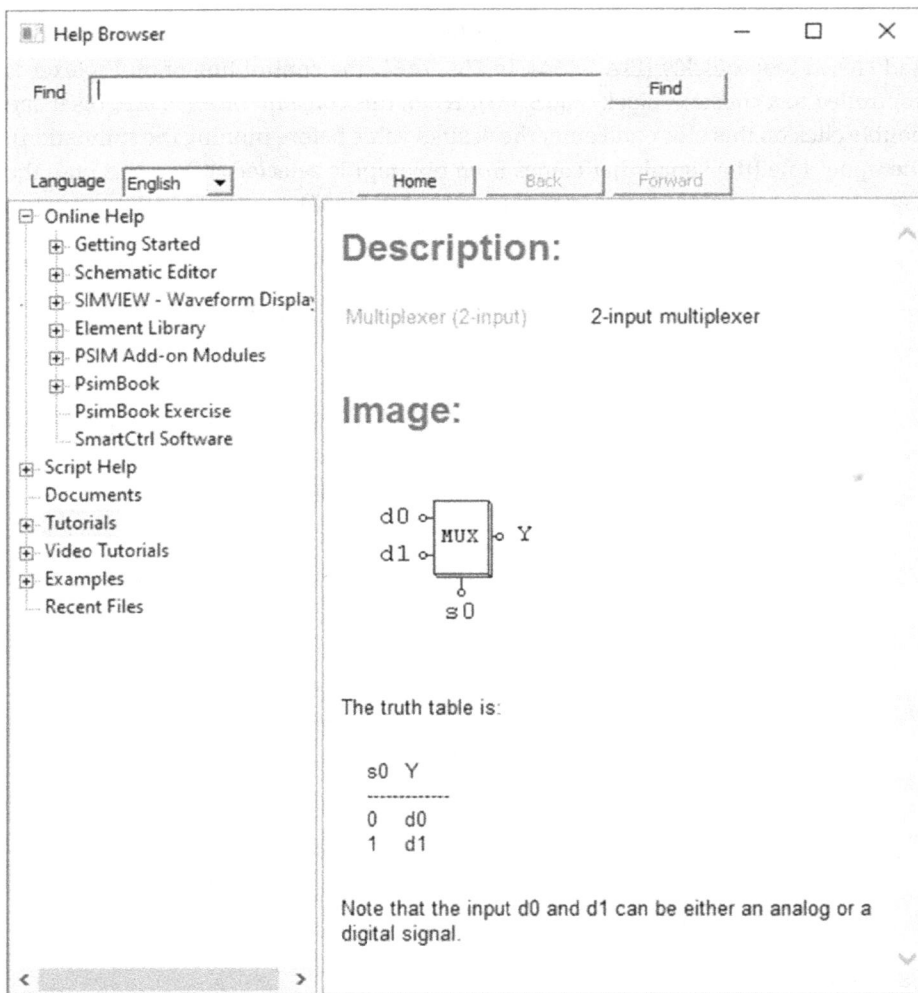

Help Browser

Find

Language English   Home   Back   Forward

- Online Help
  - Getting Started
  - Schematic Editor
  - SIMVIEW - Waveform Display
  - Element Library
  - PSIM Add-on Modules
  - PsimBook
    - PsimBook Exercise
    - SmartCtrl Software
- Script Help
  - Documents
- Tutorials
- Video Tutorials
- Examples
  - Recent Files

Description:

Multiplexer (2-input)   2-input multiplexer

Image:

d0 —
         MUX — Y
d1 —
          |
         s0

The truth table is:

```
s0  Y
-------------
0   d0
1   d1
```

Note that the input d0 and d1 can be either an analog or a
digital signal.

Fig. 5.466.

5.40 Example 39: simulation of flyback and push–pull converter

Flyback and push–pull converters are among the important DC–DC converters.
PSIM has ready examples for this converter. In order to see these examples, click the
File>Open Examples. Then search for flyback and push–pull (Figs. 5.467 and 5.468).

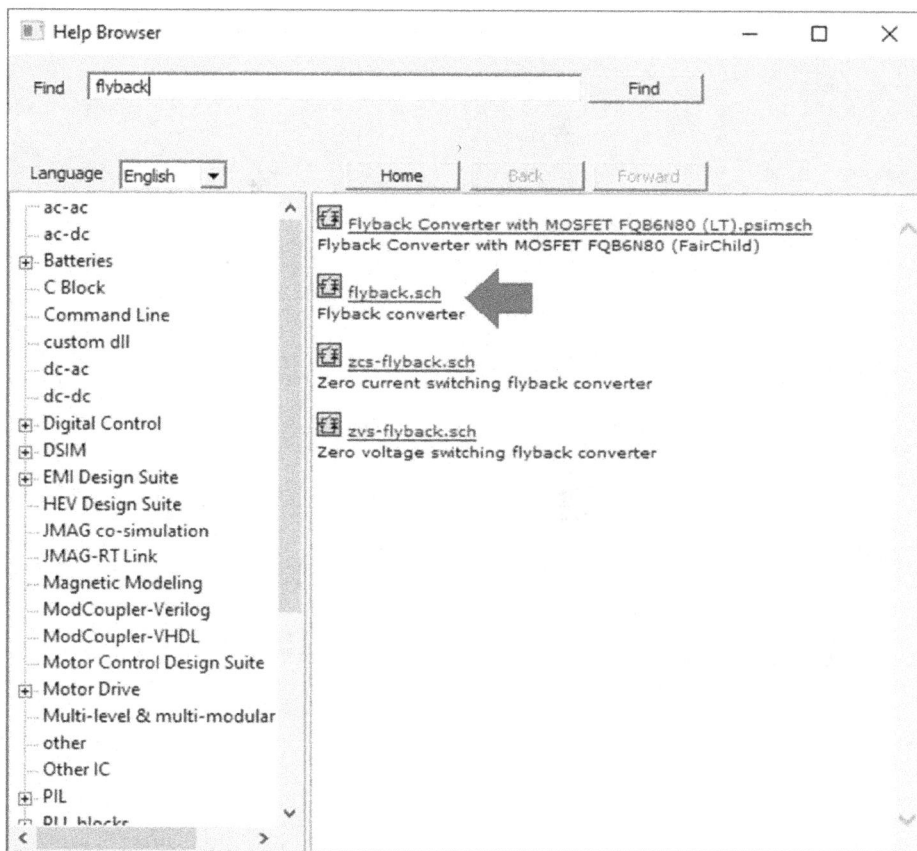

Fig. 5.467.

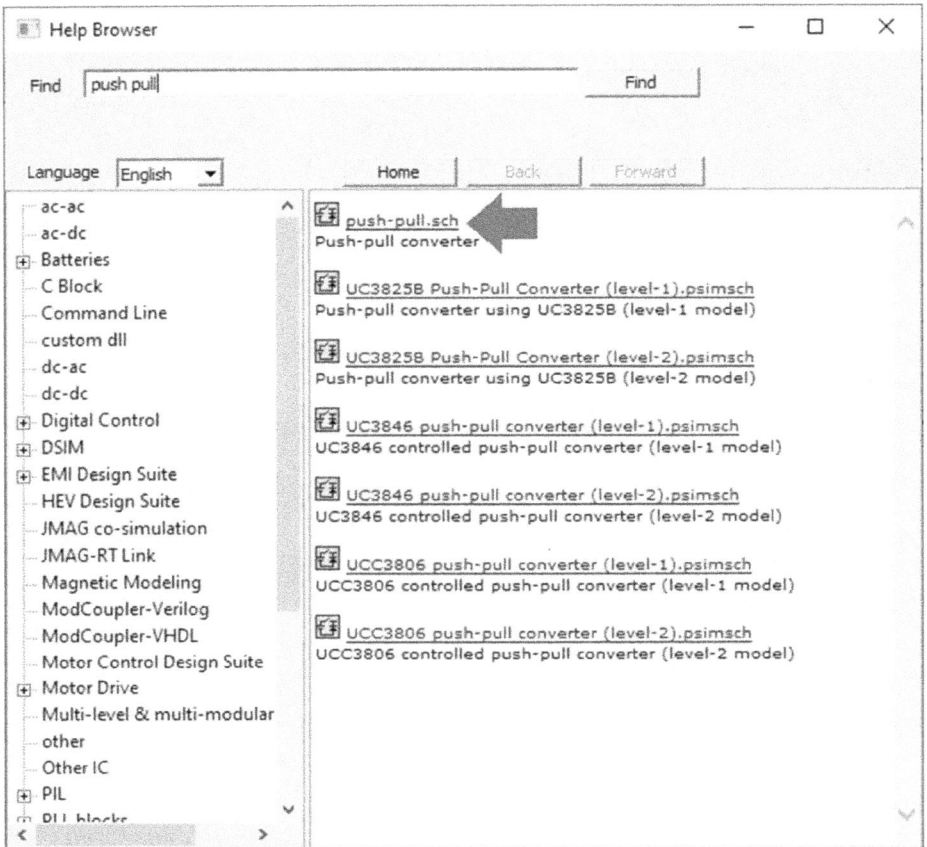

Fig. 5.468.

5.41 Example 40: single-phase inverter

In this example, we want to simulate a single-phase inverter. Assume the circuit shown in Fig. 5.469.

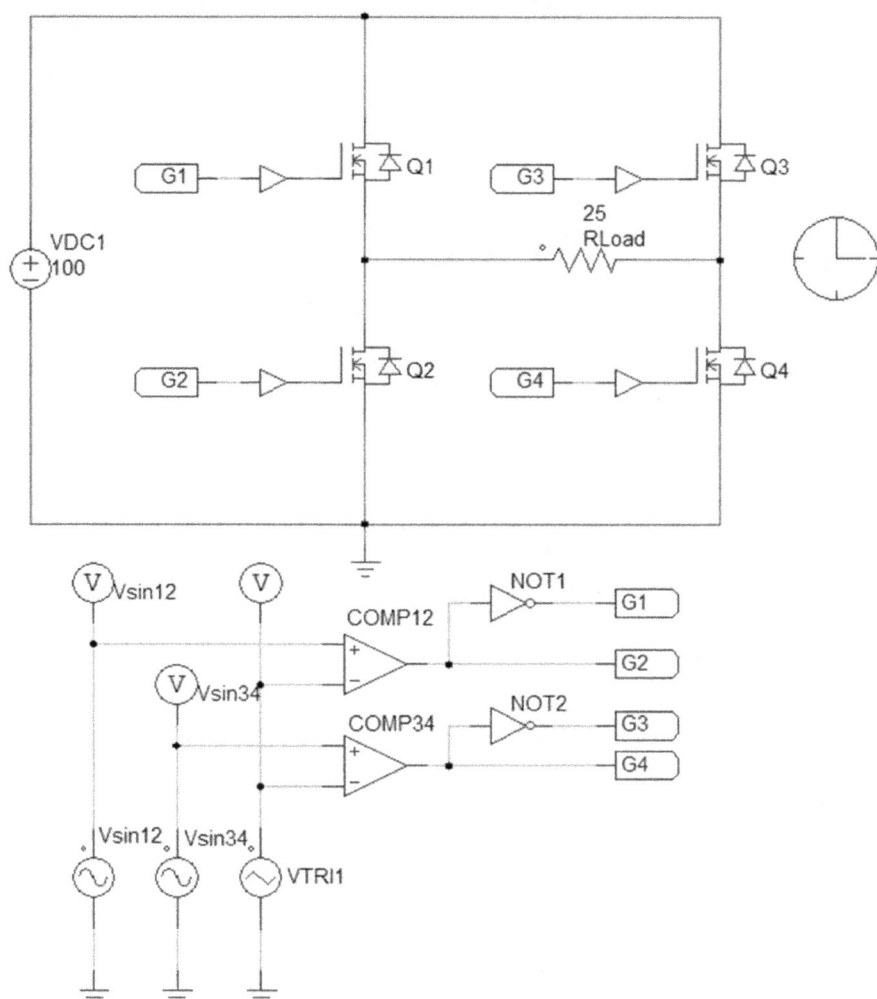

Fig. 5.469.

The settings of the blocks are shown in Figs. 5.470–5.476.

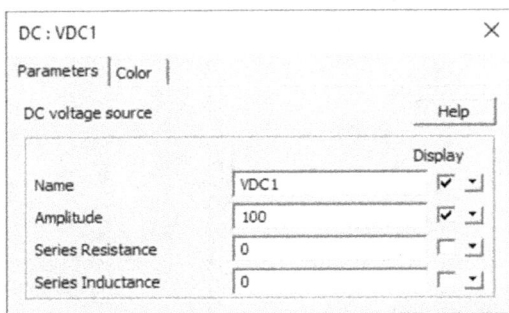

Fig. 5.470.

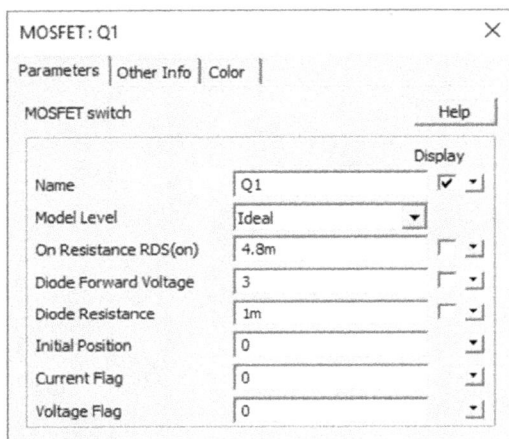

Fig. 5.471.

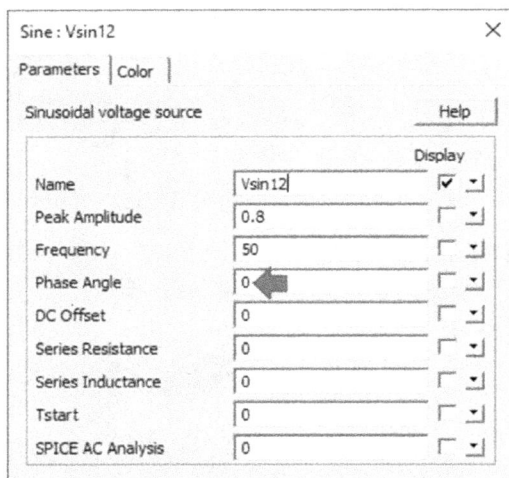

Fig. 5.472.

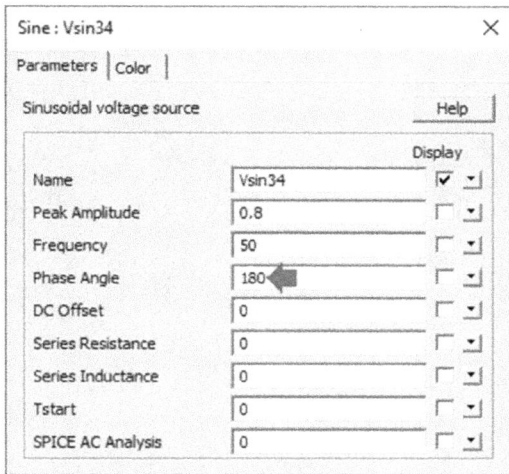

Fig. 5.473.

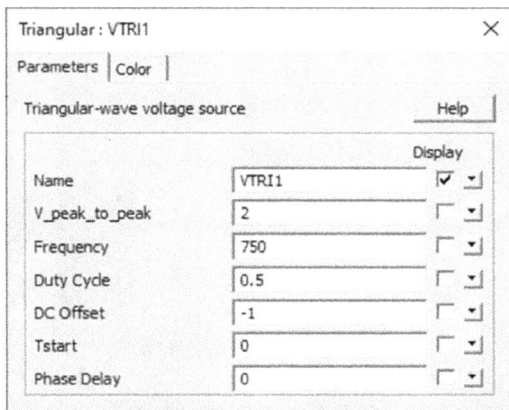

Fig. 5.474.

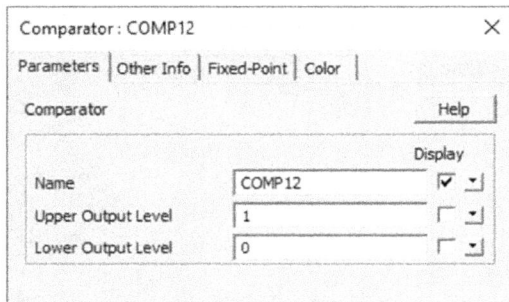

Fig. 5.475.

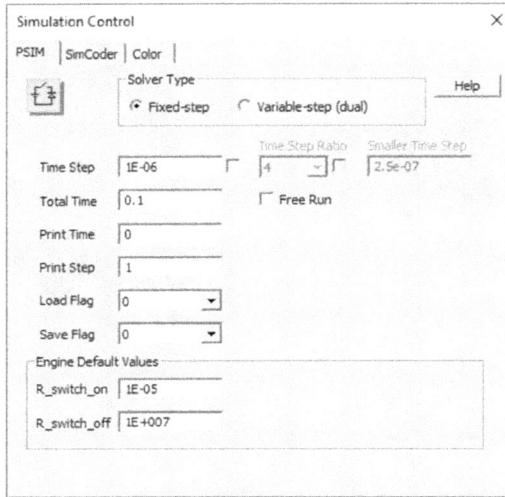

Simulation Control   ✕

PSIM | SimCoder | Color |

┌─ Solver Type ──────────────────────────────┐   Help
│ ⦿ Fixed-step   ○ Variable-step (dual)   │
└───┘

                              Time Step Ratio   Smaller Time Step
Time Step   1E-06   ▢   4   ▾ ▢   2.5e-07

Total Time   0.1   ▢ Free Run

Print Time   0

Print Step   1

Load Flag   0   ▾

Save Flag   0   ▾

┌ Engine Default Values ──────────────────
R_switch_on   1E-05

R_switch_off   1E+007

Fig. 5.476.

Run the simulation. The load voltage and current are shown in Figs. 5.477 and 5.478, respectively.

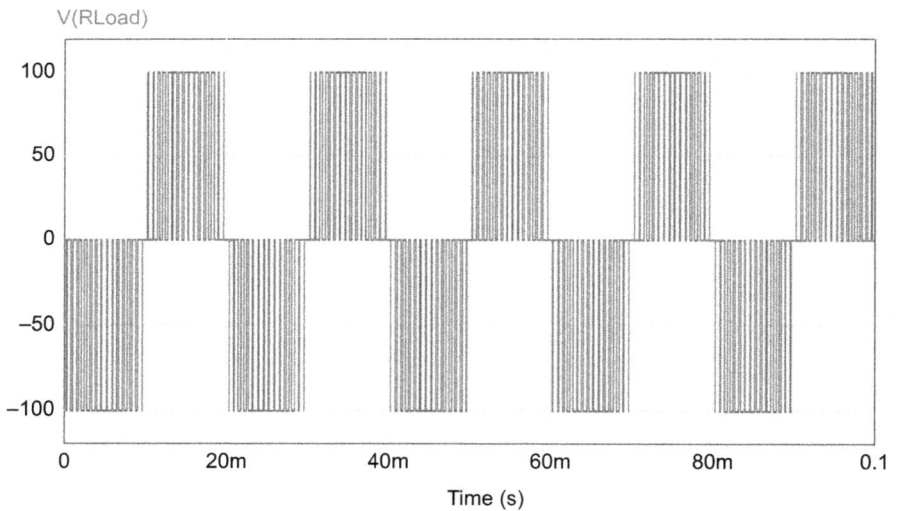

V(RLoad)

Time (s)

Fig. 5.477.

I(RLoad)

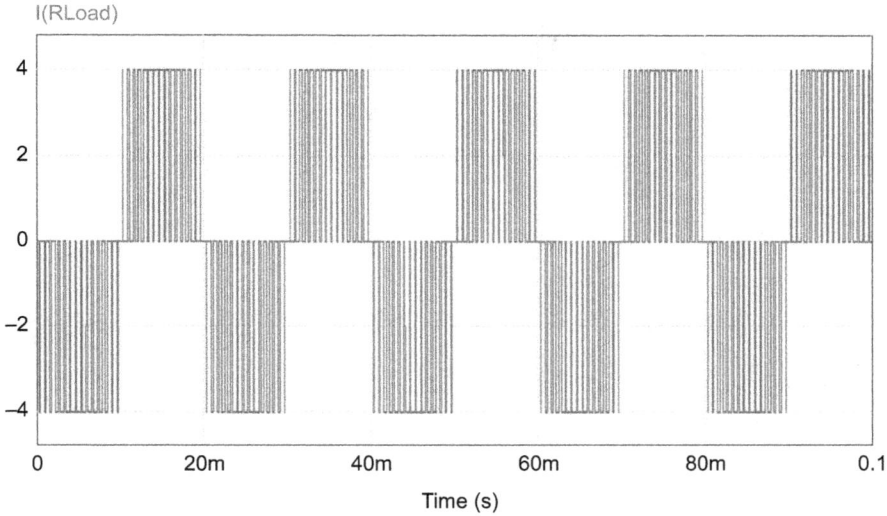

Fig. 5.478.

Let's measure the THD of load voltage. Since the load is resistive, the current is only scaled version of the load voltage $(i(t) = \frac{1}{R} \times v(t))$. So, the THD of load voltage and current are the same. Click the THD icon in order to measure the THD. .

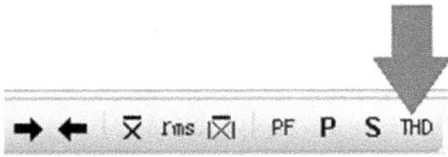

Fig. 5.479.

Enter the Fundamental Frequency. In this example, the reference signal frequency is 50 Hz, so the Fundamental Frequency box must be filled with 50 (Fig. 5.480).

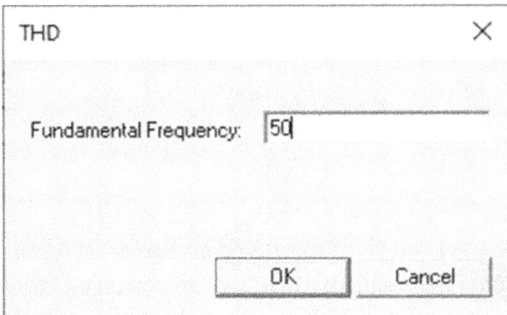

Fig. 5.480.

According to Fig. 5.481, the THD of load voltage (or current) is 76%.

Measure				☒
⋮	X1	X2	Δ	THD
Time	4.07193e-02	6.07889e-02	2.00696e-02	freq=50
V(RLoad)	-9.99616e+01	-2.84217e-14	9.99616e+01	7.65385e-01

Fig. 5.481.

You can see the frequency content of the signal by clicking the FFT icon.

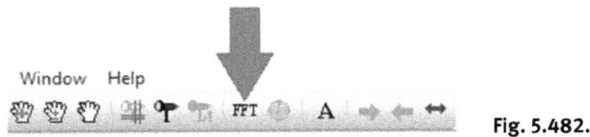

Window   Help

🖐 🖐 🖐 ⚟ ⌖ 🔍 FFT ⊙ A ⇒ ⇐ ↔   **Fig. 5.482.**

The frequency content is shown in Fig. 5.483. Note to the frequency content at high frequencies. According to Fig. 5.483, the amplitude of fundamental frequency is about 80.83 V.

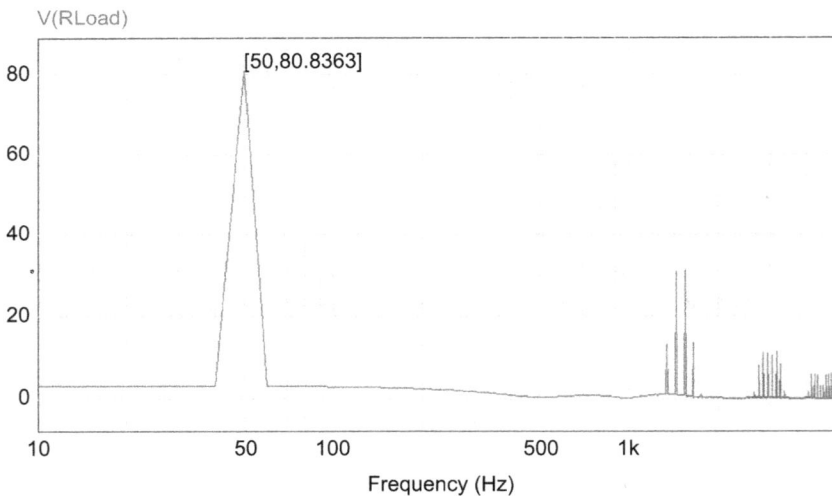

Fig. 5.483.

Let's change the load to RL type with $R = 25\ \Omega$ and $L = 50$ mH and see the waveforms. The voltage does not change (Fig. 5.484); however, the load current changes more drastically and changes to what shown in Fig. 5.485. The reason of this change is the inductor. The inductor acts as a filter and decreases the high frequency content of the load current. The THD of load current decreased to 0.039 (Fig. 5.486).

V(L1)+V(RLoad)

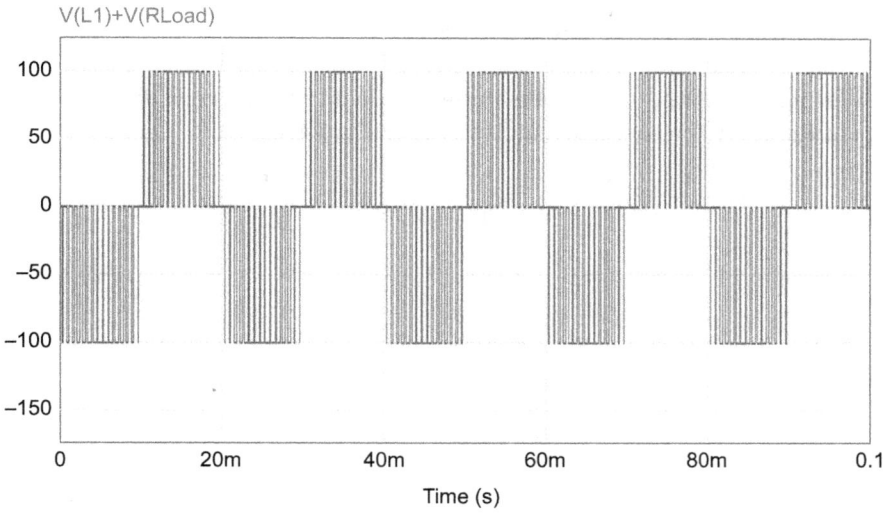

Fig. 5.484.

I(RLoad)

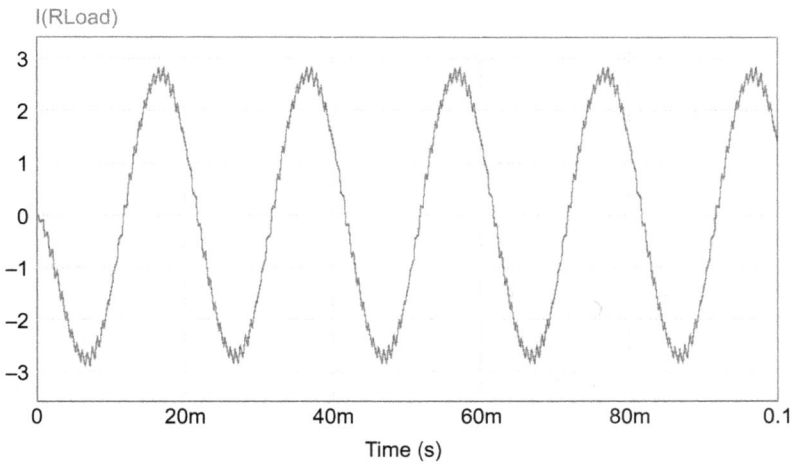

Fig. 5.485.

	X1	X2.	Δ	THD
Time	3.16629e-02	7.17540e-02	4.00911e-02	freq=50
I(RLoad)	-1.00746e-01	8.18847e-02	1.82631e-01	3.97938e-02

Measure

Fig. 5.486.

5.42 Example 41: three-phase inverter

In this example, we want to simulate a three-phase inverter. Assume the schematic shown in Fig. 5.487. The control circuit is shown in Fig. 5.488.

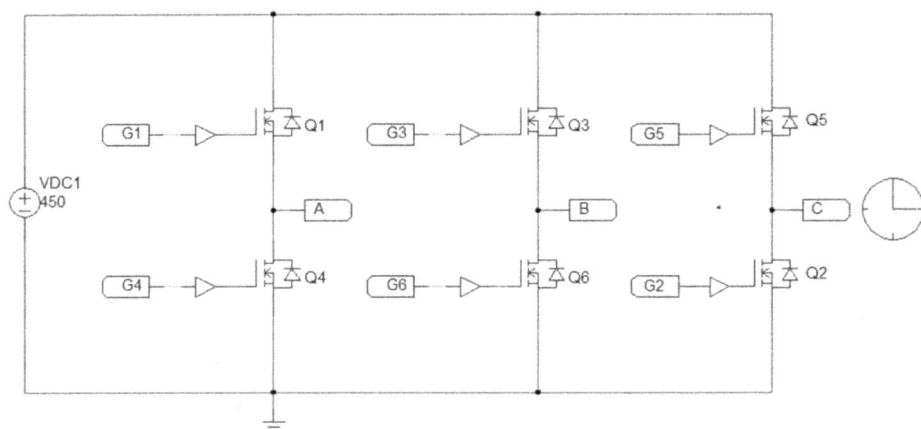

Fig. 5.487.

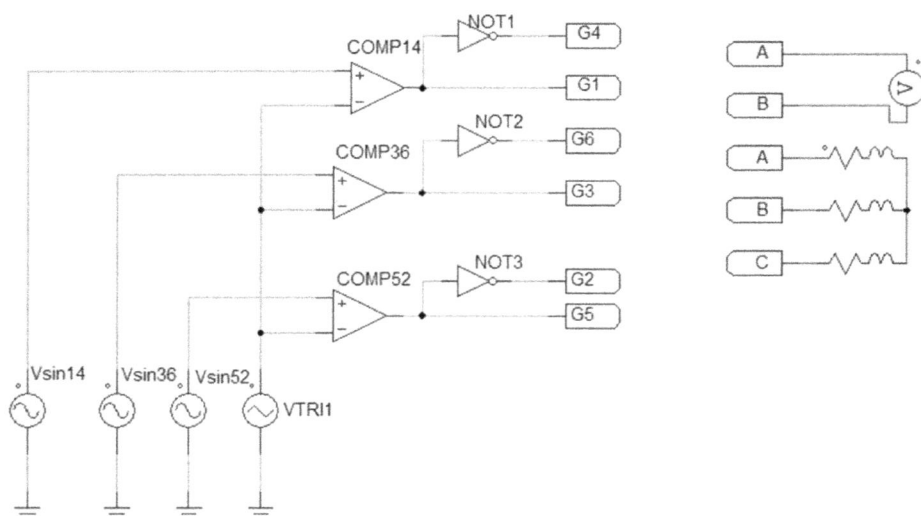

Fig. 5.488.

Settings of the used blocks are shown in Figs. 5.489–5.497.

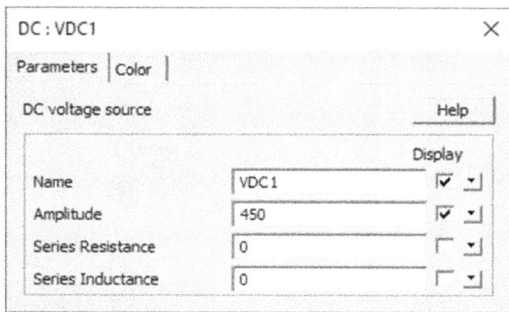

Fig. 5.489.

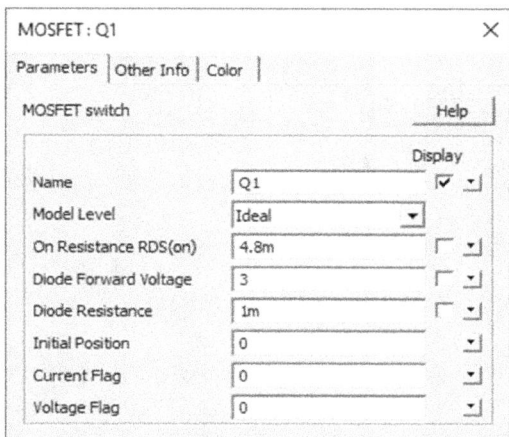

Fig. 5.490.

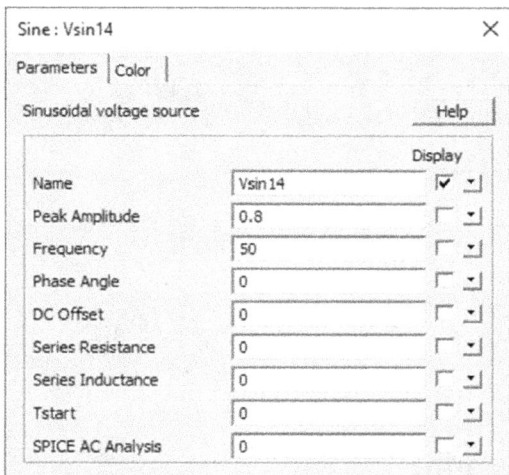

Fig. 5.491.

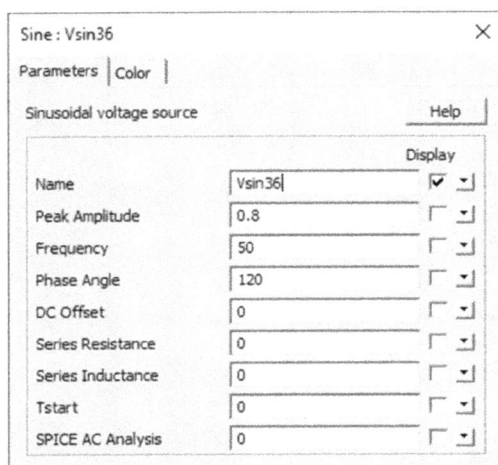

Fig. 5.492.

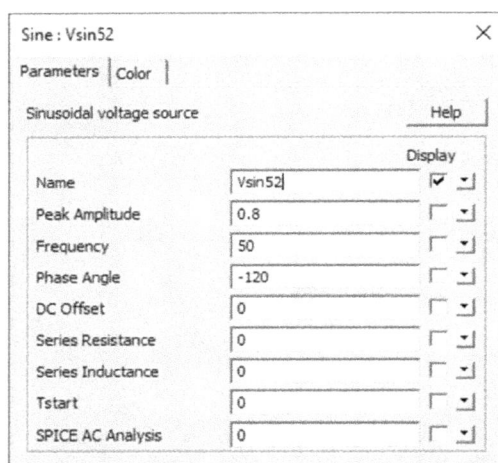

Fig. 5.493.

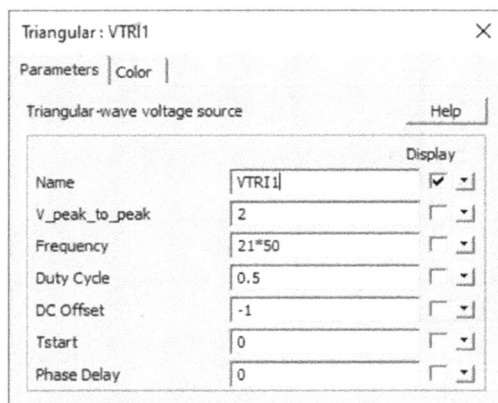

Fig. 5.494.

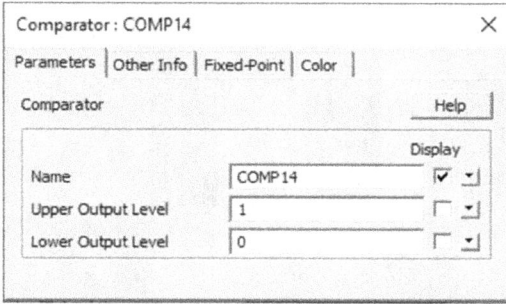

Fig. 5.495.

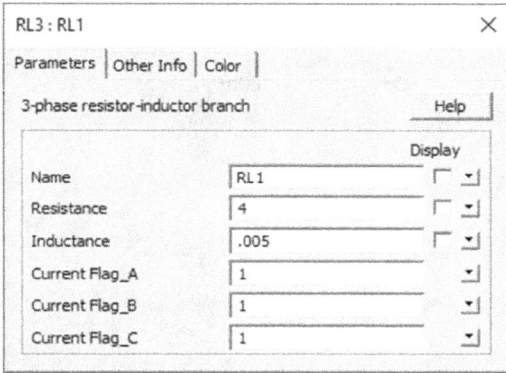

Fig. 5.496.

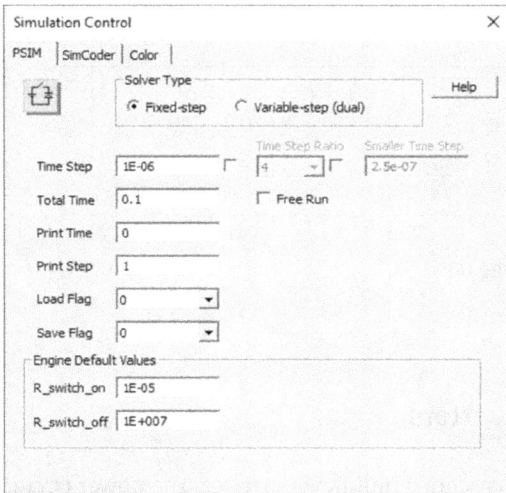

Fig. 5.497.

Run the simulation. The line–line voltage and load currents are shown in Fig. 5.498. The inductive nature of load decreased the current harmonics and caused the load current to be similar to sinusoidal waveform.

VAB

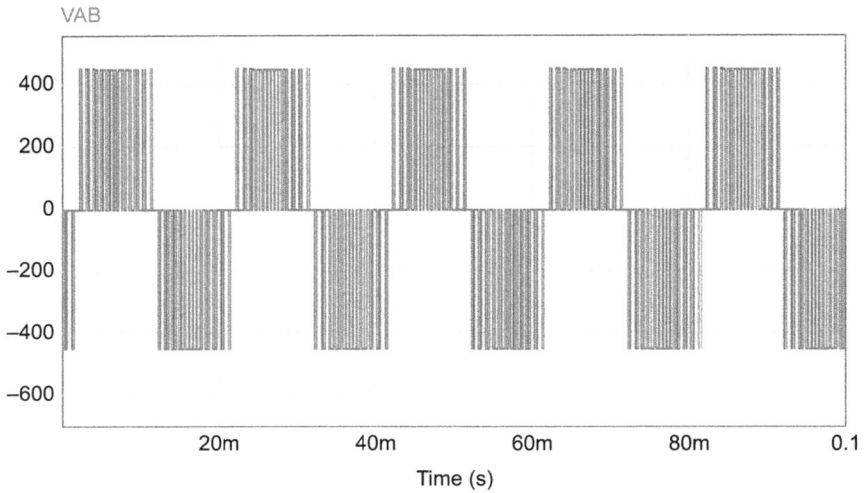

Fig. 5.498.

I(RL1a) I(RL1b) I(RL1c)

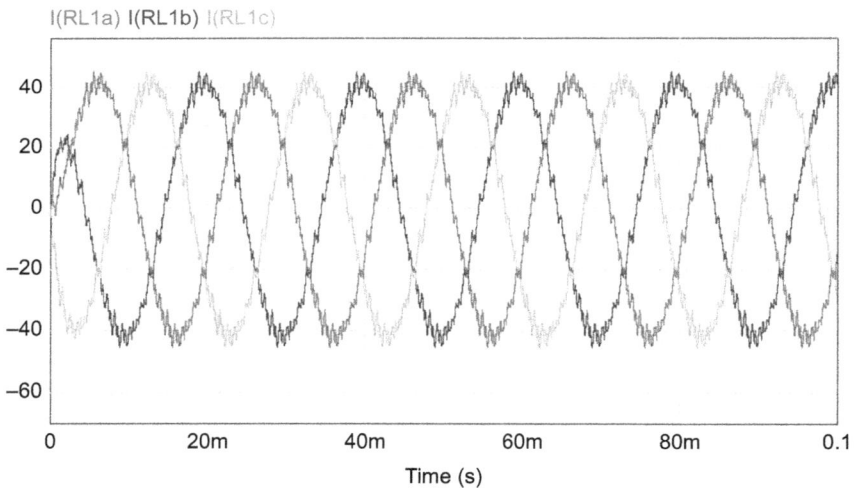

Fig. 5.499.

5.43 Example 42: cascaded inverters

In this example, we want to simulate a cascaded multilevel inverter. The power circuit of the inverter is shown in Fig. 5.500. It is series connection of two single-phase H bridge inverters. Each of H bridge inverter can produce output voltage of $-V_{dc}$, 0 and $+V_{dc}$. So, with two series H bridge, we can produce $-2V_{dc}$, $-V_{dc}$, 0, $+V_{dc}$, and $2V_{dc}$. Increasing the number of series stages helps us to decrease the THD.

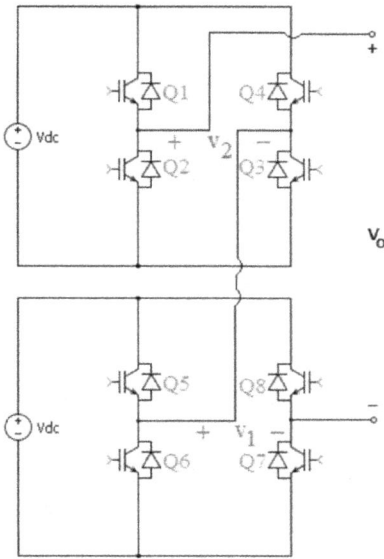

Fig. 5.500.

Assume the switching pattern shown in Tabs. 5.1 and 5.2.

Tab. 5.1: Switching pattern for Q1, Q2, Q3 and Q4.

Interval	Control signal for Q1	Control signal for Q2	Control signal for Q3	Control signal for Q4
A:$0 < \omega t < \alpha_2$	1	0	0	1
B:$\alpha_2 < \omega t < \pi - \alpha_2$	1	0	1	0
C:$\pi - \alpha_2 < \omega t < \pi + \alpha_2$	1	0	0	1
D:$\pi + \alpha_2 < \omega t < 2\pi - \alpha_2$	0	1	0	1
E:$2\pi - \alpha_2 < \omega t < 2\pi$	1	0	0	1

Tab. 5.2: Switching pattern for Q5, Q6, Q7 and Q8.

Interval	Control signal for Q5	Control signal for Q6	Control signal for Q7	Control signal for Q8
a:$0 < \omega t < \alpha_1$	1	0	0	1
b:$\alpha_1 < \omega t < \pi - \alpha_1$	1	0	1	0
c:$\pi - \alpha_1 < \omega t < \pi + \alpha_1$	1	0	0	1
d:$\pi + \alpha_1 < \omega t < 2\pi - \alpha_1$	0	1	0	1
e:$2\pi - \alpha_1 < \omega t < 2\pi$	1	0	0	1

The voltage waveforms shown in Fig. 5.501 will be obtained with this switching pattern.

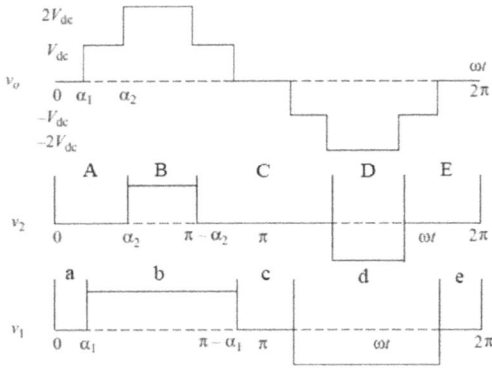

Fig. 5.501.

We want to simulate the circuit shown in Fig. 5.500 with the switching pattern of Tabs. 5.1 and 5.2 ($\alpha_1 = 20°$ and $\alpha_1 = 40°$). Assume the schematic shown in Fig. 5.502.

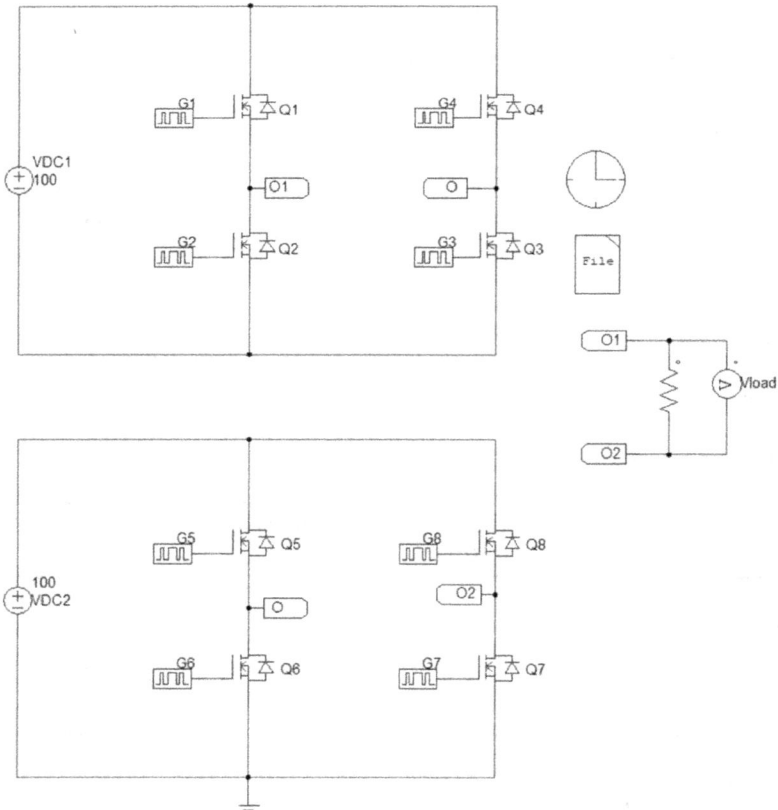

Fig. 5.502.

Settings of used elements are shown in Figs. 5.503–5.513.

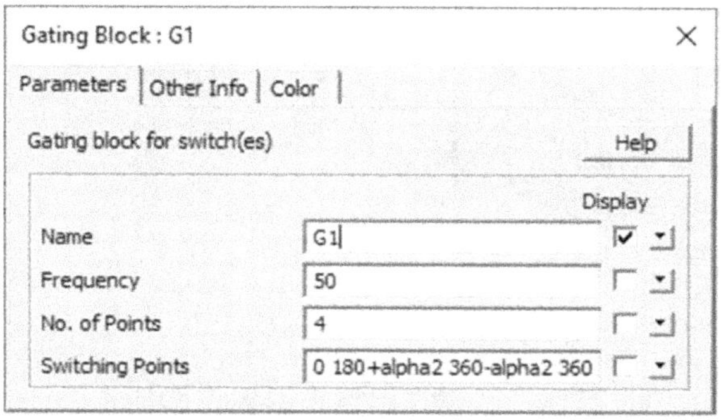

Fig. 5.503.

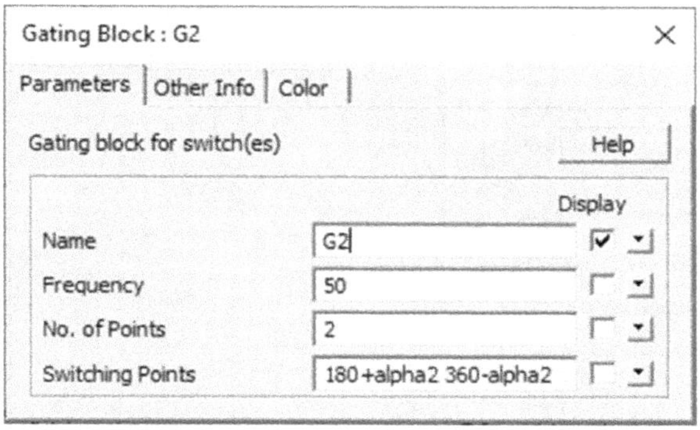

Fig. 5.504.

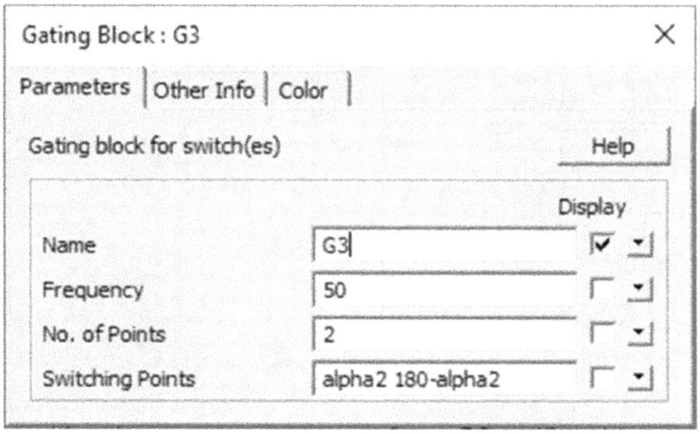

Fig. 5.505.

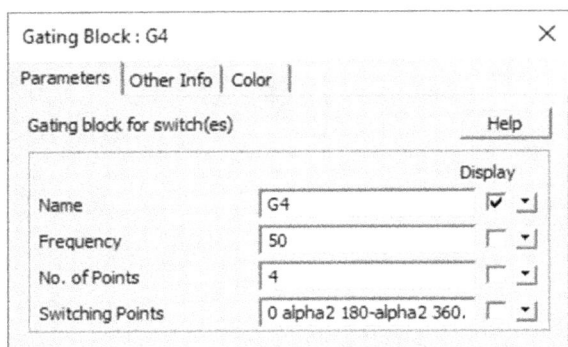

Fig. 5.506.

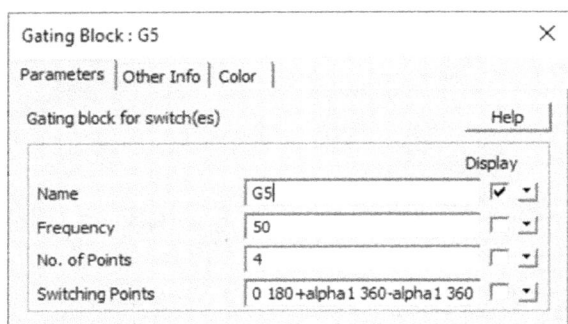

Fig. 5.507.

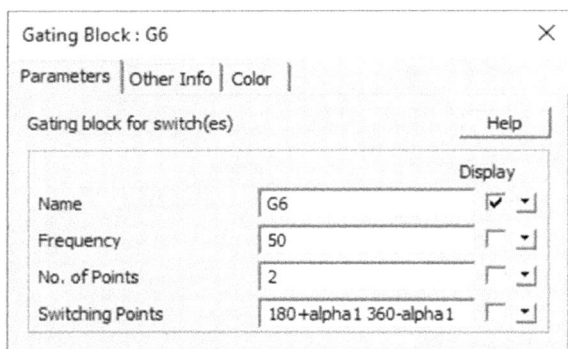

Fig. 5.508.

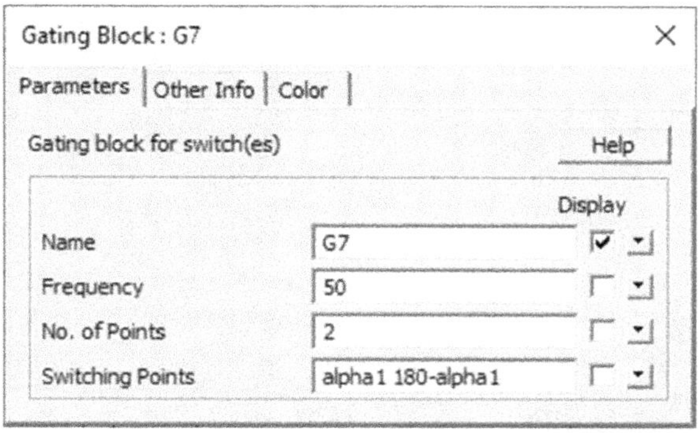

Gating Block : G7

Parameters | Other Info | Color

Gating block for switch(es)   Help

		Display
Name	G7	☑ ▾
Frequency	50	☐ ▾
No. of Points	2	☐ ▾
Switching Points	alpha1 180-alpha1	☐ ▾

Fig. 5.509.

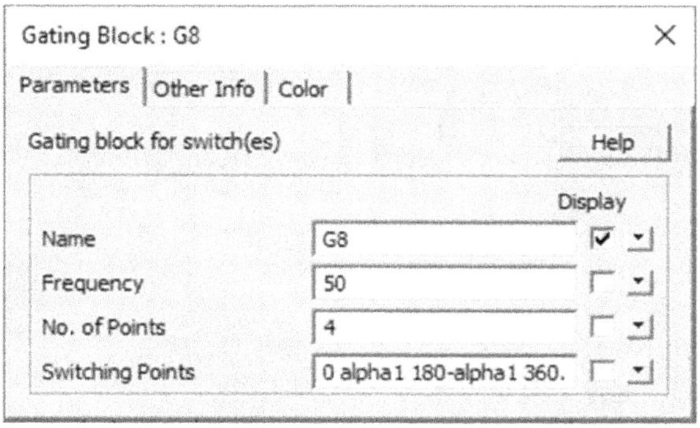

Gating Block : G8

Parameters | Other Info | Color

Gating block for switch(es)   Help

		Display
Name	G8	☑ ▾
Frequency	50	☐ ▾
No. of Points	4	☐ ▾
Switching Points	0 alpha1 180-alpha1 360.	☐ ▾

Fig. 5.510.

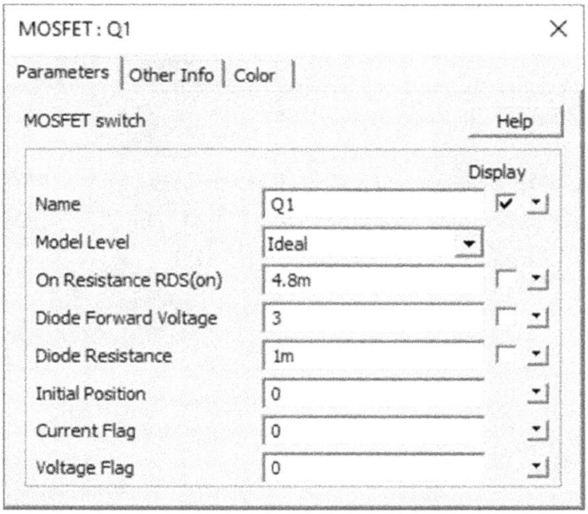

MOSFET : Q1

Parameters | Other Info | Color

MOSFET switch   Help

		Display
Name	Q1	☑ ▾
Model Level	Ideal ▾	
On Resistance RDS(on)	4.8m	☐ ▾
Diode Forward Voltage	3	☐ ▾
Diode Resistance	1m	☐ ▾
Initial Position	0	▾
Current Flag	0	▾
Voltage Flag	0	▾

Fig. 5.511.

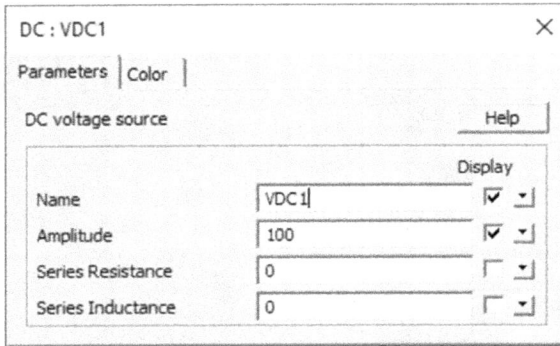

Fig. 5.512.

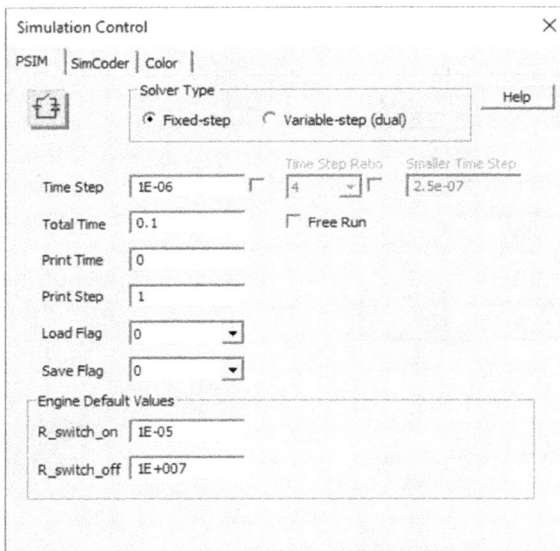

Fig. 5.513.

Note that α_1 and α_2 are defined in a Parameter File block (Fig. 5.514). The Parameter File block can be found in Other section of Elements menu (Fig. 5.515).

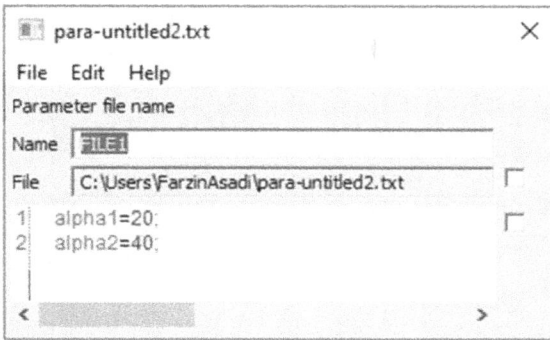

Fig. 5.514.

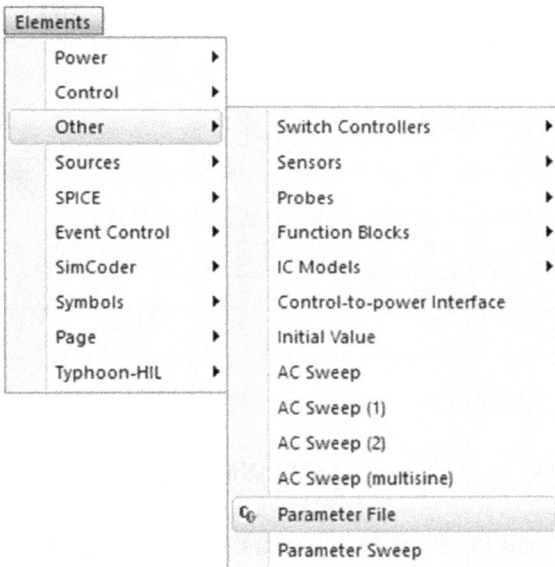

Fig. 5.515.

Run the simulation. The load voltage is shown in Fig. 5.516. The output voltage has −200, −100, 0, 100, and 200 levels. According to Fig. 5.517, the THD of this waveform is 0.19.

Vload

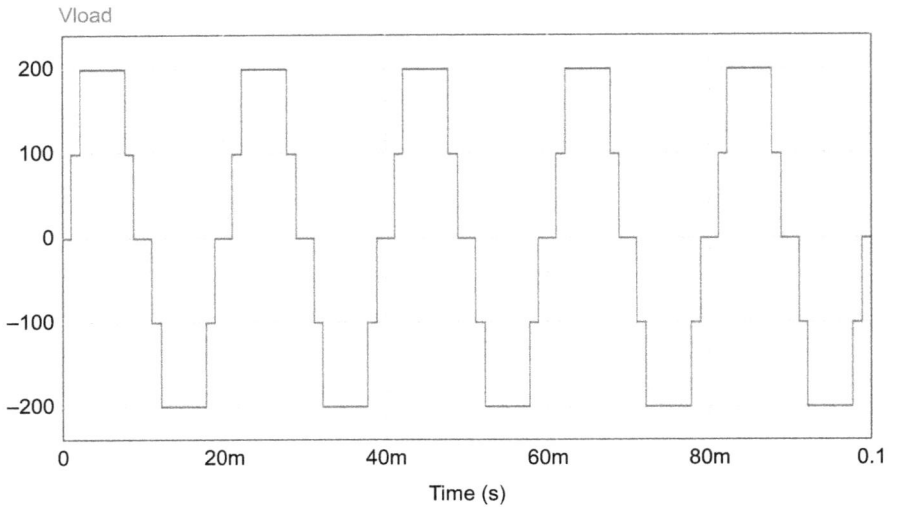

Fig. 5.516.

Measure	X1	X2	Δ	THD
Time	1.87935e-02	3.88631e-02	2.00696e-02	freq=50
Vload	-9.98084e+01	-9.98084e+01	0.00000e+00	1.91058e-01

Fig. 5.517.

5.44 Example 43: multilevel inverters and SV PWM block

PSIM has some ready-to-use examples for multilevel inverters. In order to see them, click the File>Search Examples and type multi-level in the Find box.

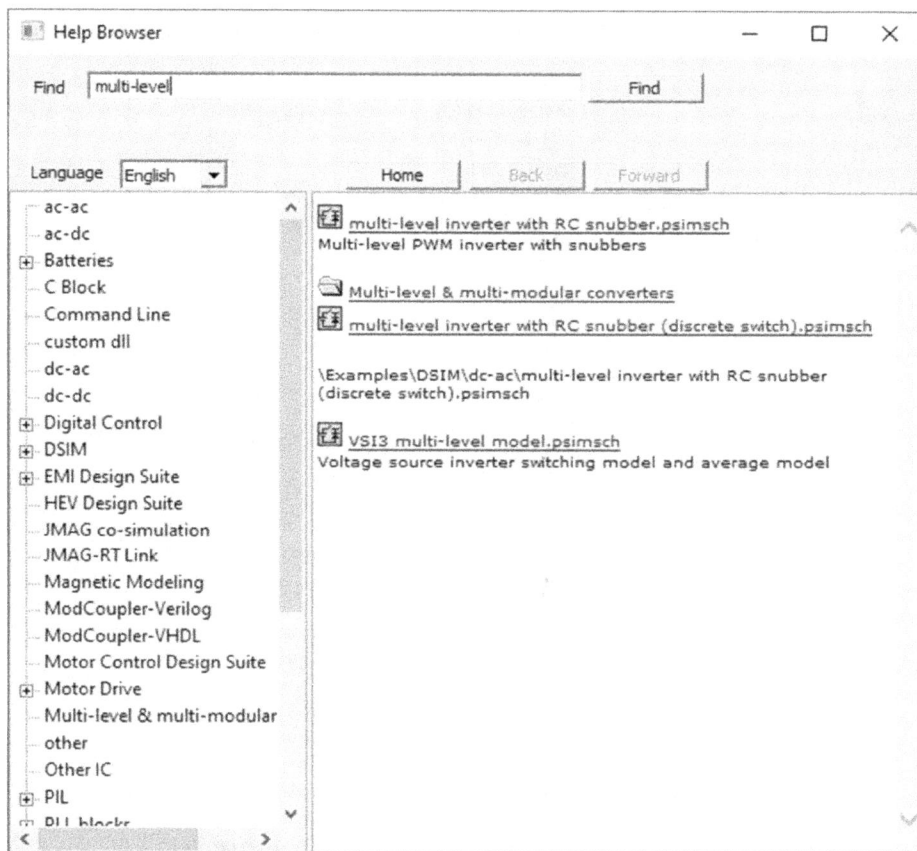

Fig. 5.518.

PSIM has a ready-to-use space vector pulse width modulator (SV PWM) block.

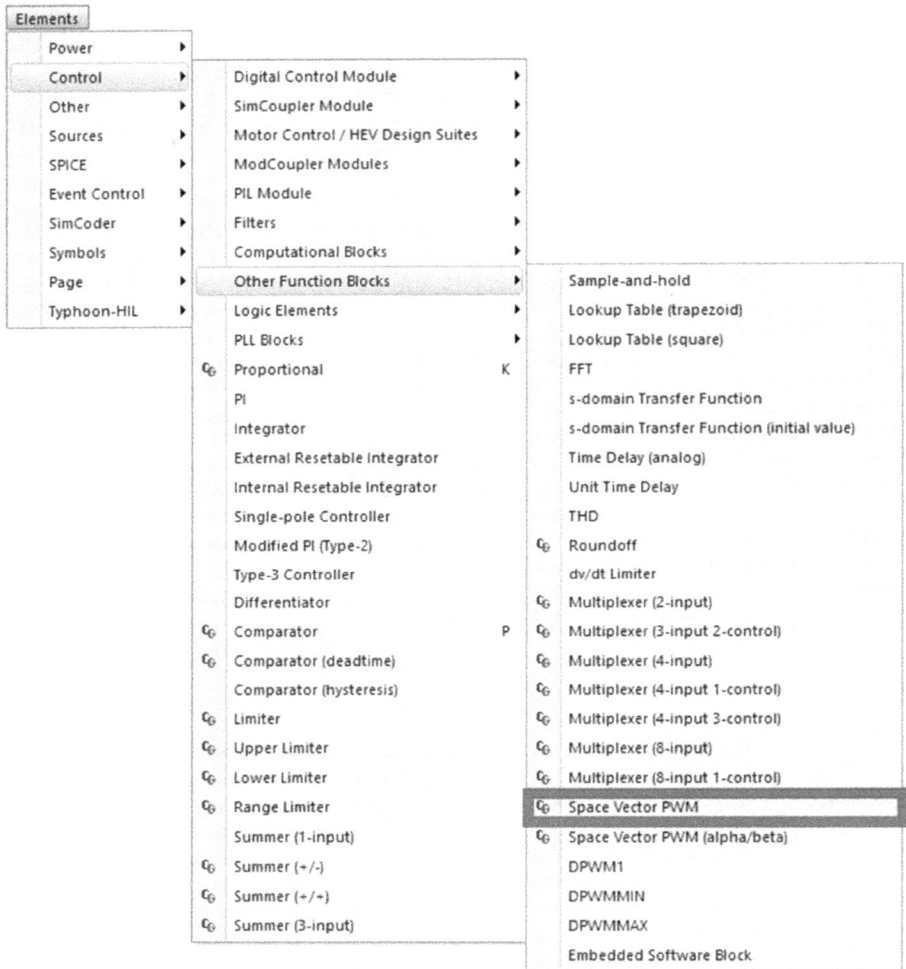

Elements		
Power ▶		
Control ▶	Digital Control Module ▶	
Other ▶	SimCoupler Module ▶	
Sources ▶	Motor Control / HEV Design Suites ▶	
SPICE ▶	ModCoupler Modules ▶	
Event Control ▶	PIL Module ▶	
SimCoder ▶	Filters ▶	
Symbols ▶	Computational Blocks ▶	
Page ▶	Other Function Blocks ▶	Sample-and-hold
Typhoon-HIL ▶	Logic Elements ▶	Lookup Table (trapezoid)
	PLL Blocks ▶	Lookup Table (square)
	⌁ Proportional K	FFT
	PI	s-domain Transfer Function
	Integrator	s-domain Transfer Function (initial value)
	External Resetable Integrator	Time Delay (analog)
	Internal Resetable Integrator	Unit Time Delay
	Single-pole Controller	THD
	Modified PI (Type-2)	⌁ Roundoff
	Type-3 Controller	dv/dt Limiter
	Differentiator	⌁ Multiplexer (2-input)
	⌁ Comparator P	⌁ Multiplexer (3-input 2-control)
	⌁ Comparator (deadtime)	⌁ Multiplexer (4-input)
	Comparator (hysteresis)	⌁ Multiplexer (4-input 1-control)
	⌁ Limiter	⌁ Multiplexer (4-input 3-control)
	⌁ Upper Limiter	⌁ Multiplexer (8-input)
	⌁ Lower Limiter	⌁ Multiplexer (8-input 1-control)
	⌁ Range Limiter	⌁ Space Vector PWM
	Summer (1-input)	⌁ Space Vector PWM (alpha/beta)
	⌁ Summer (+/-)	DPWM1
	⌁ Summer (+/+)	DPWMMIN
	⌁ Summer (3-input)	DPWMMAX
		Embedded Software Block

Fig. 5.519.

In order to see how SV PWM block is used in simulations, click the File>Search Examples and type svpwm in the Find box.

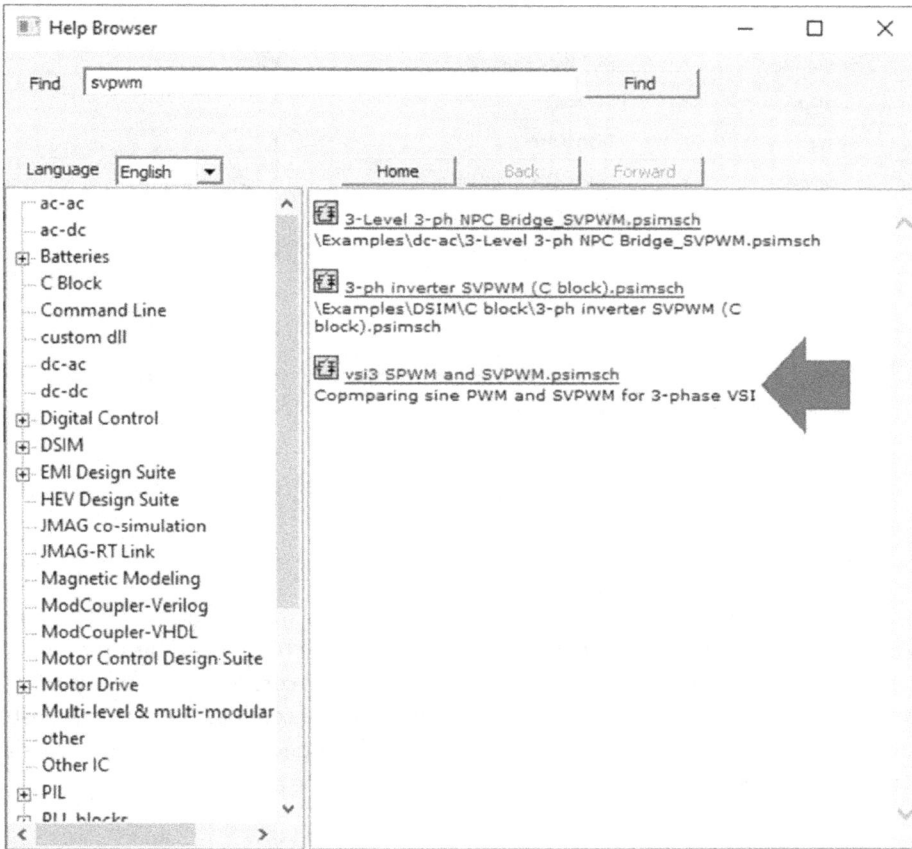

Fig. 5.520.

5.45 Example 44: magnetic components

You can simulate complex magnetic circuits easily with the aid of PSIM's magnetic blocks. Simulation of nonlinear effects such as hysteresis, saturation and flux leakage is quite easy with the aid of blocks inside the Magnetic Elements section.

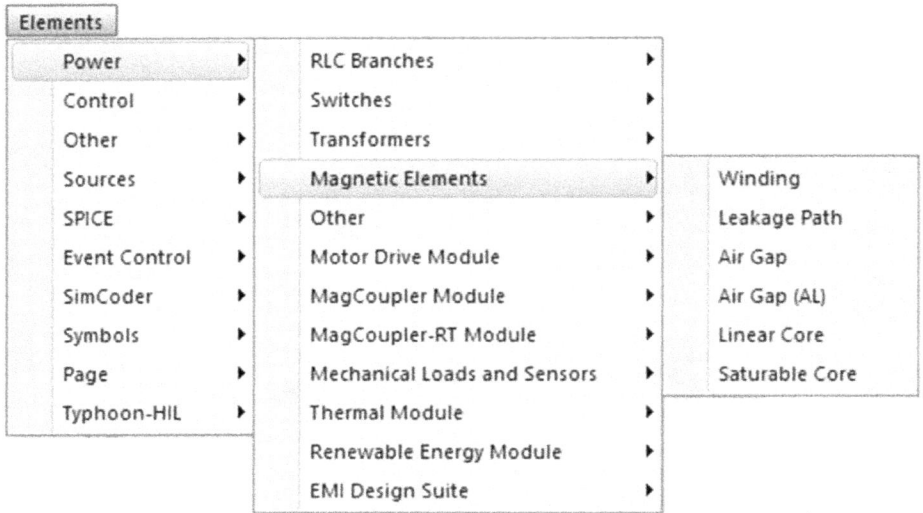

Fig. 5.521.

Consider the configuration shown in Fig. 5.522. This figure shows a ferromagnetic core whose mean path length is 40 cm. There is a small gap of 0.05 cm in the structure of the otherwise whole core. The cross-sectional area of the core is 12 , the relative permeability of the core is 4,000, and the coil of wire on the core has 400 turns. Resistance of wire and input DC current are 10 Ω and 0.6 A, respectively. Fringing in the air gap is ignored. We want to calculate the flux density in the air gap.

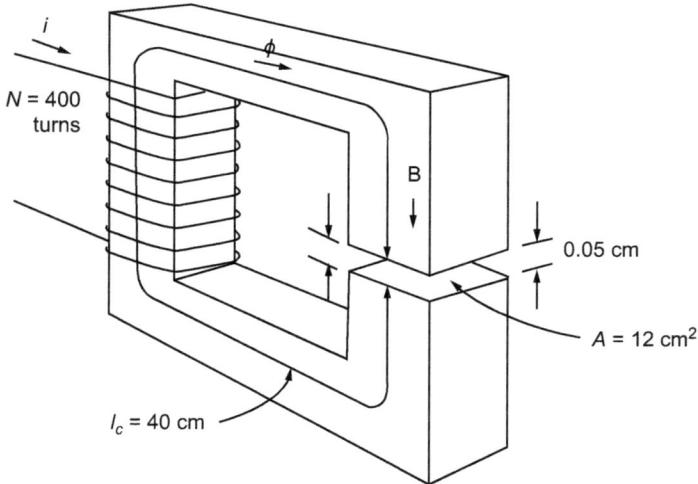

Fig. 5.522.

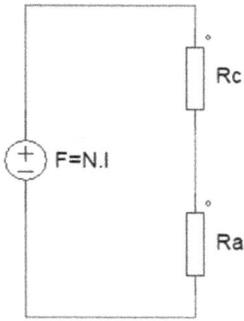

Fig. 5.523.

The reluctance of core $(\mathcal{R}_c)$ is

$$\mathcal{R}_c = \frac{l_c}{\mu_r \mu_0 A_c} = \frac{0.4\,m}{(4{,}000) \times (4\pi \times 10^{-7}) \times 0.0012\,m^2} = 66{,}298\,A.turn\,/\,Wb$$

The reluctance of air gap $(\mathcal{R}_a)$ is

$$\mathcal{R}_a = \frac{l_a}{\mu_r \mu_0 A_c} = \frac{0.0005\,m}{(4\pi \times 10^{-7}) \times 0.0012\,m^2} = 331{,}490\,A.turn\,/\,Wb$$

The total reluctance is

$$\mathcal{R}_t = \mathcal{R}_a + \mathcal{R}_c = 397{,}788\,A.turn\,/\,Wb$$

The MMF distribution can be calculated easily:

$$\mathcal{F}_c = \frac{\mathcal{R}_c}{\mathcal{R}_t} = \frac{\mathcal{R}_c}{\mathcal{R}_a + \mathcal{R}_c} \times N \times I = \frac{66{,}298}{397{,}788} \times 400 \times 0.6 = 40\,A.turn$$

$$\mathcal{F}_a = \frac{\mathcal{R}_a}{\mathcal{R}_t} = \frac{\mathcal{R}_a}{\mathcal{R}_a + \mathcal{R}_c} \times N \times I = \frac{331{,}490}{397{,}788} \times 400 \times 0.6 = 200\,A.turn$$

The flux density of air gap is

$$B = \frac{\varphi}{A} = \frac{\dfrac{N.I}{\mathcal{R}_t}}{A} = \frac{\dfrac{400 \times 0.6}{397{,}788}}{0.0012} = 0.5\,T$$

The inductance can be calculated as

$$L = \frac{N^2}{\mathcal{R}_t} = \frac{400^2}{397{,}788} = 0.40\,H$$

The magnetic circuit of Fig. 5.522 can be analyzed with the aid of schematic shown in Fig. 5.524.

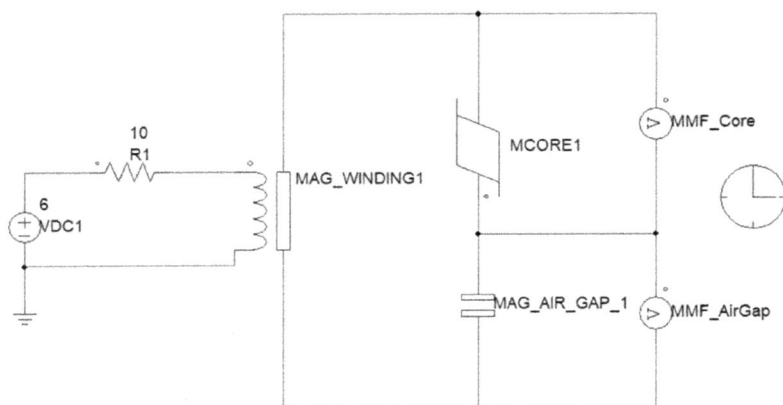

Fig. 5.524.

Settings of used blocks are shown in Figs. 5.525–5.532.

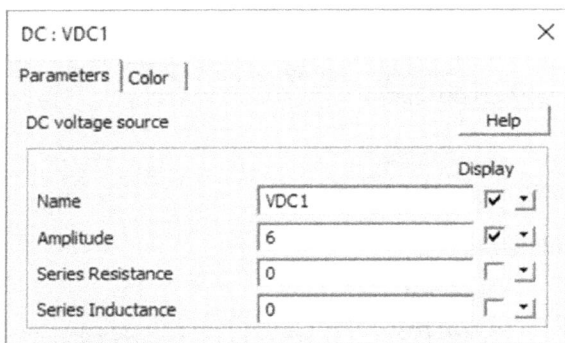

Fig. 5.525.

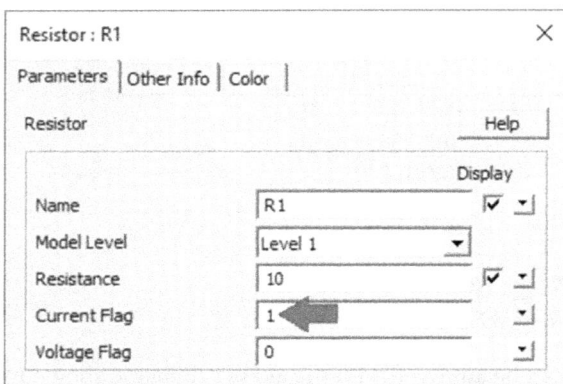

Fig. 5.526.

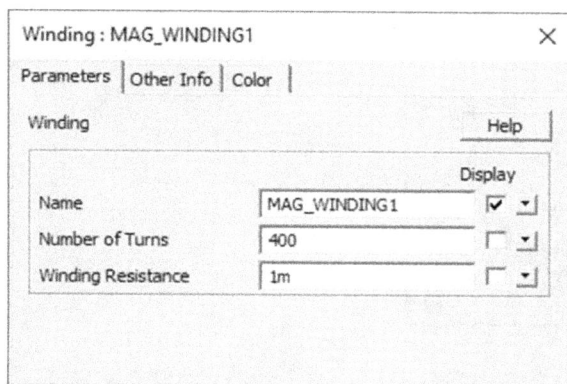

Fig. 5.527.

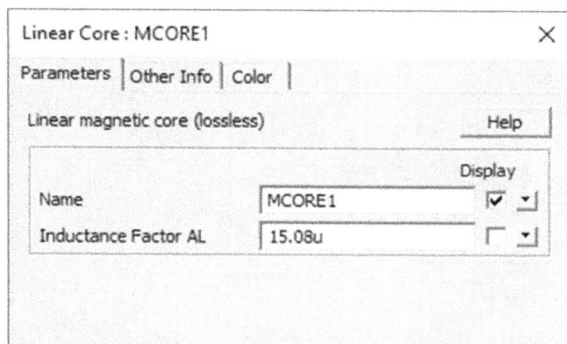

Fig. 5.528.

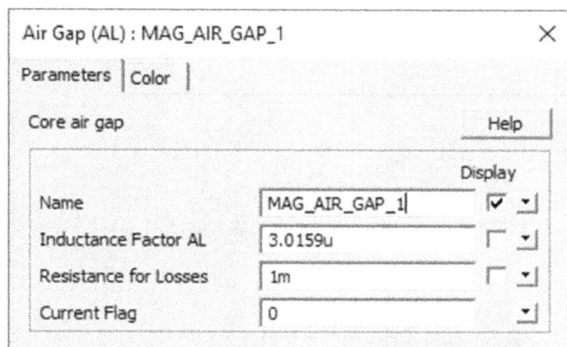

Fig. 5.529.

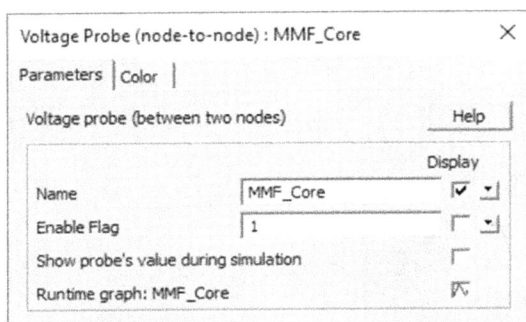

Fig. 5.530.

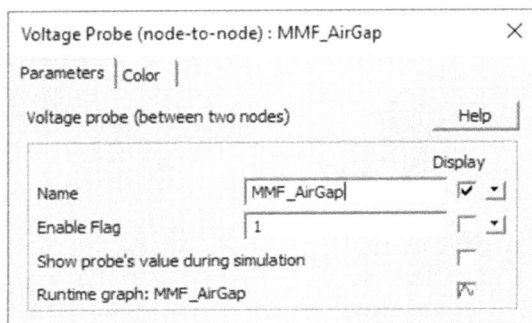

Fig. 5.531.

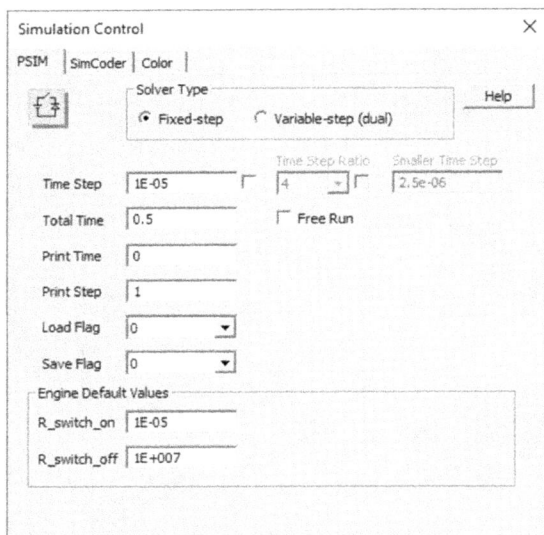

Fig. 5.532.

Note that lossless magnetic core and air gap blocks use inductance factor. The inductance factor definition for theses blocks are shown in Figs. 5.533 and 5.534. The inductance factor of used blocks is calculated in Fig. 5.535.

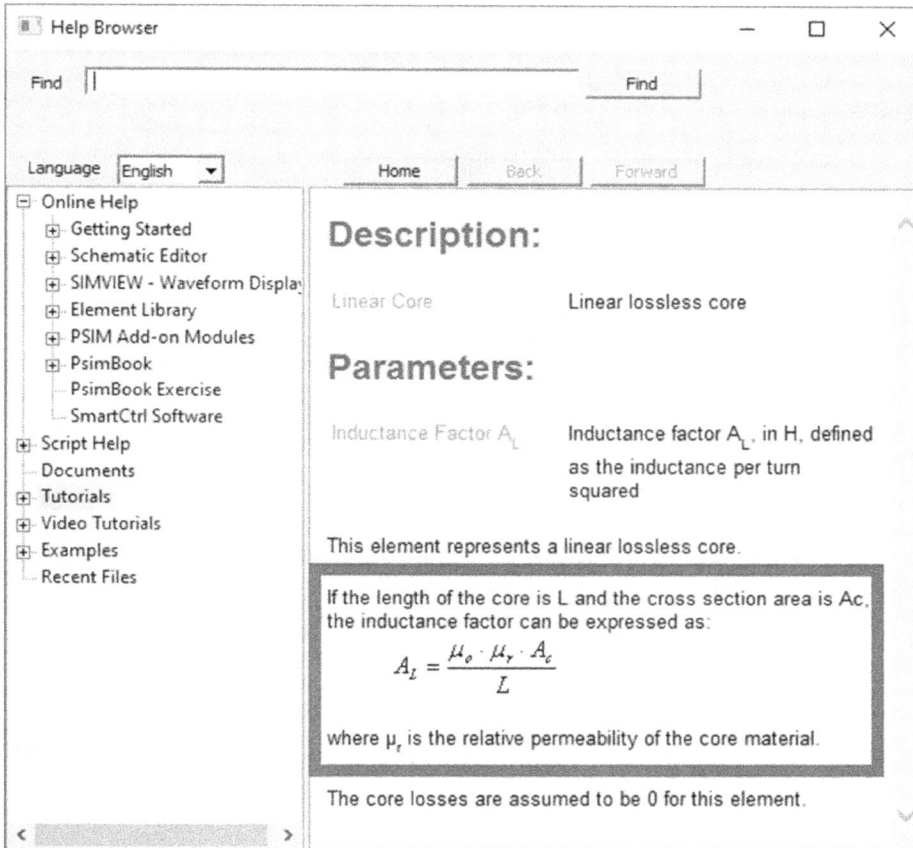

Fig. 5.533.

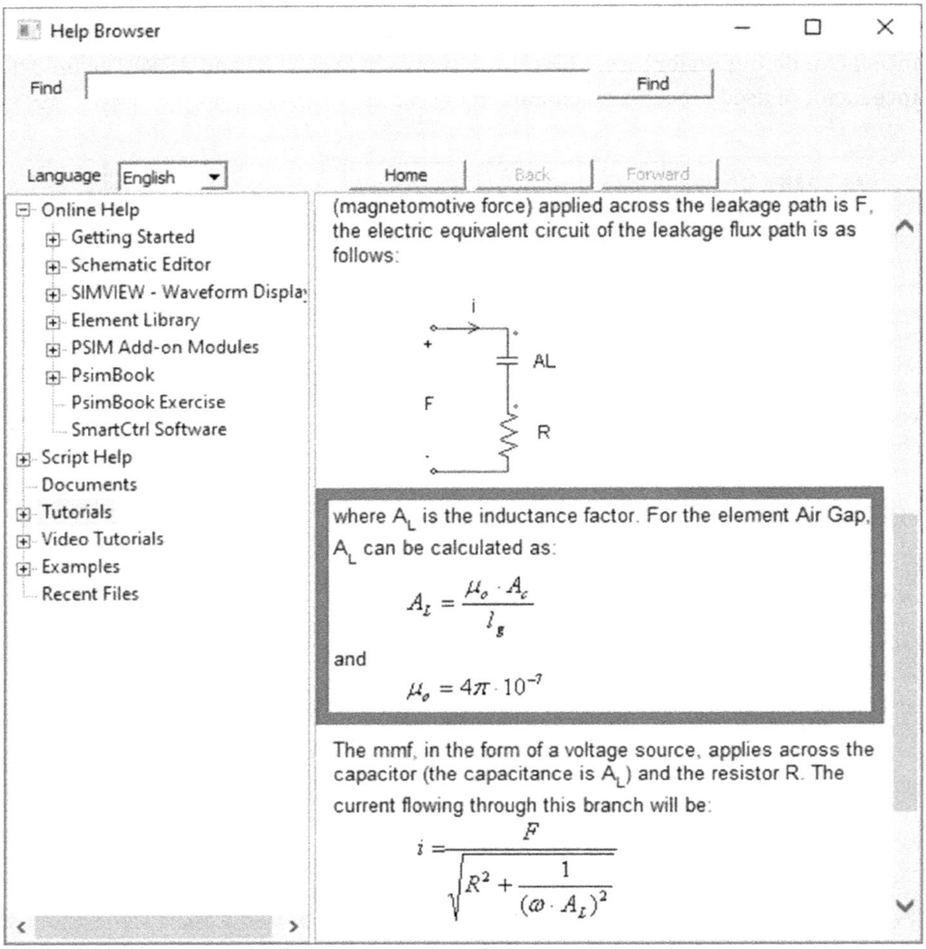

Help Browser — □ ×

Find [   ] Find

Language [English ▼]   Home   Back   Forward

- Online Help
  - Getting Started
  - Schematic Editor
  - SIMVIEW - Waveform Display
  - Element Library
  - PSIM Add-on Modules
  - PsimBook
    - PsimBook Exercise
    - SmartCtrl Software
- Script Help
  - Documents
- Tutorials
- Video Tutorials
- Examples
  - Recent Files

(magnetomotive force) applied across the leakage path is F, the electric equivalent circuit of the leakage flux path is as follows:

where A_L is the inductance factor. For the element Air Gap, A_L can be calculated as:

$$A_L = \frac{\mu_o \cdot A_c}{l_g}$$

and

$$\mu_o = 4\pi \cdot 10^{-7}$$

The mmf, in the form of a voltage source, applies across the capacitor (the capacitance is A_L) and the resistor R. The current flowing through this branch will be:

$$i = \frac{F}{\sqrt{R^2 + \frac{1}{(\omega \cdot A_L)^2}}}$$

Fig. 5.534.

```
Command Window                              ⊙

 >> u0=4*pi*1e-7;
 >> ur=4000;
 >> Ac=12e-4;
 >> Lcore=0.4;
 >> LAirGap=0.0005;
 >> CoreInductanceFactor=ur*u0*Ac/Lcore

CoreInductanceFactor =

    1.5080e-05

 >> AirGapInductanceFactor=u0*Ac/LAirGap

AirGapInductanceFactor =

    3.0159e-06

fx >>
```

Fig. 5.535.

Run the simulation. The simulation results (circuits current, core magnetic motive force, and air gap magnetic motive force) are shown in Figs. 5.536–5.538.

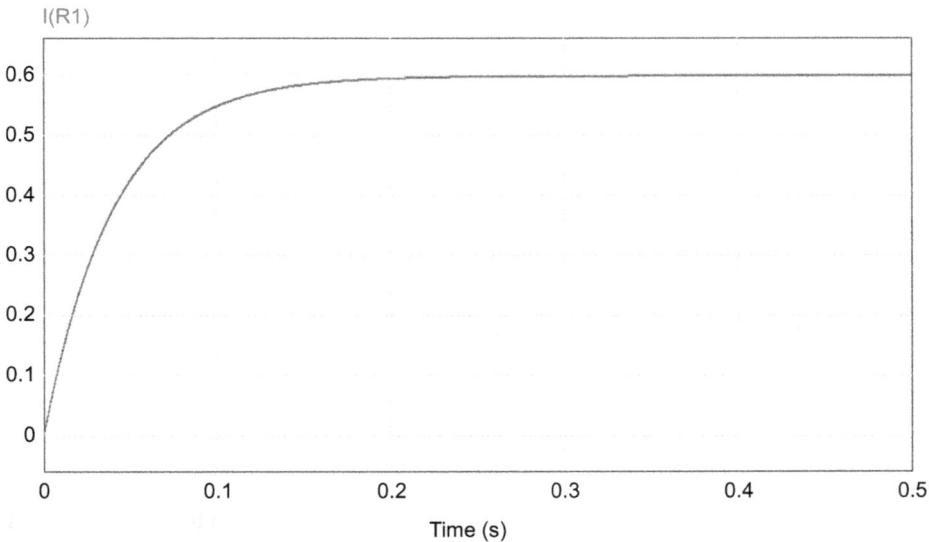

Fig. 5.536.

MMF_Core

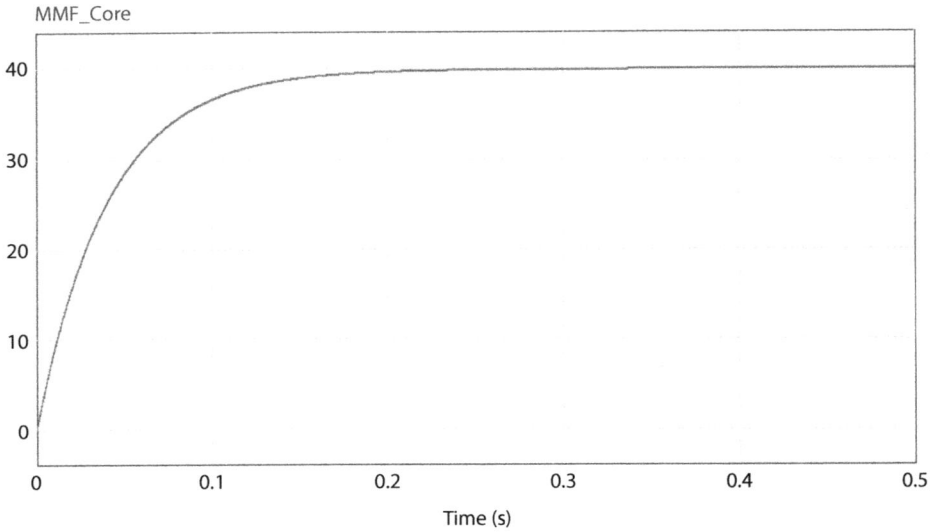

Fig. 5.537.

MMF_AirGap

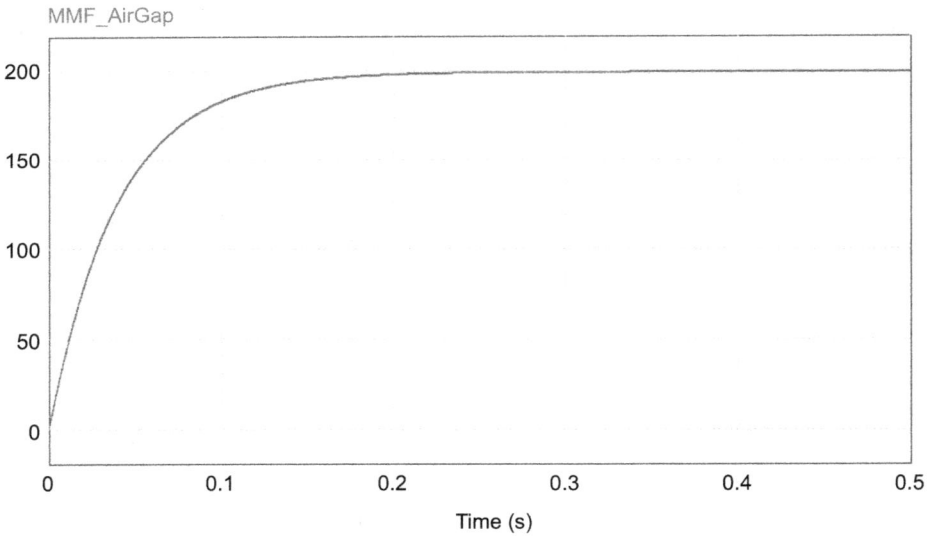

Fig. 5.538.

We used a linear core (without saturation) in the schematic shown in Fig. 5.524, PSIM has a block called Saturable Core (Fig. 5.539) which can be used to simulate the effect of core saturation.

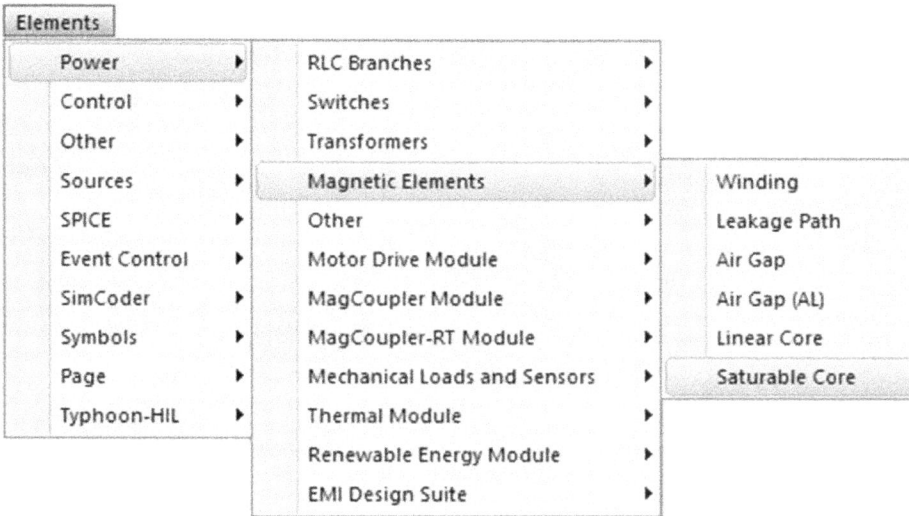

Elements		
Power ▶	RLC Branches ▶	
Control ▶	Switches ▶	
Other ▶	Transformers ▶	
Sources ▶	Magnetic Elements ▶	Winding
SPICE ▶	Other ▶	Leakage Path
Event Control ▶	Motor Drive Module ▶	Air Gap
SimCoder ▶	MagCoupler Module ▶	Air Gap (AL)
Symbols ▶	MagCoupler-RT Module ▶	Linear Core
Page ▶	Mechanical Loads and Sensors ▶	Saturable Core
Typhoon-HIL ▶	Thermal Module ▶	
	Renewable Energy Module ▶	
	EMI Design Suite ▶	

Fig. 5.539.

The settings of Saturable Core block are shown in Fig. 5.540. The B-H Curve tool in the Utilities menu can be used to set the block according to the manufacturer data sheet. Click the Help button in Fig. 5.540. The help page of Saturable Core block will appear. Scroll down until you see the tutorial link (Fig. 5.541). This tutorial shows how to define saturable core parameters.

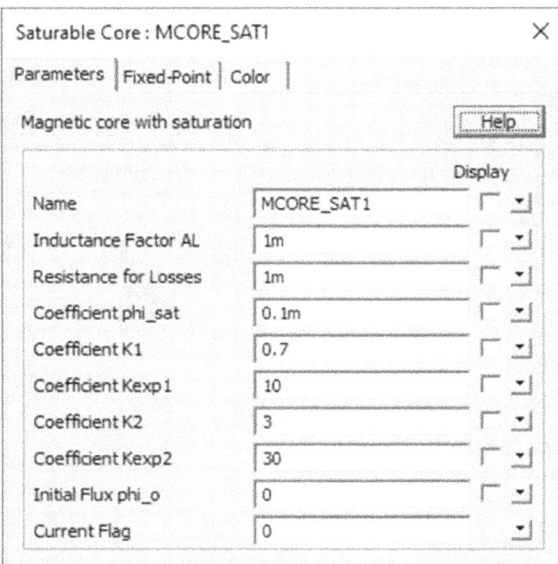

Saturable Core : MCORE_SAT1		×
Parameters ⎸ Fixed-Point ⎸ Color ⎸		
Magnetic core with saturation		Help
		Display
Name	MCORE_SAT1	▢ ▾
Inductance Factor AL	1m	▢ ▾
Resistance for Losses	1m	▢ ▾
Coefficient phi_sat	0.1m	▢ ▾
Coefficient K1	0.7	▢ ▾
Coefficient Kexp1	10	▢ ▾
Coefficient K2	3	▢ ▾
Coefficient Kexp2	30	▢ ▾
Initial Flux phi_o	0	▢ ▾
Current Flag	0	▾

Fig. 5.540.

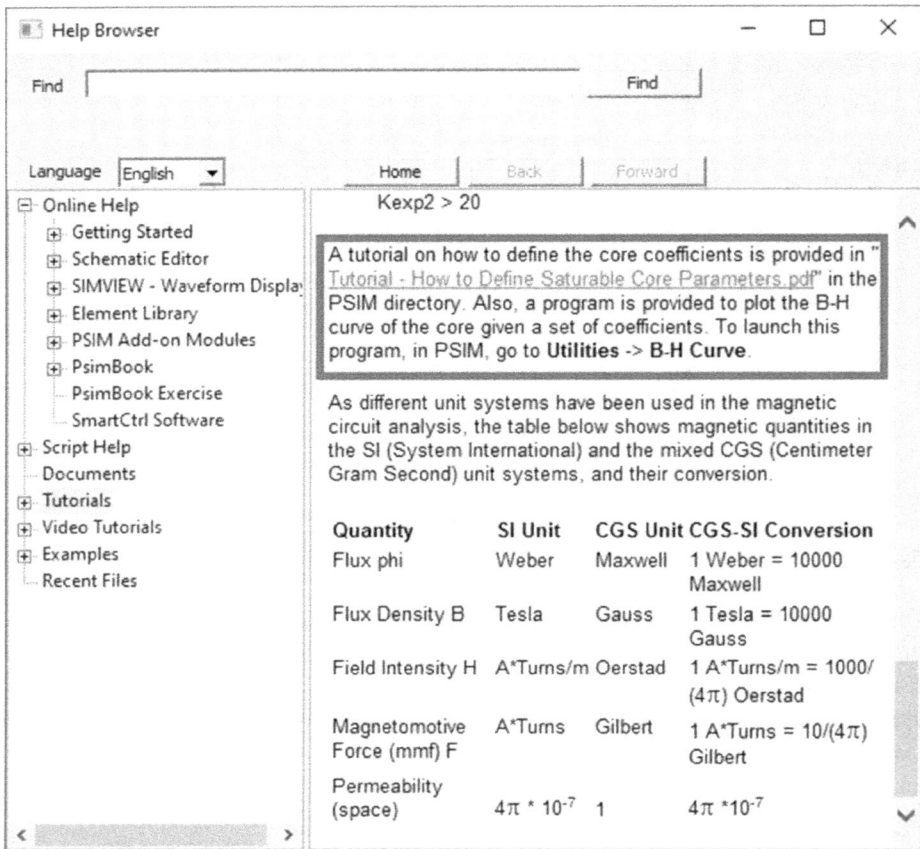

Fig. 5.541.

PSIM has some ready examples for magnetic components. In order to see the examples, click the File>Open Examples and then go to the Magnetic Modeling folder.

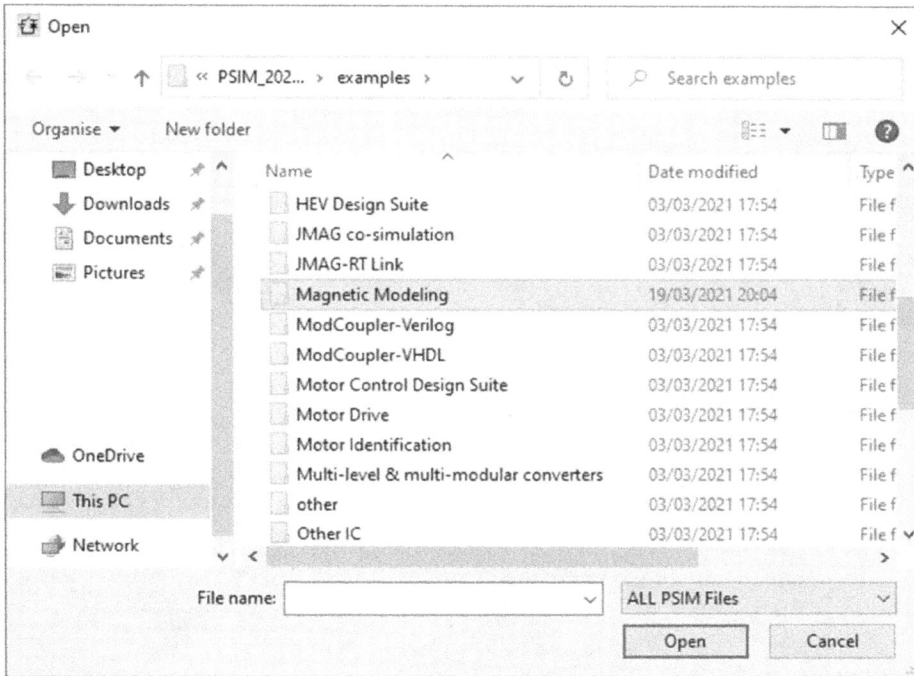

Fig. 5.542.

5.46 Example 45: code generation and thermal modules

PSIM can generate code for different targets boards. SimCoder™ generates high quality, consistent C code from a PSIM control schematic automatically – free from human error.

The tutorials in the Help> Tutorials> Code Generation section (Fig. 5.543) and [28, 29] are good references to learn the details of SimCoder module.

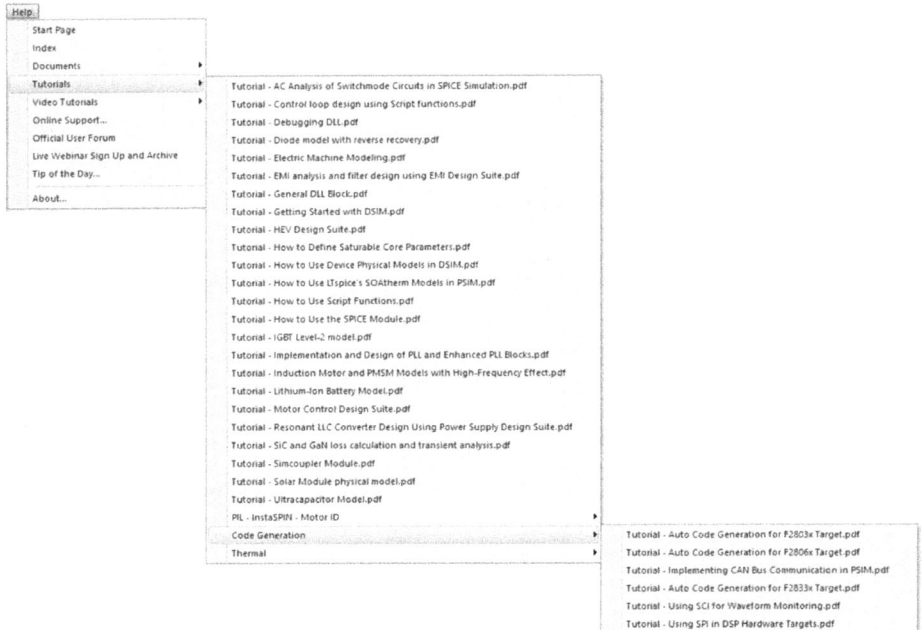

Fig. 5.543.

The Thermal Module is another add-on option to PSIM. Its purpose is to simulate the losses of semiconductor devices and inductors quickly from manufacturer device datasheets.

The tutorials in the Help> Tutorials> Thermal section (Fig. 5.544) and [30] are good references to learn the details of this module.

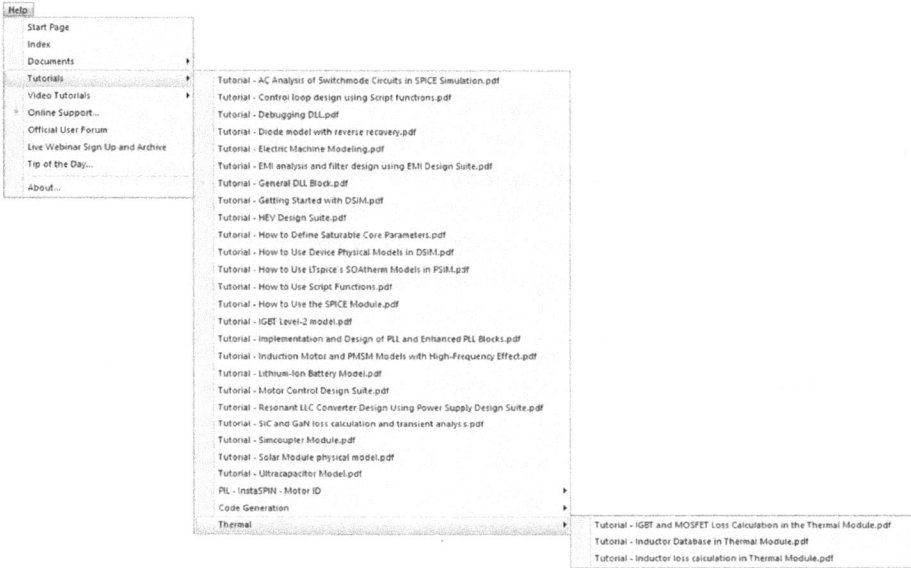

Fig. 5.544.

Chapter 6
Electrical machines

6.1 Introduction

Electrical machines and electric motor drives are important part of any industry. PSIM provides commonly used electric machine models, mechanical load models, and control blocks (such as Maximum-Torque-Per-Ampere Control and Field Weakening Control blocks). You can create and use custom-built machine or load models for greater flexibility.

This chapter is a brief introduction to motor drive module of PSIM.

6.2 Simulation of a loaded DC motor

A model of different types of motors (Fig. 6.1), mechanical loads, and sensors (Fig. 6.2) is available in PSIM. This simplifies the simulation of motor drive circuits.

Fig. 6.1.

https://doi.org/10.1515/9783110740653-006

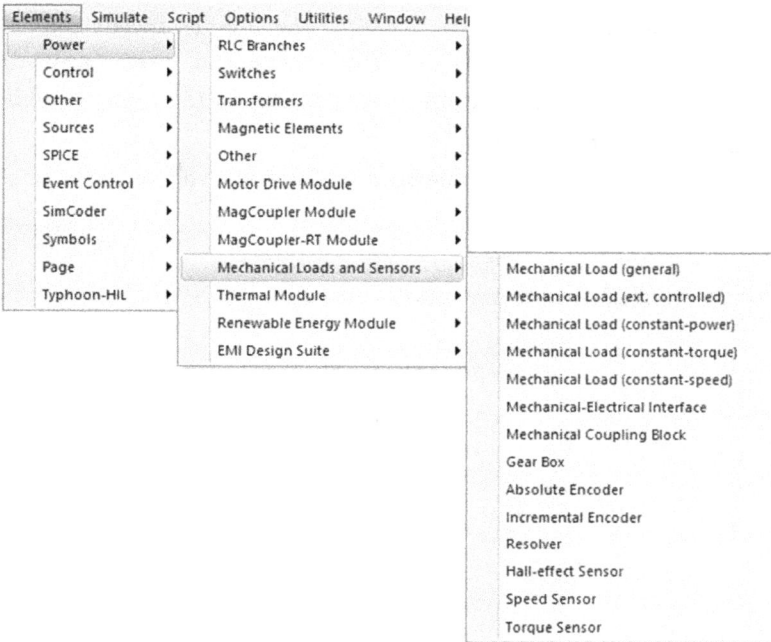

Fig. 6.2.

In this example, we want to simulate a DC motor with a load. Assume the schematic shown in Fig. 6.3. This schematic is composed of a DC Machine (Fig. 6.1), Speed Sensor, and Mechanical Load (general) blocks (Fig. 6.2).

Note that the Speed Sensor block measures the speed in revolutions per minute (RPM). If you multiply the output of speed sensor with , you will $\frac{2\pi}{60}$=0.1047 obtain the speed in Rad/s.

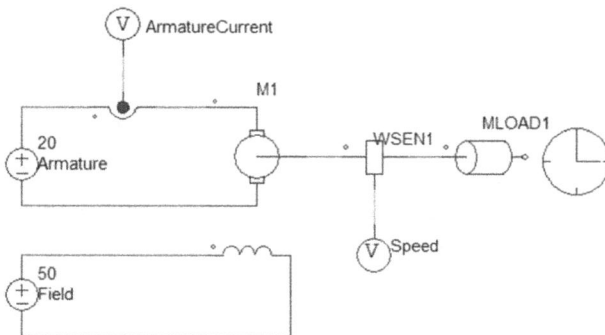

Fig. 6.3.

Settings of blocks are shown in Figs. 6.4–6.7. The torque–speed characteristic for mechanical load (general) is. The mechanical load (general)
$T=\text{sign}(\omega_m)\left(T_C+K_1|\omega_m|+K_2|\omega_m|^2+K_3|\omega_m|^3\right)$ is of resistive type. That is, the load torque is
always against the direction of the rotation.

By choosing the coefficients properly, one can model various types of mechanical
load, such as:

a) Conveyor (Tload = $K_1|\omega_m|$)

b) Fans, centrifugal pumps, and compressors (Tload = $K_2|\omega_m|^2$)

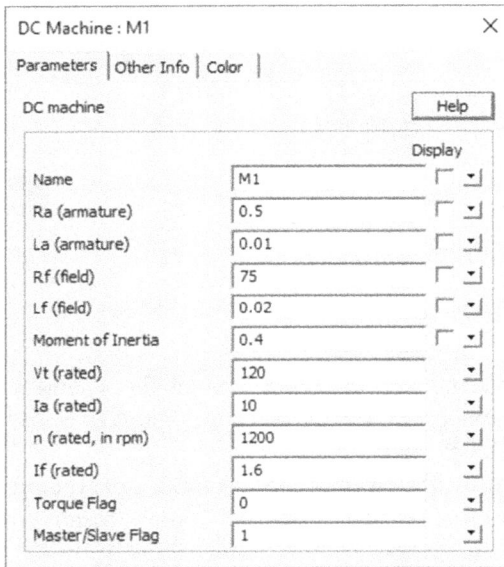

Fig. 6.4.

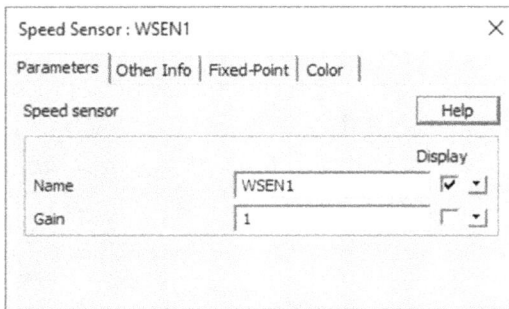

Fig. 6.5.

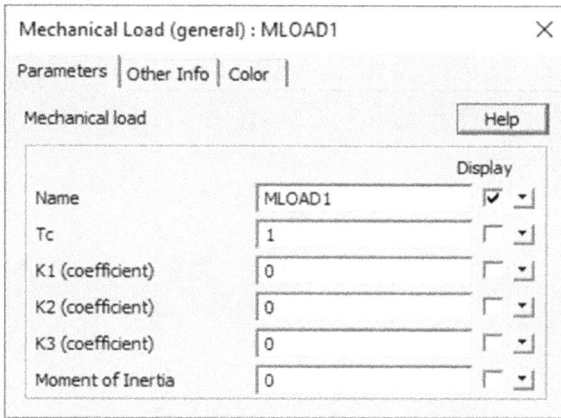

Fig. 6.6.

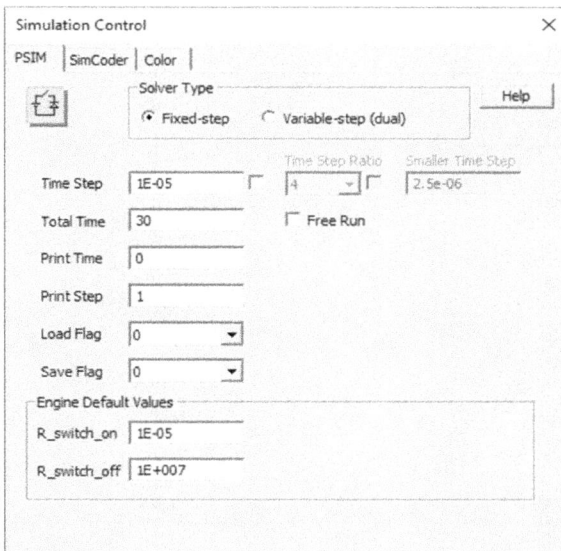

Fig. 6.7.

Run the simulation. The results (load speed and armature current) are shown in Figs. 6.8 and 6.9. Note the peak in the armature current. Such a peak is expected because at low speeds (during the time that machine shaft starts rotation), the back electromotive force (BEMF) is small and the only factor that restricts the current is the armature current. As the shaft speed increases, the BEMF increases and the armature current reaches its steady-state value.

Speed

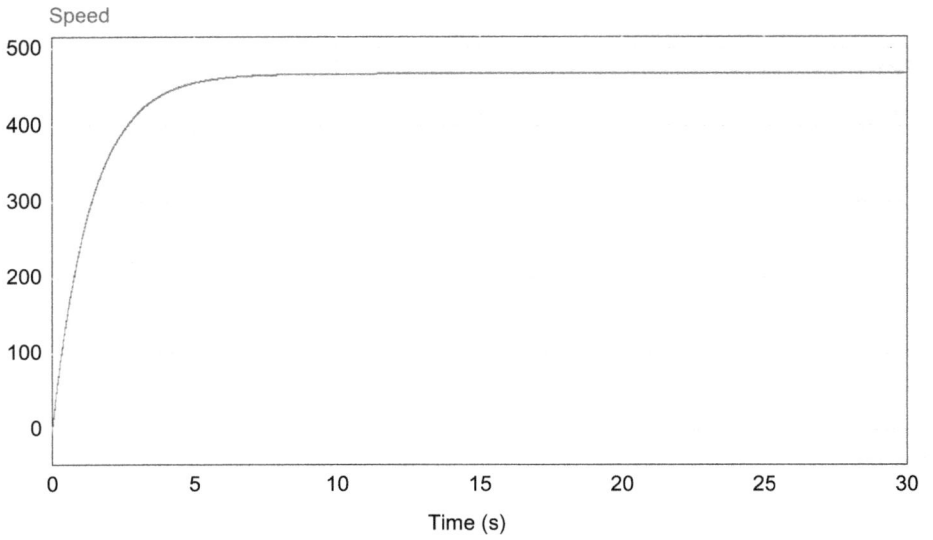

Fig. 6.8.

ArmatureCurrent

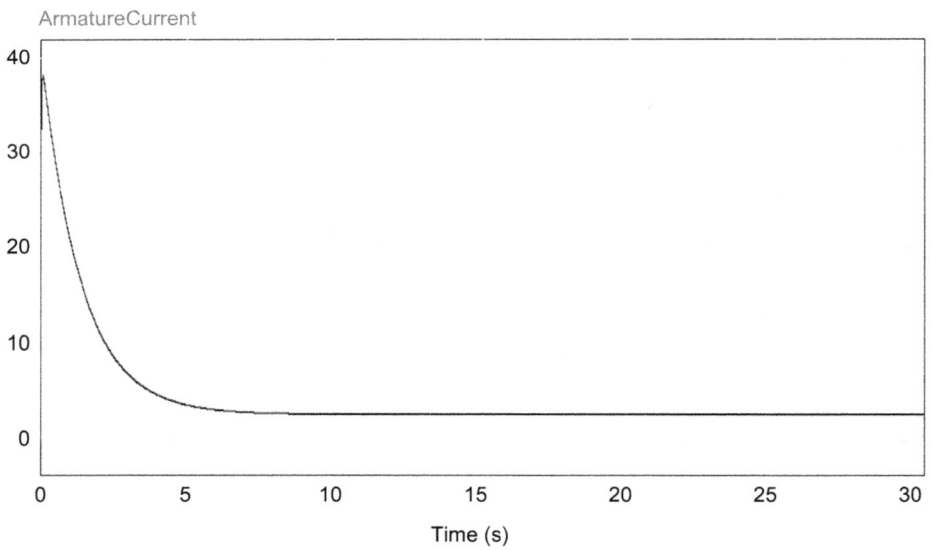

Fig. 6.9.

6.3 Reference direction of mechanical systems

In a motor drive system, in order to formulate equations for the mechanical system, a position notation needs to be defined. Take the motor drive system shown in Fig. 6.10 as an example.

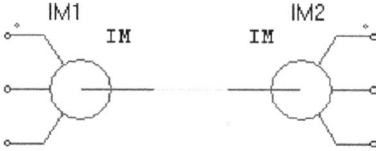

Fig. 6.10.

The system consists of two induction machines, IM1 and IM2, connected back to back. One operates as a motor, and the other as a generator. From the point of view of the first machine IM1, the mechanical equation can be written as:

$$\left(J_1+J_2\right)\frac{d\omega_{\mathrm{m}}}{dt}=T_{\mathrm{em1}}-T_{\mathrm{em2}}$$

where J_1 and J_2 are the moments of inertia, and T_{em1} and T_{em2} are the developed torques of the machine IM1 and IM2, respectively. From the point of view of the second machine IM2, however, the mechanical equation can be written as

$$\left(J_1+J_2\right)\frac{d\dot{\omega}_{\mathrm{m}}}{dt}=T_{\mathrm{em1}}-T_{\mathrm{em2}}$$

These two equations are equally valid, but will produce opposite mechanical speed (i.e. $\dot{\omega}_{\mathrm{m}}=-\omega_{\mathrm{m}}$).

In order to avoid this ambiguity, in PSIM, the concept "reference direction" is used in the mechanical system so that the mechanical equation can be uniquely defined.

In a mechanical system, one element is designated as the master unit (this element is considered to operate in the master mode), and the rest of the elements are in the slave mode. Elements that can be master units are: Electric machines, mechanical-to-electrical interface blocks, and gearboxes.

The master unit defines the reference direction of the mechanical system. The direction is defined as the direction from the shaft node of the master unit, along the shaft, to the rest of the mechanical system.

Once the reference direction of the mechanical system is defined, the speed and torque reference of the mechanical system can be defined. For example, if we use the right-hand method, with the thumb pointing in the reference direction of the mechanical system, by rotating the right hand, the fingers will point to the positive direction of the speed and the torque.

Moreover, each mechanical element has its own reference direction. Figure 6.11 shows the reference direction of each mechanical element, as indicated by the arrow.

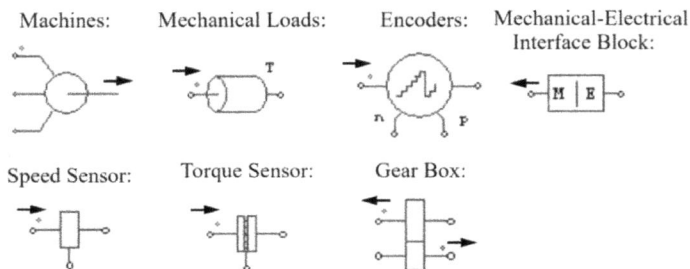

Fig. 6.11.

The reference direction of each element and the reference direction of the overall mechanical system determine how the element interacts with the mechanical system.

For example, if the reference direction of a machine is along the same direction as the reference direction of the mechanical system, the developed torque of the machine will contribute to the shaft rotation in the positive direction. However, if the reference direction of the machine is opposite to that of the mechanical system, the developed torque will contribute to the shaft rotation in the negative direction.

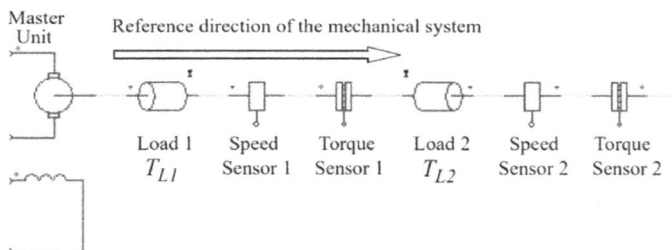

Fig. 6.12.

In this mechanical system, the machine on the left is the master unit. The reference direction of the mechanical system is from left to the right along the mechanical shaft. Comparing this direction with the reference direction of each element, Load 1, Speed Sensor 1, and Torque Sensor 1, will be along the reference direction, and Load 2, Speed Sensor 2, and Torque Sensor 2 will be opposite to the reference direction of the mechanical system.

Therefore, if the speed of the machine is positive, Speed Sensor 1 reading will be positive, and Speed Sensor 2 reading will be negative.

Similarly, the two constant-torque mechanical loads, with the amplitudes of TL1 and TL2, interact with the machine in different ways. Load 1 is along the reference direction, and the loading torque of Load 1 to the master machine will be TL1. On the other hand, Load 2 is opposite to the reference direction, and the loading torque of Load 2 to the machine will be -TL2.

6.4 Measurement of motor efficiency

You can measure the efficiency of the system with the aid of schematic shown in Fig. 6.13.

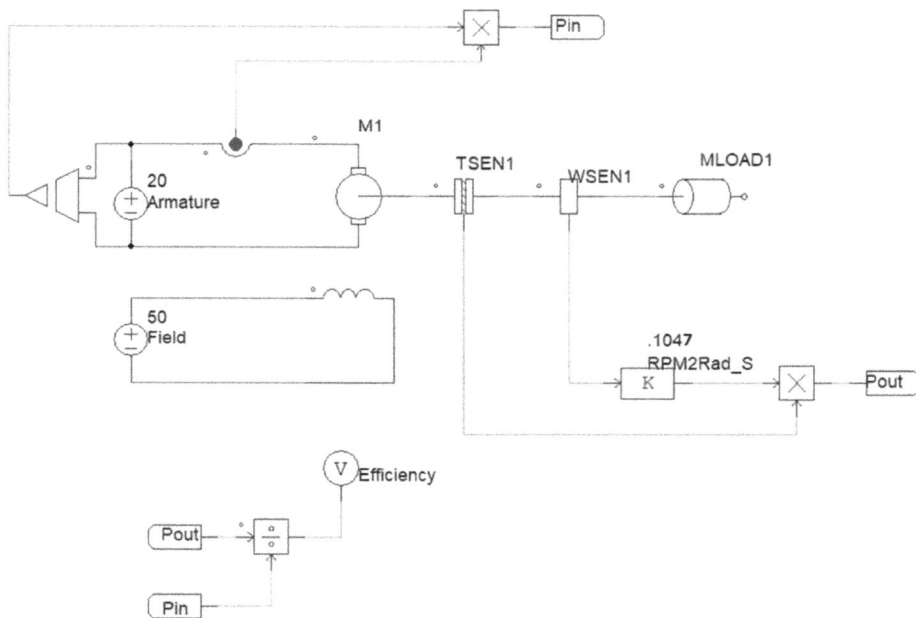

Fig. 6.13.

After running the simulation, the result shown in Fig. 6.14 will be obtained. According to Fig. 6.15, the efficiency of the system is 93%.

Efficiency

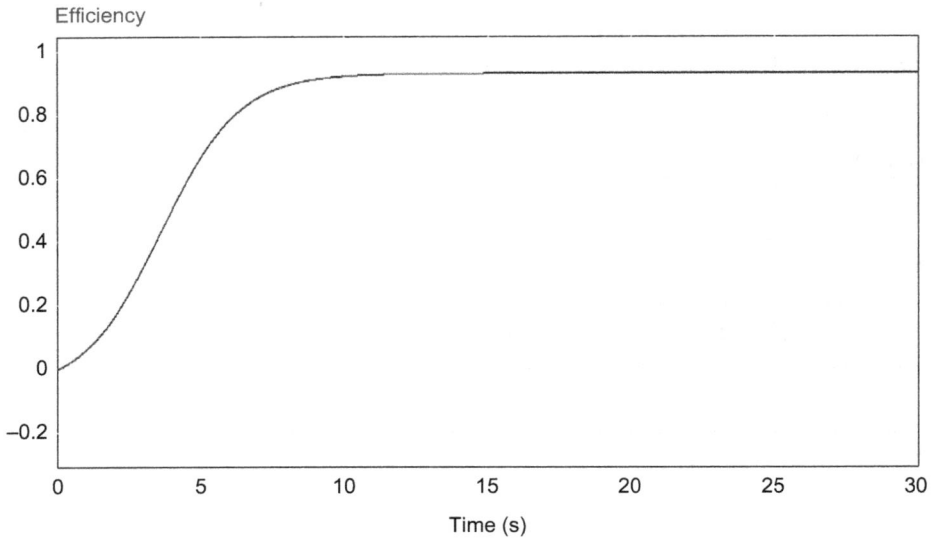

Fig. 6.14.

Measure				x
		X1	X2	Δ
Time		2.64665e+01	2.79908e+01	1.52425e+00
Efficiency		9.32473e-01	9.32473e-01	3.24290e-08

Fig. 6.15.

6.5 Generator simulation

In this section, we will study the simulation of a DC generator. We want to couple two DC machines together and use one of them as motor and the other one as generator to feed the load. We need a coupling block (Fig. 6.16) in order to couple the machines shaft.

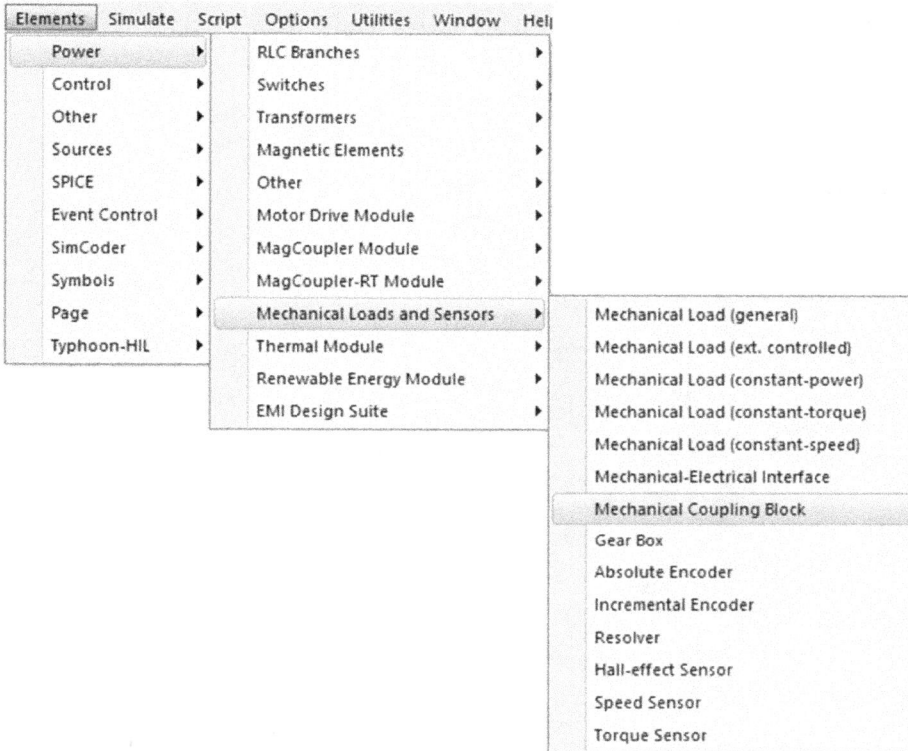

Fig. 6.16.

Assume the schematic shown in Fig. 6.17.

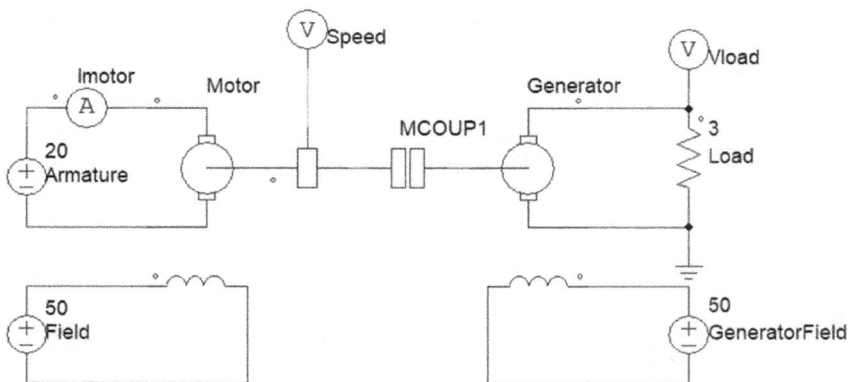

Fig. 6.17.

Settings of the used blocks are shown in Figs. 6.18 and 6.19.

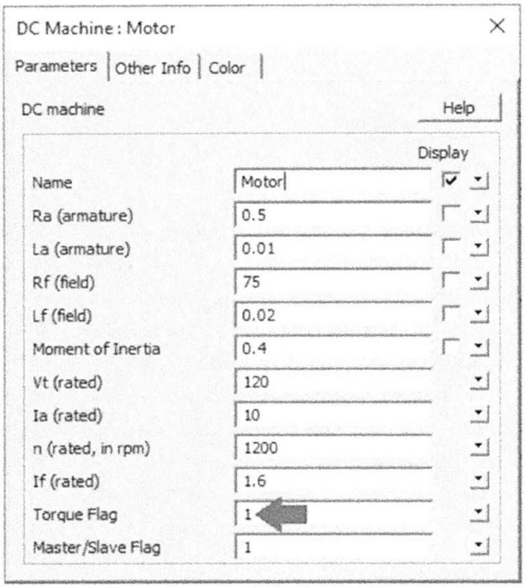

Fig. 6.18.

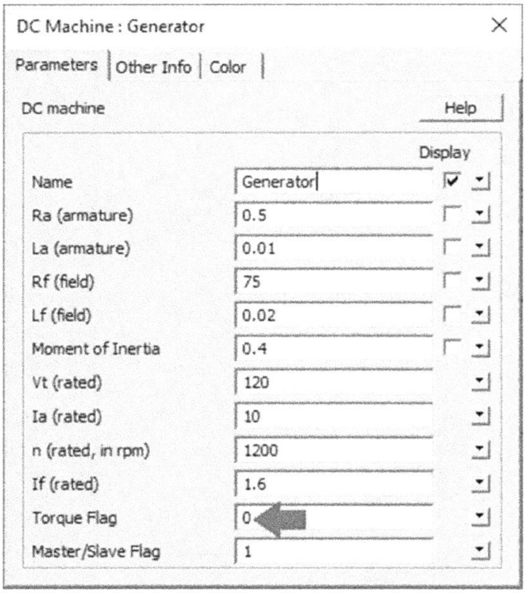

Fig. 6.19.

Run the simulation. The load voltage is shown in Fig. 6.20. The DC machine generated about 15 V across the 3 Ω load.

Vload

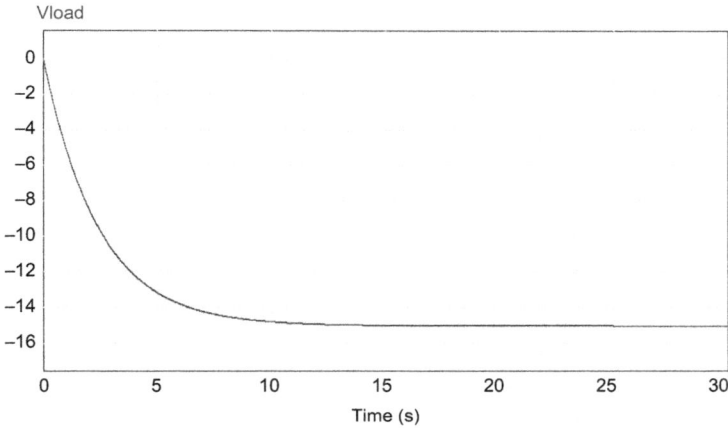

Fig. 6.20.

There is another way to simulate a generator: Using the Mechanical Load (constant-speed) block (Fig. 6.21). You can rotate the shaft of a generator with a Mechanical Load (constant-speed) block. The rotation speed is determined by the user.

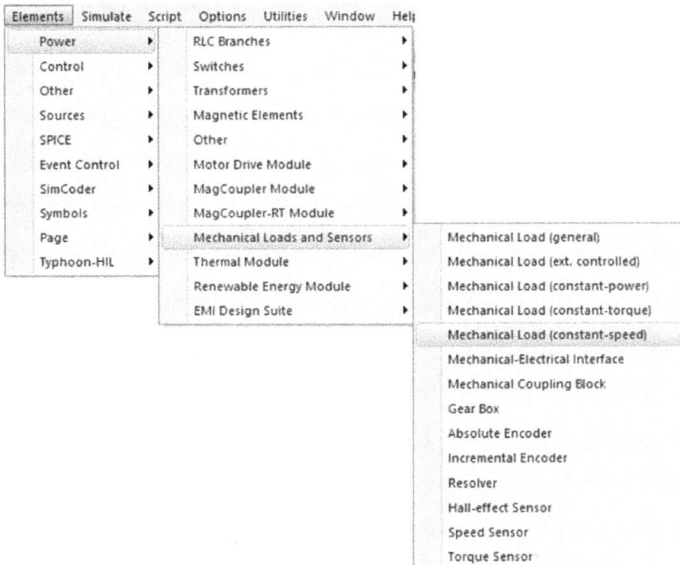

Fig. 6.21.

Assume the schematic shown in Fig. 6.22.

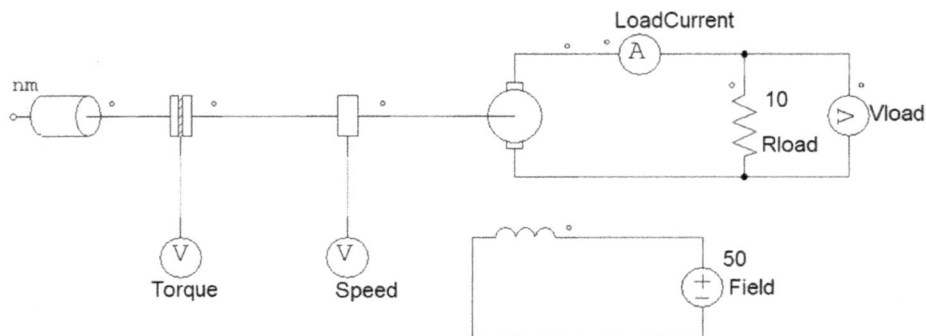

Fig. 6.22.

Settings of Mechanical Load (constant-speed) block are shown in Fig. 6.23.

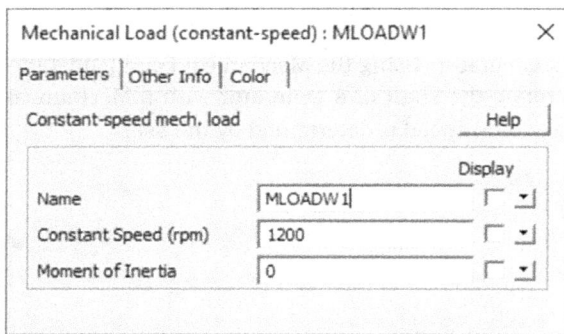

Fig. 6.23.

Run the simulation. The 10 Ω load voltage reached 45.6 V.

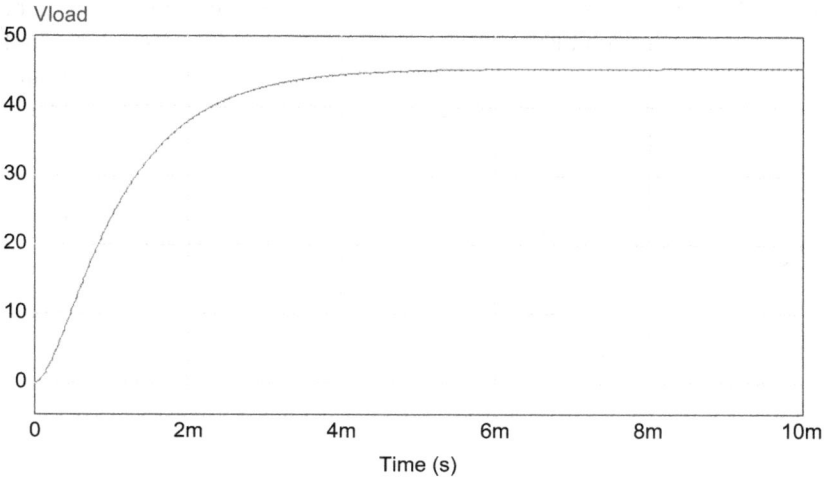

Fig. 6.24.

Double click on the Mechanical Load (constant-speed) block and set the speed to 2,400 RPM. Run the simulation. In this case, the result shown in Fig. 6.25 will be obtained. The load voltage reached 91.25 which is twice the voltage for 1,200 RPM.

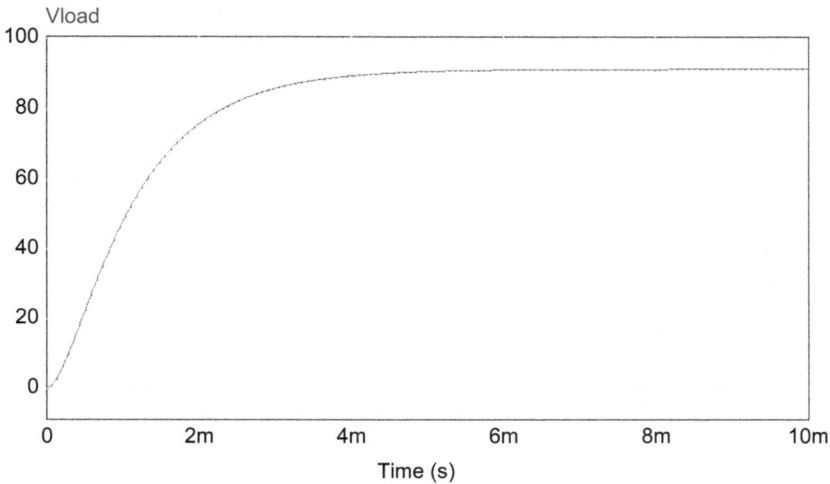

Fig. 6.25.

6.6 Efficiency of generator

The efficiency of generator system shown in Fig. 6.22 is measured in this section. Assume the schematic shown in Fig. 6.26. The input mechanical power is measured with the aid of $P_{in} = \omega \times \tau$. Settings of Speed Sensor are shown in Fig. 6.27. The Gain box is equal to 0.1047, so the output of the block speed is in Rad/s. The load power is measured with a Wattmeter block.

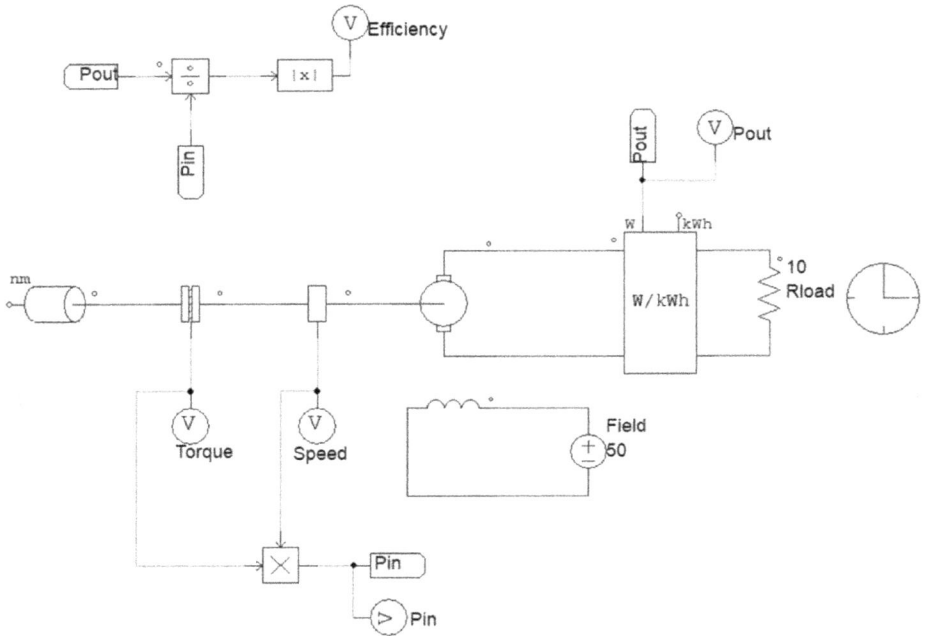

Fig. 6.26.

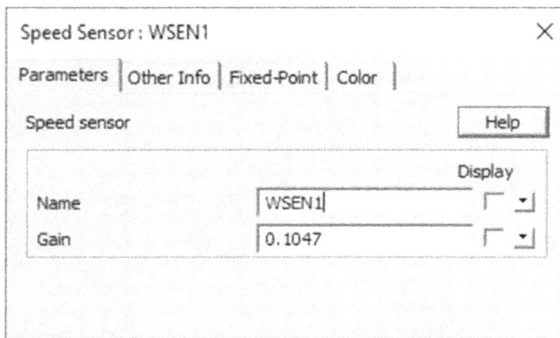

Fig. 6.27.

Since the input power is negative, we need to use an absolute value block (Fig. 6.28) after the division to see a positive value on the graph.

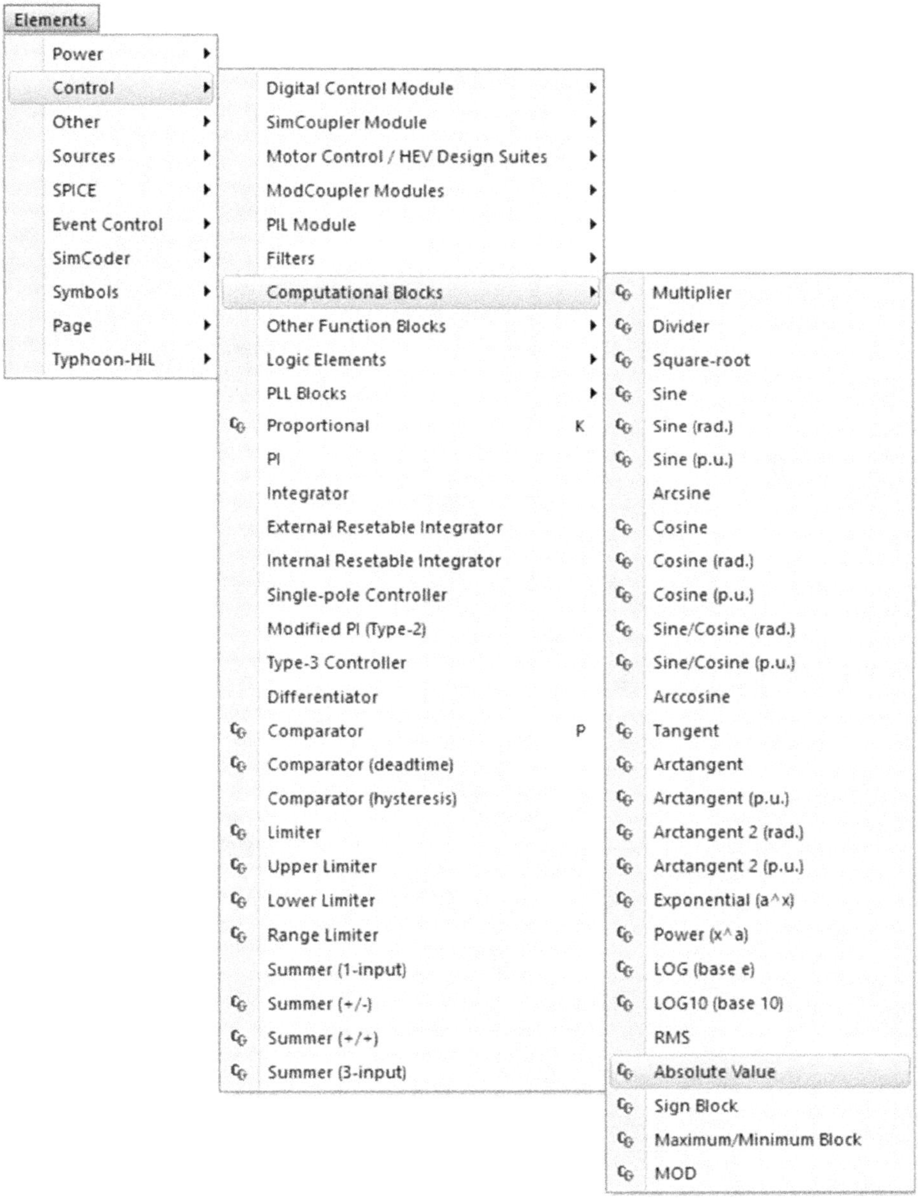

Fig. 6.28.

Run the simulation. The result shown in Fig. 6.29 is obtained. Zoom into the steady-state region. As shown in Fig. 6.30, the steady-state value is around 0.95. So, the efficiency is around 95%.

Efficiency

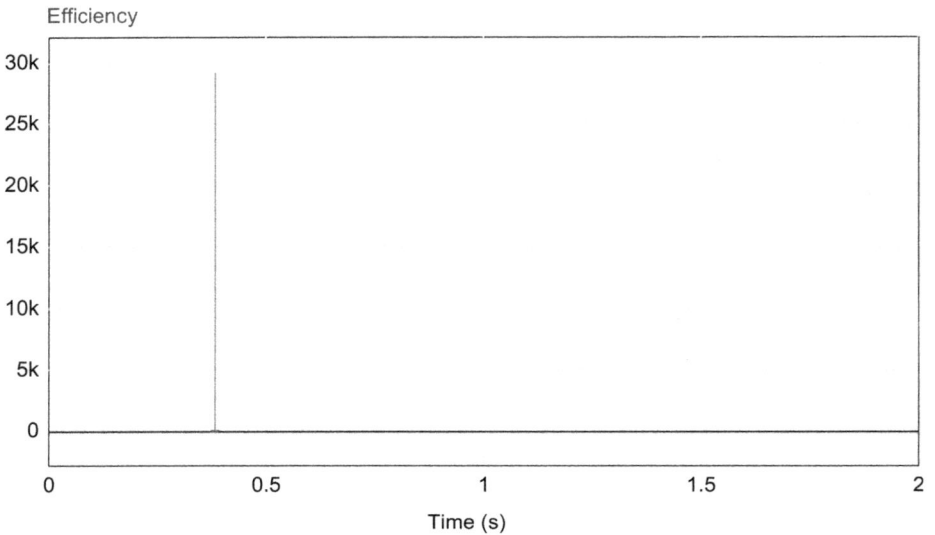

Fig. 6.29.

Efficiency

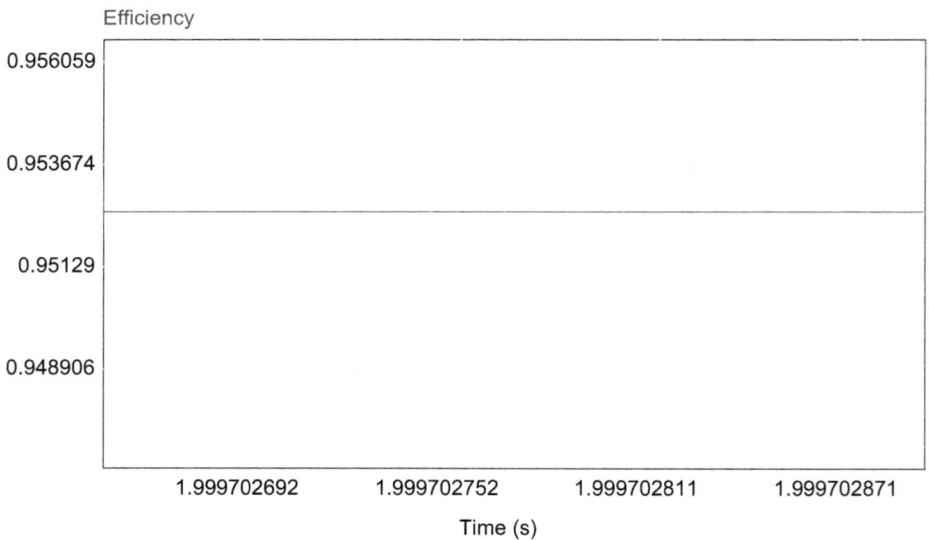

Fig. 6.30.

You can see the input and output powers as well. According to Fig. 6.31, the steady-state input power and output power are around 218 and 208 W, respectively.

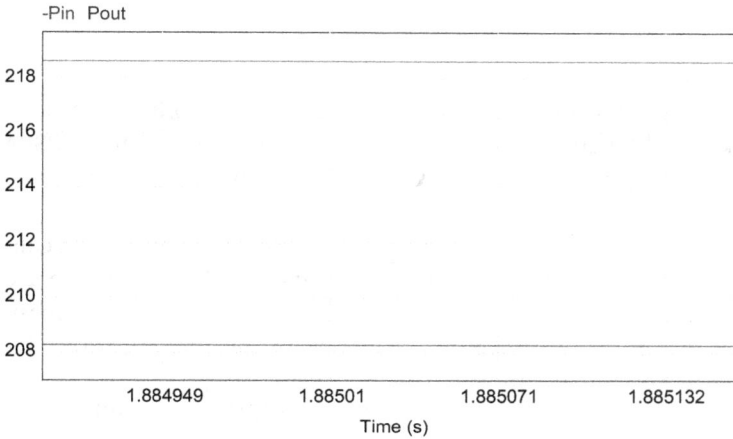

Fig. 6.31.

PSIM has many examples related to electrical machines. In order to see these examples, click the File>Open Examples and see the folders shown in Fig. 6.32.

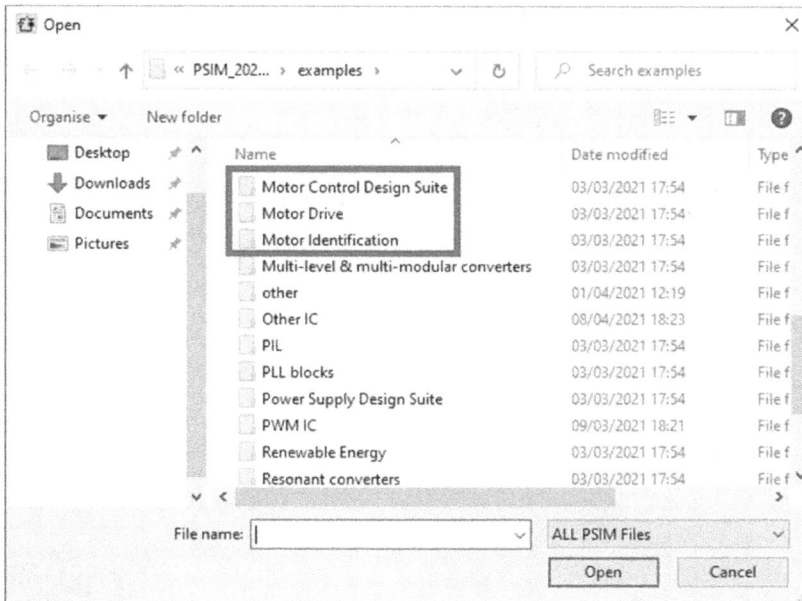

Fig. 6.32.

Chapter 7
SimCoupler™

7.1 Introduction

Simulink® is one of the most powerful simulation programs in the world. It has many ready-to-use blocks for different branches of engineering. SimCoupler fuses PSIM with Simulink by providing an interface for co-simulation. Part of a system can be implemented and simulated in PSIM and the rest in Simulink.

For instance, assume that you want to simulate a converter with a fuzzy logic controller. In this case, you can draw the power circuit schematic in PSIM and then use the Simcoupler and MATLAB's Fuzzy Logic Toolbox™ in order to simulate the fuzzy logic controller. So, with the aid of SimCoupler, all the blocks of Simulink are available for you and there is no need to start your work from zero.

In this chapter, we will study the co-simulation between PSIM and Simulink. First of all, we will show how to set up the Simcoupler. Then, we will study an example.

7.2 Setting up the SimCoupler

You need to set up the SimCoupler before using it in the Simulink environment. Click the SimCoupler Setup in order to set up the SimCoupler.

Fig. 7.1.

https://doi.org/10.1515/9783110740653-007

The window shown in Fig. 7.2 will appear. Check the box and click OK. PSIM set up the SimCoupler and after successful set up shows the message shown in Fig. 7.3.

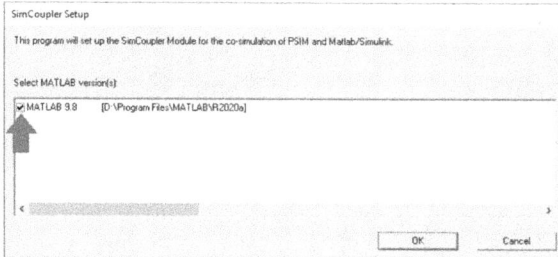

Fig. 7.2.

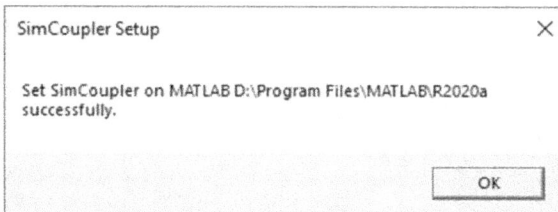

Fig. 7.3.

Now go to the Simulink. A new section with title of "S-function SimCoupler" is added to Simulink Library Browser (Fig. 7.4). Open a new simulation file in the Simulink environment and add the SimCoupler block to it. Double click on the block. The window shown in Fig. 7.5 will appear. Use the Browse button in order to add the PSIM schematic file.

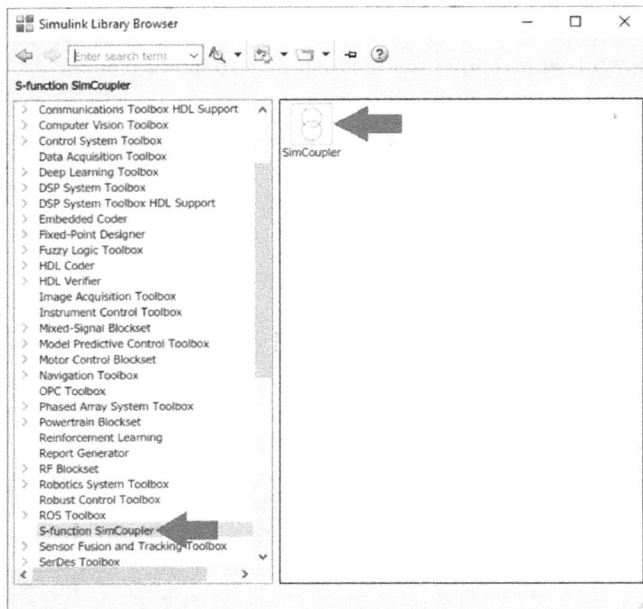

Fig. 7.4.

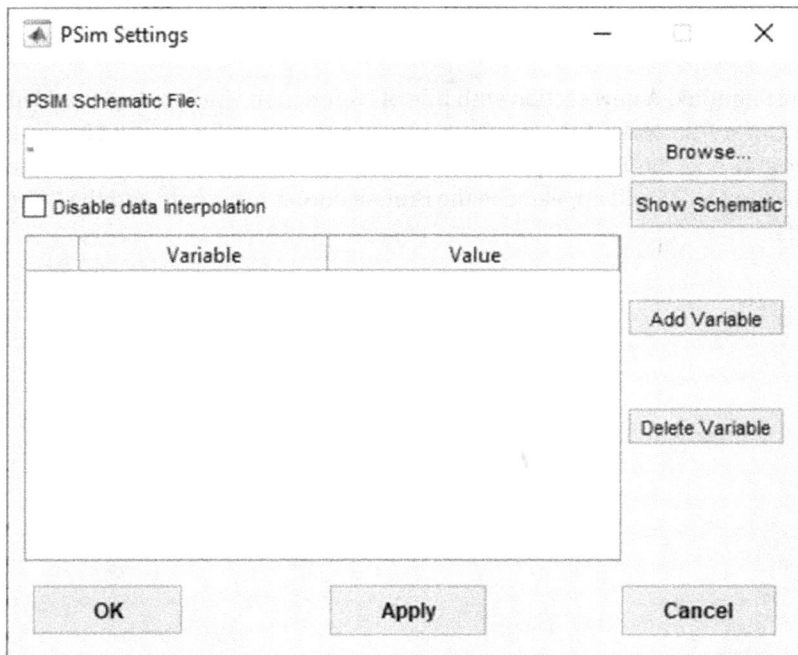

Fig. 7.5.

7.3 Preparing the PSIM model

In this section, we want to simulate a closed-loop flyback converter. The power circuit is designed in PSIM and the controller will be implemented in the Simulink environment.

The PSIM schematic must have the In Link Node and Out Link Node blocks (Fig. 7.6) in order to communicate with Simulink.

The In Link Node is used in PSIM to provide interface to Simulink. An In Link Node receives a value from Simulink, and the node behaves as a voltage source in PSIM.

The Out Link Node is used in PSIM to provide interface to Simulink. An Out Link Node passes a value from PSIM to Simulink.

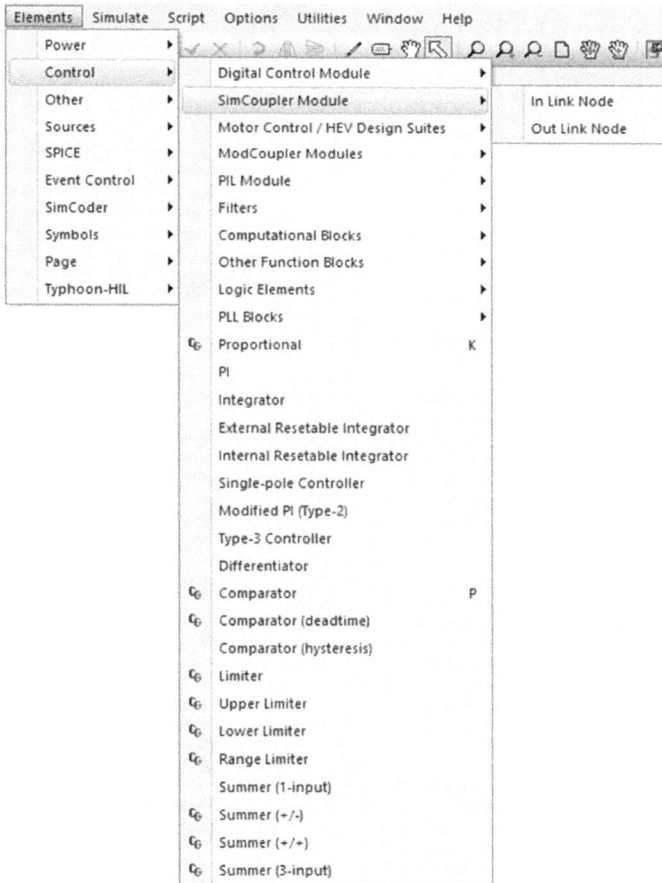

Fig. 7.6.

Assume the schematic shown in Fig. 7.7.

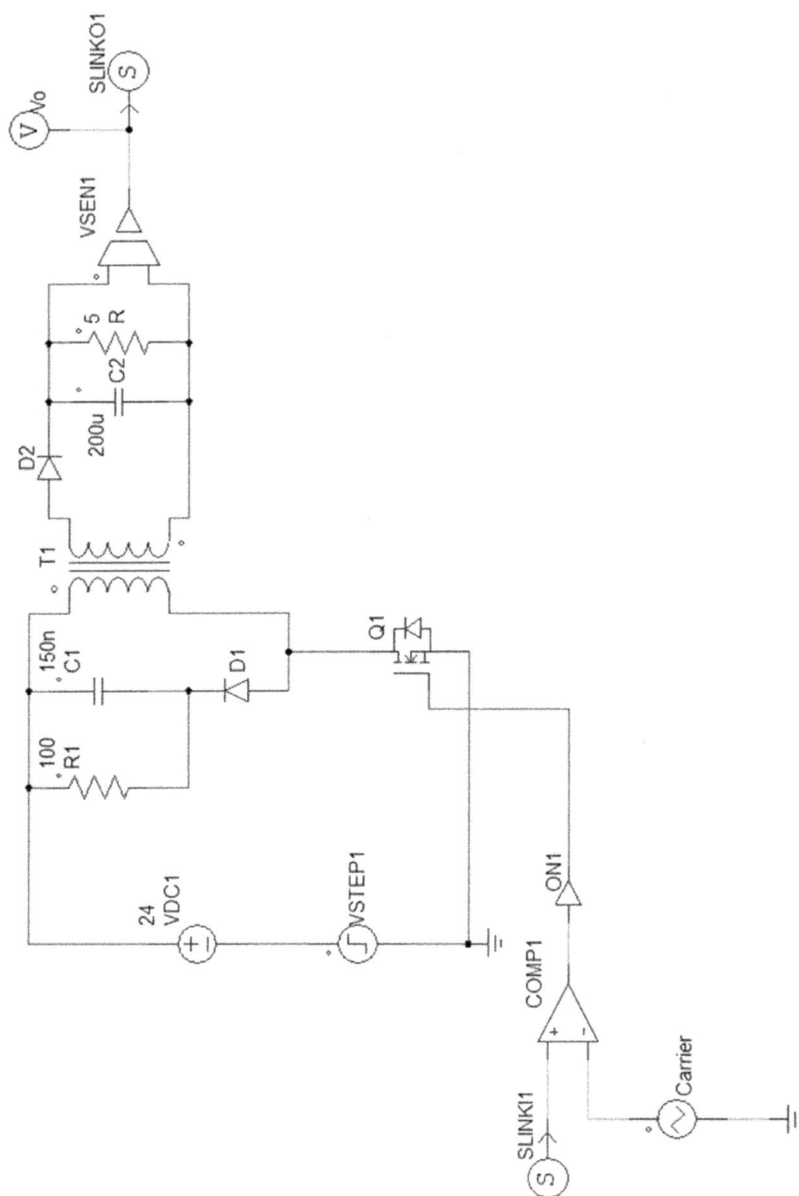

Fig. 7.7.

The settings of blocks are shown in Figs. 7.8–7.15.

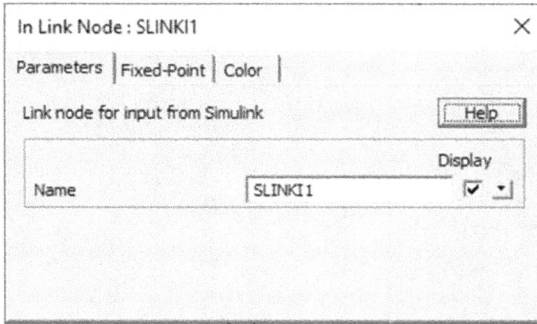

Fig. 7.8.

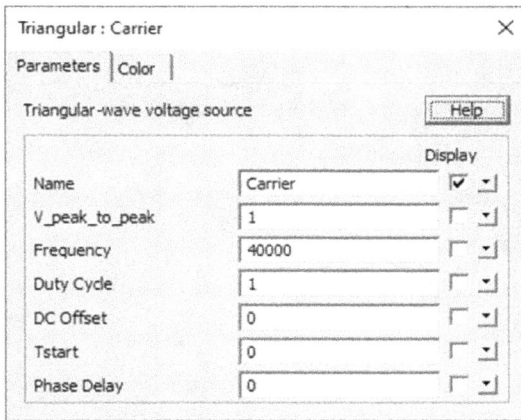

Fig. 7.9.

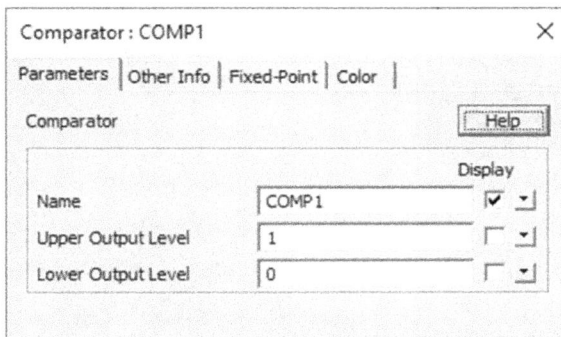

Fig. 7.10.

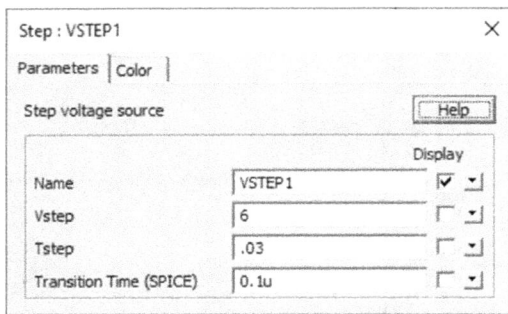

Fig. 7.11.

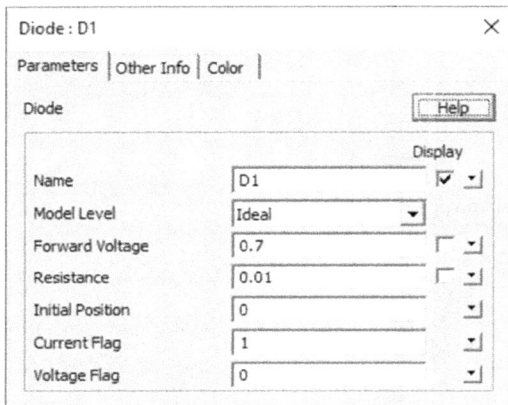

Fig. 7.12.

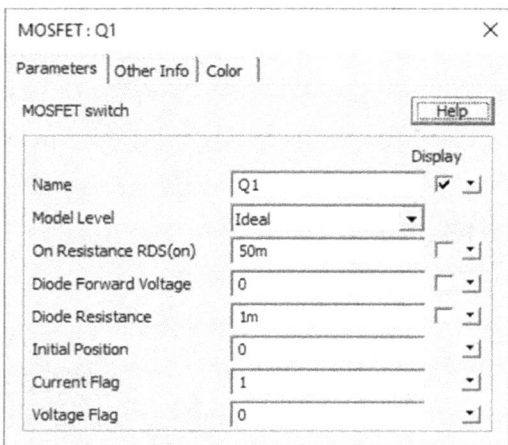

Fig. 7.13.

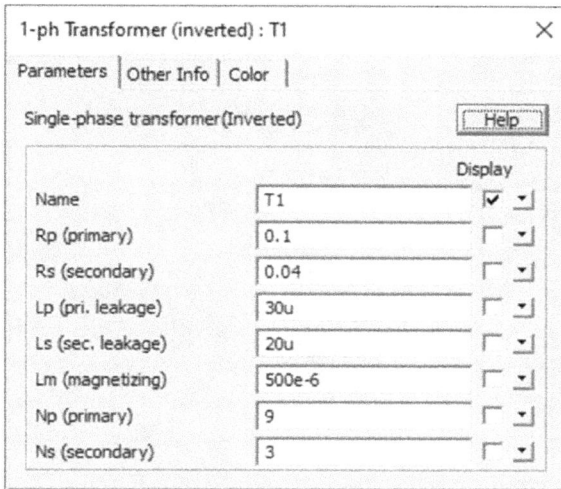

Fig. 7.14.

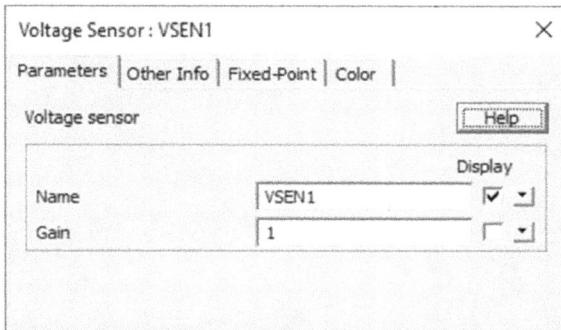

Fig. 7.15.

7.4 Preparing the Simulink model

After preparing the PSIM schematic, save it in desired location and go to Simulink. Add a SimCoupler block to the Simulink model. Double click the SimCoupler block and click the Browse button in order to select the schematic that you draw in the previous step. Then click the OK button. After clicking the OK button, Simulink adds the input output ports to the block.

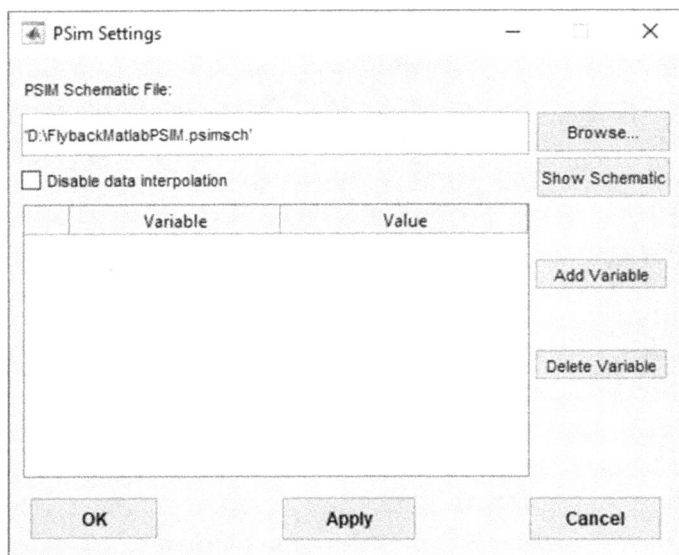

Fig. 7.16.

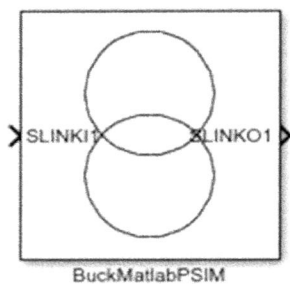

BuckMatlabPSIM

Fig. 7.17.

Draw the closed-loop control system shown in Fig. 7.18.

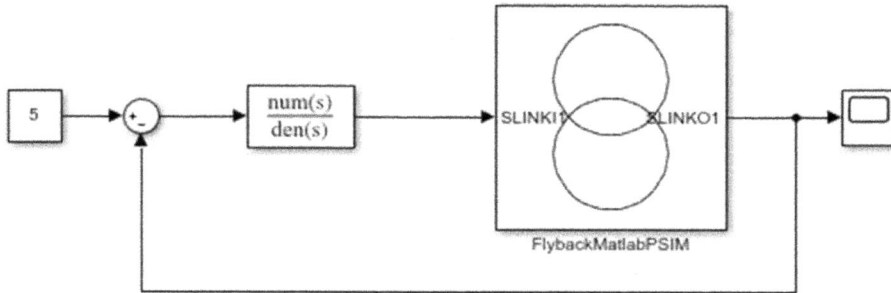

Fig. 7.18.

The settings of used blocks are shown in Figs. 7.19 and 7.20.

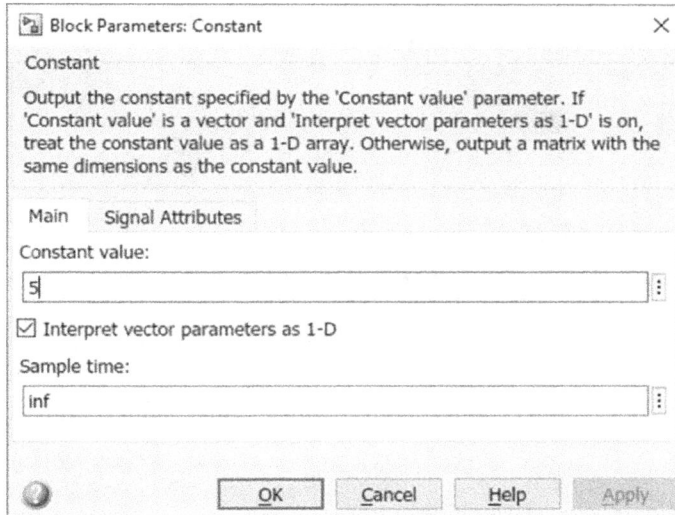

Fig. 7.19.

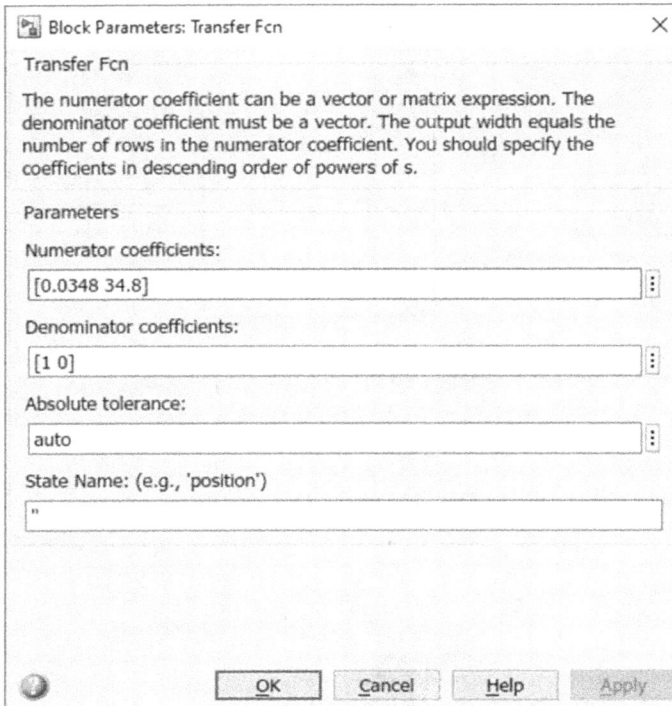

Fig. 7.20.

Now go to MODELING tab of Simulink and click the Model Settings (Fig. 7.21). You can use the Ctrl+E shortcut key as well. The window shown in Fig. 7.22 will appear. Use the desired solver in the Solver selection section. The switching frequency of this circuit is 40 kHz (switching period is 25 μs). So, time steps of less than $\frac{25\mu s}{10}=2.5\mu s$ are good. Let's set it to 1 μs. Keep in mind that selection of suitable solver is an important step in co-simulations.

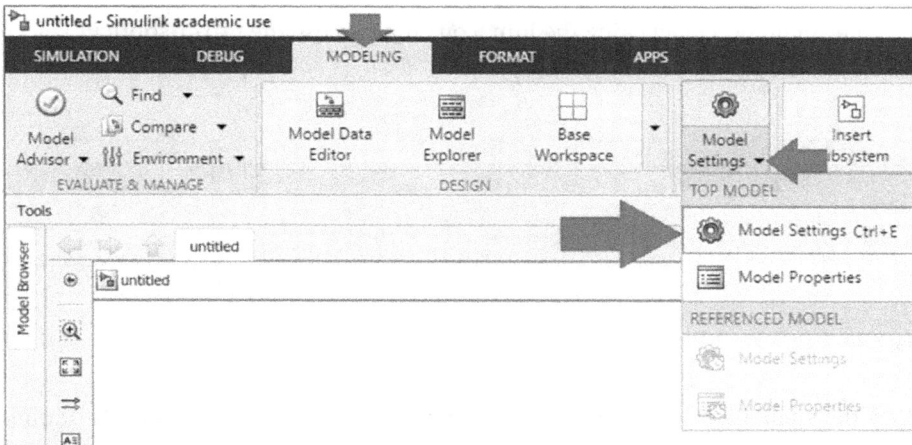

Fig. 7.21.

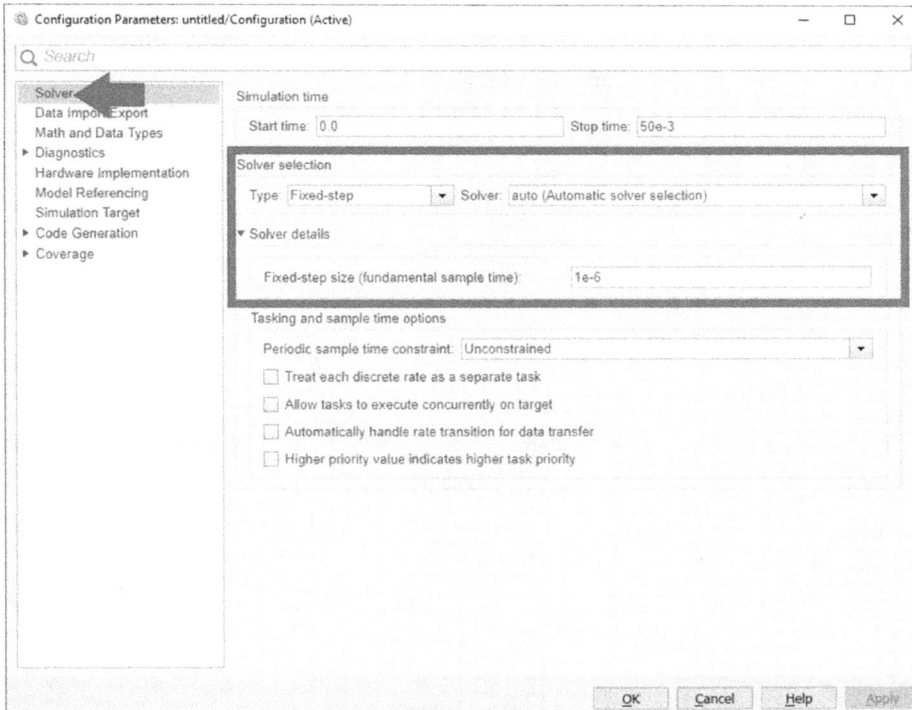

Fig. 7.22.

Run the simulation by clicking the Run icon. You can use the Ctrl+T shortcut key as well.

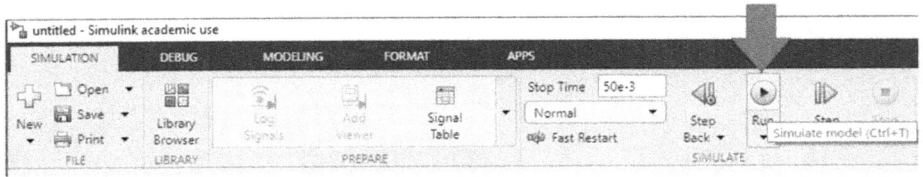

Fig. 7.23.

Once the simulation is completed, you can analyze the waveforms in PSIM or Simulink (Figs. 7.24 and 7.25).

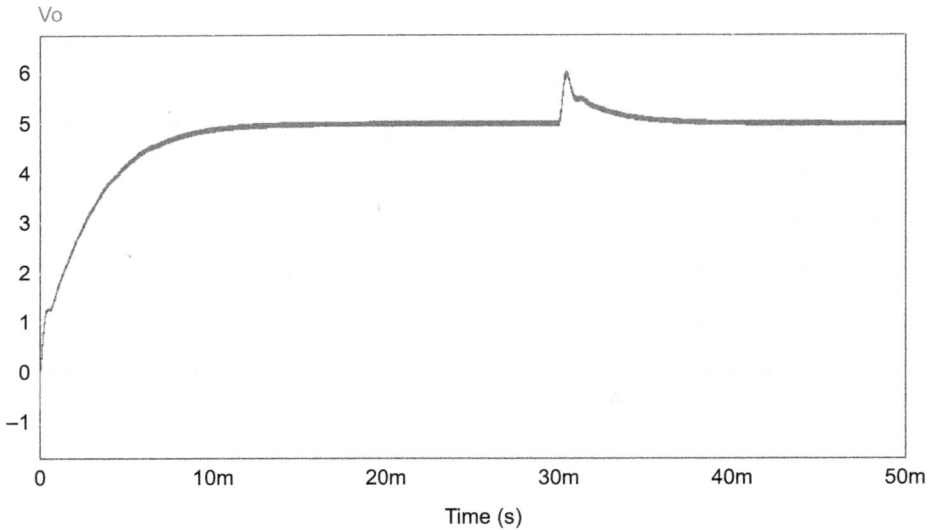

Fig. 7.24.

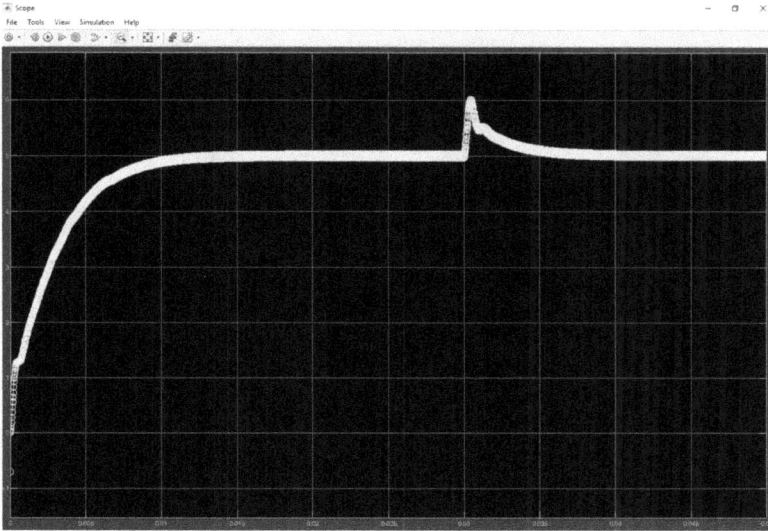

Fig. 7.25.

If you zoom in the waveform shown in Simulink (Fig. 7.26), you can see that the difference between any consecutive points of the graph is 1 μs (remember that we asked Simulink to do the simulation with 1 μs steps).

0.02127

0.021271

Fig. 7.26.

Chapter 8
SmartCtrl

8.1 Introduction

With a user-friendly interface, simple workflow, and easy-to-understand display of control loop stability, SmartCtrl is an indispensable tool in your design and simulation arsenal. It seamlessly integrates with PSIM, allowing you to expand the possibilities of your power electronics design.

SmartCtrl is introduced in this chapter. Control problem of a buck converter is studied as an example in this chapter.

8.2 Power stage of the converter

Assume the schematic shown in Fig. 8.1. There is no feedback loop in this converter. The small resistors in series with the inductor and capacitor show the equivalent series resistance of components. The switching frequency of this converter is 100 kHz and the duty cycle is 50%.

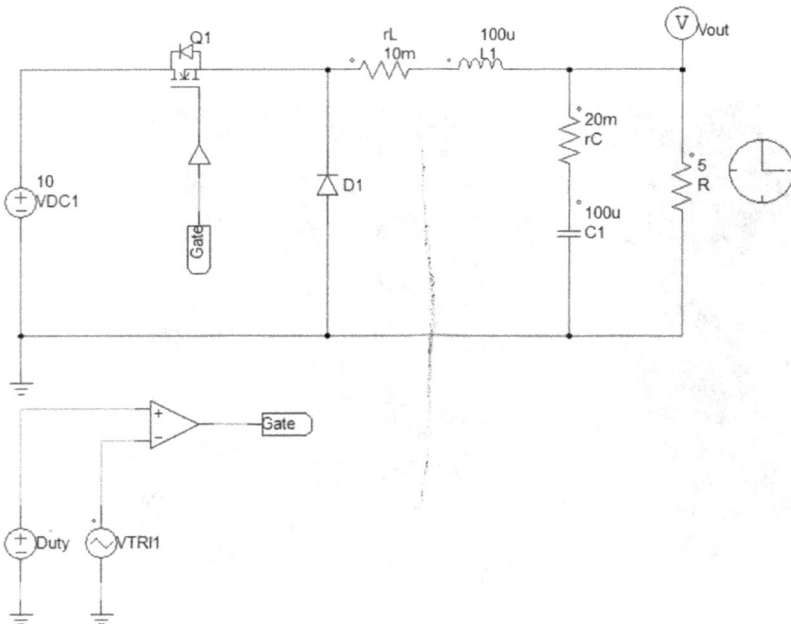

Fig. 8.1.

https://doi.org/10.1515/9783110740653-008

The settings of the blocks are shown in Figs. 8.2–8.6.

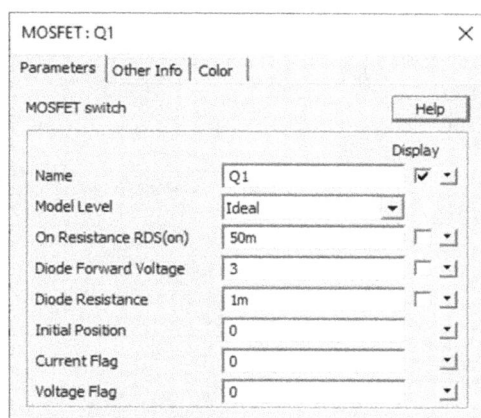

MOSFET : Q1   ✕

Parameters | Other Info | Color

MOSFET switch   Help

		Display
Name	Q1	☑ ▾
Model Level	Ideal ▾	
On Resistance RDS(on)	50m	☐ ▾
Diode Forward Voltage	3	☐ ▾
Diode Resistance	1m	☐ ▾
Initial Position	0	▾
Current Flag	0	▾
Voltage Flag	0	▾

Fig. 8.2.

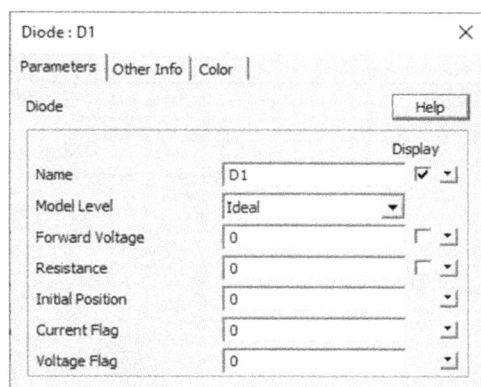

Diode : D1   ✕

Parameters | Other Info | Color

Diode   Help

		Display
Name	D1	☑ ▾
Model Level	Ideal ▾	
Forward Voltage	0	☐ ▾
Resistance	0	☐ ▾
Initial Position	0	▾
Current Flag	0	▾
Voltage Flag	0	▾

Fig. 8.3.

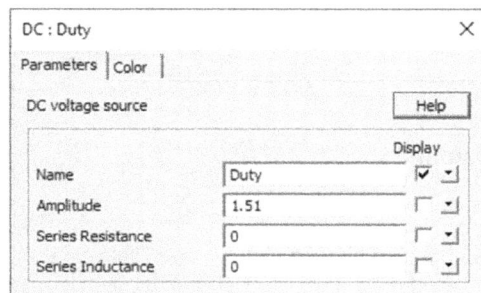

DC : Duty   ✕

Parameters | Color

DC voltage source   Help

		Display
Name	Duty	☑ ▾
Amplitude	1.51	☐ ▾
Series Resistance	0	☐ ▾
Series Inductance	0	☐ ▾

Fig. 8.4.

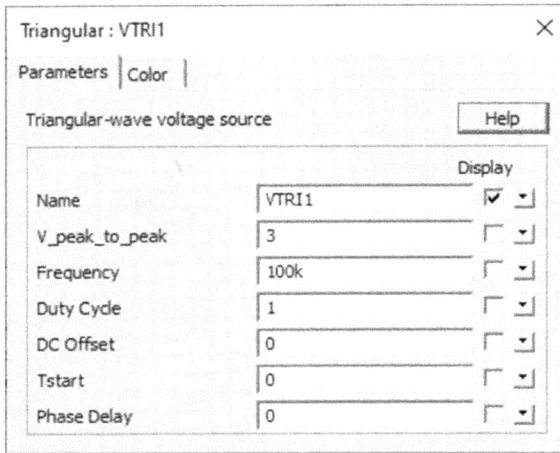

Fig. 8.5.

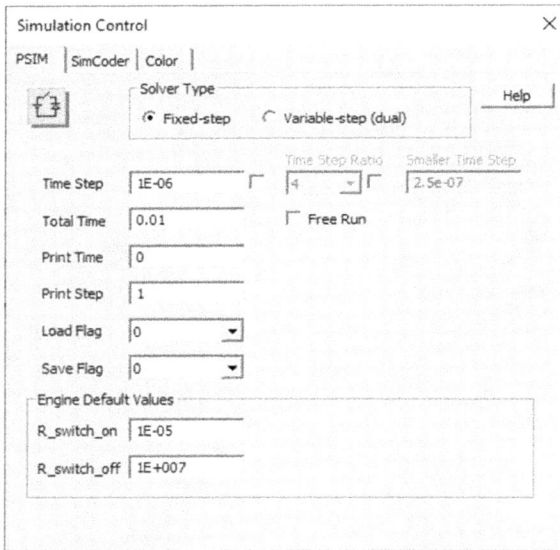

Fig. 8.6.

Run the simulation. The output voltage is around 5 V.

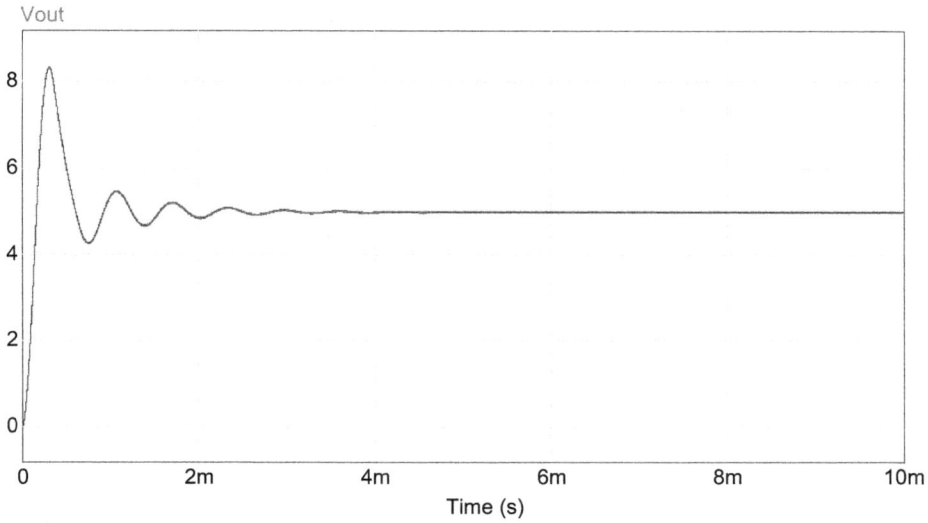

Fig. 8.7.

The inductor current is shown in Fig. 8.8. The zoomed view of inductor current in steady state is shown in Fig. 8.9. According to Fig. 8.9, the minimum and maximum of inductor currents are around 0.875 and 1.25 A, respectively. So, the converter operates in continuous current mode.

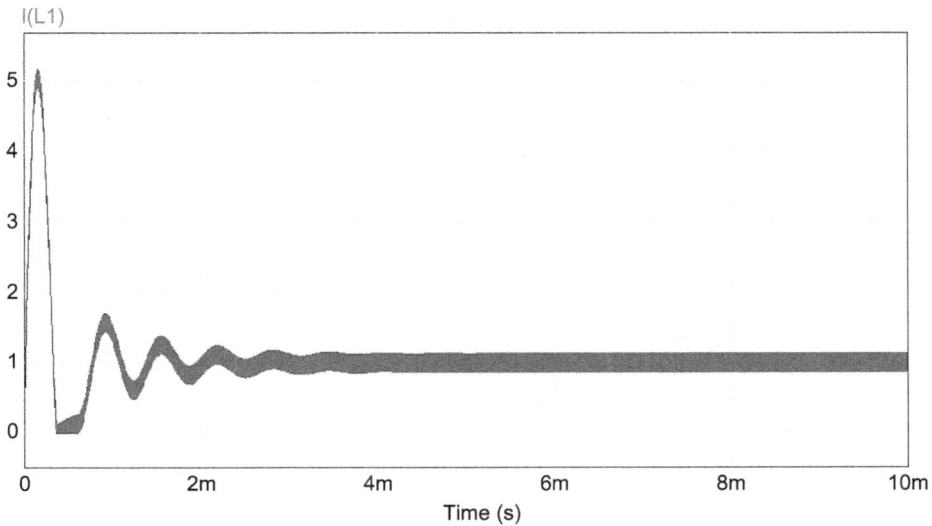

Fig. 8.8.

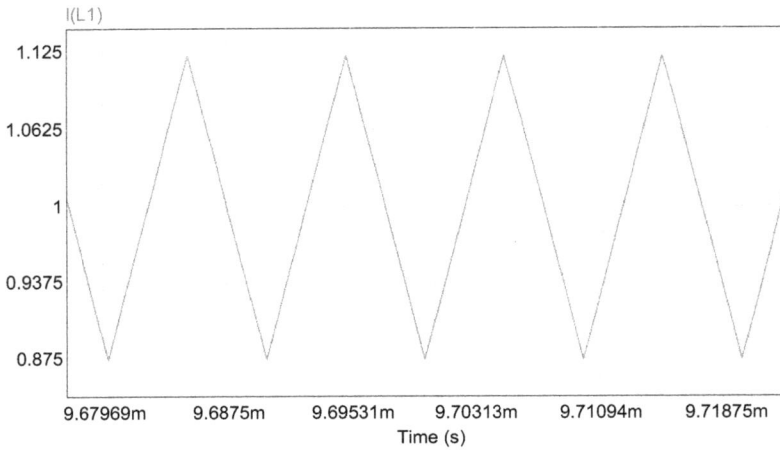

Fig. 8.9.

8.3 Obtaining the frequency response of the power stage

Required blocks are added to the schematic in order to obtain the frequency response of control-to-output voltage.

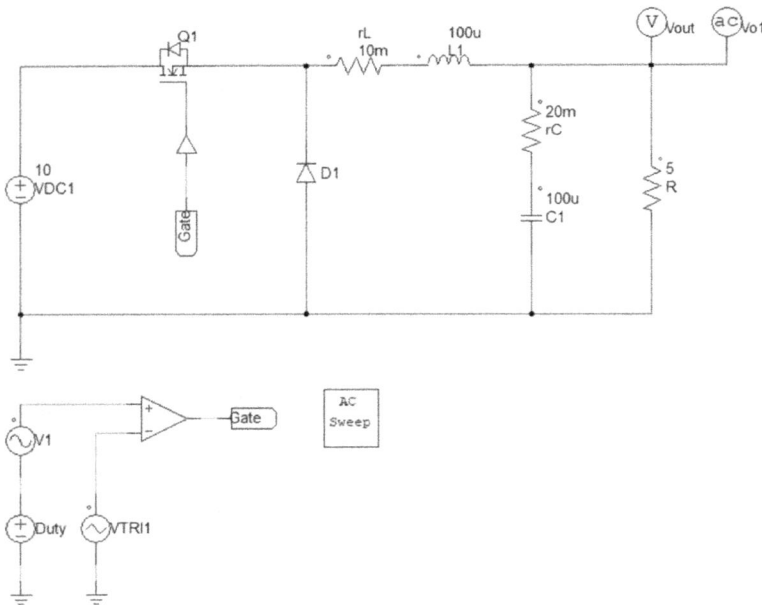

Fig. 8.10.

Settings of AC Sweep block are shown in Fig. 8.11.

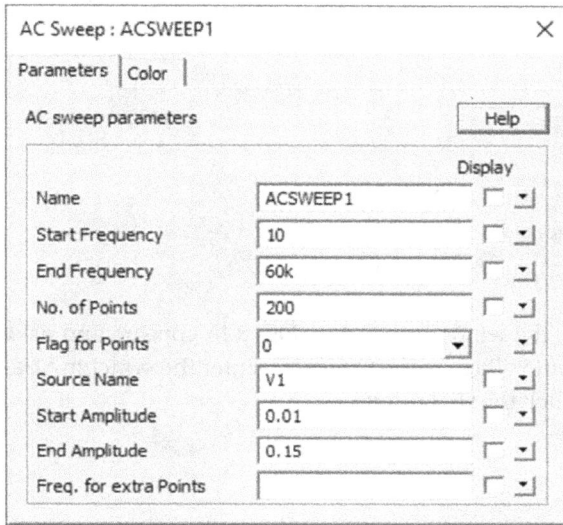

AC Sweep : ACSWEEP1		×
Parameters Color		

AC sweep parameters   Help

		Display
Name	ACSWEEP1	⊓ ▾
Start Frequency	10	⊓ ▾
End Frequency	60k	⊓ ▾
No. of Points	200	⊓ ▾
Flag for Points	0 ▾	▾
Source Name	V1	⊓ ▾
Start Amplitude	0.01	⊓ ▾
End Amplitude	0.15	⊓ ▾
Freq. for extra Points		⊓ ▾

Fig. 8.11.

Run the simulation. The result shown in Fig. 8.12 will be obtained. The controller will be designed based on the obtained frequency response.

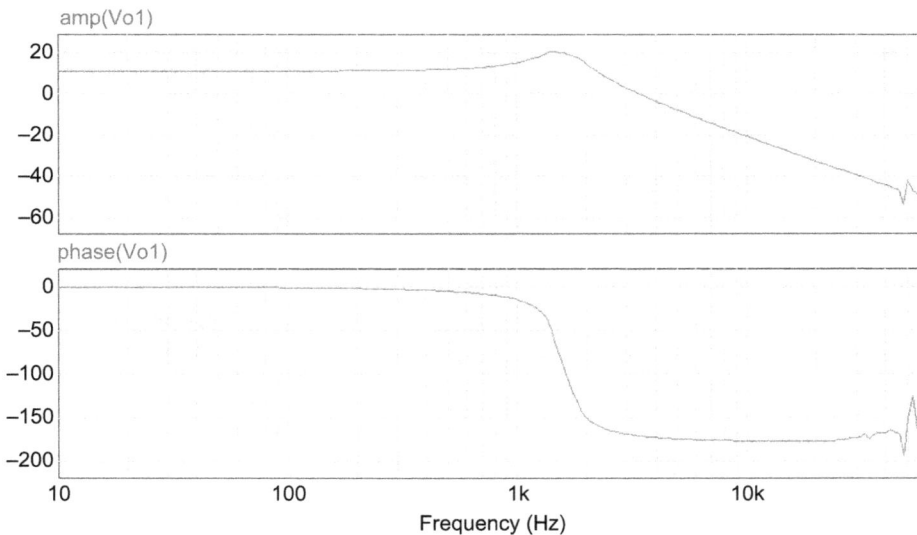

Fig. 8.12.

8.4 Designing the controller

Now, click the SmartCtrl icon in order to run the SmartCtrl.

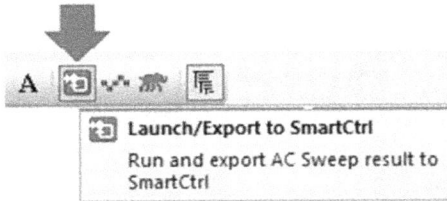

Launch/Export to SmartCtrl

Run and export AC Sweep result to
SmartCtrl

Fig. 8.13.

After clicking the SmartCtrl icon, the window shown in Fig. 8.14 appears and asks the switching frequency and output voltage of the converter. Enter the switching frequency and output voltage, and click the OK button.

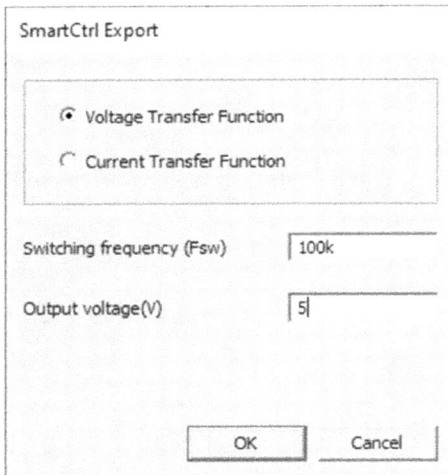

SmartCtrl Export

⦿ Voltage Transfer Function

○ Current Transfer Function

Switching frequency (Fsw)   100k

Output voltage(V)   5

OK   Cancel

Fig. 8.14.

After clicking the OK button, the window shown in Fig. 8.15 will appear. PSIM reads the obtained frequency response (Fig. 8.12) automatically and shows it in this window. Click OK to move forward.

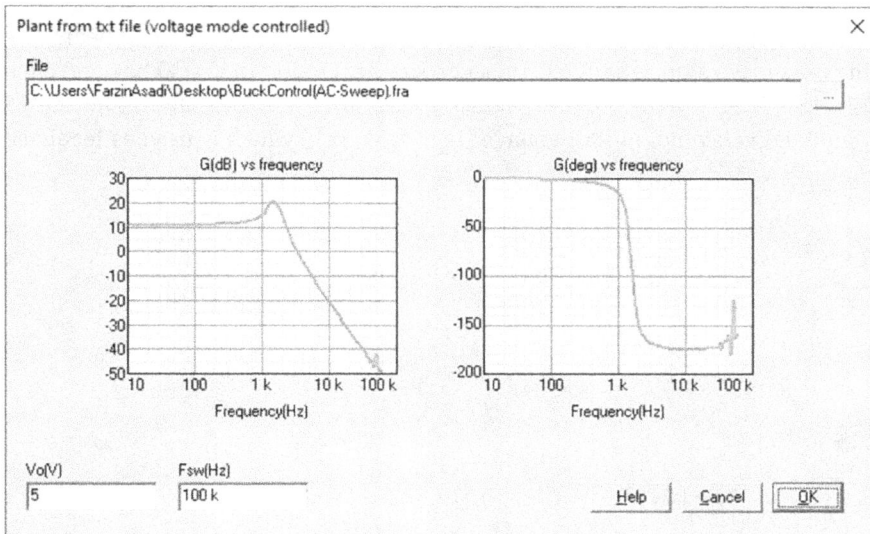

Fig. 8.15.

After clicking the OK button, the window shown in Fig. 8.16 will appear. Select the desired sensor type. The resistive voltage divider is the simplest and most commonly used type of voltage sensors.

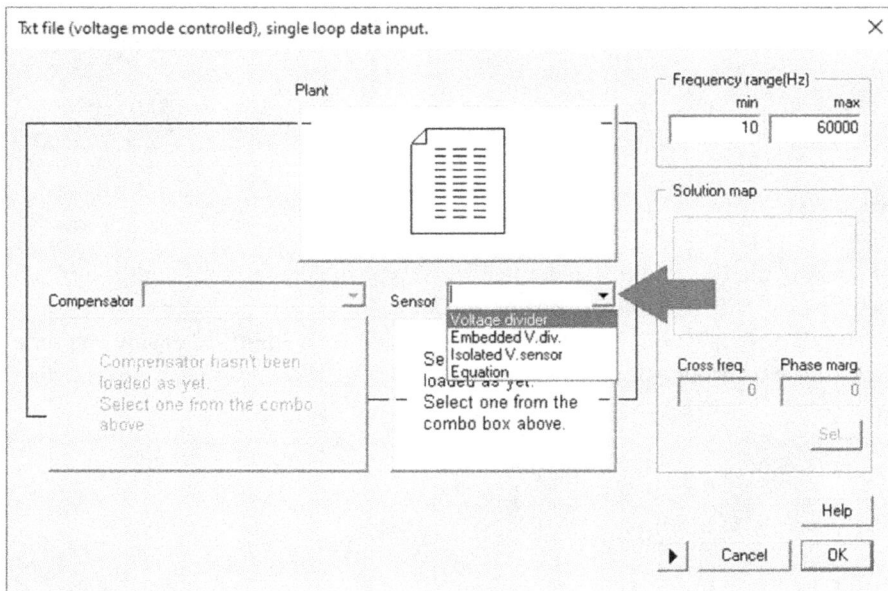

Fig. 8.16.

After clicking the voltage divider sensor, the window shown in Fig. 8.16 appears and asks the reference value, where 2.5 V is a good value. Enter it to V_{ref} (V) box and click the Calculate Gain = V_{ref}/V_o from V_{ref} button. Then click the OK button. The Vref is the attenuated version of output voltage $\left(\left(R_b/(R_b+R_b)\right)V_o\right)$ which is used as feedback signal.

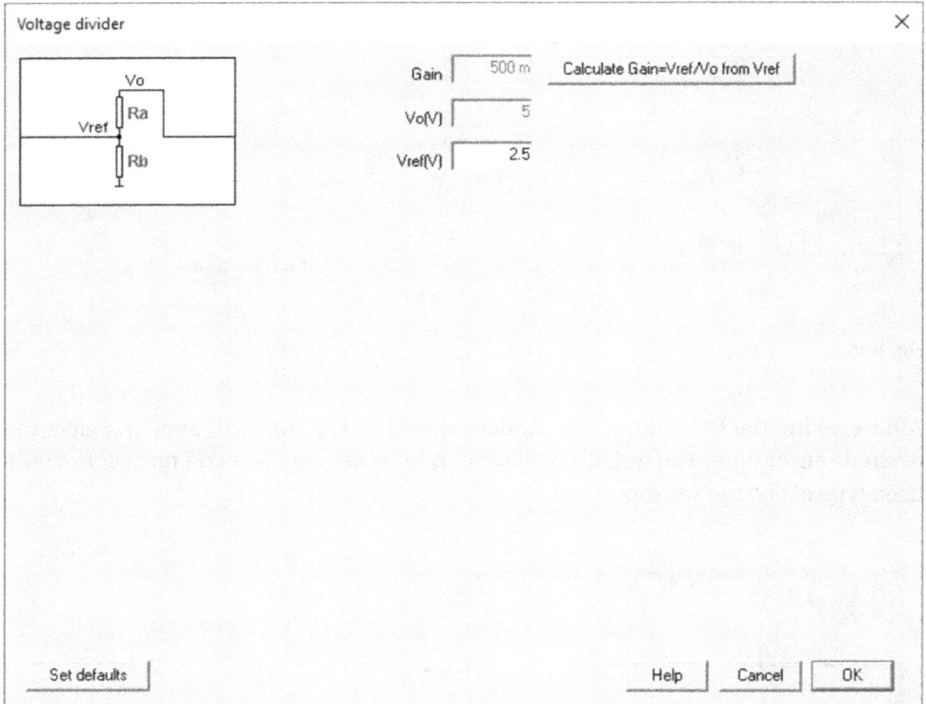

Fig. 8.17.

After clicking the OK button, select the desired type of compensator (controller). For instance, select the Type 2 compensator (Fig. 8.18).

You can design Type 2, Type 3, and PI controller with SmartCtrl easily. The schematic of these controllers is shown in Fig. 8.19.

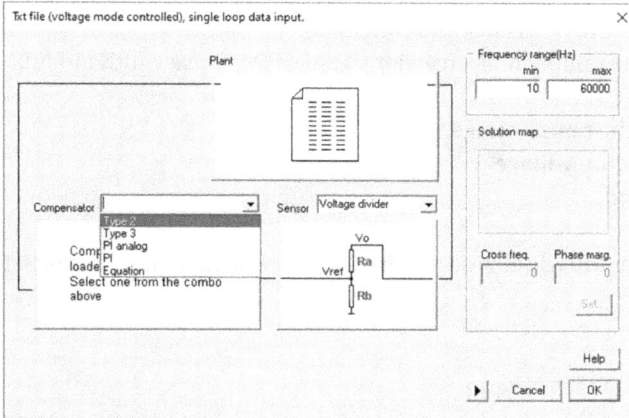

Fig. 8.18.

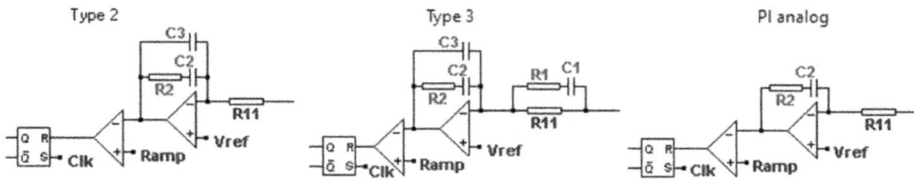

Fig. 8.19.

After clicking the Type 2 controller, the window shown in Fig. 8.20 will appear.

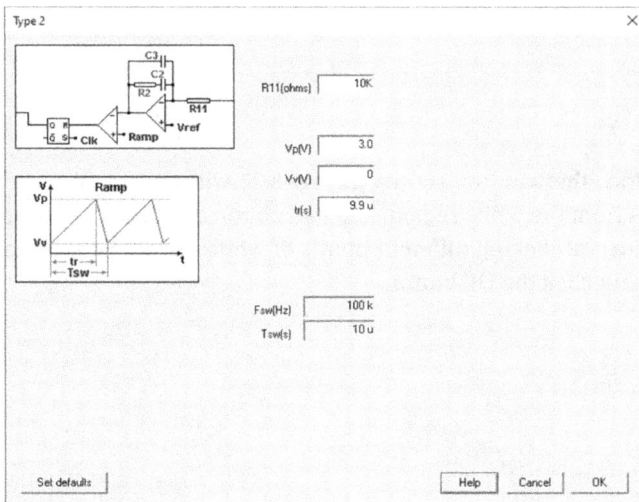

Fig. 8.20.

Enter the values in the boxes and click OK. Note that:

- **Vp(V):** Peak value of the ramp voltage (carrier signal of the pulse width modulation modulator)
- **Vv(V):** Valley value of the ramp voltage
- **Tr(s):** Rise time of the ramp voltage
- **Tsw(s):** Switching period

After clicking the OK button, the window shown in Fig. 8.21 will appear. Click the Set . . . button.

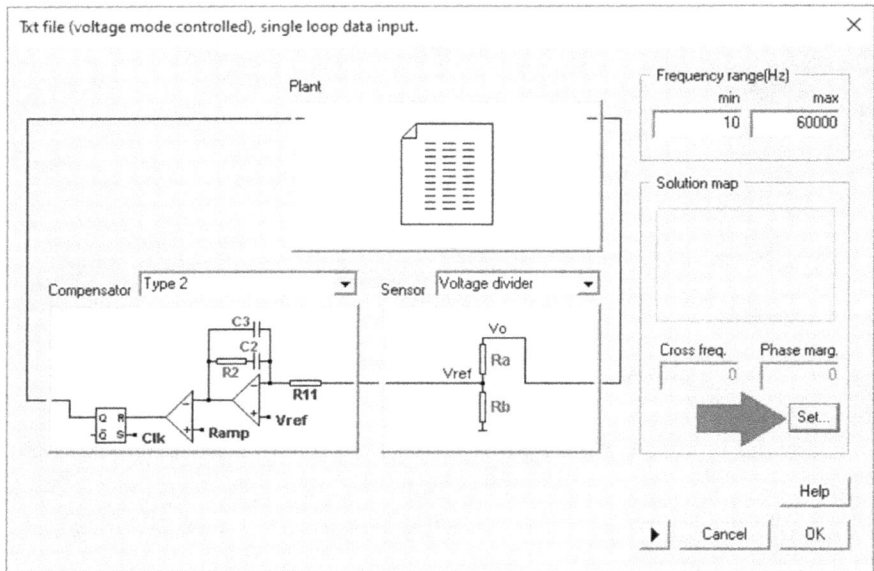

Fig. 8.21.

After clicking the OK button, the window shown in Fig. 8.22 will appear. You are allowed to select the points from the white region. You can obtain different crossover frequency and phase margins by clicking different points of white region. Select one of the white region points and click the OK button.

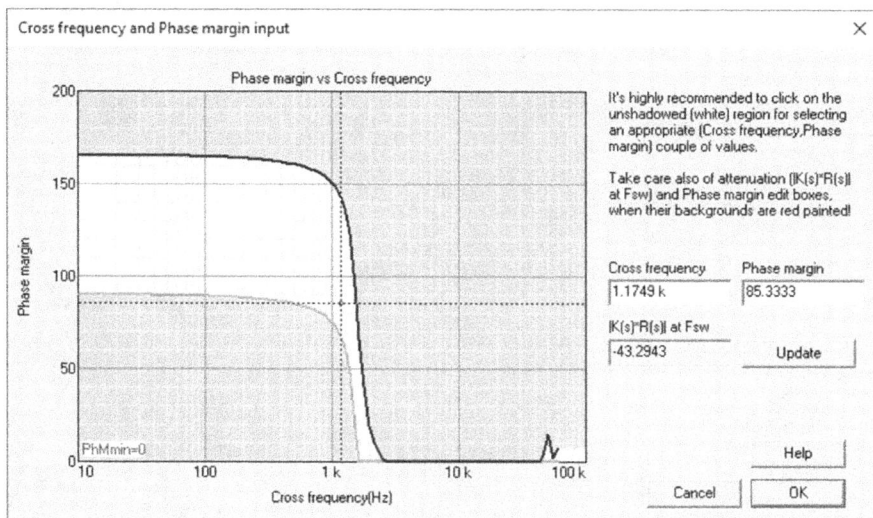

Fig. 8.22.

After clicking the OK button, the window shown in Fig. 8.23 will appear. Click OK button.

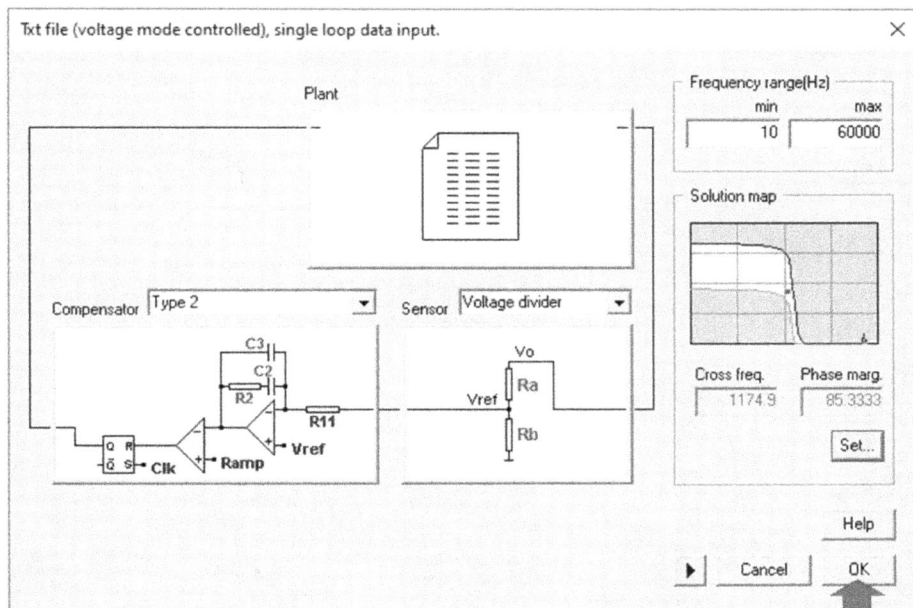

Fig. 8.23.

After clicking the OK button, the window shown in Fig. 8.24 will appear.

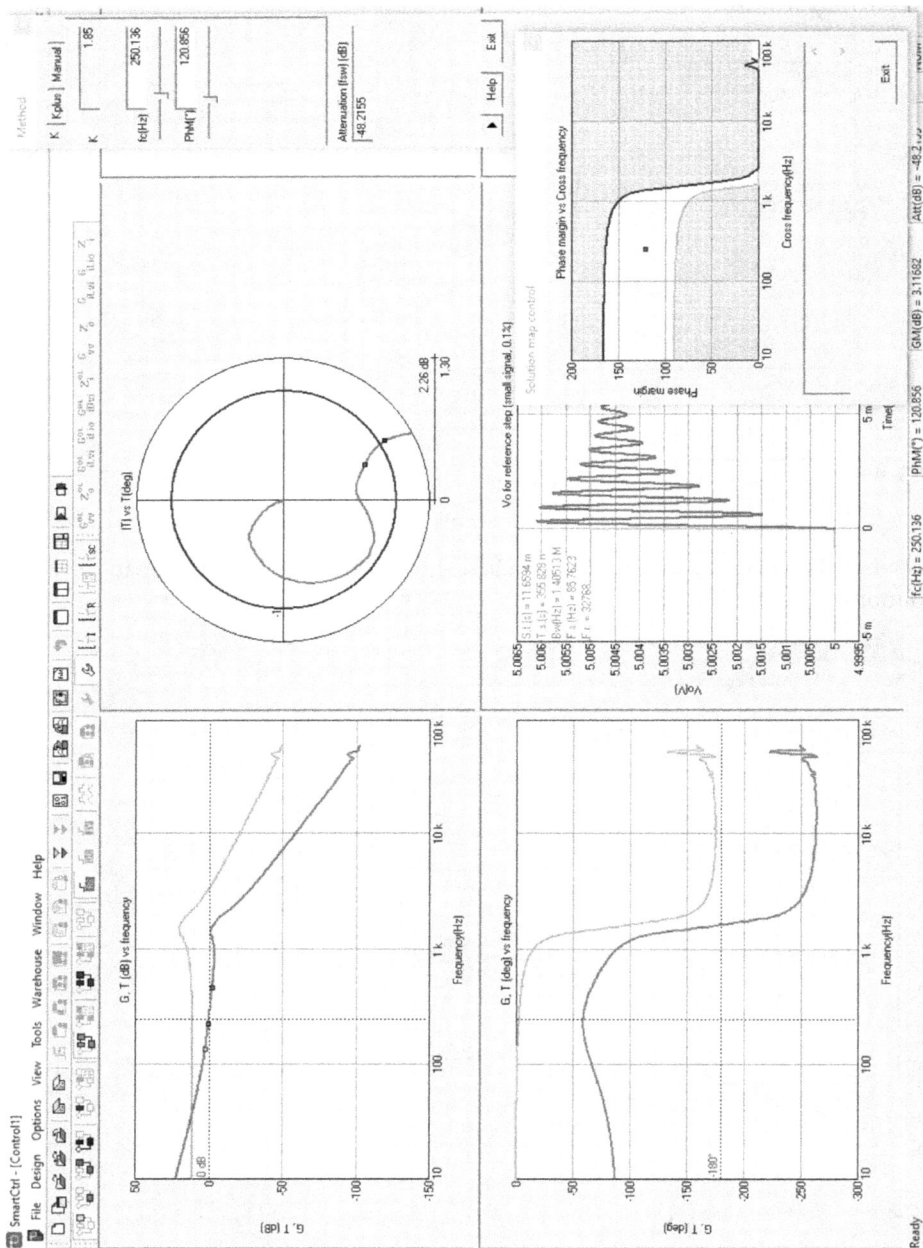

Fig. 8.24.

You can double click on any shown graph, and PSIM shows the zoomed view of that graph to you. You can click the See all panels icon in order to see all the graphs again.

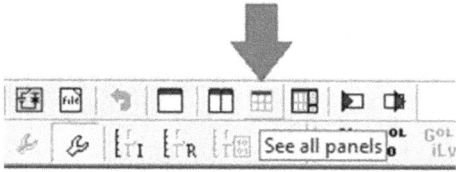

Fig. 8.25.

SmartCtrl can show Bode plot of different transfer functions simultaneously. You can see the Bode plot of different loop transfer functions with the aid of icons shown in Fig. 8.26.

Fig. 8.26.

You can see the Bode plot of input impedance, output impedance, and audio susceptibility with the aid of icons shown in Fig. 8.27. Click on different points of allowed white region or push the mouse left button and move the cursor over the white region and see the change of Bode plots.

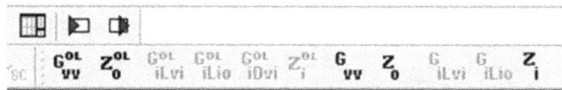

Fig. 8.27.

So, you need to try different points of allowed white space in order to see whether or not the design criteria are satisfied. For instance, assume that your design criteria are overshoot of less than 10% and settling time of less than 1 ms for step input.

In this case, you need to pay attention to the step response graph that is shown on the screen (Fig. 8.28). You need to move your cursor over different points of allowed white region and see whether or not the design criteria are satisfied.

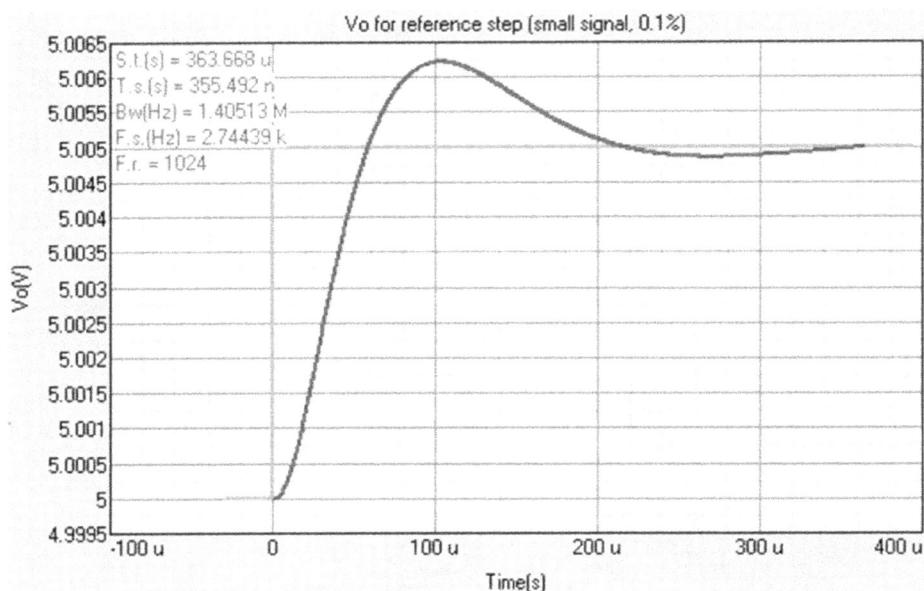

Fig. 8.28.

8.5 Parameters of designed controller

After reaching the point that satisfies the design criteria, click the Output data.

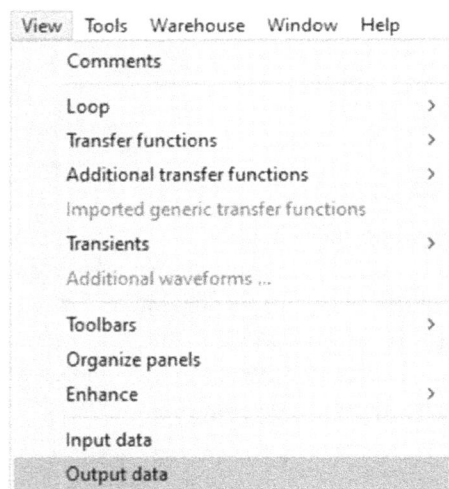

Fig. 8.29.

After clicking the Output data, the window shown in Fig. 8.30 will appear. Required values of components are given in this window (R_a and R_b are output voltage divider resistors of Fig. 8.17). According to Fig. 8.30, the equation of designed controller is

$$C(s) = \frac{0.0017914s + 1}{5.2127 \times 10^{-7} s^2 + 0.00105328 s} .$$

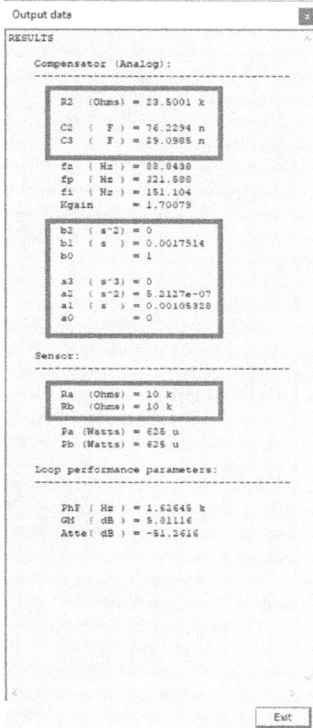

Fig. 8.30.

8.6 Exporting the parameters of designed controller

You can export the values shown in Fig. 8.30 to a text file. In order to do that, click the File> Generate report> To Notepad (Fig. 8.31). After clicking the To Notepad, the parameters will be exported to Notepad.

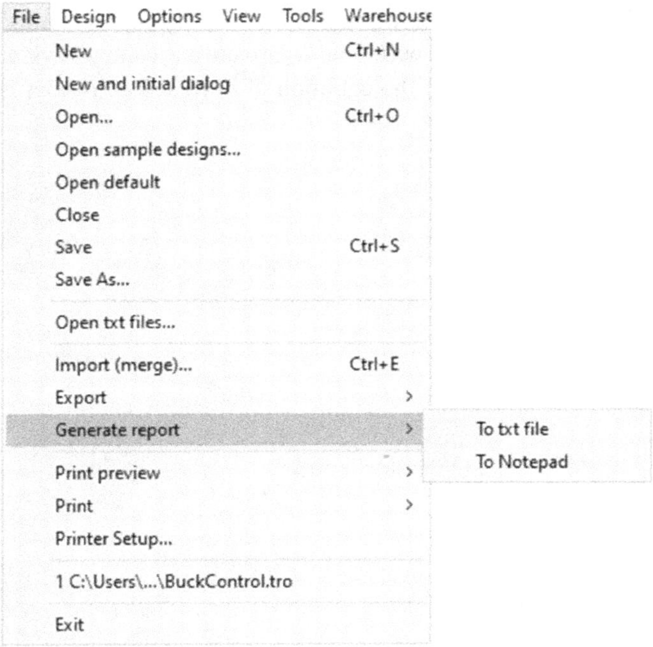

Fig. 8.31.

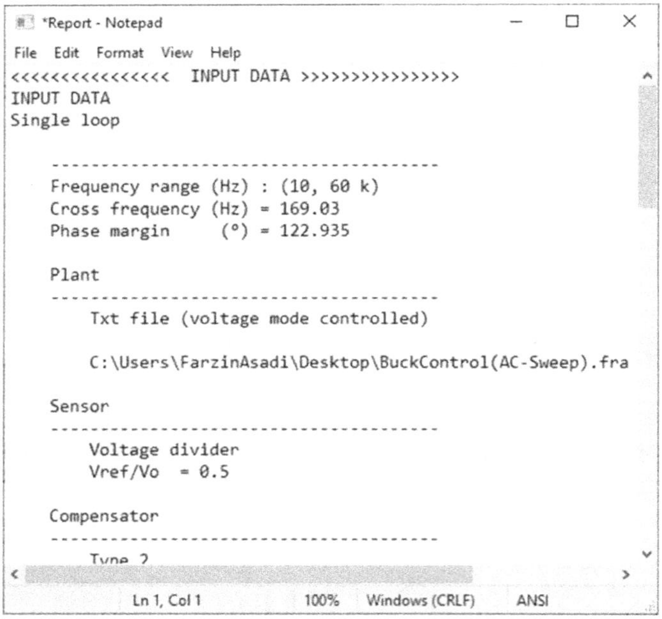

Fig. 8.32.

You can click the Generate report icon as another way to get the parameter values in Notepad.

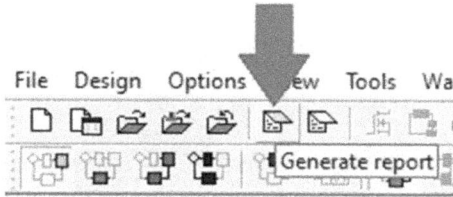

Fig. 8.33.

8.7 Addition of the controller to the power stage

You need to export the controller which is designed in the SmartCtrl to PSIM and test it. In order to test the designed controller, go to PSIM schematic and add a sub circuit block to it. The exported controller will be placed inside the added sub circuit.

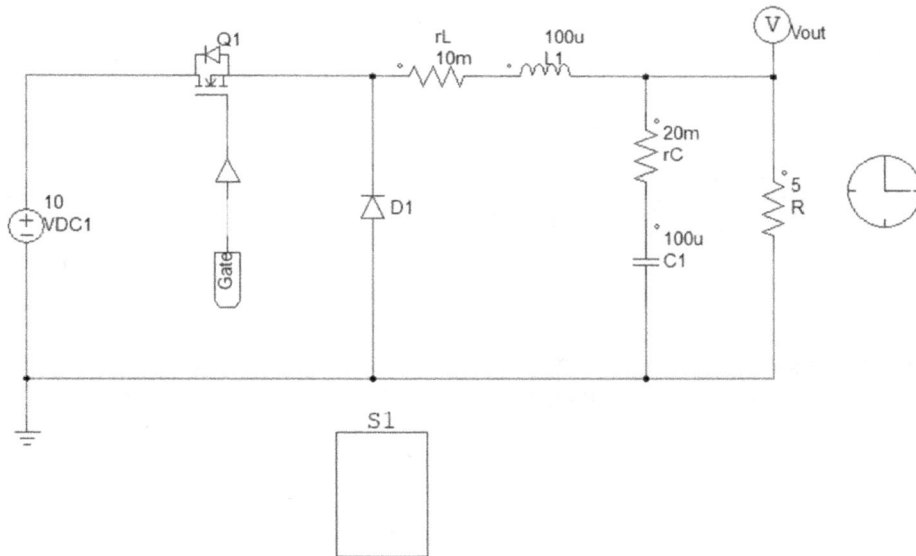

Fig. 8.34.

Now go to the SmartCtrl environment and click the File> Export> To PSIM> Schematic (Fig. 8.35). Instead of using File menu, you can use the shortcut icon shown in Fig. 8.36.

Fig. 8.35.

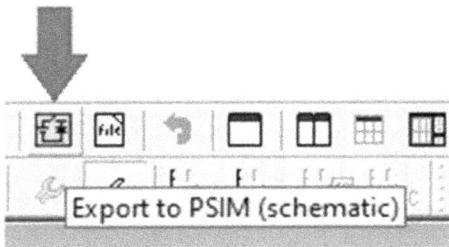

Fig. 8.36.

SmartCtrl opens the Selecting schematic file window. Set the desired path and enter the desired name for the file which will be exported. Then click the Open button.

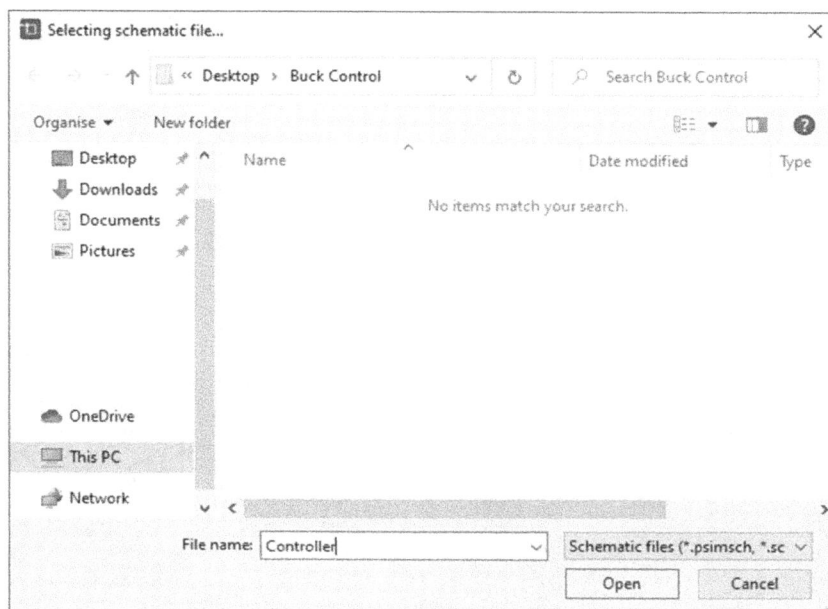

Fig. 8.37.

PSIM opens the window shown in Fig. 8.38. Select the Components (R1, C1, . . . are given) and click OK button.

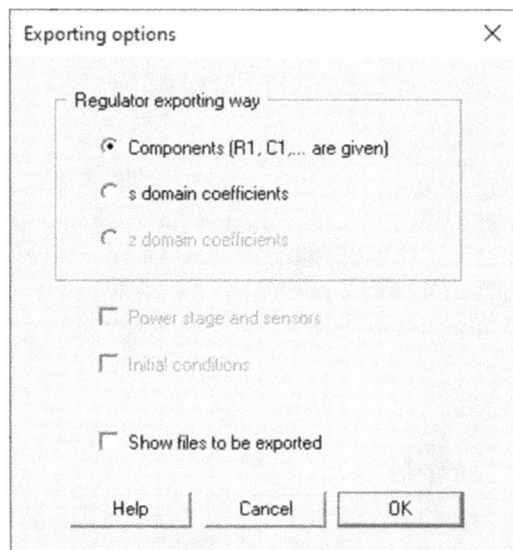

Fig. 8.38.

After click the OK button, the schematic of designed controller will be exported to the file that you determined in Fig. 8.37. PSIM automatically opens this file for you.

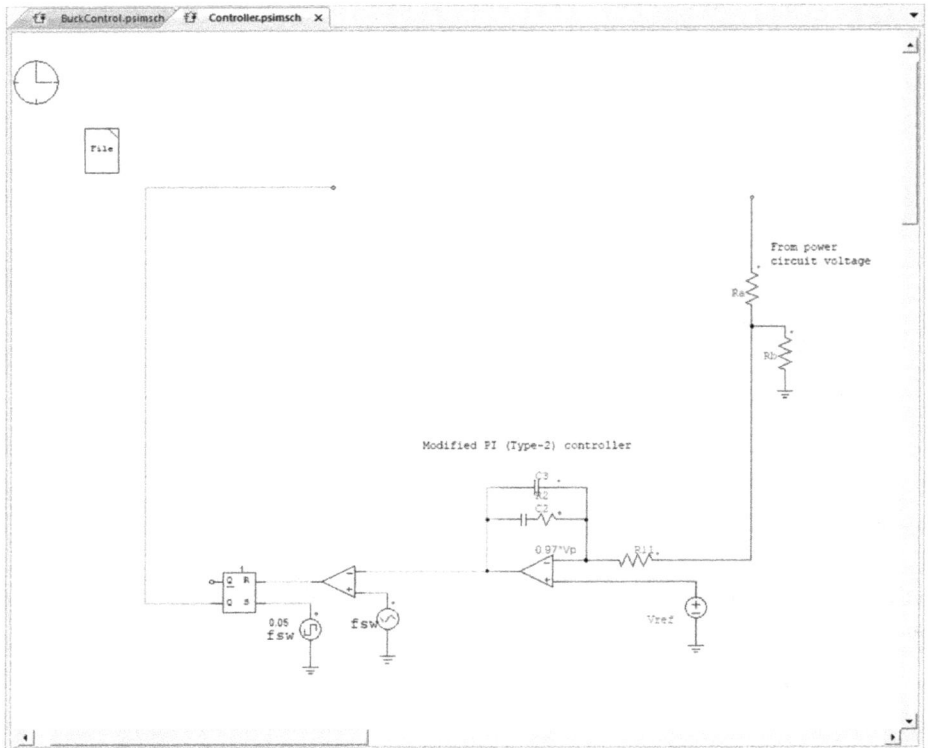

Fig. 8.39.

Select the components of controller. In order to do this, press the mouse left button on an empty point of the schematic and draw a rectangle around the components without releasing the mouse left button (Fig. 8.40). When the rectangle surrounds all the components (except of Simulation Control block), release the mouse left button and press the Ctrl+C. This copies the components into the Windows clipboard.

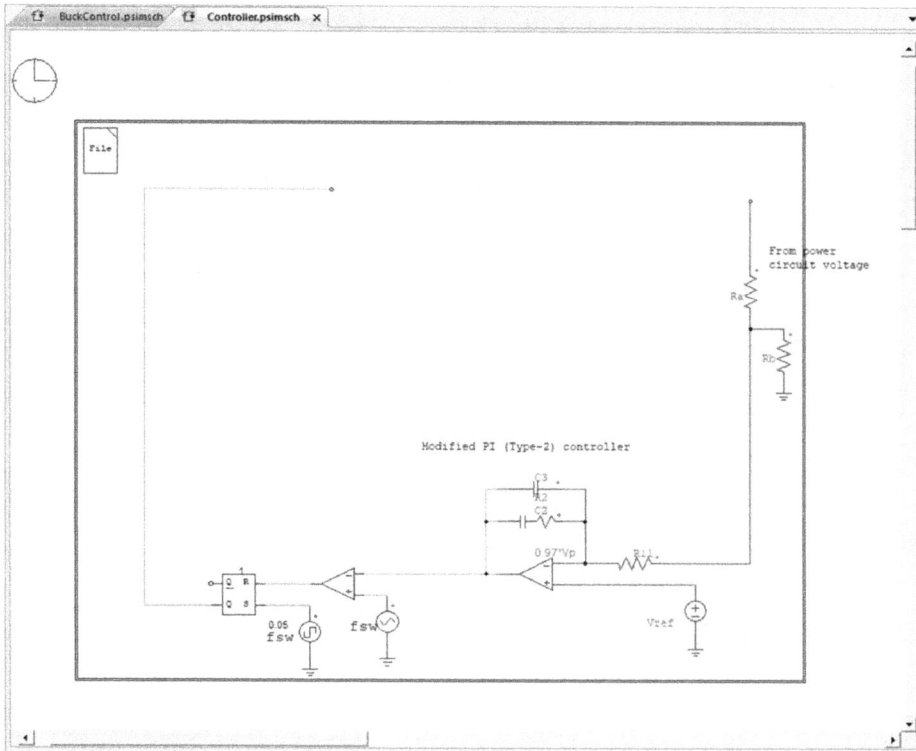

Fig. 8.40.

Now, open the sub circuit block (Fig. 8.34) by double clicking on it. Press Ctrl+V in order to paste the copied components. Connect the input and output ports to the input and output of controller circuit (Fig. 8.41).

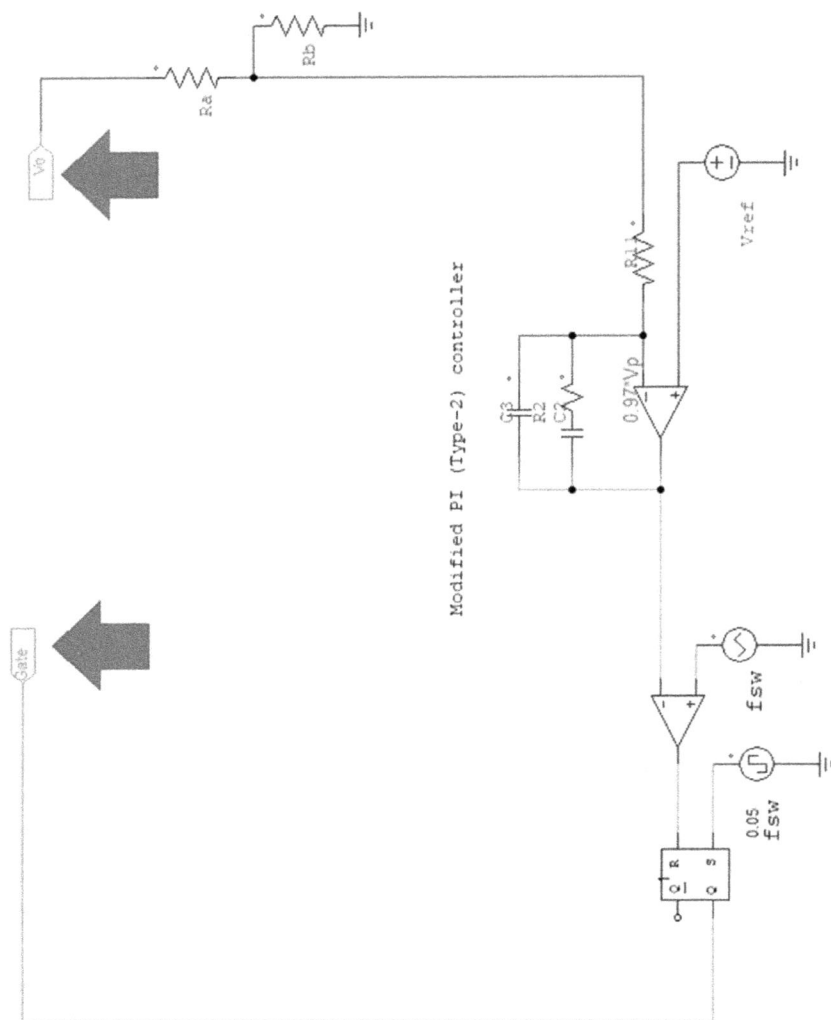

Fig. 8.41.

Now connect the sub circuit block to the rest of circuit.

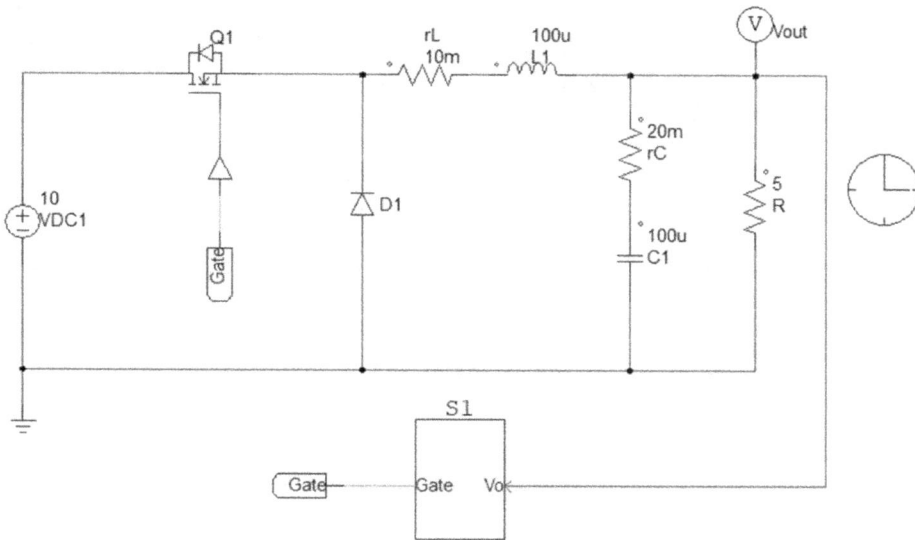

Fig. 8.42.

Run the simulation. The result shown in Fig. 8.43 is obtained. The steady-state voltage value is 5 V.

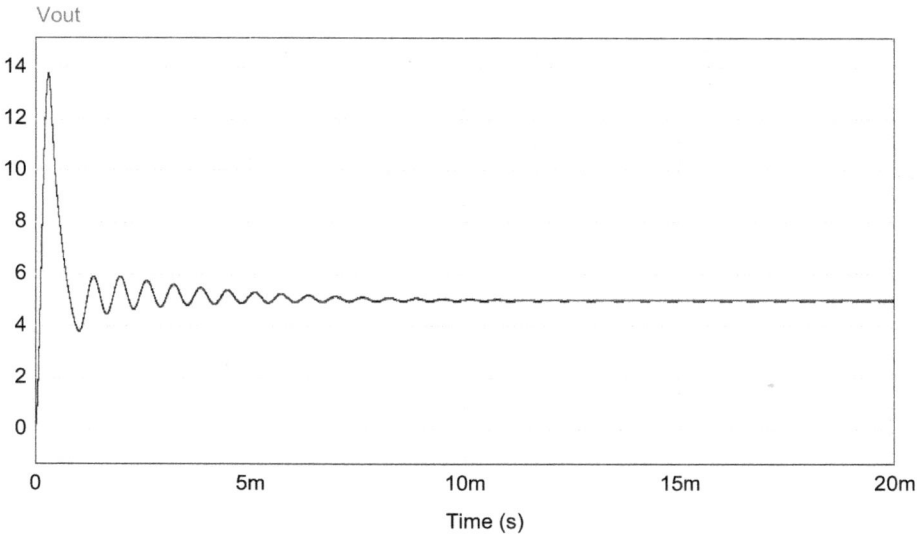

Fig. 8.43.

8.8 Testing the controller

In this section, we want to see the performance of designed controller. For instance, let's see the behavior of closed loop system for a step change in the reference signal.

In order to do this test, add a step source in series with the V_{ref}. This step source makes a jump in the reference signal of controller.

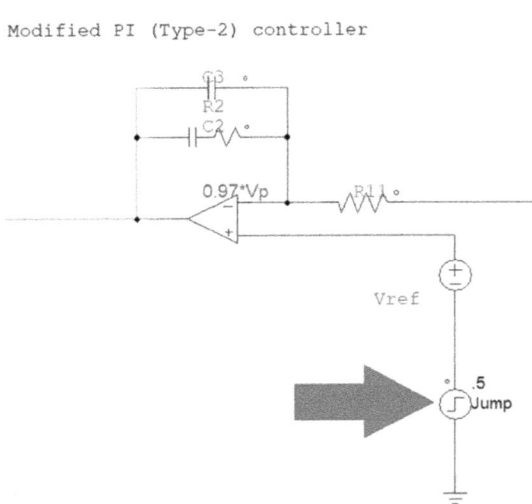

Fig. 8.44.

Settings of added step source are shown in Fig. 8.45. According to Fig. 8.45, the value of step voltage source is 0.5 for 15 ms $< t$. So, the reference signal of control system is $2.5 + 0 = 2.5$ V for $t < 15$ ms and $2.5 + 0.5 = 3$ V for $t > 15$ ms. Since the output voltage sensor has a gain of 0.5, the converter must follow 5 V for $t < 15$ ms and 6 V for $t > 15$ ms.

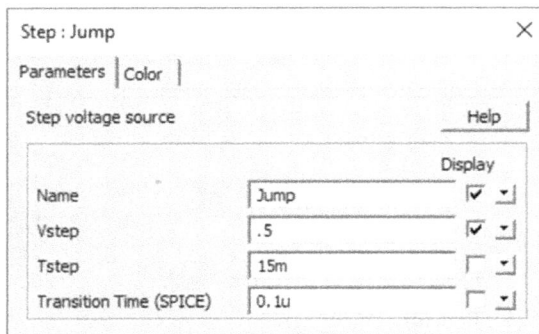

Fig. 8.45.

Run the simulation. The simulation result is shown in Fig. 8.46. According to Fig. 8.46, the converter follows the reference with zero steady-state error. However, we see overshoot when the reference of control system changes. If such an overshoot is not allowed, then you need to return to SmartCtrl environment and redesign the controller.

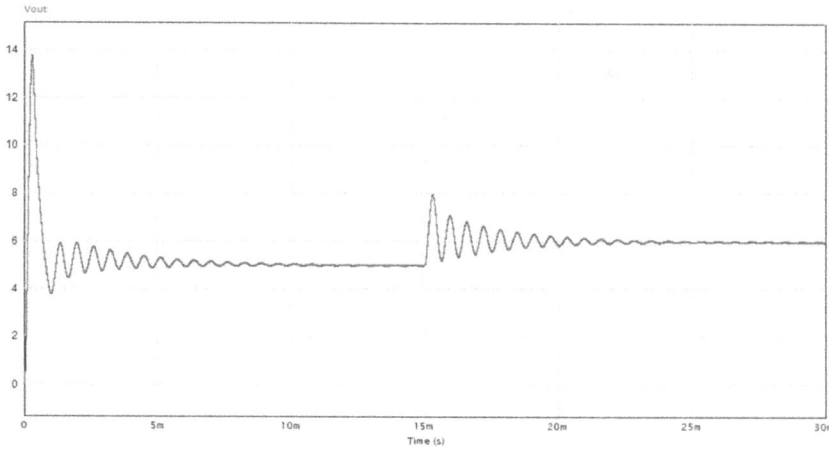

Fig. 8.46.

Assume that you want to study the effect of input voltage changes on the output voltage of converter. Put a step voltage source in series with the input voltage source in order to do such a test.

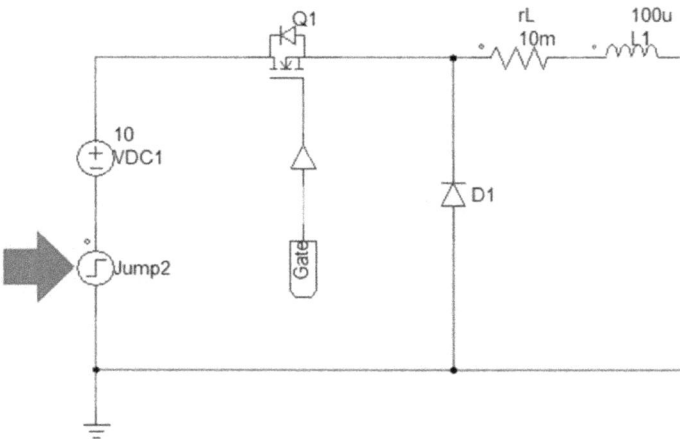

Fig. 8.47.

Assume that you want to study the effect of output load changes on the output voltage of converter. Add the blocks shown in Fig. 8.48 in parallel with the output load. The step voltage source controls the status of MOSFET. When the MOSFET is turned on, the output load is $\frac{5\times5}{5+5}=2.5\Omega$. When the MOSFET is turned off, the output load is .

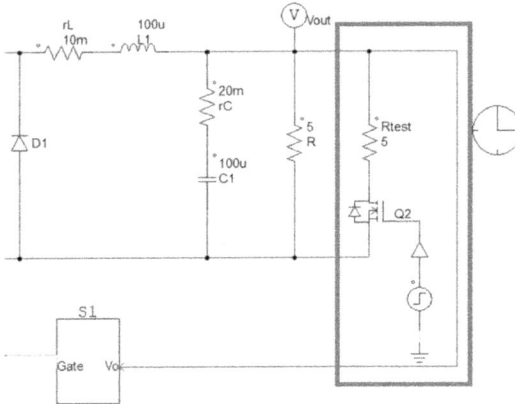

Fig. 8.48.

The schematic shown in Fig. 8.48, decreases the output load. If you want to study the effect of increase in the load, you need use the schematic shown in Fig. 8.49. When the MOSFET is turned on, the output load is 5Ω. When the MOSFET is turned off, the output load is 5 + 5 = 10Ω.

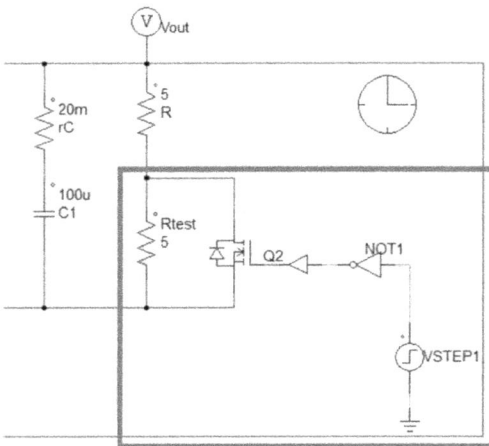

Fig. 8.49.

8.9 Designing with another type of controller

Sometimes you cannot obtain the desired response with the selected controller. In such cases, changing the controller type is one of the options that can help you.

For instance, assume that we want to design a Type 3 controller for the studied buck converter. In order to do this, click the Design> Modify data (Fig. 8.50) or click the shortcut icon shown in Fig. 8.51.

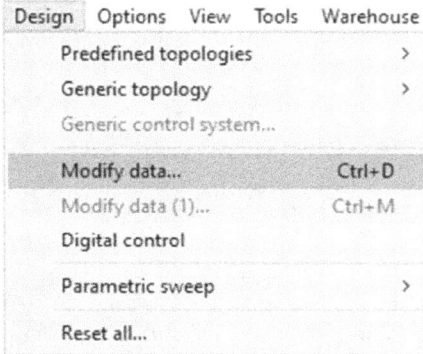

Fig. 8.50.

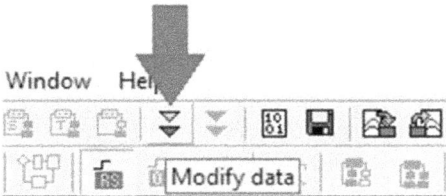

Fig. 8.51.

The window shown in Fig. 8.52 will appear. Now, you can change what you want. Select the Type 3 controller.

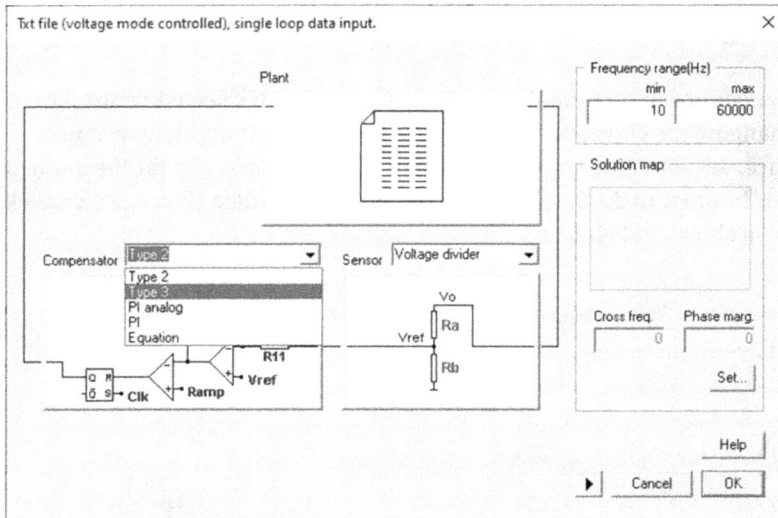

Fig. 8.52.

After selecting the Type 3 controller, the window shown in Fig. 8.53 will appear. Enter the ramp signal characteristics. This is the ramp signal which is used in the pulse width modulator circuit. After entering the values, click the OK button.

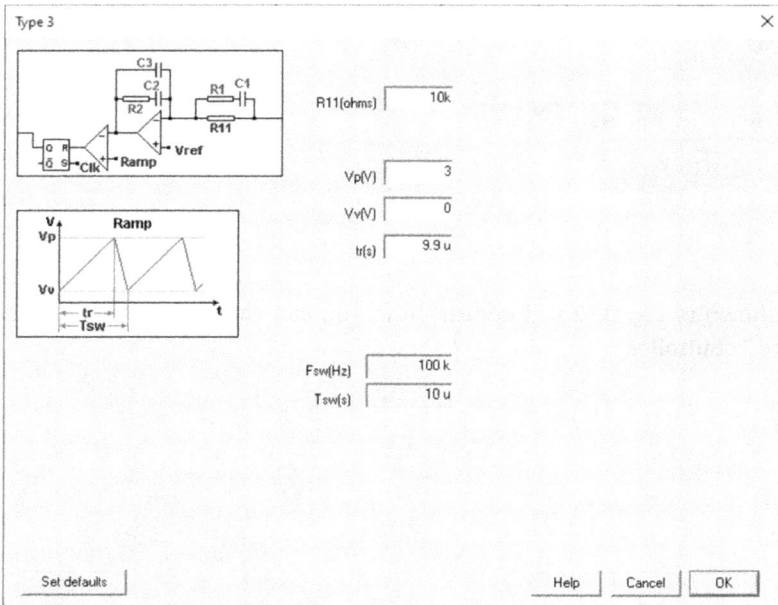

Fig. 8.53.

Now click the Set button and select a point from the white region. Then click the OK button.

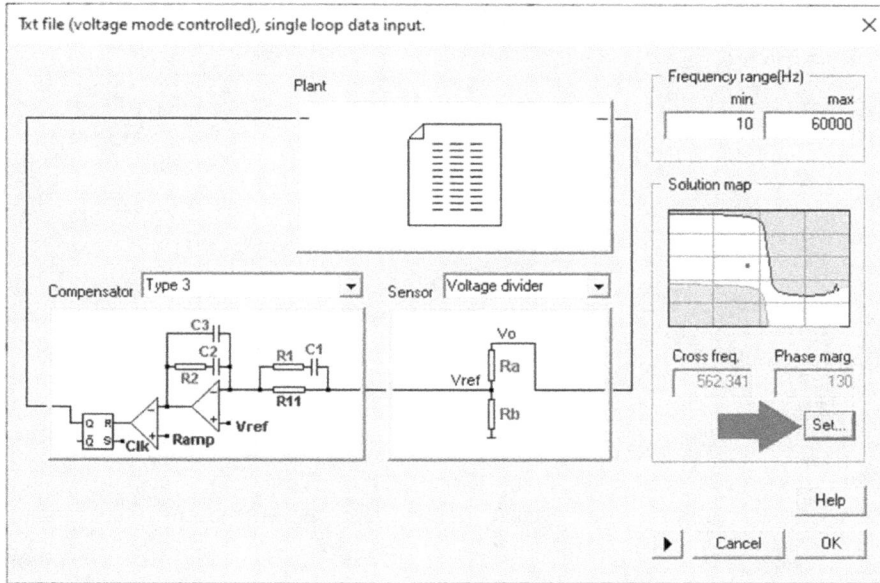

Fig. 8.54.

After clicking the OK button, the message shown in Fig. 8.55 will appear. Click Yes to continue.

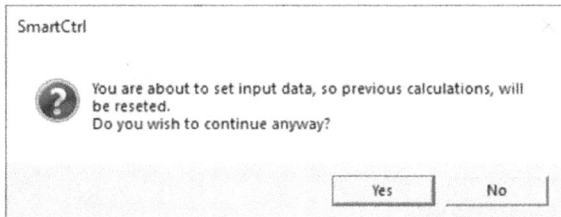

Fig. 8.55.

After clicking the Yes button, SmartCtrl opens the Bode graph of different transfer function and you can design a Type 3 controller. For instance, for values shown in Fig. 8.56, the simulation result shown in Fig. 8.57 is obtained. In this simulation, the reference signal of control system changed from 2.5 to 3 V at 15 ms. The settling time and percent of overshoot for this controller is less than the Type 2 controller (compare Fig. 8.57 with Fig. 8.46).

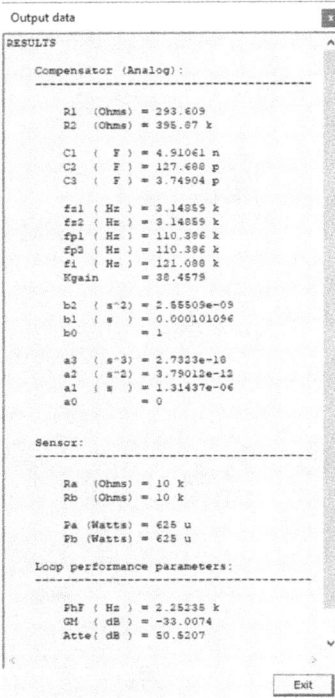

Fig. 8.56.

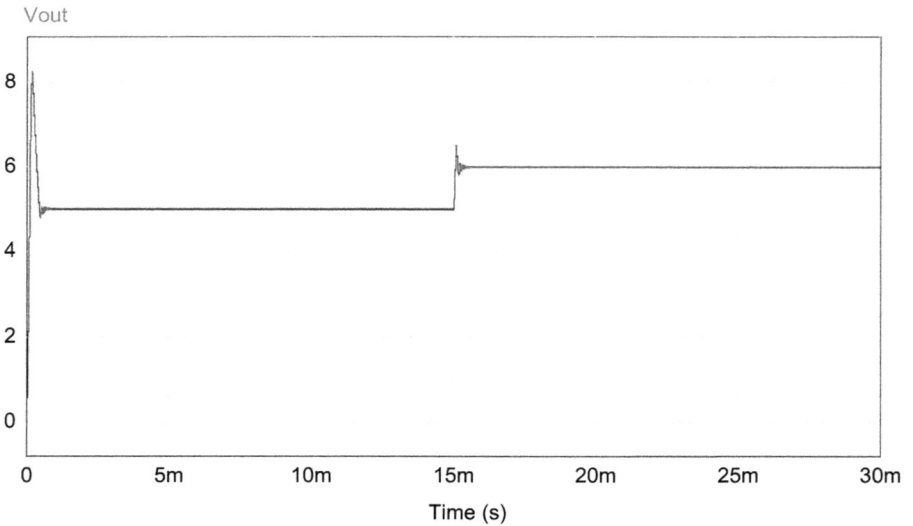

Fig. 8.57.

Let's test the controller. Assume that the output load changes from 5 to 2.5 Ω at 20 ms. The simulation result is shown in Fig. 8.58. Figure 8.59 is the zoomed big view of Fig. 8.58.

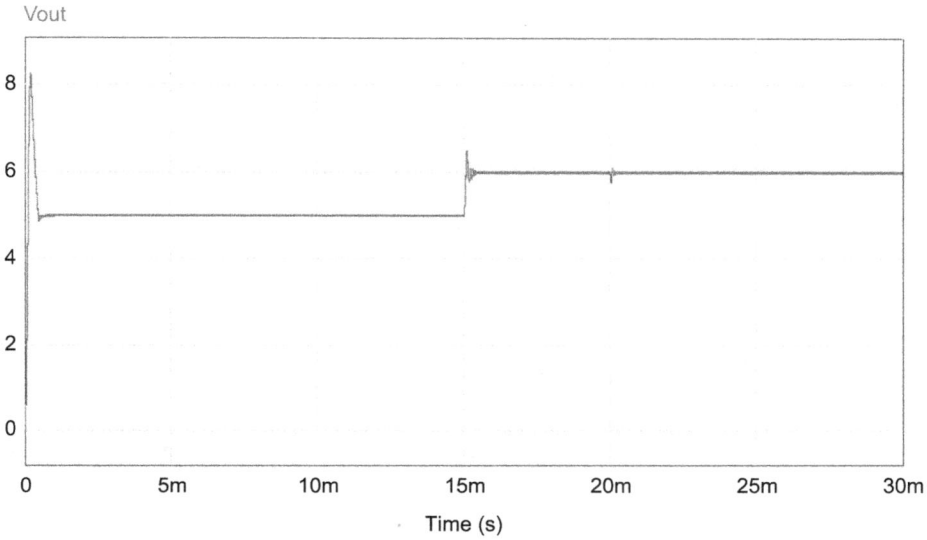

Fig. 8.58.

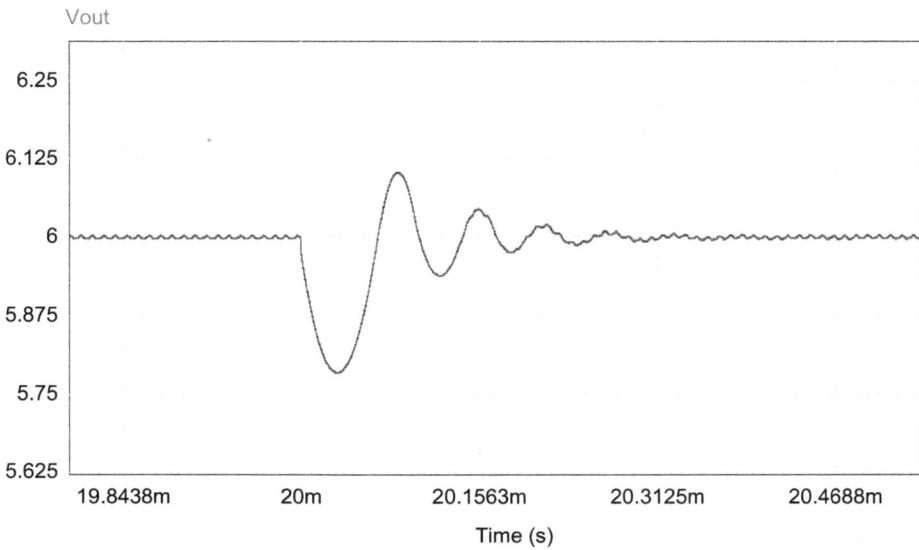

Fig. 8.59.

As another test, assume that the output load changes from 5 to 10 Ω at 20 ms. The simulation result is shown in Fig. 8.60. Figure 8.61 is the zoomed big view of Fig. 8.60.

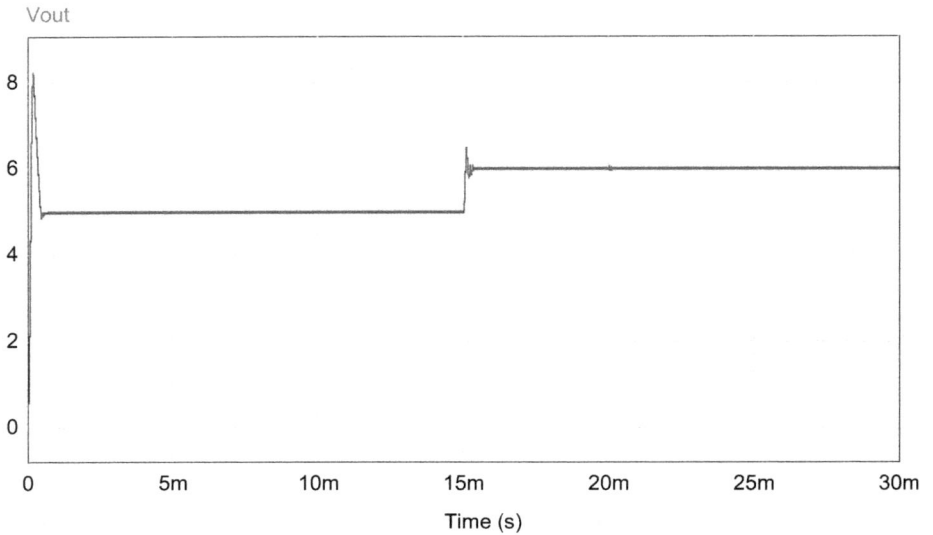

Fig. 8.60.

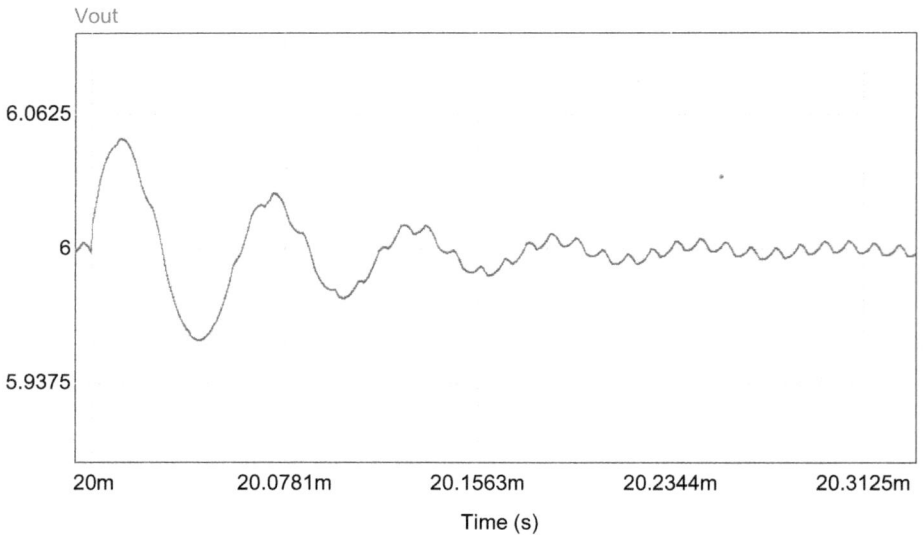

Fig. 8.61.

8.10 Automatic design of power stage and controller

SmartCtrl can design the power stage and required controller for most of commonly used converters automatically. In order to do that, click the New and initial dialog.

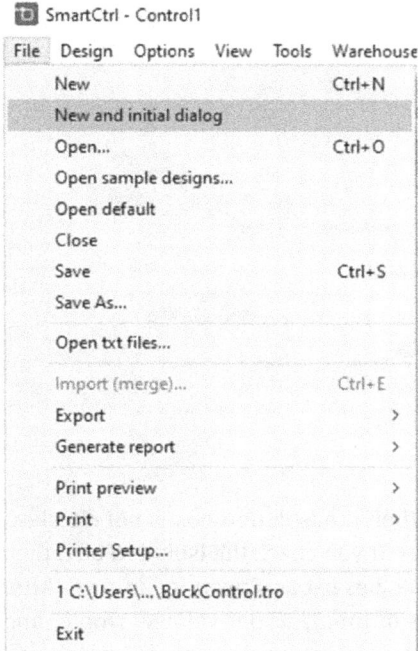

Fig. 8.62.

The window shown in Fig. 8.63 will appear. Click the DC–DC power stage and control circuit design.

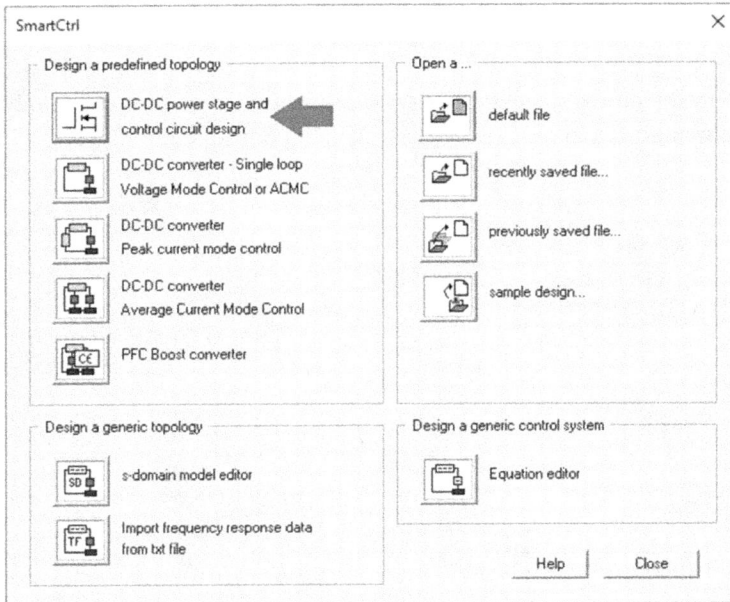

Fig. 8.63.

The window shown in Fig. 8.64 will appear. When the Isolation box is not checked, you can design buck and buck–boost topologies. If you check the Isolation box, then Flyback and Forward will be added to the topologies list as shown in Fig. 8.65. After selection of desired topology, enter the values of input/output voltage, ripple, and so on to the boxes and click the OK button. SmartCtrl design the required circuit and display it (Fig. 8.66). Value of V_{ref} for schematic shown in Fig. 8.66 is 2.5 V.

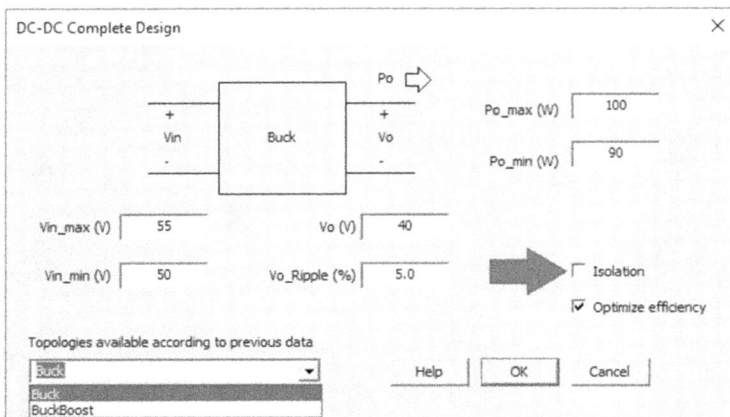

Fig. 8.64.

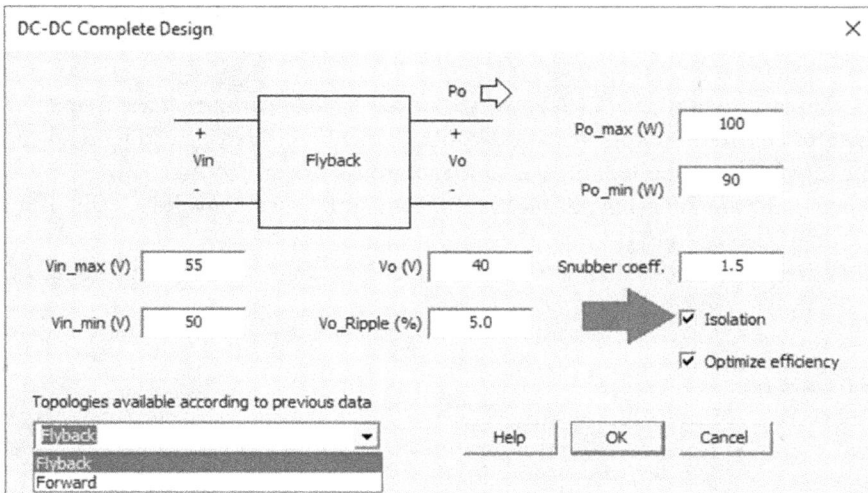

Fig. 8.65.

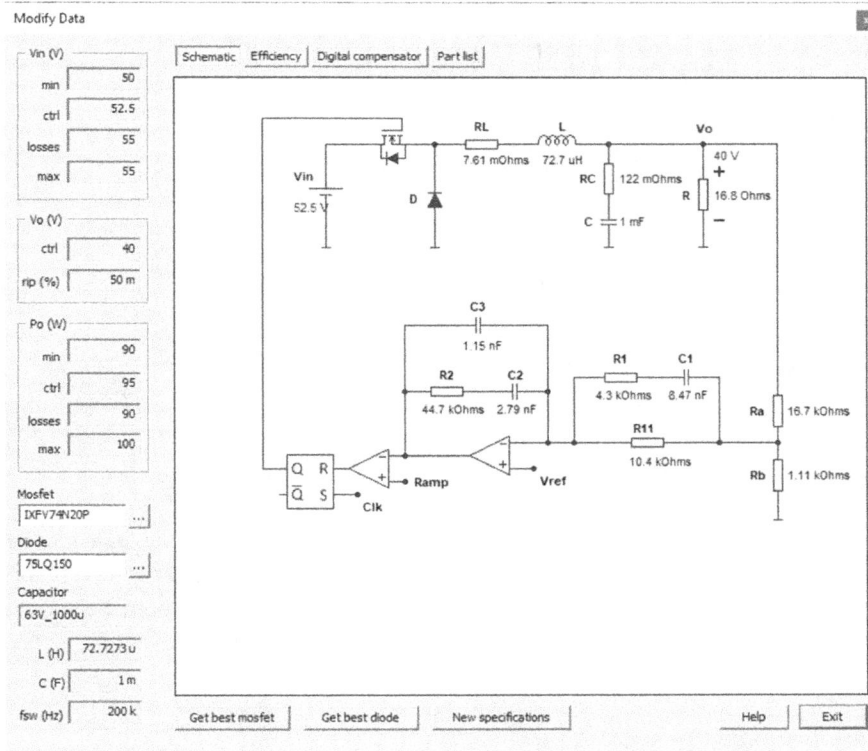

Fig. 8.66.

8.11 Controller design for common type of DC–DC converters

In this chapter, we used PSIM's AC Sweep analysis in order to extract the frequency response of control-to-output for studied buck converter. Then SmartCtrl read the obtained frequency response and used it for controller design process.

When you want to design a controller for commonly used converters, there is no need to do AC Sweep in PSIM.

Click New and initial dialog in order to design a controller without doing AC Sweep analysis.

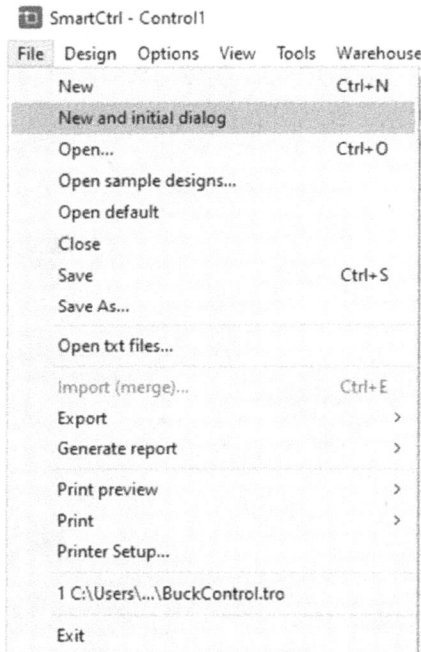

Fig. 8.67.

The window shown in Fig. 8.68 appears. The three buttons shown in Fig. 8.68 can be used to design the control loop. For instance, click the DC–DC converter – Single loop Voltage Mode Control or ACMC. The window shown in Fig. 8.69 will appear. Select the type of converter and desired control technique from the Plant drop down list.

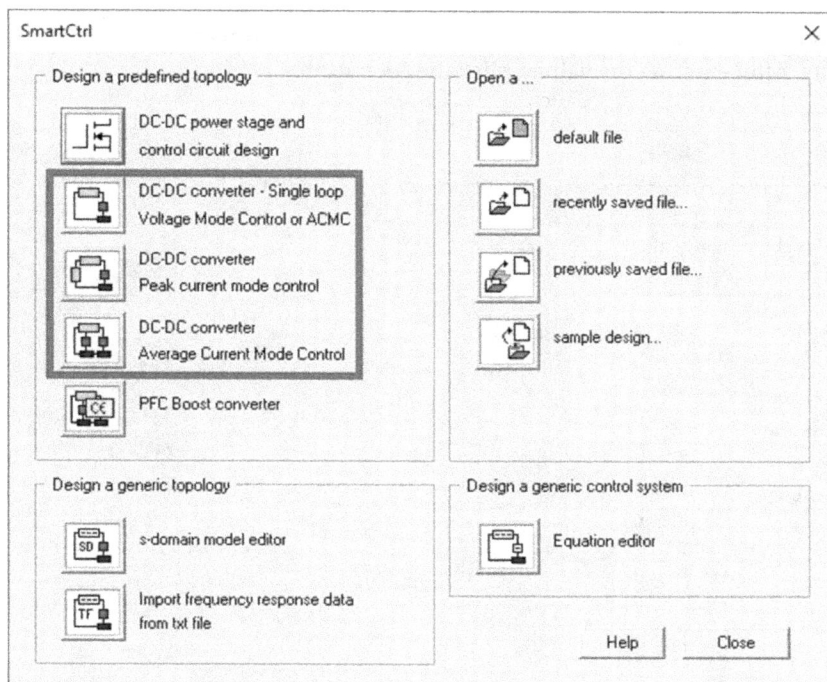

Fig. 8.68.

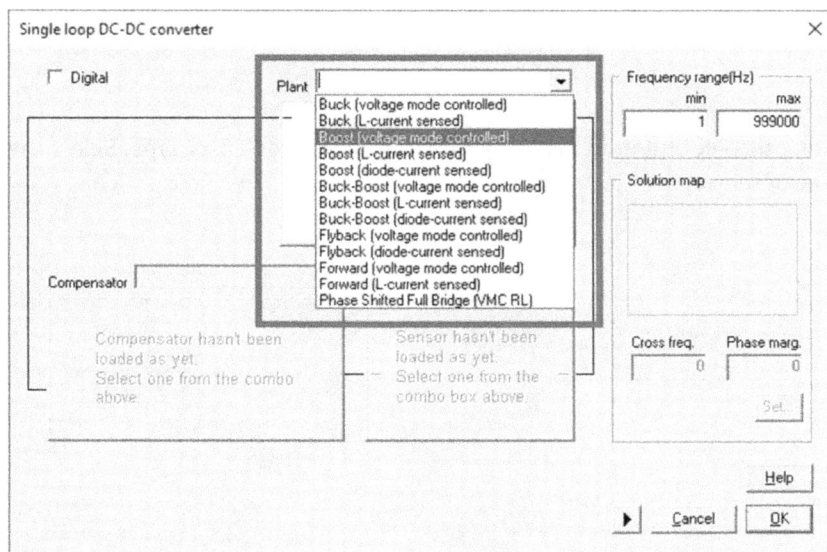

Fig. 8.69.

After converter type is selected, SmartCtrl opens a window for you and asks the values of elements. After entering the values, click the OK button.

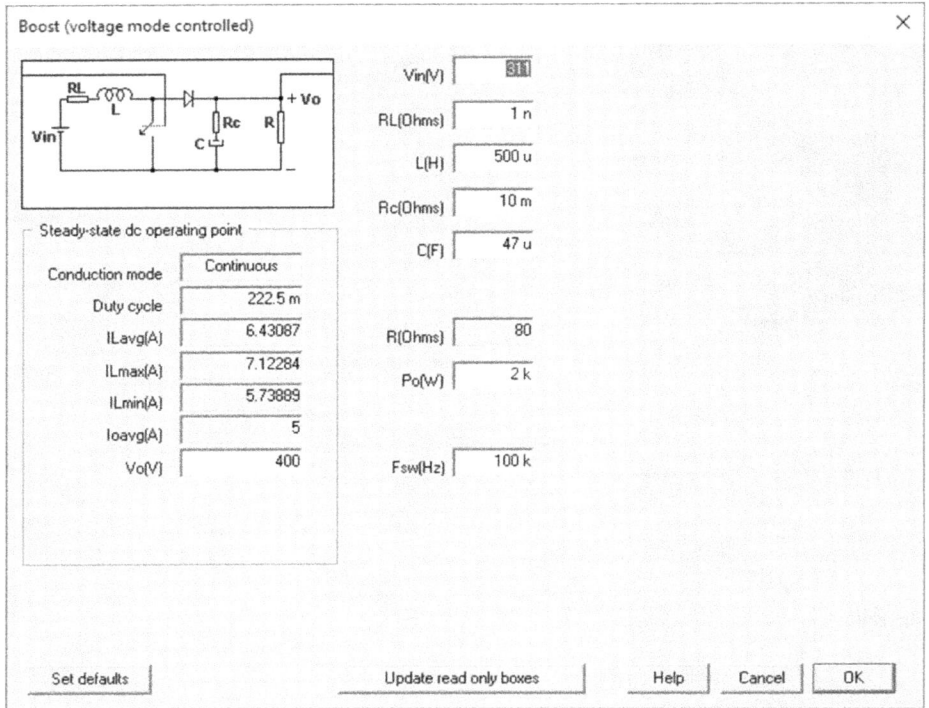

Fig. 8.70.

After clicking the OK button in Fig. 8.70, SmartCtrl asks the Sensor type. Select the desired type of sensor.

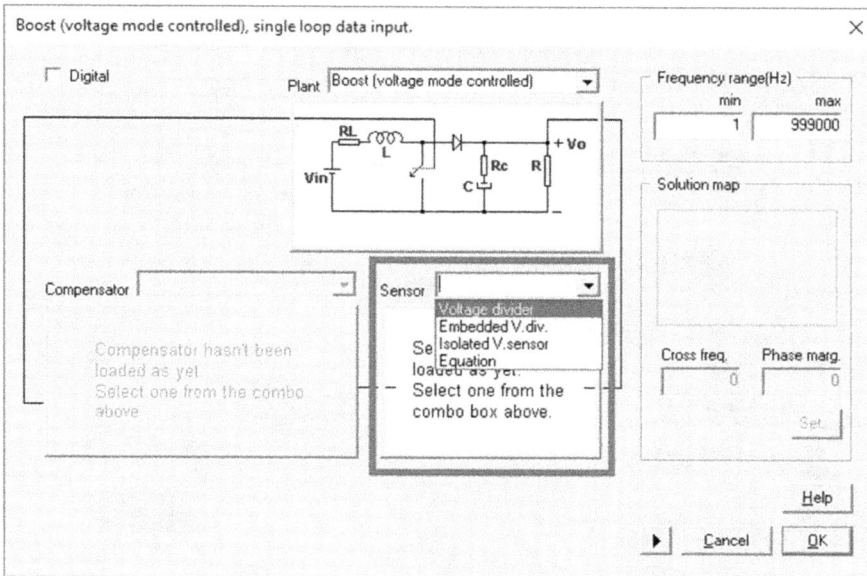

Fig. 8.71.

Once the desired type of sensor is selected, the Compensator box becomes actives. Select the desired type of compensator from this box.

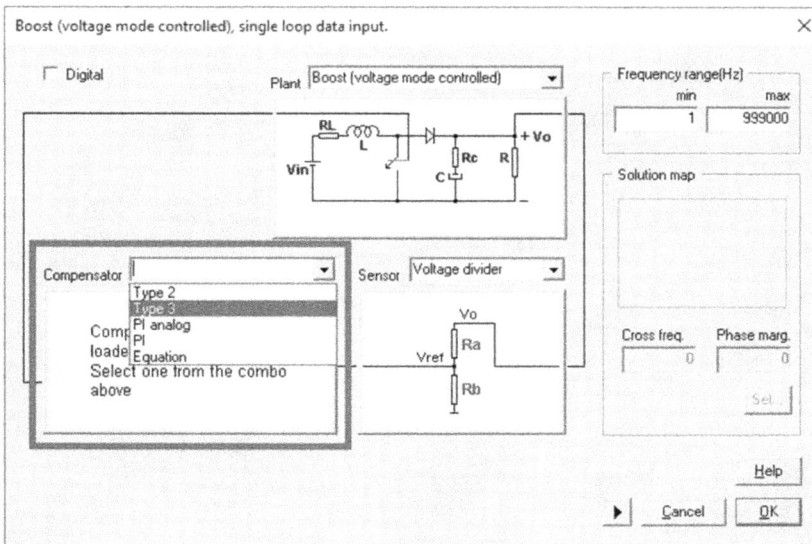

Fig. 8.72.

If you want to design a digital controller, you must check the Digital box.

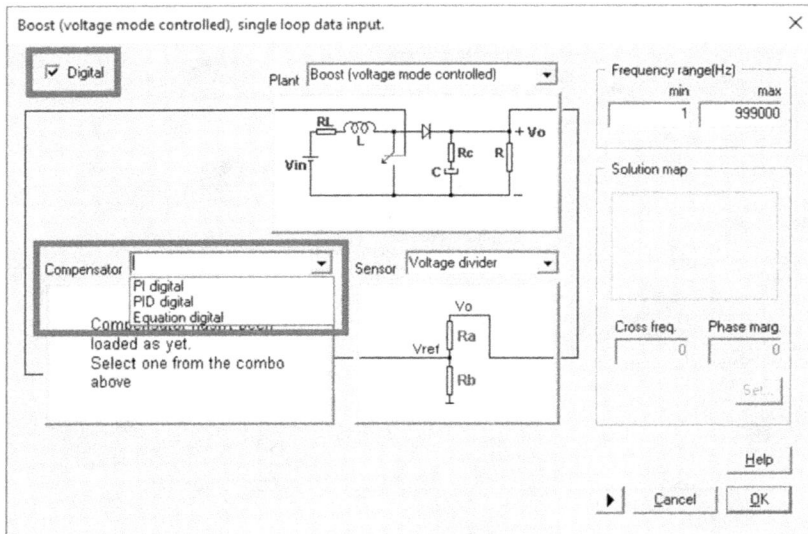

Fig. 8.73.

After desired type of controller is selected, click the Set button (Fig. 8.74) and click one of the points in the white region (Fig. 8.75). Then click the OK button.

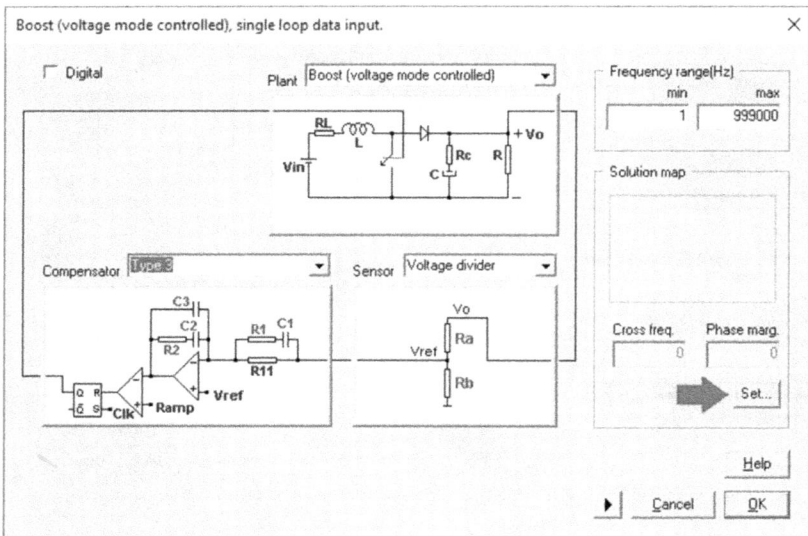

Fig. 8.74.

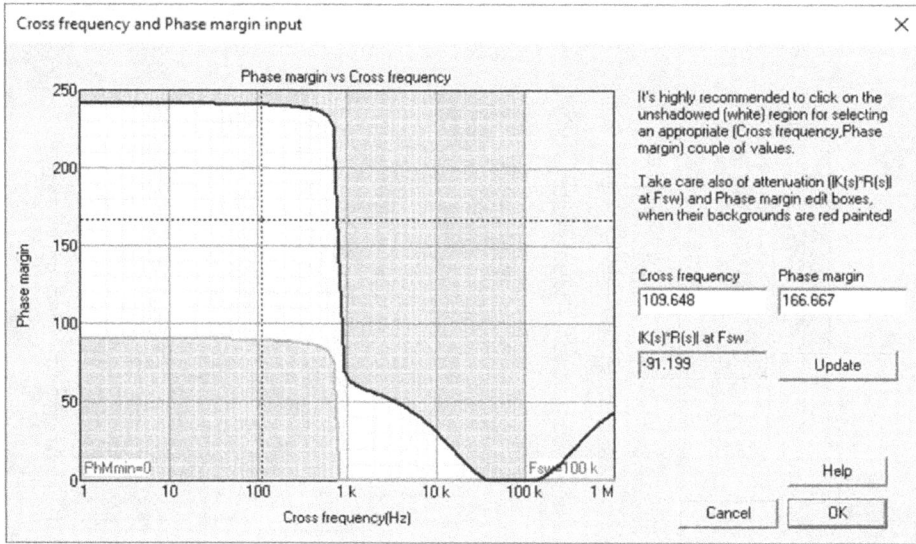

Fig. 8.75.

After clicking the OK button, SmartCtrl shows different Bode plots (Fig. 8.76) and you can start the design.

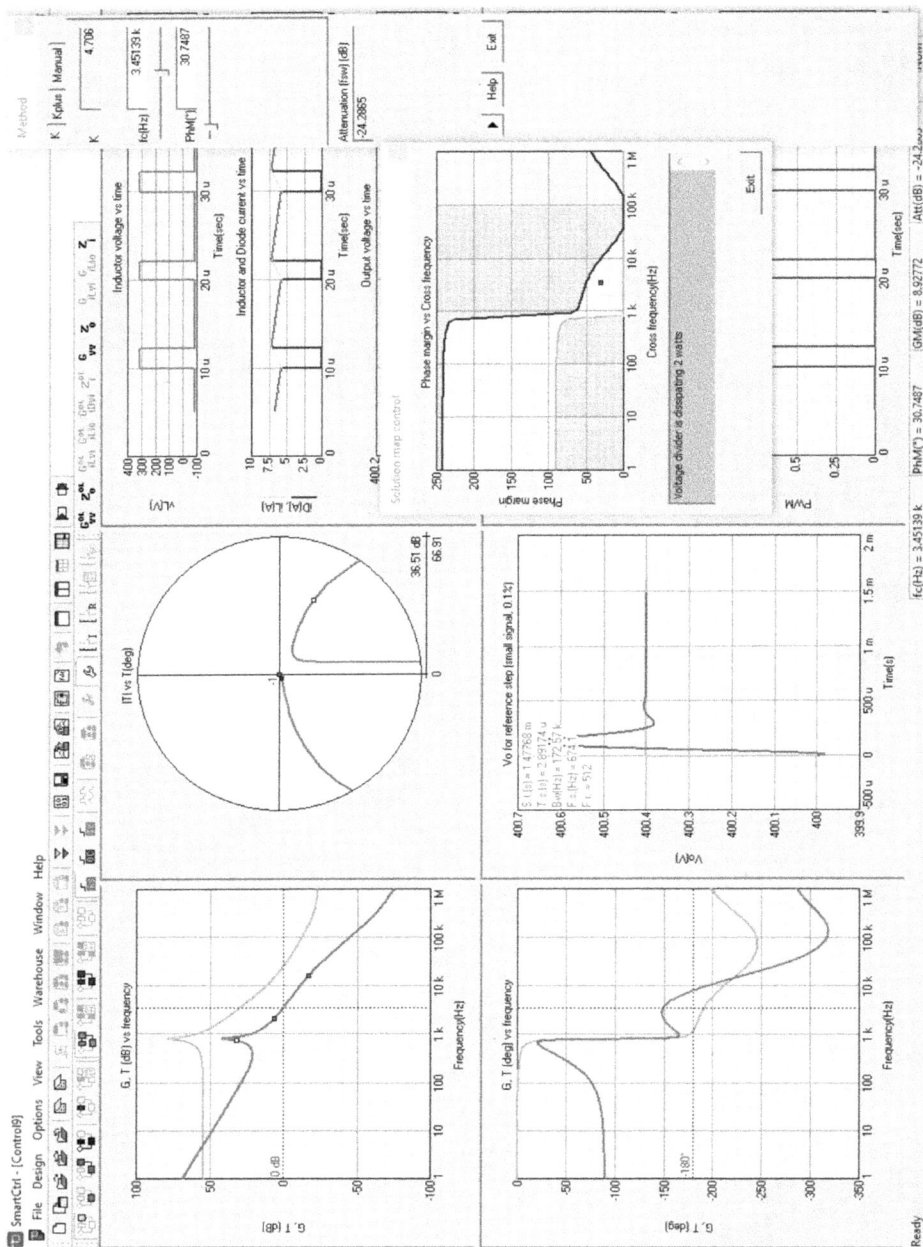

Fig. 8.76.

8.12 Sample designs for SmartCtrl

In the example folder of SmartCtrl, you can find different examples for different types of converters. In order to see the examples, click the Open sample design icon (Fig. 8.77). The window shown in Fig. 8.78 will appear. Open the examples folder in order to see the sample designs. Note that SmartCtrl user guide is available in the doc folder (Fig. 8.78) as well.

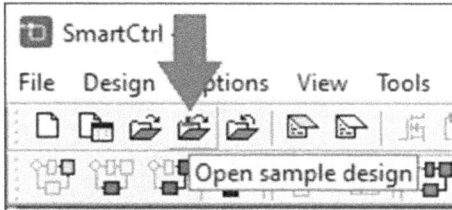

Fig. 8.77.

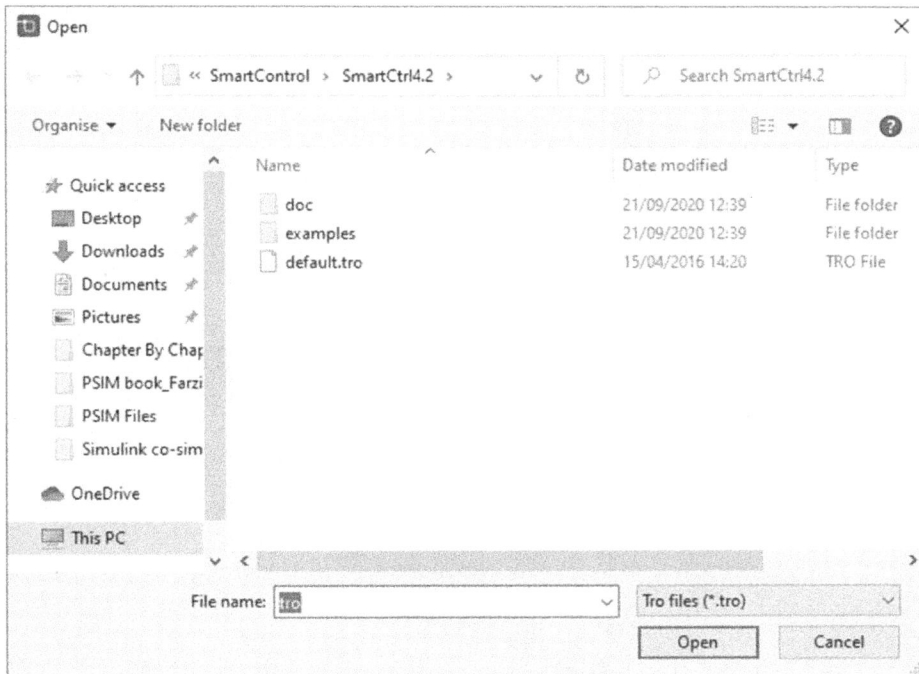

Fig. 8.78.

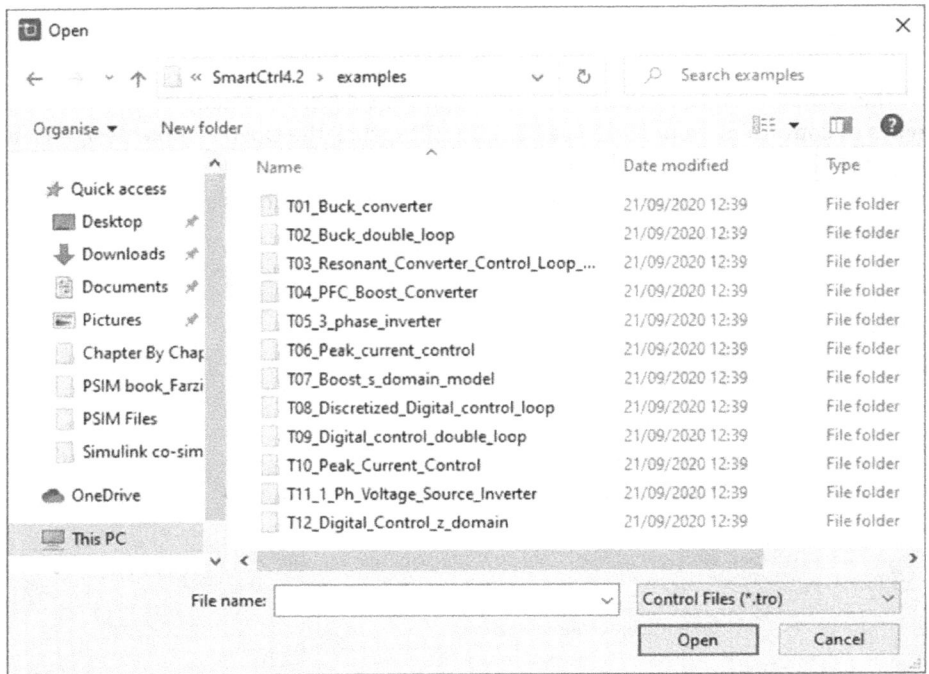

Fig. 8.79.

References for further study

Plenty of good power electronics references are available in the market. [1] is a good textbook with plenty of numeric examples. It is quite useful for a first course on power electronics, since the concepts are described clearly. [2-4] are more advanced books and suggested to advanced readers. [5] and [6] are good references for readers interested in DC-DC converters. [7-9] studied the dynamics and control of DC-DC converters. [10] is a good reference for readers interested in inverters.

[1]   D. Hart, Power Electronics, Mc Graw Hill, 2011.
[2]   N. Mohan, T. Undeland, Power Electronics Converters, Applications and Design, Wiley, 2002.
[3]   M. Rashid, Power Electronics Devices, Circuits and Applications, Pearson, 2013.
[4]   R. Erikson, D. Maksimovic, Fundamentals of Power Electronics, Springer, 2001.
[5]   S. Ang, A. Oliva, Power Switching Converters, Taylor & Francis, 2005.
[6]   Marian K. Kazimierczuck, Pulse Width Modulated DC-DC Power Converters, John Wiley, 2012.
[7]   F. Asadi, K. Eguchi, Dynamics and Control of DC-DC Converters, Morgan&Claypool (2018)
[8]   F. Asadi, Computer Techniques for Dynamic Modeling of DC-DC Power Converters, Morgan&-Claypool (2019).
[9]   F. Asadi, K. Eguchi, On the Extraction of Input and Output Impedance of PWM DC-DC Converters, Balkan Journal of Electrical and Computer Engineering (2019).
[10] F.L. Luo, H. Ye, Advanced DC/AC Inverters: Applications in Renewable Energy, CRC Press, 2013.
[11] Supporting documents for Powersim products (Visited at: 15.07.2021): https://powersimtech.com/resources-support/more-resources/
[12] PSIM Academic License Options (Visited at: 15.07.2021): https://www.youtube.com/watch?v=8NvtWHMVeYQ&t=271s
[13] PSIM Tutorials (Visited at: 15.07.2021): https://powersimtech.com/resources/tutorials/
[14] How to use PSIM help (Visited at: 15.07.2021): https://powersimtech.com/resources/tutorials/examples-tutorials-how-to-use-psim-help/
[15] Getting start with PSIM (Visited at: 15.07.2021): https://www.youtube.com/watch?v=LyS8ptNo4ww
[16] IGBT and MOSFET Loss Calculation in Thermal Module (Visited at: 15.07.2021): https://psim.powersimtech.com/hubfs/PDF%20Tutorials/Level-2%20and%20SPICE%20Model%20Simulation,%20Loss%20Calculation/Tutorial-IGBT-and-MOSFET-Loss-Calculation-in-the-Thermal-Module.pdf
[17] Using the Curve capture tool (Visited at: 15.07.2021): https://www.youtube.com/watch?v=Buz3S77QQQc
[18] How to Define Saturable Core Parameters (Visited at: 15.07.2021): https://psim.powersimtech.com/hubfs/PDF%20Tutorials/General%20PSIM%20Use/Tutorial-How-to-Define-Saturable-Core-Parameters-11.1.pdf
[19] Solar Module Physical Model (Visited at: 15.07.2021): https://psim.powersimtech.com/hubfs/PDF%20Tutorials/Inverters/Tutorial%20-%20Solar%20Module%20physical%20model.pdf
[20] Ultracapacitor Model Tool (Visited at: 15.07.2021): https://psim.powersimtech.com/hubfs/PDF%20Tutorials/General%20PSIM%20Use/Tutorial-Ultracapacitor-Model.pdf
[21] Using PSIM Free Run Mode (Visited at: 15.07.2021): https://powersimtech.com/resources/tutorials/using-psim-free-run-mode-runtime-variables-with-a-full-bridge-rectifier/
[22] How to analyze waveforms with PSIM (Visited at: 15.07.2021): https://powersimtech.com/resources/tutorials/how-to-analyze-waveforms-with-psim/
[23] A Guide to Coupled Inductors (Visited at: 15.07.2021): https://www.coilcraft.com/en-us/edu/series/a-guide-to-coupled-inductors/

https://doi.org/10.1515/9783110740653-009

[24] How to use the Subcircuit block in PSIM (Visited at: 15.07.2021): https://powersimtech.com/resources/tutorials/how-to-use-psim-subcircuit/

[25] AC Sweep troubleshoot tips (Visited at: 15.07.2021): https://powersimtech.com/resources/tutorials/ac-sweep-troubleshooting-tips/

[26] Introduction to PSIM multi-sine AC sweep block (Visited at: 15.07.2021): https://www.youtube.com/watch?v=tZzdXWTUrww&t=2s

[27] Buck Converter Output Impedance (Visited at: 15.07.2021): https://www.youtube.com/watch?v=21y4yBGlQWM

[28] Buck Converter Input Impedance (Visited at: 15.07.2021): https://www.youtube.com/watch?v=i6RBcBUfuxY

[29] Intro to the Simplified C Block (Visited at: 15.07.2021): https://www.youtube.com/watch?v=HreH4Cx-a94&t=4s

[30] How to Define Saturable Core Parameters (Visited at: 15.07.2021): https://psim.powersimtech.com/hubfs/PDF%20Tutorials/General%20PSIM%20Use/Tutorial-How-to-Define-Saturable-Core-Parameters-11.1.pdf

[31] Get started with PSIM code generation using F28379D (Visited at: 15.07.2021): https://www.youtube.com/watch?v=uL3JpALXORM

[32] Intro to Auto-Code Generation for F2833x DSP (Visited at: 15.07.2021): https://www.youtube.com/watch?v=hpKZ83exiJ0

[33] Introduction to thermal module in PSIM (Visited at: 15.07.2021): https://www.youtube.com/watch?v=4m393CqxaZM

[34] Co-simulation of Simulink & PSIM with SimCoupler (Visited at: 15.07.2021): https://www.youtube.com/watch?v=B5f1pbWdKCg

[35] Buck inner current loop design with SmartCtrl (Visited at: 15.07.2021): https://www.youtube.com/watch?v=gbsPRwNg1HM

[36] Using SmartCtrl to design the outer voltage loop of a Buck with SmartCtrl (Visited at: 15.07.2021): https://www.youtube.com/watch?v=Ag7GFsPLYFg

Index

AC ammeter 119
AC Sweep (multisine) block 458
AC Sweep block 426
AC Sweep Probe 426
AC volt meter 119
add cursor to the graph 146
alpha controller block 364
apparent power factor 153
average icon 151
average power 153
average value 120

back electro motive force (BEMF) 517
back up file 84
B-H Curve 91
Bi-directional Switch 285

C generate code 511
Carrier PWM Controller (s-domain) block 407
capacitor 167
capacitor (electrolytic) 169
computational blocks 207
conduction loss 286
continuous current mode (CCM) 549
control signal 57
Control-to-power Interface 61
coupled inductor 179
current flag 51
current probe 39
current sensor 46
Curve Capture Tool 90

DC ammeter 119
DC component 120
DC generator 522
DC Load (constant-power) block 188
DC motor 515
DC volt meter 119
demo version 3
diode 234
Disable icon 76
DSIM 2

electrical machines 514
Enable icon 77
export Simview waveforms as CSV 162
exporting the Simview waveforms 165

fast Fourier transform (FFT) analysis 159
Filter block 202
Fit to Page 33
fixed step solver 100
Flip horizontal 23
Flip vertical 23
Flyback converter 474
Free-run mode 109
frequency response of converter 424

generator efficiency 528
grid points 81
ground icon 25

half-wave rectifier 333
help system 11
hybrid electric vehicle 1

inductor 167
In Link Node 535
inductance factor 505
initial condition 257
input impedance of buck converter 442
input mechanical power 528
inverter 476

JMAG 1

Level 1 model 171
Level 2 model 171
Library browser 6
limiter block 419
LTspice 1

magnetic circuit 499
Math Function (voltage source)
    block 298
Math function block 226
mechanical load 514
Mechanical Load (constant-speed) 525
mechanical load (general) 516
menu bar 4
Model Level 171
Model Settings 542
ModelSim 1
motor efficiency 521
multilevel inverter 488

https://doi.org/10.1515/9783110740653-010

multiplexer 472
multiplier block 47

New (worksheet) 66
new file 17
Nonlinear Element block 196

Op amp 247
open loop buck converter 388
Optocoupler 248
Out Link Node 535
output impedance of buck converter 450

PAGE block 65
password protection 70
performance of controller 570
perturbation 424
Photovoltaics 92
PI controller 459, 554
power factor 153
power stage design 579
Powersim 1
progress bar 8
Push Button Switch 285
pushpull converter 474

R_switch_off 272
R_switch_on 272
RC circuit 254
reactive power 318
reference direction 519
remove the schematic password 72
rheostat block 176
RMS block 212
RMS icon 153
Rotation icon 22
resistor 167

s2z Converter 88
saturable inductor 178
SimCoder 511
SimCoupler 532
simplified C code 468
simulation control block 100
Simulation Message window 9
Simview 128

single-ended primary-inductor converter (SEPIC)
    converter 409
single-phase diode bridge 348
single-phase full-wave controlled rectifier 363
single-phase full-wave rectifier 342
single-phase half-wave controlled rectifier 358
single-phase inverter 476
SmartCtrl 546
Solar Module (physical model) 93
space vector pulse width modulator 498
Speed Sensor block 515
split the screen 140
subcircuit block 398
switches 234

THD measurement 289
Thermal Module 512
three-phase controlled rectifier 380
three-phase diode bridge 350
three-phase four-wire source 351
three-phase inverter 484
three-phase uncontrolled rectifier 351
thyristor 358
TL431 248
Transfer Function block 220
Transformer 243
Type 2 554
Type 3 554
Typhoon 2

Ultracapacitor Model Tool 97
Unit Converter 99

VA/Power Factor Meter block 323
VAR meter block 310
variable step solver 100
video tutorial 12
voltage probe 28, 51
voltage sensor 46

Wattmeter block 300
wind turbine 1
wire tool 24

z-domain Transfer Function block 223
Zero Order Hold 471

www.ingramcontent.com/pod-product-compliance
Lightning Source LLC
Chambersburg PA
CBHW060941210326

41598CB00031B/4692